HYDROLOGY

An Advanced Introduction to
Hydrological Processes and Modelling

Other Titles of Interest

BISWAS
Hydrology — A History, 2nd Edition
United Nations Water Conference: Summary and Main Documents
Water Development and Management (4 volumes)

GOLUBEV & BISWAS
Interregional Water Transfers: Problems and Prospects

INTERNATIONAL COMMISSION ON IRRIGATION AND DRAINAGE
Application of Systems Analysis to Irrigation, Drainage and Flood Control

RAUDKIVI
Loose Boundary Hydraulics, 2nd Edition

TEBBUTT
Principles of Water Quality Control, 2nd Edition

YALIN
Mechanics of Sediment Transport, 2nd Edition

Related Pergamon Journal
Water Supply and Management

HYDROLOGY

An Advanced Introduction to Hydrological Processes and Modelling

ARVED J. RAUDKIVI
Ph.D., Dipl.-Ing. (hons.), F.I.C.E., F.N.Z.I.E.
Professor of Civil Engineering
University of Auckland
New Zealand

PERGAMON PRESS

OXFORD · NEW YORK · TORONTO · SYDNEY · PARIS · FRANKFURT

U.K.	Pergamon Press Ltd., Headington Hill Hall, Oxford OX3 0BW, England
U.S.A.	Pergamon Press Inc., Maxwell House, Fairview Park, Elmsford, New York 10523, U.S.A.
CANADA	Pergamon of Canada, Suite 104, 150 Consumers Road, Willowdale, Ontario M2J 1P9, Canada
AUSTRALIA	Pergamon Press (Aust.) Pty. Ltd., P.O. Box 544, Potts Point, N.S.W. 2011, Australia
FRANCE	Pergamon Press SARL, 24 rue des Ecoles, 75240 Paris, Cedex 05, France
FEDERAL REPUBLIC OF GERMANY	Pergamon Press GmbH, 6242 Kronberg-Taunus, Pfederstrasse 1, Federal Republic of Germany

---

First edition 1979

British Library Cataloguing in Publication Data

Raudkivi, Arved Jaan
Hydrology.
1. Hydrology
I. Title
551.4'8 GB661.2 79-40857

ISBN 0-08-024261-8

In order to make this volume available as economically and as rapidly as possible the author's typescript has been reproduced in its original form. This method has its typographical limitations but it is hoped that they in no way distract the reader.

Printed and bound in Great Britain by
William Clowes (Beccles) Limited, Beccles and London

Photo by Bill Reaves - Texas Parks & Wildlife Department

Photo by Otago Catchment Board, New Zealand

CONTENTS

PREFACE

"A net is a collection of holes held together by a piece of string"

In the following text a collection of topics are strung together in the hope that these will form a sufficiently well-defined net, an interconnected framework of the elements of applied hydrology. However, since hydrology is the study of all aspects of water, no one book will satisfy the needs of everybody. The text is orientated towards discussion of the hydrological processes and methods of estimation of the various quantities involved. It has grown out of twenty years of teaching hydrology and water resources engineering to civil engineering students, and has been guided by the conviction that it is important for engineers to understand the physics of the processes involved before they get engrossed in the methodology of calculations. Gross errors in design and planning can be avoided only if the designer understands the hydrological processes and their interactions. The book has been written for senior and postgraduate students, but it is hoped that it will also be of value to the practitioner. The topics are developed to a stage from where the student should be able to advance on his own. Some sections have material which may be considered too specialized, and there are topics which have been omitted or only briefly referred to, for example, methods of river gauging and field measurement in general, field surveys and hydrological mapping. Of particular importance among the omitted material, and complexity, are the remote sensing techniques, such as the many forms of aerial photography by aircraft and satellites, microwave, nuclear and chemical methods of mapping, magnetic surveys and the associated data handling and data processing methods. The more specialized sections of the book may be omitted in the initial study and for the techniques of measurement the reader is referred to the existing literature. Techniques of problem solving which exploit the pocket calculator or the computer are adaptations of the principles and therefore not part of this treatment. I have also refrained from including pages of problems. These tend to be artificial and sterile. I believe that the teacher should set problems which are related to local conditions. This creates interest and shows the relevance of the study. In the same spirit the student could formulate his own problems, a process which is at least as instructive as the solving of ready-made problems. There is a place for a book of problems in which the majority of the problems are solved as examples and the solutions are explained in detail. Such a book would be a useful aid to teaching but, in my opinion, would form a separate volume.

The book draws together contributions from many authors and I wish to thank them all. I have attempted to give source references to the material used, both as acknowledgement and as a source for further study. Any omissions which might be found are unintentional and I offer my apologies in advance. I would also like to thank all the individuals and organizations who helped me with information, data, references and comments.

The support provided during the closing stages of this work by the Deutsche Forschungsgemeinschaft, SFB 80, of the Federal German Republic is gratefully acknowledged.

Finally, I would like to record my appreciation of the help given with proof reading by my family and to thank my wife for typing the manuscript in its many modified versions.

A.J.R.

Auckland, N.Z. Jan. 1978

Chapter 1

INTRODUCTION

The word hydrology is derived from the Greek words *hydor*, meaning water, and *logos*, meaning science. In this broad sense hydrology is concerned with all water on the Earth, its occurrence, distribution and circulation, its physical and chemical properties, its effects on the environment and on life of all forms. It is hard to think of a discipline of science which could not come under this definition. The botanist studying the movement of moisture through the plant, the medical scientist studying the role of water in human body, etc. could all come under the braod definition of hydrology. Many branches of hydrology are scientific disciplines in their own right, such as meteorology and hydrometeorology (study of atmospheric water); oceanography; hydrography (study of surface waters) which is further subdivided into potamology (potamos-river), dealing with flow in streams and rivers, limnology (limne-lake) is concerned with fresh water lakes, reservoirs, etc. and cryology (kruos-frost) with snow and ice; geohydrology dealing with water in the ground, and many more.

In customary usage, however, hydrology has come to mean studies of precipitation and runoff, that is, it has been linked with problems associated with design and management of water resources projects, such as water supply, flood control, or recreational use of water. In fact, hydrologists are expected to supply the basic data on which the design as well as management will be based. The most sophisticated methods of optimization and management studies are still only as good as the basic data supplied by the hydrologists and much of the data is very difficult to obtain and interpret.

1.1 Water Resources of the World.

It is hardly necessary to state that water is one of the most important minerals and vital for all life. It has played an important role in the past and in the future it will play the central role in the well-being and development of our society. This most precious resource is sometimes scarce, sometimes plentiful and always very unevenly distributed, both in space and time.

Estimates of the total amount of water vary. Table 1.1, due to Lvovich, indicates the order of size of the resource and how it is distributed.

Towards the end of the last glacial period, about 18 000 years ago, the ocean level has been estimated to have been some 105-120 m lower than at present. The difference is equivalent to 40 x 10^6 km^3 of water. If this water was stored in the form of ice then the total water equivalent of the polar caps and glaciers must have been about three times that at present. During the last century there appears to have been a puzzling increase in the total water equivalent in oceans and as ice. Measurements indicate an average rise of ocean level of 1.2 mm per annum or about 430 km^3/year; some estimates of this increase are even as high as 1750 km^3/year. An explanation is that this water comes from exploitation of groundwater in excess of recharging, but 430 km^3/year averaged over the total land areas of 134 x 10^6 km^2, not covered by water, means a lowering of the groundwater table by 3.2 mm/year, or a third of a metre in the century and there is little evidence to support this on world wide scale. Indeed, changes in the sea level could more readily be ascribed to changes in the volume of the oceans, caused by continental drifts and warping of land masses. According to Fairbridge (1961), variations of ±100 m with respect to present levels have occurred in the last 300 000 years.

TABLE 1.1 Quantity and Distribution of Water

	Area covered 10^6 km^2	Volume in 10^3 km^3	% of total volume
Oceans	360	1 370 323	93.93
Total groundwater, incl. zones of		64 000	4.39
active water exchange		(4 000)	(0.27)
Polar ice and glaciers	16	24 000	1.65
Lakes		230	0.016
Soil moisture		75	0.005
Atmospheric water	510	14	0.001
Rivers		1.2	0.0001
		1 458 643	100

The total fresh water amounts to 88.32 x 10^3 km^3 or less than 6% and only about ½% is readily available in lakes and rivers. The atmospheric water content is equivalent to less than 3 cm of water and the total amount of water in growing matter (the biomass) is less than 10 km^3. A more illuminating picture is obtained when the water masses involved in the processes of the hydrosphere - as the global circulation is referred to - are associated with their turnover times, Fig. 1.1. The fresh water resources of Continents are shown in Table 1.2 and the per capita volume of runoff in streams and rivers is shown in Table 1.3.

It is useful to reflect that Europe and Asia together accommodate about 76% of the world population but have only 27% of the total fresh water runoff. About two-third of the Earth's surface is arid or semi-arid where the extent of agricultural and industrial development depends primarily on the availability of water. Of the total land surface of 140 x 10^6 km^2, only about 10% is arable and of this about 10^6 km^2 is at present irrigated. Few people realize that 1 m^3 of water is required to grow 1 to 3½ kg of dry matter by agricultural cropping, or to make about 14 kg of paper, 36 kg of steel, etc. If we allow for a total consumptive use of water for all purposes of 1000 m^3 per head per year then Table 1.2 shows that Europe and Asia are close to the population limit set by availability of fresh water. In order, however, to make use of all the available water it must be stored and distributed. For example, the Indian subcontinent is at present not short of water, that is if the water was distributed evenly throughout the year over the entire continent. But to achieve this redistribution we would require storage and distribution systems on a scale not yet known to man. Another example is the basin of the river Rhine. The annual runoff is about 69 km^3/year and the population is about 50 million, that is 1400 m^3/year per capita. The total use of water is approaching 25 km^3/year or about 30% of the total runoff and this is about the fraction of the runoff that can be controlled at reasonable cost.

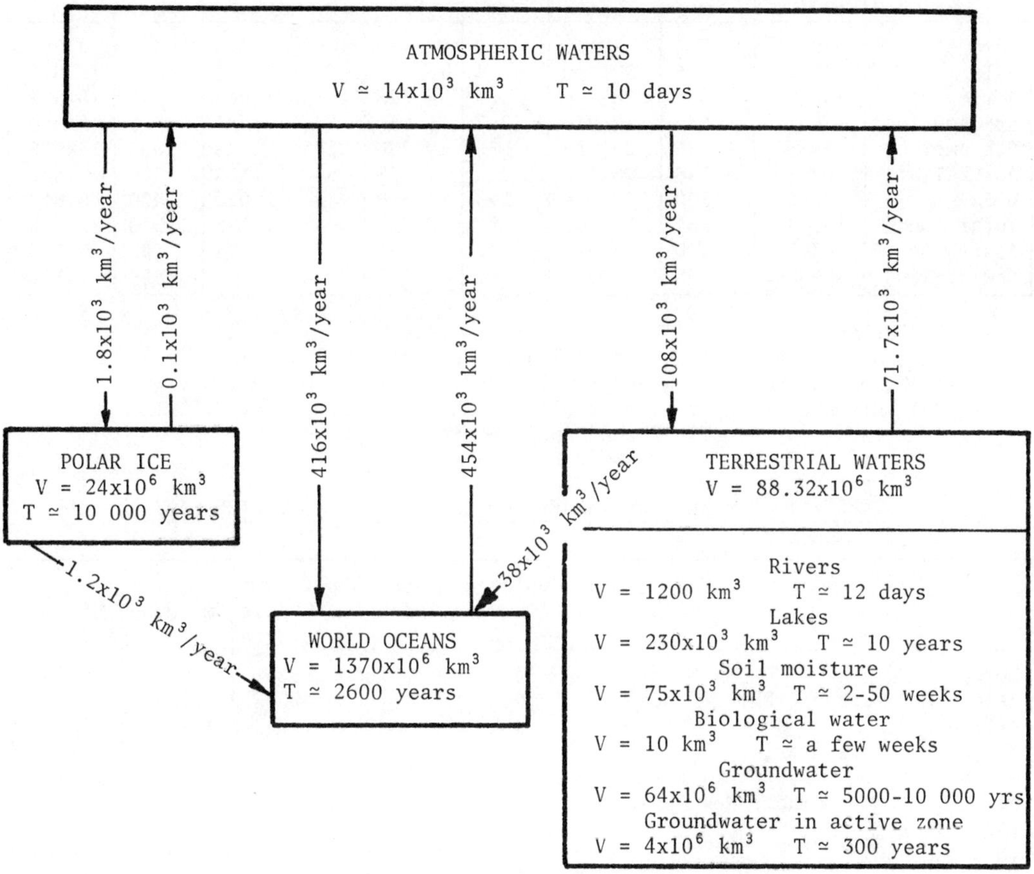

Fig. 1.1. Principal static and dynamic characteristics of the hydrosphere, V is volume in km^3 and T is the average turnover period in years.

However, it is not only the quantity but also the quality of water that is important. The quality aspect in a narrow sense refers to the pollution of fresh water by domestic, industrial and agricultural wastes. Not only may water returned to a river be unfit for use but a much greater volume of the river flow is made unfit for other uses. Mineral oils, for example, make water unfit for drinking in a ratio of $1:10^6$, one gram of radio-active strontium - 90 spoils a reservoir, i.e. $1:10^{15}$. The sewage discharge annually is of the order of 430 km^3 and it spoils about 5 500 km^3. This is more than 30% of the total runoff of rivers. But water quality is also important for recreational use, for maintenance of the ecological balance, etc. Indeed, water quality today is a subject of its own right and for this reason will not be further discussed here.

TABLE 1.2 Fresh Water Resources of Continents, after Lvovich (1973)

	Area	Precipitation		Runoff Total		Runoff Subsurface		Evaporation	
	$x10^6 km^2$	mm	km^3	mm	km^3	mm	km^3	mm	km^3
Europe[1]	9.8	734	7 165	319	3 110	109	1 065	415	4 055
Asia	45.0	726	32 690	293	13 190	76	3 410	433	19 500
Africa	30.3	686	20 780	139	4 225	48	1 465	547	16 555
Nth America[2]	20.7	670	13 910	287	5 960	84	1 740	383	7 950
Sth America	17.8	1 648	29 355	583	10 380	210	3 740	1 065	18 975
Australia[3]	8.7	736	6 405	226	1 965	54	465	510	4 440
USSR	22.4	500	10 960	198	4 350	46	1 020	300	6 610
Total land[4]	132.3	834	110 305	294	38 830	90	11 885	540	71 468
Australia	7.7	440	3 390	47	362	7	54	393	3 028
New Zealand	0.265	2 059	546	1 481	387			599	159

1. Incl. Iceland.
2. Excl. Canadian Archipelago and including Central America.
3. Incl. Tasmania, New Guinea and New Zealand. For New Guinea, Aitken et al. (1972) estimate precipitation at 3150 mm and total runoff at 2110 mm.
4. Excl. Antarctica, Greenland and Canadian Archipelago.

TABLE 1.3 Freshwater Runoff per Capita, after Lvovich (1973)

	Population (1969) in 10^6	Ann. Runoff km^3 Total	Ann. Runoff km^3 Stable Portion	m^3/Head	Stable Portion
Europe	642	3100	1325	4850	2100
Asia, incl. Japan & Philippines	2040	13190	4005	6465	1960
Africa incl. Madagascar	345	4225	1905	12250	5500
North & Central America	334	5960	2380	17844	7125
South America	188	10380	3900	55213	20745
Australia, New Guinea, New Zealand	18	1965	495	109000	27500
(Australia	12.45	362		2908	)
(New Zealand	3	387	150	129000	56000)
All land areas	3567	38830	14010	10886	3928

1.2 Transport Processes for Energy and Matter

In the framework of hydrology one cannot talk of water alone. Instead we have to consider the flows and storage of both energy and matter on global scale. We really have to look upon the hydrosphere as a giant heat engine. Hydrology involves the budget of radiant energy exchange, the balance of energy or heat and the balance of water (Miller, 1965). Heat and water are separately conserved but their cycles mesh when water changes phase. The energy sources are solar radiation, terrestrial and tidal energy. The earth intercepts about 1.7×10^{14} kW of solar radiation of which a little less than half (47%) is absorbed and converted into heat at ambient temperature, about 23% goes into the hydrological cycle and 30% is returned to space. The heat flow from the interior of the earth has been estimated at about 3×10^{10} kW and the tidal energy, from potential and kinetic energy of the Earth, Moon and Sun system, at 3×10^9 kW. Most of the heat in the hydrological cycle is consumed by evaporation. The atmospheric heat engine is diagrammatically shown in Fig. 1.2. It is the equivalent of the heart in mammals. However, it is even more important

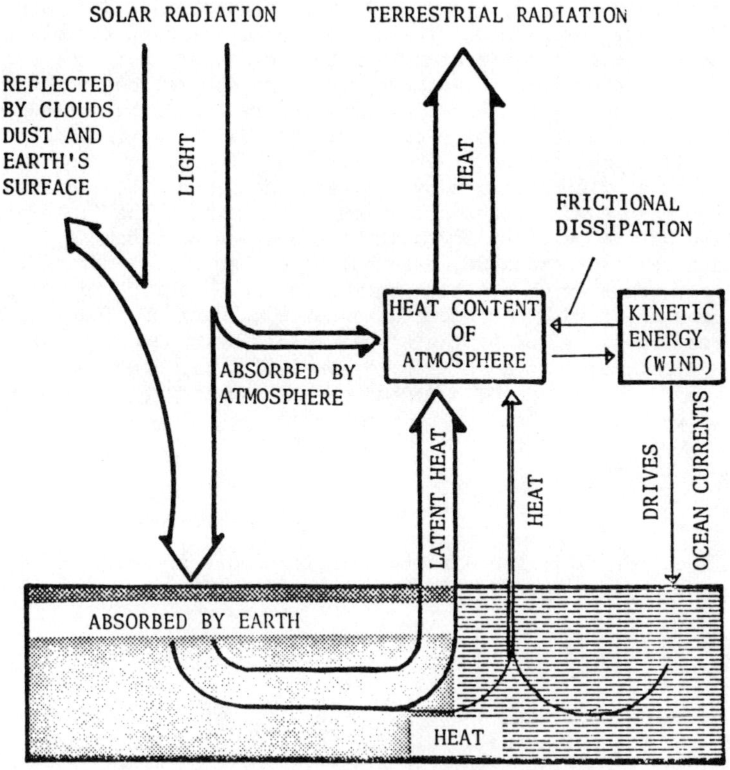

Fig. 1.2. Schematic representation of the atmospheric heat engine. The thickness of the arrows indicates the relative amounts of energy flow.

to realize that in hydrology we are dealing with a dynamic system involving the transport of energy, momentum and mass. The transport may be by convection, by conduction or by radiation and the laws of conservation of mass and energy have to be satisfied. Before dealing with heat budget and heat radiation it may be helpful to review briefly the concept of transport by convection and conduction.

Convection is transport by flow, for example, mass transport by flowing water. Velocity is momentum per unit mass and hence, momentum is transported by convection. Likewise, the moving masses transport heat, that is energy, by convection. The energy equation assembles the terms for the various forms of energy: heat energy, shaft or mechanical work, work done by shear stresses on the boundaries and the terms from the Bernoulli equation (pressure energy, kinetic energy and potential energy). All these terms have to be accounted for in the energy balance but only heat can be transported, the other terms describe the transformation between work done and capacity to do work. Convection of conservative quantities by large scale motion (usually horizontal) is called *advection*. Convection may also result from buoyancy effects which will be discussed in the section on meteorology together with elements of thermodynamics.

Conduction, like convection, may be of mass, energy or momentum and it may be in any direction relative to the mean flow, or in the absence of it. In connection with mass and momentum it is customary to talk of diffusion but in connection with energy it is referred to as heat conduction. The mass transport by diffusion may be by molecular diffusion or turbulent diffusion, or both, although turbulent diffusion is usually orders of magnitude greater than the molecular one; for example, the diffusion of water vapour into the atmosphere by random molecular motion alone would be very small compared to that resulting from the random turbulent eddying of the air flow. Transport of momentum by convection or diffusion in a direction normal to that of flow is by the action of viscosity and turbulence. For example, consider a large body of fluid bounded by a plane boundary and let this boundary start to slide. As a result of viscous action more and more of the fluid starts to move in the direction of the boundary movement - conduction of momentum. In turbulent flow with a mean velocity gradient, a turbulent exchange of lumps of fluid between layers leads to a transfer of momentum normal to the velocity - diffusion of momentum. The diffusion of mass and momentum and the conduction of heat are analogous processes and may be described by the so-called diffusion equation as follows: *the time rate of transport of a quantity Q per unit area, normal to the direction of transport, is proportional to the gradient of the quantity Q per unit volume of fluid (concentration) in the direction of transport*

$$\frac{\partial Q}{\partial t} \propto \frac{\partial^2}{\partial s^2}\left(\frac{Q}{\forall}\right) A\, \Delta s \qquad 1.1$$

where A is an area normal to the direction of transport s, and $\forall$ is volume. Thus, for mass transport per unit area (diffusion)

$$\frac{\partial Q}{\partial t} = - D \frac{dc}{ds} \qquad 1.2$$

and is known as the Fick's law, where D is the coefficient of diffusion or diffusivity, and c is the concentration ($Q/\forall$), for example, of water vapour. The minus sign indicates that the transport is in the direction of decreasing concentration. For heat conduction the gradient term would be dT/ds, the temperature gradient, D is replaced by k, the thermal conductivity, and $\partial Q/\partial t$ would be the rate of transport of heat per unit area, i.e. heat flux.

The rate of transfer of momentum per unit area has the dimension of stress. Thus, the diffusion of momentum in laminar flow in y-direction is

$$\tau_{yx} = - \mu \frac{du}{dy} \qquad 1.3$$

where u is velocity in x-direction and the constant of proportionality is now the coefficient of viscosity. (Here the minus sign is usually omitted and gradient du/dy is inserted with its sign). The shear stress in the xy-plane is equivalent to momentum flux in the y-direction. Likewise, for turbulent motion

$$\tau_{yx} = \rho \varepsilon \frac{\partial \bar{u}}{\partial y} \qquad 1.4$$

where ε is the momentum exchange coefficient (kinematic eddy viscosity) or in terms of the Reynolds stress

$$\tau_{yx} = - \rho\, \overline{u'v'} \qquad 1.5$$

where $u'v'$ is the time average value of the product of the turbulent velocity fluctuations; $u = \bar{u} + u'$, $v = \bar{v} + v'$.

The processes of convection and conduction of energy are essentially independent of temperature, although heat conduction depends on temperature gradient. However, *thermal radiation* is proportional to the absolute temperature of the radiator. Thus, processes dominate, and vice versa at the high temperatures like that of sun (6000°K). At the Earth's surface we have an intermediate condition where all processes share in the energy budget.

1.3 Energy from the Sun

Radiation of thermal energy from the sun is the primary source of energy for the global heat engine, the hydrologic cycle. Radiant energy cannot be stored but some may be absorbed as heat energy by the atmosphere and the earth. The processes involved are complex and not yet fully explained. The atmospheric adsorption is a selective process with respect to wave lengths of the radiated energy, composition of the atmosphere, and temperature. It also depends on reflection and on both molecular and particulate scattering. Adsorption by surfaces depends on the absorptive properties of surfaces. The intensity of thermal radiation - radiant energy flux - is described in terms of power density, Wm^{-2}. Meteorological texts in the non SI units use the unit ly/min (1 Langley = 1 cal/cm^2, hence 1 ly/min = 697.8 Wm^{-2}). Only a very brief discussion of radiation physics is possible here and the reader has to be referred to basic texts of physics.

Adsorption is described by the Kirchhoff's law. Consider i-bodies of surface area A_i each in vacuum in a large thermally perfectly insulated space. At thermal equilibrium the bodies emit thermal radiation at rates A_iR_i, where R is the intensity of radiation, and receive radiation from the space at $a_iA_iR_s$, where a_i is the *absorptivity* of body i. Thus, for conservation of energy

$$a_1A_1R_s = A_1R_1 \text{ , } a_iA_iR_s = A_iR_i$$

or

$$\frac{R_1}{a_1} = \frac{R_2}{a_2} = \cdots = \frac{R_i}{a_i} = R_s \qquad 1.6$$

and it follows that *the ratio of radiation intensity (emissive power, emittance, radiant-flux density) to absorptivity at thermal equilibrium is the same for all bodies.*

When all incident radiant energy is absorbed (a = 1) we speak of a perfect radiator or black-body, and R_b is a maximum. The ratio of R/R_b is called the emissivity, ε, and is equal to a, under thermal equilibrium. *Radiation* is described by the Stefan-Boltzmann's law and the Planck's law. Analysis of a light from a source with a spectrometer leads to spectra as shown in Fig. 1.3. The ordinate R_λ ($Wm^{-2}\mu m^{-1}$) is the spectral radiancy defined so that $R_\lambda d\lambda$ (Wm^{-2}) is the rate of energy radiated per unit surface area in the interval of wave lengths λ to $\lambda + d\lambda$ (μm). The area under the spectral curve is the radiancy R(Wm^{-2})

$$R = \int_0^\infty R_\lambda \, d\lambda \qquad 1.7$$

Every material has a family of typical spectral radiancy curves, one for each temperature. However, if the light is emitted from a cavity in a heated solid, the cavity radiator, then the light-emitting properties are independent of any particular material and vary in a simple way with temperature. The radiation from

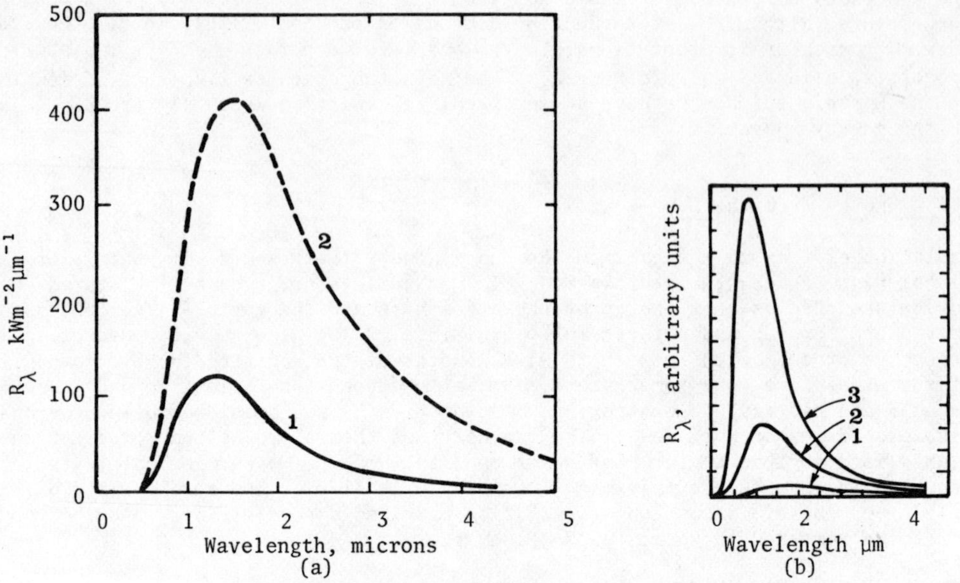

Fig. 1.3. (a) Spectral radiancy of tungsten (1) at 2000 °K, and of a cavity radiator (2) at the same temperature. (b) Spectral radiancy for cavity radiation at 2000, 3000 and 4000 °K, shown as 1, 2 and 3, respectively.

the cavity is always more intense than the radiation from the outside wall, c.f. Fig. 1.3 where the ratio of radiancies is 0.259. The cavity radiancy for any material is

$$R_c = \sigma T^4 \qquad 1.8$$

where $\sigma = 5.67 \times 10^{-8}$ Wm^{-2}K^{-4} is a universal constant - the Stefan-Boltzmann constant. The radiancy of the outer surface is given by $R = \varepsilon R_c$ where ε is the emissivity and depends on the material and temperature. The spectral radiancy curve R_λ for the cavity radiation varies with temperature but is independent of material or size of the cavity. For black-body radiation

$$\frac{R_b}{T^4} = \int_0^\infty \frac{R_{b\lambda}}{T^5}\, d(\lambda T) = \sigma \qquad 1.9$$

and this is known as the Stefan-Boltzmann's law.

Planck first modified Wien's formula for the cavity radiation spectrum by adding a minus one in the denominator. The formula then became

$$R_\lambda = \frac{c_1}{\lambda^5} \frac{1}{\exp(c_2/\lambda T) - 1}$$

where c_1 and c_2 are constants. This formula gave a perfect fit to the experimental points of the R_λ versus λ curve. He then proposed a theory in terms of atomic

Percentage of total energy below λ as function of λT

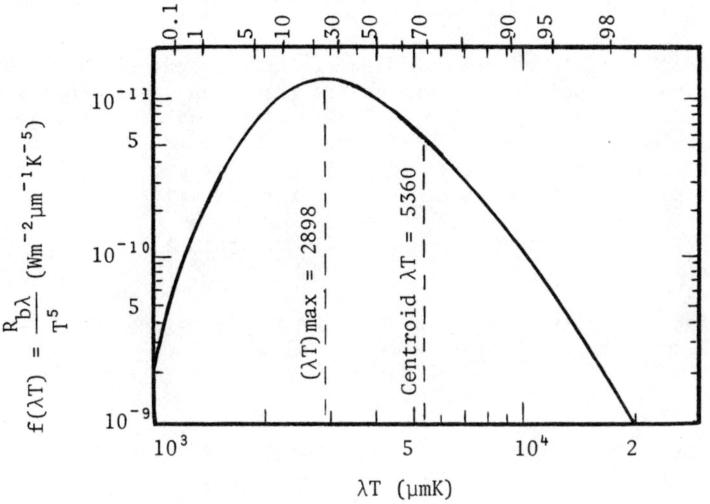

Fig. 1.4. Spectrum of thermal radiation from a black body, after McAdams (1954)

processes at the cavity walls. The basic assumption was that the atoms of the walls behave like tiny electromagnetic oscillators which, with their characteristic frequencies, emit electromagnetic energy into the cavity and absorb electromagnetic energy from the cavity. He assumed that

1. an oscillator's energy is given by $E = nh\nu$ (later modified to $E = (n + \frac{1}{2})h\nu$), where ν is the oscillator frequency, $h = 6.625 \times 10^{-34}$Js is the Planck's constant, n is a number (now known as the quantum number), and
2. the oscillators radiate energy in jumps or quanta when they change from one frequency to other.

His theoretical expressions for the empirical constants were

$$c_1 = 2\pi c^2 h \quad \text{and} \quad c_2 = \frac{hc}{k}$$

where c is the speed of light ($2.998 \times 10^8 ms^{-1}$) and k is the Boltzmann constant (k is the ratio of the universal gas constant $R = 8.314\ J\ mol^{-1}K^{-1}$ to the Avogadro's number (number of molecules per mole) $N_0 = 6.023 \times 10^{23}\ mol^{-1}$. Thus, $k = 1.380 \times 10^{-23} JK^{-1}$). The theory was presented on Dec. 14, 1900 and quantum physics dates from that day. Planck was awarded the Nobel Prize for this in 1918. Planck still treated the radiation within the cavity as an electromagnetic wave, but this concept was later replaced by Einstein by his photon theory.

Planck's law may thus be written as

$$\frac{R_{b\lambda}}{T^5} = f(\lambda T) = \frac{2\pi\ hc^2\lambda^{-5}T^{-5}}{e^{ch/k\lambda T} - 1} \qquad 1.10$$

when plotted with λT as abscissa the peak of the spectrum of thermal radiation occurs at

$$\lambda T = 2898\ \mu m\ K \qquad 1.11$$

which is the Wien's displacement law. Fig. 1.4 shows such a thermal radiation spectrum for a black body.

Assuming the sun and earth to be black bodies at 6000 °K and 300 °K respectively, it is possible to calculate their radiation spectra (Fig. 1.5) using eqns 1.9, 1.10 and 1.11. At the respective surfaces $R_b = 74.7\ MWm^{-2}$ and $0.46 kWm^{-2}$. Only a fraction of the energy radiated by the sun is intercepted by the earth, Fig. 1.5. The average intensity of solar radiation received at the outer limit of the earth's atmosphere on a unit area normal to the incident radiation is called the *solar constant*, R_{bo}. The currently accepted value of it is

$$R_{bo} = 1.354 \simeq 1.4\ kWm^{-2} \qquad 1.12$$

The solar radiation intensity on a *horizontal surface* is called *insolation* and at the outer limit of the atmosphere the insolation

$$I_o = R_{bo} \sin \alpha \qquad 1.13$$

where α is the angle between the direction of radiation and the horizontal surface. It is given from spherical trigonometry by

$$\sin \alpha = \sin \delta \sin \phi + \cos \delta \cos \phi \cos \tau \qquad 1.14$$

where δ is the declination, ϕ the local latitude and τ is the sun's hour angle. Figure 1.6 shows daily insolation amounts outside the earth's atmosphere. The values are tabulated in Table 4.1. The spectrum of incident solar radiation at the earth's surface is quite different from that shown in Fig. 1.5 because of losses in traversing the atmosphere due to reflection, scattering and absorption.

Reflection is a non-selective process. It occurs at all wavelengths and from all surfaces, including clouds and dust, as long as the particle size is greater than the wavelength. The ratio of reflected to incident energy, the reflection coefficient, is knwon *albedo*.

Scattering may be molecular or particulate scattering. Molecular scattering is a selective refraction process and approximately proportional to λ^{-4} (van de Hulst, 1949). The scattering of short wavelengths (violet) by air molecules is supposed to give the blue colour of the clear sky. Particulate scattering is due to dust particles and water droplets, but is not as well understood as the molecular scattering. It is also a selective process with a maximum value when the particle size is roughly equal to the wavelength. The scattering becomes negligible for λ much greater than the particle size and has an asymptotic value for λ much smaller than the particle size. A dust cloud, for example, may appear red because blue light is reflected and red is scattered.

Molecular adsorption is highly selective. It is confined to narrow spectral bands. Some of these bands are shown in Fig. 1.5. Goldberg (1954) lists 127 bands between $0.3 < \lambda < 24\ \mu m$. Figure 1.7 shows absorption of radiation at various wavelengths by O_2, O_3, H_2O and by the principal absorbing gases. We see that the atmosphere is reasonably transparent to the shortwave radiation of the sun but water vapour and carbon dioxide are seen to absorb a large proportion of the energy radiated from the earth. This keeps the earth warm and provides the energy for the atmospheric circulation. The absorption arises from excitation of the molecules of the gases in the atmosphere at particular resonant modes.

These effects are accounted for by expressing the change in local radiation inten-

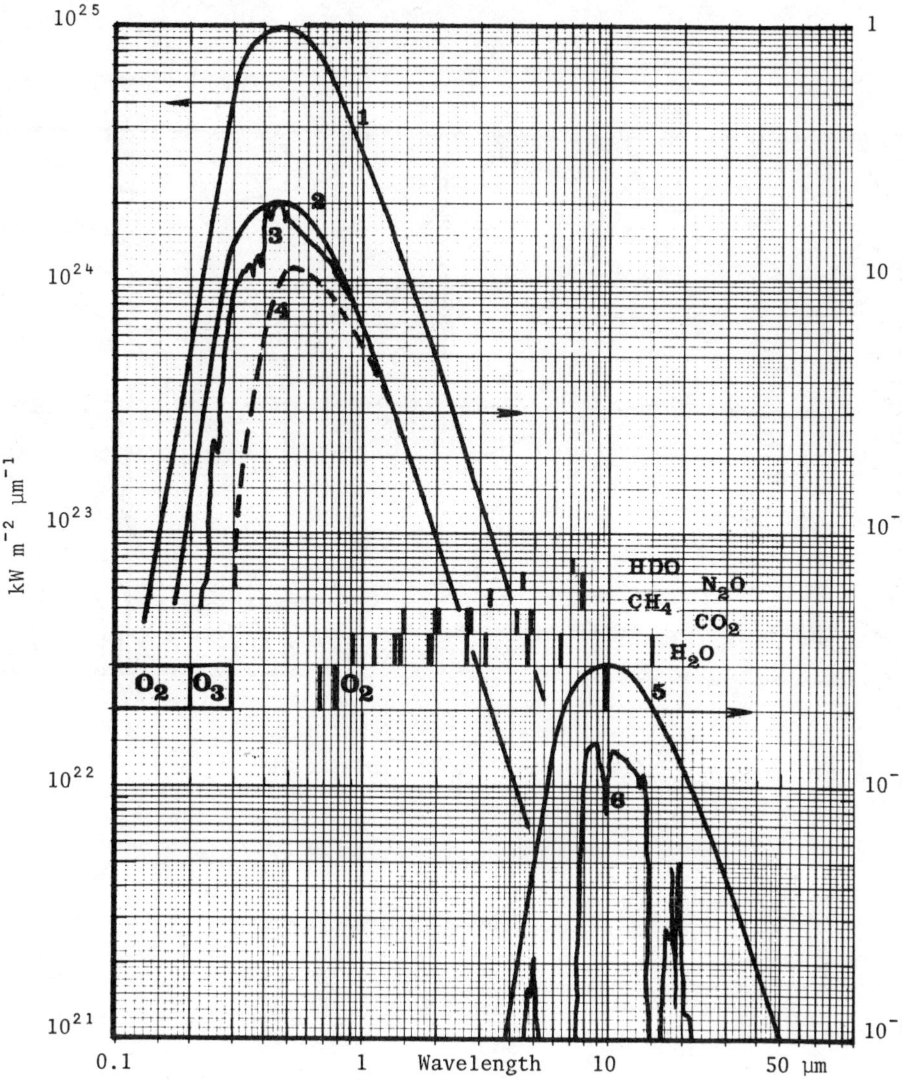

Fig. 1.5. The black-body radiation spectrum of the sun (1) at 6000 °K; the spectrum at the top of the atmosphere (2) (obtained by reducing the values of (1) by the square of the ratio of the sun's distance from earth); the measured values of (2) are shown as line (3), and line (4) joins together the peaks of radiation received at the earth's surface. Line (5) shows the black-body radiation spectrum of earth at 300 °K, R = 0.46 kWm^{-2} (at 287 °K the value of R_b of the earth is 0.39 kWm^{-2}) and line (6) indicates infrared emission to space. Note the different scales, only line (1) relates to the scale at the right hand side. The actual spectra received and radiated are modified by absorption. The locations of the very strong absorption bands are indicated on the figure.

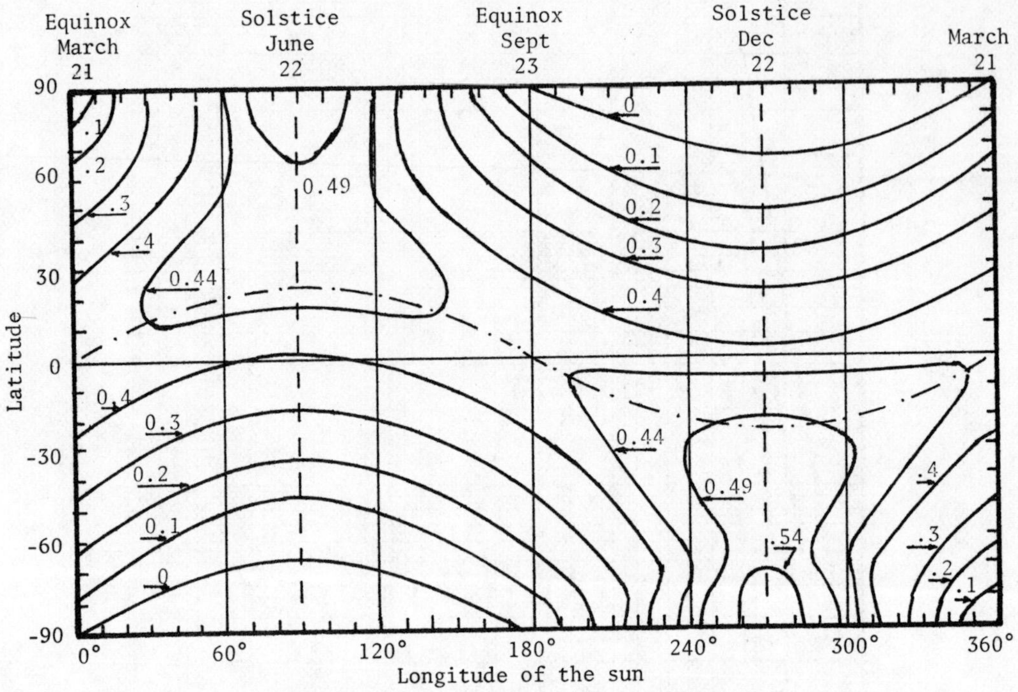

Fig. 1.6. Daily insolation amounts outside the earth's atmosphere in Wm^{-2} per day, after Milankovitch (1930). Declination of sun is shown by chain-dotted line.

sity caused by traversing a thin layer of atmosphere for monochromatic radiation as

$$-\frac{dI_\lambda}{ds}\Delta s = \rho k_{\lambda a} I_\lambda \Delta s$$

or

$$dI_\lambda = -\rho k_{\lambda a} I_\lambda ds = -a_\lambda I_\lambda \qquad 1.15$$

If at the same time radiation from the layer itself is significant then

$$dI_\lambda = a_\lambda (R_{b\lambda} - I_\lambda)$$

where Δs is the length of path through the thin layer, ρ is the density of the layer, $k_{\lambda a}$ is the absorption coefficient of the layer and a_λ is the absorptivity. Integrating eqn 1.15 over a distance s yields

$$\frac{I_{\lambda s}}{I_{\lambda o}} = \exp(-\int_o^s k_{\lambda a} \rho ds) \qquad 1.16$$

The same form of relationship is used for evaluation of scattering, except that an appropriate coefficient for scattering is used instead of $k_{\lambda a}$. The evaluation, however, of the integral for absorption or for scattering is a very difficult task because the structure of the absorption spectra is complex and there are large variations in the composition of the atmosphere with height.

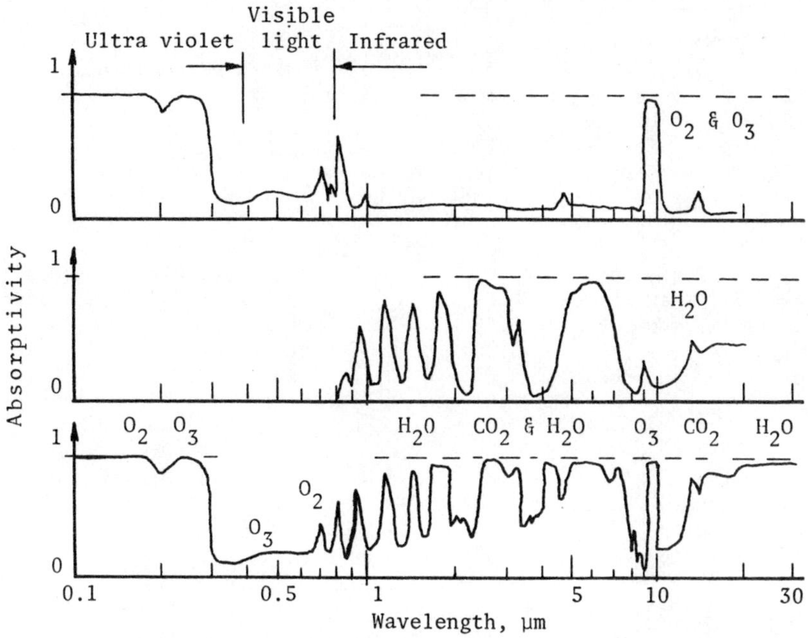

Fig. 1.7. Illustration of absorption of radiation at various wavelengths by O_2 and O_3, by water vapour and a combined absorption spectrum for the atmosphere showing the principal gases.

The budget of radiant exchange for the earth is still one of the major problems of geophysical sciences. The net rate at which solar energy is received by the earth's surface varies widely with latitude, with local albedo etc., and we will be discussing some aspects of this in the chapters to follow. Estimated mean albedo values, according to Budyko (1956), Budyko et al. (1961, 1962) are shown in Table 1.4. The value of albedo is not a constant. The solar altitude and cloud cover can substantially alter the albedo values (Raphael, 1962), particularly in the tropics where the clear sky value could be twice that of an overcast day. The albedo of clouds themselves varies with type and thickness of the clouds from 0.05 to 0.80. The albedo is also a function of the wavelength of the incoming radiation.

The global variation of the heat gives rise to energy gradients. This leads to large scale circulations in the atmosphere and the oceans, and convection of heat. Figure 1.8 summarizes the heat balance of the earth. The estimates of the percentages vary from estimator to estimator. The largest differences appear in the total emitted longwave radiation of the atmosphere. Figure 1.8 shows for this 166% whereas Strahler and Strahler (1974) show 137%, i.e. 77% of counter radiation instead of 106%. The differences arise from the difficulties with separation of longwave radiation from the earth's surface and the counter radiation from the atmosphere. However, the net longwave radiation gain of the atmosphere usually varies little, i.e. in these two cases 11% and 13%, respectively.

TABLE 1.4 Estimated Mean Values of Albedo

ZONAL VARIATIONS

Perennial snow cover in polar regions (polewards from °60)	0.60
Snow cover of longer duration in the moderate zone (below °60)	0.70
Snow cover of short duration	0.45
Coniferous forests	0.14
Deciduous forests, prairie areas during humid seasons	0.18
Prairie areas during drought periods, semi-deserts	0.30

EFFECT OF SURFACE PROPERTIES

Snow, ice, water		Agricultural areas	
Fresh, dry snow	0.80 - 0.87	Cereals	0.10 - 0.25
Clean moist snow	0.60 - 0.70	Potatoes	0.15 - 0.25
Dirty snow	0.40 - 0.50	Cotton	0.20 - 0.25
Ice on seas and oceans	0.30 - 0.40	Meadow	0.15 - 0.25
Water surfaces	0.03 - 0.10*	Tundra areas	0.15 - 0.20
Bare soil		**Forest areas**	
Dark soils	0.05 - 0.20	Coniferous forests	0.10 - 0.15
Dry clay or raw soils	0.20 - 0.35	Deciduous forests	0.15 - 0.20
Dry sand soils	0.15 - 0.45		

* At solar elevations 1°- 20° can be as high as 0.45

1.4 The Hydrologic Cycle

The above thumb-nail sketch of the various physical processes was aimed at drawing attention to the complexity of hydrology and to the interdependence of the many processes involved. It is not intended to be a treatise of these processes.

The hydrologic cycle can be represented in many different ways, with more or less detail, in pictorial or diagrammatic form as shwon in Fig. 1.9 to 1.11. Engineering problems are usually primarily concerned with one or two sub-systems of the hydrologic cycle at any one time. The problem is to find analytical relationships between the variables characterizing the inflow and outflow process and those defining the corresponding state of the system:

Variables characterizing inflow process → [Physical system] → Variables characterizing the outflow system

For example,

Rainfall excess → [Catchment] → Runoff

The catchment as a physical system responds to rainfall input. The problem is to express the runoff from the catchment as a function of catchment properties. In principle, one ought to solve the equations of momentum, energy, continuity and state, but because of inadequate knowledge of the physical behaviour, unknown system heterogeneities or anisotropies, unknown time dependence of system parameters and approximations introduced for calculations this is nearly impossible. This is also the reason why there is often a very highly variable difference between calculated and observed values. Therefore, there is a tendency to abandon the *causal deterministic* formulation in favour of the non-causal stochastic representation of the output variable which omits consideration of the dynamics of the process. Very few hydrological variables can be calculated in the form of a unique number and statistical methods - although not hydrology - form a very important part of an hydrologist's training.

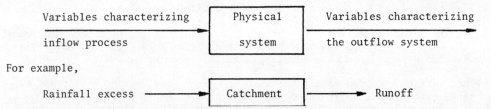

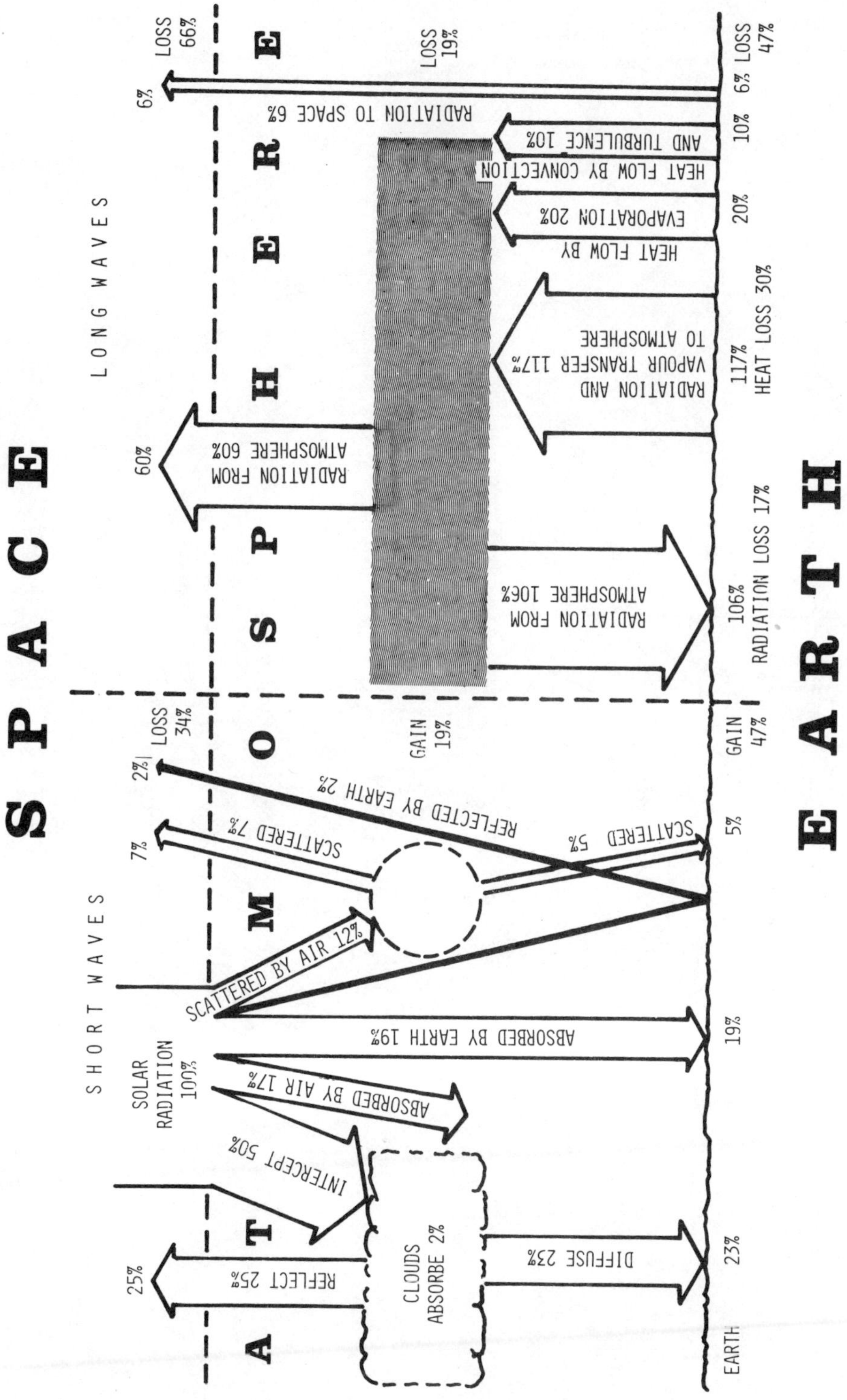

Fig. 1.8. Diagrammatic illustration of the earth's heat balance.

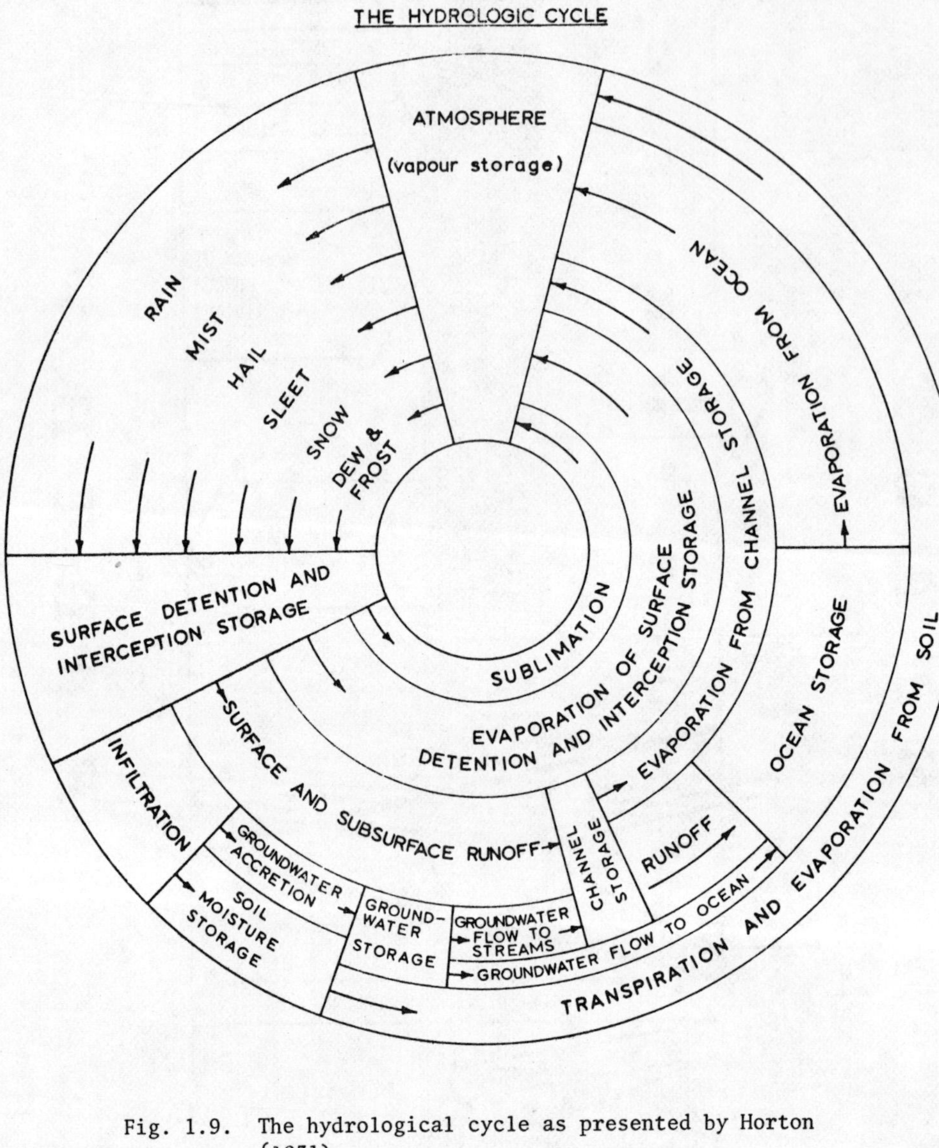

Fig. 1.9. The hydrological cycle as presented by Horton (1931)

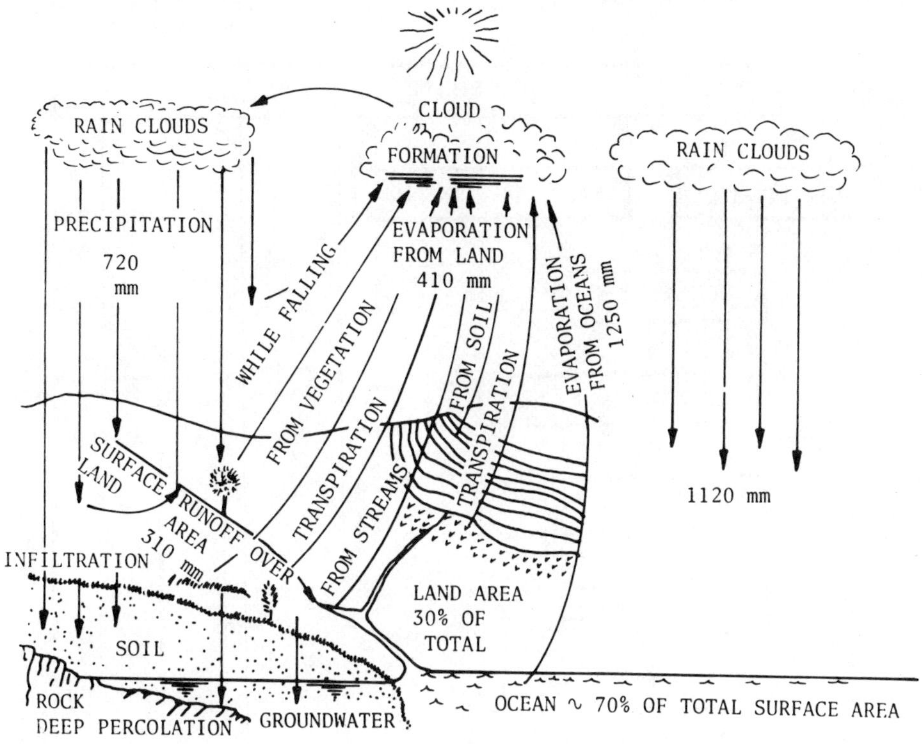

Fig. 1.10. The hydrologic cycle - pictorial representation.

1.5 Utilization and Management of Water Resources.

Engineering works aimed at utilization of water in one form or another may have profound effects on the local economic and social conditions as well as on the environment. For example, the transport of water to an arid region can totally change the regions ecological structure. It may not always be essential to maintain the status quo, and indeed, as long as the population growth continues it will be necessary to modify nature to man's benefit, but this must be done with care so that we shall not be confronted by very undesirable side effects. The effects of man-made changes are most important on large scale projects. If extensive areas of desert are irrigated and turned into green fields the reflective properties of earth are changed and this may lead to change in the total amount of heat retained in the global system. Or if, for example, all the large Russian rivers which flow into the Arctic Ocean are reversed, and diverted to the central desert areas, the arctic region would be deprived of a vast amount of heat carried by the rivers. The effects of the redistribution of such quantities of heat may have profound effects on global climate. Likewise, the cutting off of the large freshwater inflow will lead to significant changes is salinity in the Arctic Ocean and in its density induced ocean currents. The study of the environmental effects requires teamwork by specialists in all appropriate disciplines of science and no more is possible here than to mention the need for these studies.

The well being of future generations depends to a large extent on the wise manage-

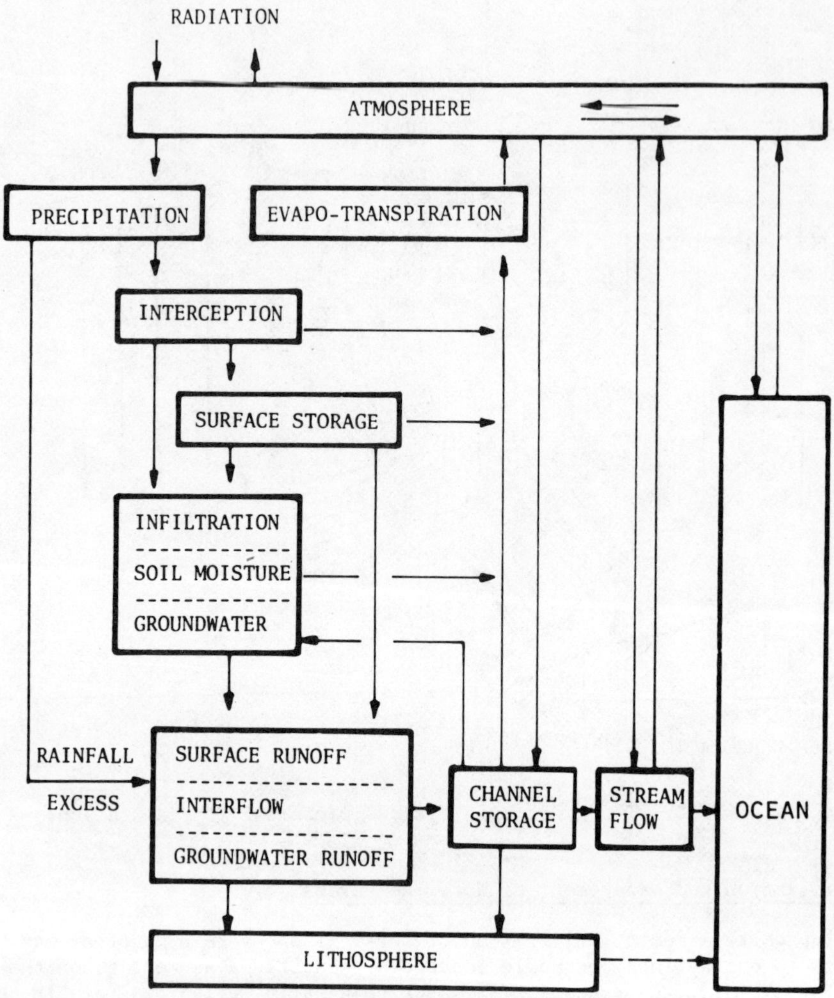

Fig. 1.11. A systems representation of the hydrologic cycle.

ment of water resources. Throughout history, the prosperity of nations correlates closely with the management of water resources. The 4000 years of history in Mesopotamia from 2400 B.C. to 1600 A.D. is a history of water resources management. The Beled Dam on the river Tigris upstream from Bagdad diverted flood waters into a basin from where it was used for irrigation. Failure to maintain the system and probable mismanagement of the upper catchment led to failure of the system, and to ruined cities and desert. The earlier irrigation works in Ceylon are dated to about 2100 B.C. By 1100 A.D. there were some 15 000 storage reservoirs behind earth embankments. Warfare and malaria destroyed all this. However, the Peruvians have continued the use of the stone-walled pre-Inca terraces to the present day.

Water resources management has established itself as an almost self-contained discipline concerned with seeking optimal solutions to problems associated with *demand*

and *availability* of water. These two components, although essential, do not alone constitute the water resources management problem. Political, social and ecological considerations play a very important role. Figure 1.12 by Klemes (1973) illustrates the interaction of water resources management with the various factors which together constitute the environment. The management problem has moved from finding an optimal solution to a particular problem to one of broader environmental planning. A major fraction of the information required for such planning has to be provided by the hydrologist.

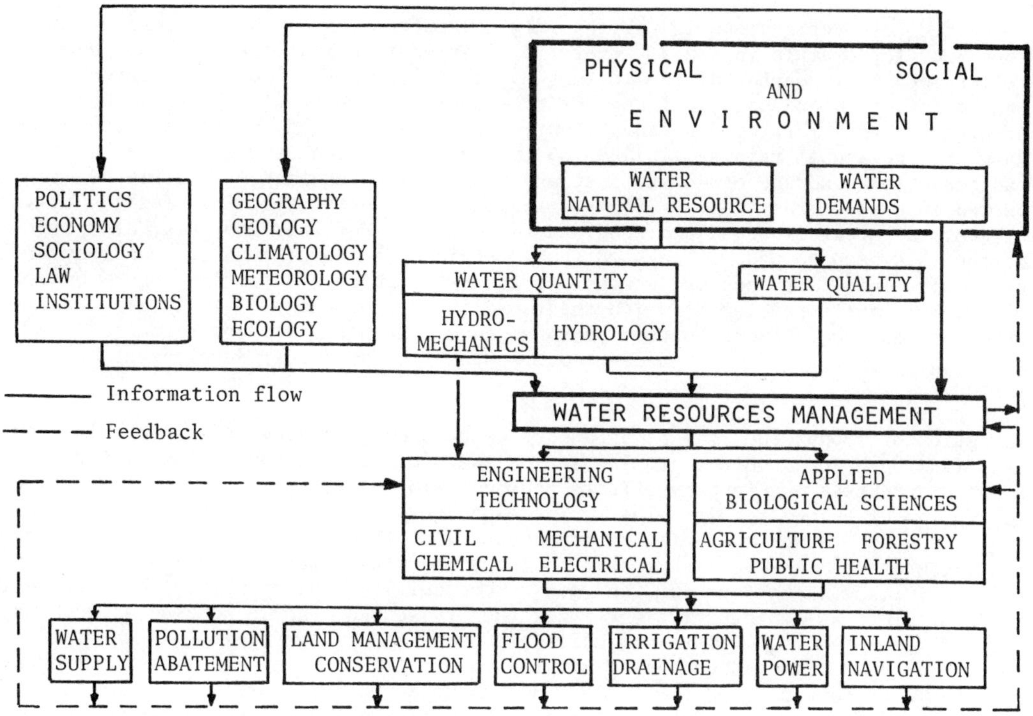

Fig. 1.12. Water resources management in context of the physical and social environment, after Klemes (1973)

For sound management it is necessary to know the distribution of available water both in space and time, as well as that of demand. The water resources planning itself can be subdivided into long term and short term planning. The aims of the long term planning are to obtain the desired distribution and optimal use of water for all purposes, for use by man, animals, plants and for recreation. The principal aspects of this planning are distribution systems, storage and the environmental effects. Probably the most difficult of these is storage. Suitable reservoir sites are very limited and involve loss of land. In addition, in warm countries the evaporation losses become very important. The implication is that the groundwater aquifers have to be used more extensively for storage of water.

Short term planning - could be separated into two kinds - *soil conservation* and *water storage*. The best returns for investment are obtained from reduction of runoff. Where land is severely over grazed or over cultivated runoff from it is very rapid, erosion is high and infiltration is low. The latter means that soil moisture storage will be small and the land becomes more arid still. Erosion means not only that soil useful for farming or forestry is lost but also that rivers will be overloaded with sediment and the deposition of it in lower reaches makes expensive protection works necessary. The Ganges delta is probably the most drastic example of this. It must be realized that soil conservation is an investment which will yield dividends in due course. River works, on the contrary, are defensive works, a continuous expenditure on protection, and yield no direct dividend. All this has been known to man for a long time, yet is still not appreciated by a vast number of people. Already in the dialogues of Plato (Critias), about 400 B.C., one can read over the effects of catchment mismanagement, loss of trees and loss of soil.

The effect of over grazing is particularly severe on grasslands with long dry seasons - as for example in India - where in addition to climatic effects the uncontrolled impact of quarter of the world's cattle is serious. Yet over grazed land can be retrieved and used in a controlled manner. Pereira (1973) shows photographs of over grazed a retrieved savanna country in the Karamoja district of Northern Uganda where the annual rainfall is 500-750 mm and evaporation 1800 mm. Over grazing had converted hundreds of km^2 to dust and thorn scrub and this was retrieved and converted into good grassland. Over grazing is at present responsible for the loss of enormous areas of grass and woodland to desert. For this the word "desertification" has been coined. The word should strictly be understood to mean loss of productive land, irrespective whether to desert or through water logging. Desertification is not a problem to which the remedies are not known. It is a problem which arises predominantly out of national, and international, power game of the countries involved. Associated are over population and under development of the region.

Selection of crops, too, may be an aspect of catchment management. Crops from arable farming have a shorter growing season and use, in most cases, less water than grasses and trees. Cropping effects the reception of rainfall, and its partition to overland flow and infiltration. The exposure of bare soil to rain, sun and wind will, however, reduce the infiltration capacity of intense rainfalls. Without protective measures the overland runoff is increased with resultant loss of *water and soil*. Tillage on the contour, and subsurface cultivation, by which the plant residuals are left as a protective cover for the dry season, should be used as much as possible. Sorghums, millets and maize are often sown in dry soil when the rains are expected, to take advantage of all the rainfall and of the release of nitrogen from the clay particles on first wetting. Further measures are ridging of the fields. *Runoff* can be reduced by construction of small dikes which prevent water from running off as overland flow into the rivers. These little flooded fields will increase the soil moisture and groundwater storage.

Although the single processes of hydrology, such as evaporation, soil moisture, stream flow, etc., are important, the emphasis of hydrological studies has shifted towards the study of all the effects on the water balance caused by a single intervention. For example, extraction of groundwater may effect plant life, evapotranspiration, stream flows, etc. The aim of these water balance models is to assess the magnitude of all the terms in the balance relationship. For every term in the water balance analytical functions are used, some of which are well defined. Others require more study and the end results depend critically on the quality of these basic relationships. The aim here is primarily to introduce these single processes and the elementary techniques.

Additional reading:

"History of Hydrology", Biswas, Asist K., North-Holland Publishing Co., 1970.
"Will there be enough water?", Maxwell, J.C., Am. Scientist, Vol. 53, 97-103, 1965.
"Water of the World", Nace, R.L., Nat. Hist., Vol. 73, No. 1, Jan. 1964.
"The Biosphere", A Scientific American Book, Freeman & Co., 1970.

"Topsoil and Civilization", Carter, V.G. and Dale, T., University of Oklahama Press Rev. Ed., 1974.
"Engineering Management of Water Quality", McGauhey, P.H., McGraw-Hill, 1968.
"Introduction to Environmental Science", Strahler, A.N. and Strahler, A.H., Hamilton Publishing Co., 1974.

Chapter 2

ELEMENTS OF METEOROLOGY

All atmospheric phenomena form the science of meteorology. The study of atmospheric water alone is called hydrometeorology. Meteorology, of course, is a science in its own right and all that can be done here is to introduce a few of the elements.

The moisture of the atmosphere may exist as true vapour (gas) in which case it is invisible and is referred to as humidity, or it may be in the form of minute drops formed upon condensation. In the latter form it is visible and is called fog or clouds, depending on the altitude. Although the lower atmosphere is made up of a mixture of gases, Table 2.1, water vapour is in many respects one of the most important constituents.

TABLE 2.1 Principal Constituents of the Atmosphere

Permanent constituents		Variable constituents	
Constituent	% by volume	Constituent	% by volume
Nitrogen (N_2)	78.084	Water vapour (H_2O)	< 4
Oxygen (O_2)	20.946	Water (liquid and ice)	< 1
Argon (A)	0.934	Ozone (O_3)	$< 0.07 \times 10^{-4}$
Carbon dioxide (CO_2)	0.033	Sulphur dioxide (SO_2)	$< 1 \times 10^{-4}$
(Global average)	(up to 0.1)	Nitrogen dioxide (NO_2)	$< 0.02 \times 10^{-4}$
Neon (Ne)	18.18×10^{-4}	Ammonia (NH_3)	Trace
Helium (He)	5.24×10^{-4}	Carbon monoxide (CO)	$\sim 0.2 \times 10^{-4}$
Kryton (Kr)	1.14×10^{-4}	Dust	$< 10^{-5}$
Xenon (Xe)	0.087×10^{-4}		
Hydrogen (H_2)	0.5×10^{-4}		
Methane (CH_4)	2.0×10^{-4}		
Nitrous oxide (N_2O)	0.5×10^{-4}		
Radon (Rn)	6×10^{-18}		

Ozone, which is mainly found in the layer from 25-60 km, and the bulk of it at the lower part of this layer, is important in that it prevents harmful ultra-violet radiation from penetrating down to the lower levels where biological processes occur. Carbon dioxide together with water vapour absorbs some of the longwave radiation from the earth and thus influences the heat balance. The amount of carbon dioxide in the atmosphere varies; it is consumed by the plants and produced by animal and industrial burning processes. It has been estimated that there has been a 10% net increase in the amount of carbon dioxide in the atmosphere in this century. This has been linked with the rise of air temperatures, particularly in the middle and high altitudes, during the same period.

Figure 2.1 illustrates a number of the features of the atmosphere. One of the main features is the number of more or less concentric shells with different but approximately constant temperature. The thickness of the troposphere averages 8 to 18 km depending on latitude and local weather, being thickest over the equator and thinnest over the poles. The variations of altitude of the tropopause are associated with surface temperatures. It takes a greater height for the higher temperatures of tropics to decrease at constant lapse rate, dT/dz, to the temperature of the stratosphere. The troposphere contains about 75% of the mass of atmosphere, all significant moisture, weather and most of dust. The lower region up to 2 km is the most disturbed. It is subject to frictional and diurnal influences, and frequently to temperature inversions. The lowest 2 m are known as the micro-layer and is associated with micro-meteorology and micro-climatology.

The rise of temperature in the stratosphere is due to selective absorption of UV-radiation by the ozone, but since the distribution of the ozone is not uniform over the globe and varies with season, the temperature in the stratosphere varies with space and time. This leads to strong seasonal circulation in this layer. The mesosphere is a transition layer between the two energy absorbing layers, the stratosphere where the ozone reaches its peak concentration and the thermosphere above which dissociation and ionization of O_2 and N_2 are accompanied by heating. The temperatures in the thermosphere should not be understood to mean that at 300 km we would feel 1000 °C. The individual particles have energies corresponding to that temperature but there are so few gas particles that the total energy transmitted to a body through collisions is very small.

The amount of water vapour present in the air may be expressed as the pressure that the vapour would exert if the other gases were absent; the vapour pressure $p_v = e_v$ = e millibars (1 millibar = 10^3dyne/cm^2 = 0.75 mm Hg = 100 Pa; 1 dyne = 10^{-5}N). For any given temperature there is a maximum amount of water vapour in the space or air. This is known as the saturation condition and the temperature is called the dew point. Under saturation conditions, water vapour (or air containing water vapour) and a plane surface of pure water at the same temperature are in equilibrium.

The ratio of $p_v = e$ to saturation vapour pressure $p_s = e_s$ at a given temperature, expressed in percentage, is known as the *relative humidity*

$$H_r = 100 \frac{e}{e_s} \qquad 2.1$$

It expresses the actual quantity of vapour in the air in terms of that at saturation at the same temperature. The vapour density is

$$\rho_v = \frac{e}{R_v T} \qquad 2.2$$

where

$$R_v = \frac{m_a}{m_v} R_a = 1.609\, R_a \doteqdot 8/5\, R$$

and m_a and m_v are the molecular weights of *dry air* and *water vapour*, respectively.

Specific humidity, H_s or q, is defined as the ratio of the mass of the water vapour

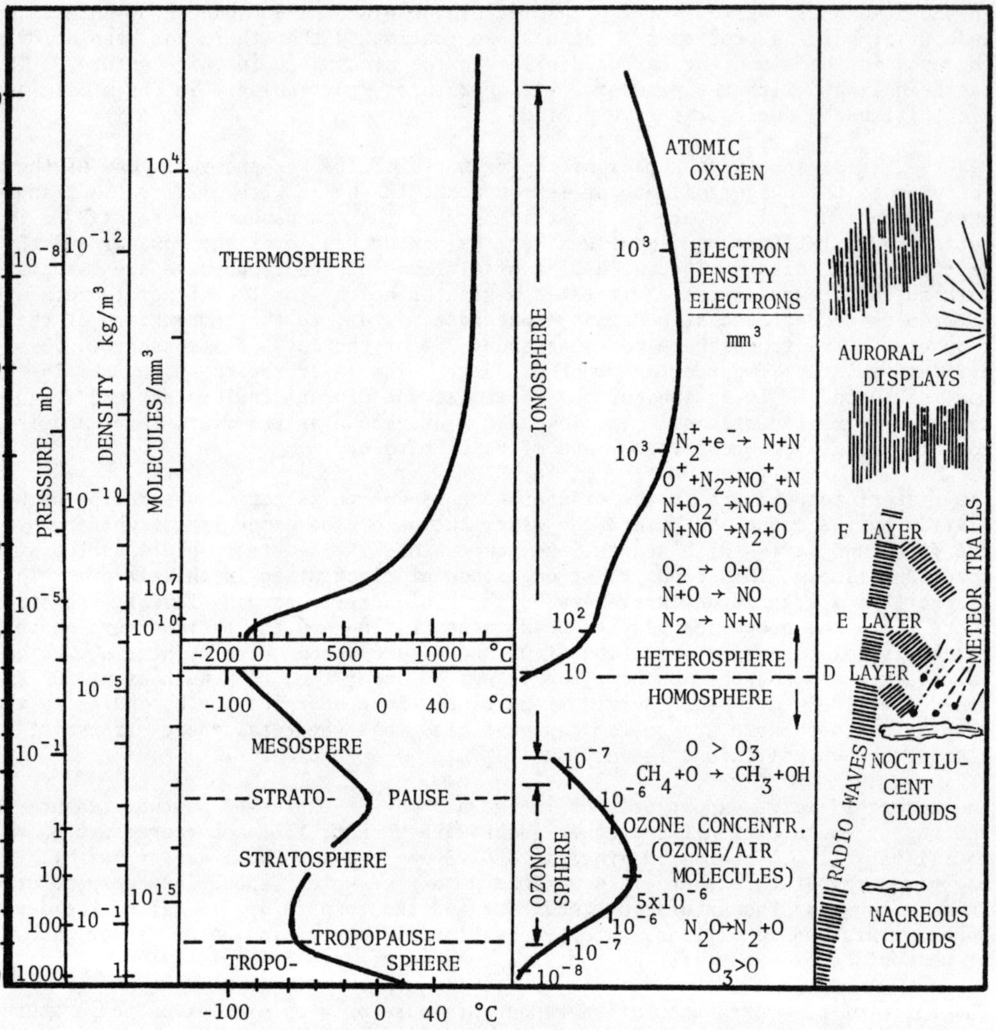

Fig. 2.1 Vertical structure of the atmosphere.

to the mass of most air containing the water vapour. The ratio of the mass of water vapour to the mass of dry air containing this vapour is known as the mixing ratio M_r or w

$$M_r = \frac{\rho_v}{\rho_a} = \frac{e}{R_v T}\,\frac{R_a T}{p - e} = \frac{em_v}{m_a R^* T}\,\frac{R^* T}{p - e} = \varepsilon\,\frac{e}{p - e}$$

$$= 0.622\,\frac{e}{p - e} \doteqdot (5/8)\,\frac{e}{p} < 0.04 \qquad 2.3$$

where $\varepsilon = m_v/m_a = 0.622$, $R^* = mR$, and p-e is the *partial pressure* of the dry air, i.e. $p = p_a + e$. Hence,

$$H_s = \frac{\rho_v}{\rho_a + \rho_v} \doteq \frac{5e}{8p - 3e} \qquad 2.4$$

Specific humidity is not affected by pressure and temperature changes unless water is added or subtracted. Table 2.2 gives some abbreviated data on saturation humidity.

Air itself is a *mixture* of gases and its behaviour is reasonably well approximated by the ideal gas laws. The behaviour of mixtures is usually explained with the aid of Dalton's law which postulates that the pressure of the mixture is the sum of the pressures of the constituents if these, one by one alone, filled the entire volume occupied by the mixture at the temperature of the mixture, that is

$$p = p_1 + p_2 + p_3 \cdots = \sum_1^k p_n$$

If V is the volume of the mixture and M_n the mass of a constituent, and m_n its molecular weight, then, if each gas separately is assumed to follow the ideal gas law $p\mathrm{v} = RT$,

$$p_n = \frac{R^*}{m_n} \frac{M_n}{V} T \ , \quad R^* = mR \ , \quad p = \Sigma p_n = \frac{R^* T}{V} \Sigma \frac{M_n}{m_n}$$

and

$$p\mathrm{v} = R^* T(\Sigma M_n/m_n)/\Sigma M_n$$

where v is the specific volume, $\mathrm{v} = 1/\rho$. Thus, if we define a mean molecular weight as

$$1/\bar{m} = (\Sigma M_n/m_n)/\Sigma M_n \qquad 2.5$$

the mixture will satisfy the ideal gas law for a single gas.

If we now look upon dry air as one constituent and water vapour as the other then

$$\frac{1}{\bar{m}} = \frac{1}{M_a + M_v} \left[\frac{M_a}{m_a} + \frac{M_v}{m_v} \right] = \frac{1}{m_a} \frac{M_a}{M_a + M_v} \left[1 + \frac{M_v/M_a}{m_v/m_a} \right]$$

where M_v/M_a = the mixing ratio M_r. Writing

$$\frac{m_v}{m_a} = \varepsilon \doteq 0.622 \sim 5/8$$

and rearranging yields

$$\frac{1}{\bar{m}} = \frac{1}{m_a} \frac{1 + M_r/\varepsilon}{1 + M_r}$$

This gives the equation of state for the moist air

$$p\mathrm{v} = \frac{R^*}{m_a} \left[\frac{1 + M_r/\varepsilon}{1 + M_r} \right] T \qquad 2.6$$

Rather than have a variable gas constant the bracket term is included with the temperature and called the virtual temperature T^*, whence

$$p\mathrm{v} = RT^* \qquad 2.7$$

TABLE 2.2a Saturation Vapour Pressure Over Water and Over Ice in mb

Tens	Temperature °C Units 0	1	2	3	4	5	6	7	8	9
41	73.777	77.802	82.015	86.423	91.034	95.855	100.89	106.16	111.66	117.40
30	42.430	44.927	47.551	50.307	53.200	56.236	59.422	62.762	66.264	69.934
20	23.373	24.861	26.430	28.086	29.831	31.671	33.608	35.649	37.796	40.055
10	12.272	13.119	14.017	14.969	15.977	17.044	18.173	19.367	20.630	21.964
+0	6.1078	6.5662	7.0547	7.5753	8.1294	8.7192	9.3465	10.013	10.722	11.474
-0	6.1078	5.623	5.173	4.757	4.372	4.015	3.685	3.379	3.097	2.837
	*6.1078**	*5.6780*	*5.2753*	*4.8981*	*4.5451*	*4.2148*	*3.9061*	*3.6177*	*3.3484*	*3.0971*
-10	2.597	2.376	2.172	1.984	1.811	1.652	1.506	1.371	1.248	1.135
	2.8627	*2.6443*	*2.4409*	*2.2515*	*2.0755*	*1.9118*	*1.7597*	*1.6186*	*1.4877*	*1.3664*
-20	1.032	0.9370	0.8502	0.7709	0.6985	0.6323	0.5720	0.5170	0.4669	0.4213
	1.2540	*1.1500*	*1.0538*	*0.9649*	*0.8827*	*0.8070*	*0.7371*	*0.6727*	*0.6134*	*0.5589*
-30	0.3798	0.3421	0.3079	0.2769	0.2488	0.2233	0.2002	0.1794	0.1606	0.1436
	0.5088	*0.4628*	*0.4205*	*0.3818*	*0.3463*	*0.3139*	*0.2842*	*0.2571*	*0.2323*	*0.2097*
-40	0.1283	0.1145	0.1021	0.09098	0.08097	0.07198	0.06393	0.05671	0.05026	0.04449
	0.1891	*0.1704*	*0.1534*	*0.1379*	*0.1239*	*0.1111*	*0.09061*	*0.08948*	*0.07975*	*0.07124*
-50	0.03935	0.03476	0.03067	0.02703	0.02380	0.02092	0.01838	0.01612	0.01413	0.01236

* Values over water at subfreezing temperatures.

TABLE 2.2b Quantity of Water Vapour Required for Saturation

(Saturation specific humidity, in g/kg)

°C	Pressure mb 1000	900	800	700	600	500	400	Abs. hum. gm^{-3}	Mix. ratio g/kg
-40	0.118	0.131	0.147	0.168	0.196	0.235	0.294		
-35	0.195	0.217	0.244	0.279	0.326	0.391	0.488		
-30	0.317	0.353	0.397	0.453	0.529	0.635	0.793		
-25	0.503	0.559	0.629	0.719	0.839	1.007	1.259		
-20	0.784	0.871	0.980	1.120	1.307	1.569	1.962		
-15	1.20	1.33	1.49	1.71	1.99	2.39	2.99		
-10	1.79	1.99	2.23	2.55	2.98	3.58	4.48		
- 5	2.63	2.92	3.29	3.76	4.39	5.27	6.59		
0	3.80	4.23	4.76	5.44	6.35	7.62	9.54	4.86	3.84
5	5.44	6.05	6.81	7.79	9.09	10.92	13.67	6.81	5.50
10	7.67	8.53	9.60	11.0	12.8	15.4		9.41	7.76
15	10.7	11.9	13.4	15.3	17.9			12.83	10.83
20	14.7	16.3	18.4	21.1				17.31	14.95
25	20.0	22.2	25.0					23.60	20.44
30	26.9	29.9	33.7					30.40	27.69
35	35.8	39.8							
40	47.3								

Since water vapour is less dense than air, vapour increases the specific volume and the virtual temperature is always higher than the observed one. With M_r < 40/1000 the virtual temperature $T^* \leq 1.0234$ T. (Meteorologists use more than a dozen different temperatures. For the definition of these the reader could refer to Berry et al. (1945) and a few are summarized in Section 2.6).

2.1 Atmospheric Thermodynamics

When an air mass ascends it is subject to a gradually decreasing pressure and it expands. As air expands it cools. The pressure change is according to the hydrostatic relationship

$$dp = -\rho g\, dz \tag{2.8}$$

For ideal dry air

$$p = \rho RT \qquad \text{or} \qquad p = \rho RT^* \tag{2.9}$$

for moist air.

The energy equation per unit mass in differential form is

$$\Delta Q - \Delta W = dE = de_i + pd\mathrm{v} + \mathrm{v}dp + gdz + d(\frac{V^2}{2}) = dH + gdz + d(\frac{V^2}{2}) \tag{2.10}$$

where $H = e_i + p\mathrm{v}$ is the enthalpy, Q the heat added and W is work done by shear stresses at boundaries or mechanical work added or extracted. In air, at low velocities, the work by shear stresses amounts to a small fraction of the total energy balance and may be ignored in this elementary treatment. The left-hand side depends on initial and final states only. By virtue of the Euler equation

$$\mathrm{v}dp + d(V^2/2) + gdz = 0$$

the energy equation becomes

$$\Delta Q - \Delta W = dH - \mathrm{v}dp = de_i + pd\mathrm{v} \tag{2.11}$$

The pressure-specific volume relationship for water is shown diagrammatically in Fig. 2.2. The kinetic theory of gases shows that the internal energy $e_i = f(T)$ only. Thus, if heat is added at constant volume $pd\mathrm{v} = 0$ and

$$\Delta Q = de_i = c_v dT \tag{2.12}$$

where the specific heat at constant volume as defined here is not necessarily a constant and is to be determined by experiment. More generally

$$de_i = \frac{\partial e_i}{\partial \mathrm{v}} d\mathrm{v} + \frac{\partial e_i}{\partial T} dT = \frac{\partial e_i}{\partial \mathrm{v}} d\mathrm{v} + c_v dT$$

i.e. $e = f(\mathrm{v},T)$ but for an ideal gas $\partial e/\partial \mathrm{v} = 0$ and is also very close to zero for real gases under usual conditions. Hence,

$$\Delta Q = c_v dT + pd\mathrm{v} \tag{2.13}$$

In texts of thermodynamics, it is shown that

$$c_p - c_v = R \tag{2.14}$$

and differentiation of $p\mathrm{v} = RT$ yields

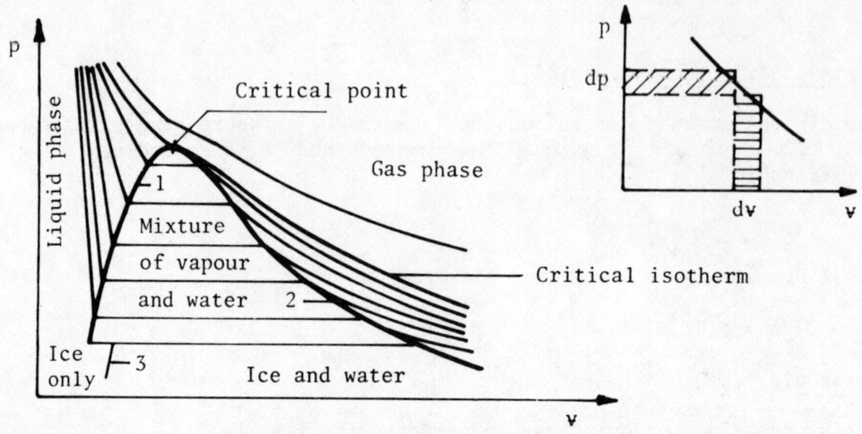

Fig. 2.2. The pressure-specific volume diagram for water. 1 - saturated liquid line, 2 - saturated vapour line, and 3 - saturated ice line for $T < 0$ °C. Critical point is at $T = 647$ °K, $p = 21.8$ MPa (218 atm), and $v = 2.50\ cm^3/g$. Water vapour at $T > T_c$ cannot be pressed into liquid.

$$p\,dv + v\,dp = R\,dT \qquad 2.15$$

Substituting from eqns 2.14 and 2.15 in eqn 2.13 leads to

$$\Delta Q = c_p dT - v\,dp$$

which upon substitution of $dp = -\gamma\,dz$ becomes

$$\Delta Q = c_p dT + g\,dz \qquad 2.16$$

If the process is assumed to be adiabatic with $\Delta Q = 0$, the rate of cooling is

$$-\frac{dT}{dz} = \frac{g}{c_p} = \alpha = \text{lapse rate.} \qquad 2.17$$

This relationship would apply to an ascending parcel for dry air and is approximately 0.01 °C/m (5.4 °F/1000 ft). Between $dp = -\rho g\,dz$ and $p/\rho = RT$, $dp/dz = -gp/RT$ and substituting this for dz in dT/dz yields

$$\alpha = \frac{gp}{RT}\frac{dT}{dp} \qquad 2.18$$

2.1.1 Changes of phase and latent heat.

When changes of phase occur (evaporation, condensation, melting, freezing, sublimation) a quantity of heat must be added or taken away while the temperature remains constant , the *latent heat* of phase change. By the first law of thermodynamics

$$\Delta Q = de_i + e_s dv \qquad 2.19$$

where e_s is the saturation water vapour pressure. The latent heat for change from phase 1 to phase 2 is

$$L_{12} = \int_2^1 \Delta Q$$

where e_s and T remain constant during the integration. Thus,

$$L_{12} = e_{i2} - e_{i1} + e_s(v_2 - v_1) \qquad 2.20$$

The internal energy e_i is f(T) only for an ideal gas or vapour but for liquids and solids it depends on specific volume as well.

If phase 1 is water and 2 vapour then L is the latent heat of evaporation and e_s is the saturation vapour pressure. Likewise for sublimation, ice to vapour. For ice to water L is the latent heat of melting and e_s is the equilibrium pressure of the mixture. A triple point is possible

$$L_{sublimation} = L_{melting} + L_{evaporation}$$

obtained by combining the single statements. For evaporation

$$L_{ev} = e_{iv} - e_{iw} + e_s(v_v - v_w) \qquad 2.21$$

but $v_{water} << v_{vapour}$ and may be ignored, and replacing v_v from $pv = RT$ by

$$v_v = \frac{m_v R^* T}{p} = \frac{R_v T}{e_s}$$

yields

$$L_{ev} = e_{iv} - e_{iw} + R_v T$$

and

$$\frac{dL_{ev}}{dT} = \frac{de_{iv}}{dT} - \frac{de_{iw}}{dT} + R_v \qquad 2.22$$

But from eqn 2.13 for v = const.

$$(c_v)_{vapour} = \left(\frac{dQ}{dT}\right)_v = \frac{de_{iv}}{dT} \qquad 2.23$$

where the last ratio is by virtue of eqn 2.12. From eqn 2.11 for $\Delta W = 0$ the specific heat of water is

$$c_w = \frac{dQ}{dT} = \frac{de_{iw}}{dT} + e_s \frac{dv_w}{dT} \doteqdot \frac{de_{iw}}{dT} \qquad 2.24$$

since dv_w/dT is negligibly small. Thus, eqn 2.22 upon substitution from eqns 2.23 and 2.24 becomes

$$\frac{dL_{ev}}{dT} \doteqdot c_v|_v - c|_w + R_v = c_p|_v - c|_w \qquad 2.25$$

since $c_p - c_v = R$.

It follows that the rate of change of the latent heat of evaporation with absolute temperature is equal to the difference between the specific heat at constant pressure of the vapour and the specific heat of water. Numerically

$$\frac{dL_{ev}}{dT} = -\ 2.3697\ \text{kJ kg}^{-1}\,{}^{\circ}\text{C}^{-1}$$

a rate of change of about 0.1% at 0 °C

$$L_{ev} = 2500.78\ \text{kJ kg}^{-1}\ \text{at 0 °C}$$

$$L_{sub} = 2834\ \text{kJ kg}^{-1}\ \text{at 0 °C}$$

$$L_{melt} = 334\ \text{kJ kg}^{-1}\ \text{at 0 °C}$$

$$L_{ev} = 2501 - 2.37(T - 273)\ \text{kJ kg}^{-1} = 2264\ \text{kJ kg}^{-1}\ \text{at 100 °C}$$

These latent heats are large compared to other known substances and they have an important influence on microclimate.

2.1.2 Entropy and the Poisson equation

The latent heat can be related to other thermodynamic variables with the aid of the entropy concept. The first law of thermodynamics

$$dQ = c_v dT + p d\mathrm{v} = c_p dT - \mathrm{v} dp$$

integrated around a closed curve

$$\oint dQ = \oint c_v dT + \oint p d\mathrm{v}$$

shows that $\oint dQ \neq 0$. For a perfect gas $c_V dT$ is an exact differential and its integral over a closed curve is zero. The integral $p d\mathrm{v}$ is the work done in the process and is the area within the curve on the p-v diagram, i.e.,

$$\oint dQ = \oint p d\mathrm{v} = \text{work} \neq 0$$

Hence, neither dQ nor $p d\mathrm{v}$ are exact differentials and their integrals depend on the path chosen.

However, dividing through by temperature T, and observing that $p\mathrm{v} = RT$ yields

$$\frac{dQ}{T} = c_p \frac{dT}{T} + \frac{\mathrm{v}dp}{T} = c_p \frac{dT}{T} - R\frac{dp}{p} = c_p d(\ell n\ T) - Rd(\ell n\ p) \qquad 2.26$$

Now, both terms on the right are exact differentials, hence dQ/T must be an exact differential also. The ratio

$$dS = \frac{dQ}{T} \qquad 2.27$$

is called the entropy of the system and is a property of the system. The units of entropy are J °C^{-1}. For an adiabatic process dQ = 0 and

$$0 = c_p \frac{dT}{T} - R\frac{dp}{p} = \frac{dT}{T} - \frac{R}{c_p}\frac{dp}{p} = d(\ell n\ T) - \kappa d(\ell n\ p)$$

$$= d(\ell n\ T) - d(\ell n\ p^{\kappa})$$

where

$$\kappa = R/c_p = 0.286 \text{ for dry air.}$$

Thus,

$$T = \text{const } p^{\kappa}$$

or

$$\frac{T}{\theta} = \left(\frac{p}{1000}\right)^{\kappa} \qquad 2.28$$

which is known as the Poisson equation. In meteorology θ is called the potential temperature - it may be looked upon as the temperature a sample of gas would have if compressed or expanded adiabatically from a given state p and T to a pressure of 1000 mb. The θ constant lines are called (dry) adiabats.

Example 2.1 Plot the T and θ isotherms on the p-v plane for dry air at T = 300 °K using mb and cm^3/g.

Here the basic relationships are $pv = RT$ and $T/\theta = (p/1000)^{\kappa}$. The mean molecular weight of dry air is m = 28.966 g/mol and R* = 8.3144 J/mol °K = mR. Hence, R = 0.2870 J/g °K and

$$p = \frac{RT}{v} = \frac{0.28704 \times 300}{v} \left(\frac{J}{cm^3}\right) = \frac{86.112 \times 10^6}{v} \left(\frac{Nm}{m^3}\right) = \frac{8.6112}{v} 10^5 mb$$

since $mb = 100\ Nm^{-2}$. For dry air κ = 0.286 and

$$v(cm^3/g) = \frac{R\theta}{1000^{\kappa}} \frac{p^{\kappa}}{p} = \frac{11.94176 \times 10^4}{p^{0.714}}$$

v cm^3/g	700	800	861.1	900	1000	1500	2000
p mb	1230.2	1076.4	1000	956.8	861.1	574.1	430.6
$p^{0.714}$	160.8	146.2	138.7	134.4	124.6	93.3	76.0
v cm^3/g	724.7	817.0	861.1	888.7	958.2	1279.8	1571.6

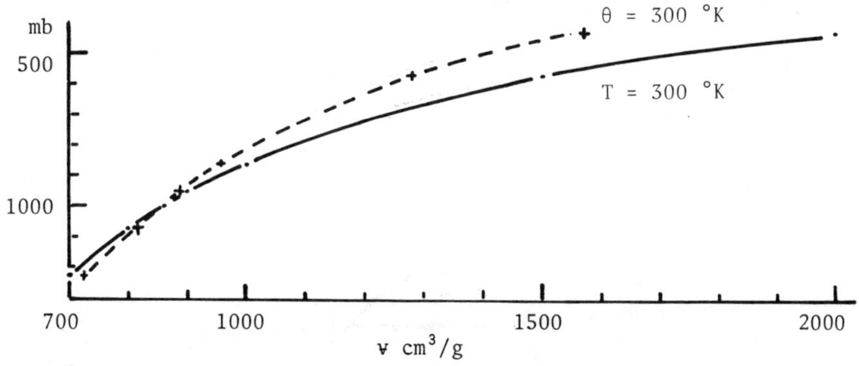

Ex. 2.1 The T and θ isotherms on the p-v plane.

2.1.3 The latent heat - pressure - temperature - specific volume relationships

Logarithmic differentiation of eqn 2.28 yields

$$c_p \frac{d\theta}{\theta} = c_p \frac{dT}{T} - R \frac{dp}{p}$$

or

$$c_p d(\ell n\ \theta) = c_p d(\ell n\ T) - Rd(\ell n\ p)$$

which is identical to the entropy expression, eqn 2.26, i.e.

$$dS = c_p d(\ell n\ \theta)$$
$$S = c_p\ \ell n\ \theta + \text{const}$$

From eqn 2.27 it follows that

$$L_{12} = \int_1^2 dQ = \int_1^2 Tds$$

and since a *reversible phase change* is isothermal

$$L_{12} = T(S_2 - S_1) \qquad 2.29$$

The term in brackets, $S_2 - S_1$, can be evaluated with the aid of eqns 2.19 and 2.27. These give

$$dQ = TdS = de_i + e_s dv$$

and integrating over a closed path yields

$$\oint TdS = \oint e_s dv$$

because the line integral of de_i over a closed curve is zero. From Fig. 2.3 the integral $\oint e_s dv \doteq (v_2 - v_1)\Delta e_s$ or

$$\oint TdS = (v_2 - v_1)\Delta e_s \qquad 2.30$$

From the exact differential d(TS)

$$\oint d(TS) = \oint TdS + \oint SdT = 0$$

it follows that

$$\oint TdS = - \oint SdT$$

From Fig. 2.3 it is seen that

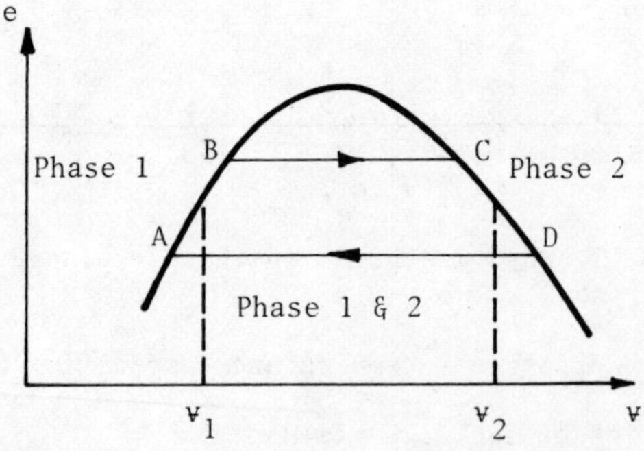

Fig. 2.3 Work done, $\int e dv$, on vapour pressure-specific volume diagram.

$$-\int_B^A SdT = -S_1\Delta T, \text{ since } e = f(T), \quad -\int_B^C SdT = 0,$$

$$-\int_D^A SdT = 0, \quad -\int_C^D SdT = -S_2\Delta T$$

where integrals BC and DA are zero because the process is isothermal. Thus,

$$\oint TdS = (S_2 - S_1)\Delta T$$

and using eqn 2.30

$$(S_2 - S_1)\Delta T = (v_2 - v_1)\Delta e_s$$

Introduction of this in eqn 2.29 yields

$$L_{12} = (v_2 - v_1)T\frac{de_s}{dT}$$

or

$$\frac{de_s}{dT} = \frac{L_{12}}{T(v_2 - v_1)} \qquad 2.31$$

which is the Clausius-Clapeyron equation. The relationship is diagrammatically shown in Fig. 2.4.

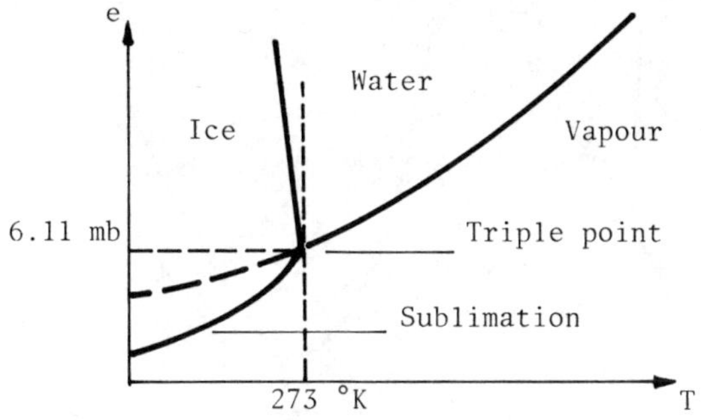

Fig. 2.4 The vapour pressure-temperature relationship for water.

If phase 2 is vapour and 1 water then $v_2 >> v_1$ and using $pv = RT$ with $R^* = m_v R_v$, where m is the molecular weight,

$$\frac{de_s}{dT} \simeq \frac{L_{12}}{Tv_2} = \frac{m_v L_{12} e_s}{R^* T^2}$$

or

$$\ell n\ e_s = -\frac{m_v L_{12}}{R^* T} = \text{const}$$

For T = 273 °K, e_s = 6.11 mb is constant and hence,

$$\ln \frac{e_s}{6.11} = \frac{m_v L_{ev}}{R^*} \left(\frac{1}{273} - \frac{1}{T}\right) \qquad 2.32$$

(Empirical expressions for e_s are given in Section 2.6). The same expression is also valid for L_{subl}. Since the specific volume of ice is only slightly larger than that of water, the right-hand side of the Clapeyron equation is large in magnitude. If these two volumes were equal the melting line would be vertical, but v_1 is slightly larger than v_2 and the slope is slightly negative. Along the melting curve water and ice are in equilibrium.

Liquid water can exist between 0 and -39 °C, so-called supercooled water. The dashed line shows the evaporation line from supercooled water. The saturation vapour pressure over supercooled water is higher than that over ice at the same temperature because $L_{subl} > L_{ev}$. Thus, if one introduces a crystal of ice into a cloud of supercooled water droplets, the vapour will rapidly condense on the ice, leading to snowflakes.

The e-v and e-T relationships can be combined into a three-dimensional plot, Fig. 2.5. Each area where the isotherms are horizontal is an area of equilibrium between two or more phases.

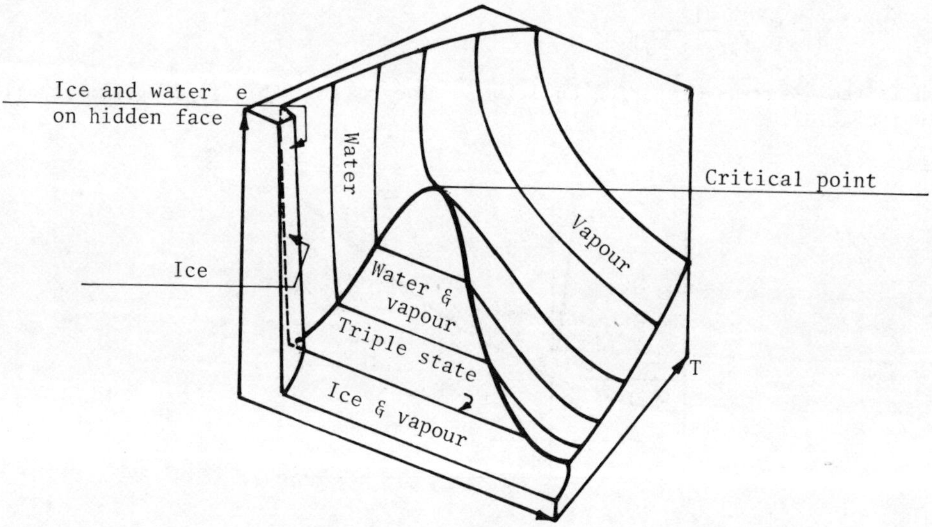

Fig. 2.5. Diagrammatic presentation of the vapour pressure-specific volume-temperature relationship for water.

The temperature changes in dry air and adiabatic processes are given by the Poisson's equation and the lapse rate expression. The potential temperature remains constant (and entropy remains constant). With moist air, however, if condensation occurs the latent heat of condensation is released. Hence, the rate of cooling of saturated air - as it expands adiabatically - is less than that of dry air, because part of the cooling is compensated for by the latent heat released. If we make the assumption that none of the condensed water falls out in the form of rain, hail or snow and all of it is carried along in the same upward moving current, then a *reversible condensation* process is possible, i.e. on descent the latent heat of condensation is available for evaporation. But if the condensation products fall out the mass and composition of the air parcel change, the heat of condensation cannot be returned when the parcel is warmed and the process is irreversible.

The real atmospheric problems lie between these extremes, but even with fairly moist air, say, 5g of water per kg of dry air, the error caused by ignoring the non-reversible features is small, i.e. from $\Delta Q = c_v dT$ the ratio

$$\frac{M_w}{M_a}\frac{c_w}{c_v}\frac{\Delta T}{\Delta T} = \frac{5}{1000}\frac{1.0}{0.17}\frac{1}{1} = \frac{1}{34} \sim 3\%$$

An approximate treatment of the irreversible process is known as the *pseudo-adiabatic* process.

Example 2.2 Determine for an air mass at 973 mb, 23 °C and with relative humidity of 86% the mixing ratio, the specific humidity, the dew point and virtual temperatures, the gas constant for moist air and the potential temperature.

By eqn 2.3 the mixing ratio is

$$M_r = w = 0.622\,\frac{e}{p - e}$$

At 23 °C the saturation vapour pressure e_s can be read from Table 2.2 as 28.086 mb or it could be estimated from

$$e_s = 6.11 \times 10^{7.5t/(t+237.3)} = 28.1 \text{ mb.}$$

The vapour pressure is

$$e_s = 0.86 \times 28.086 = 24.15 \text{ mb}$$

and the mixing ratio is

$$w = 0.622 \times 24.15/(973 - 24.15) = 0.0158$$

By eqn 2.4 the specific humidity is

$$H_s = \rho_v/(\rho_a + \rho_v) = w/(1 + w) = 0.01558.$$

For a vapour pressure of 24.15 mb the dew point temperature (interpolated from Table 2.2) is 20.5 °C. From eqn 2.6 the virtual temperature is

$$T^* = 296[(1 + 0.0158/0.622)/(1 + 0.0158)] = 298.8 \text{ °K}$$

or alternatively $R_{ma} = R[(1 + w/\varepsilon)/(1 + w)] = 0.28704 \times 1.00947 = 0.28976$ J/g °K. The potential temperature is given by eqn 2.28 as

$$\theta = 296(1000/973)^{0.286} = 298.33 \text{ °K}$$

if $\kappa = 0.286$ for dry air is used.

2.1.4 Processes during the ascent and descent of a parcel of air.

Before development of the expressions for the pseudo-adiabatic process it will be helpful to discuss what happens to a parcel of air as it is lifted, see Fig. 2.6. On ascent the air parcel cools at the dry adiabatic lapse rate of 1 °C/100 m (eqn 2.18), until saturation is reached. The saturation or condensation level depends on initial temperature and vapour content. Note, that the temperature of the condensation level is different from that of dew point. The dew point temperature is for cooling at constant pressure.

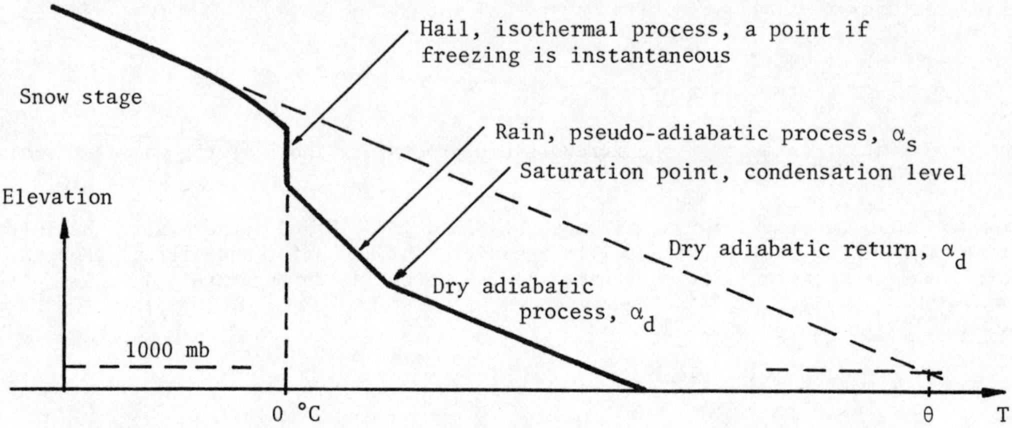

Fig. 2.6. Diagrammatic illustration of processes during the ascent of a parcel of air.

After saturation the air cools at the saturation - adiabatic rate α_s, (yet to be derived). Under the reversible assumption the droplets are carried along with the upward moving air (for example, orographic lift) until a stage is reached where through nucleation and coalescence the droplets have grown to such a size that they fall out under the action of gravity. This is the rain stage. As the parcel of air continues to rise and to cool (still moving at α_s) it comes to the limit of the rain stage when 0 °C is reached and the water begins to freeze. A large amount of heat - the heat of fusion 334 J/g - is given up by the water to the surrounding air. During this time when water changes to ice, there is essentially no further cooling of the air. The heat of fusion has a nullifying effect on the expansional cooling, but the air parcel continues to rise. This rapid freezing at essentially isothermal conditions produces hail. Compare this with heat of condensation liberated in the rain stage, which merely slows down the rate of adiabatic change. This is because most of the water freezes at the one temperature of 0 °C, while water vapour condenses to liquid a little at a time over a range of temperatures.

Snow stage. After the hail stage there is still some water vapour left in the air. The temperatures are now below 0 °C, and the remaining vapour condenses directly into ice crystals, i.e. crystallization or sublimation. This process continues as the air keeps on rising until all water vapour is condensed and then the process is again dry-adiabatic. Sublimation gives off heat in the same way as condensation but the amount is greater because it combines both the heat of condensation and the heat of fusion. However, the temperature and amounts of moisture in the snow stage are very low and the rate of cooling is faster than in the rain stage. The ice carried along from the hail stage has only a slight effect, i.e. only a very small amount of heat is given off.

During a "reversible process" the rain, hail, and snow are carried along through the successive stages. If at any time the rising air parcel starts to fall, it will follow through the same processes, but in reverse direction. The snow would evaporate, the hail would melt and water would evaporate, each of these processes absorbing from the element of air the same amount of heat given to it during condensation, fusion and crystallization. When the rain, hail and snow precipitate we have the pseudo-adiabatic process which is not reversible. The rain stage differs only in the loss of a small amount of heat which would have been given off by the liquid had it not fallen out. The hail stage is non-existent because no liquid (only vapour) remains to be frozen. The snow stage, like the rain stage, again differs only by a small amount.

When this parcel of air starts moving downward, and is compressed, the heat it had gained through condensation etc., is retained because there is no water to melt or evaporate. The parcel follows a dry-adiabatic course downwards and arrives at the initial starting point at a higher temperature than it had when it started its ascent. At 1000 mb pressure this temperature is the potential temperature.

2.1.5. Energy equation and lapse rate in saturated air.

The lapse rate for the pseudo-adiabatic process of saturated air can be derived from the first law of thermodynamics, but the derivation is not as simple as for dry air. Simplifying assumptions have to be made in order to arrive at reasonably convenient expressions for routine calculations. For an adiabatic expansion the increments dp, dT and dw are all negative, where w is the mixing ratio. The condensation (and loss of) - dw grams of water vapour releases - Ldw joules of heat.

In the air and water vapour mixture $w \ll 1$, so that the amount of heat taken up by the remaining vapour may be neglected by comparison with that of air and the first law of thermodynamics may be approximated to

$$\Delta Q = c_p dT - v dp$$

or

$$- L dw = c_p dT - \frac{RT}{p - e_s} d(p - e_s)$$

on substitution for v from $pv = RT$. Observing that $w = M_r = \rho_v/\rho_a = \varepsilon e_s/(p - e_s)$ where $\varepsilon = 0.622 \simeq 5/8$, eqn 2.3, and $e_s = f(T)$ only shows that the equation is a relationship between p and T for the pseudo-adiabatic process. A very small error is made by neglecting e_s. Thus,

$$- L \frac{dw}{T} = c_p \frac{dT}{T} - R \frac{dp}{p} \qquad 2.33$$

Substituting from $p = \rho RT$ for T, and $dp = - \rho g\, dz$, yields

$$- \frac{L}{c_p} \frac{dw}{dz} = \frac{dT}{dz} + \frac{g}{c_p} \qquad - \frac{dT}{dz} = \frac{g}{c_p} + \frac{L}{c_p} \frac{dw}{dz}$$

or

$$- \frac{dT}{dz} = \alpha_{\text{dry air}} + \frac{L}{cp} \frac{dw}{dz} = \alpha_s$$

This, however, is not a convenient form for use. A more convenient approximate form can be obtained as follows:
From mixing ratio definition $w \sim \varepsilon\, e_s/p$ and logarithmic differentiation of this

$$\frac{dw}{w} = \frac{dp}{e_s} - \frac{dp}{p}$$

From $dp = - \rho g\, dz$ and $p = \rho RT$

$$\frac{dp}{p} = - \frac{g}{RT} dz$$

where R is that of dry air. Substitution of these in eqn 2.33 yields

$$- Lw \left[\frac{de_s}{e_s} + \frac{g}{RT} dz\right] = c_p dT + g dz$$

Multiplying de_s/e_s by dT/dT, dividing through by dz and rearranging yields

$$\alpha_s = -\frac{dT}{dz} = g\left[\frac{1 + \varepsilon L e_s/pRT}{c_p + (\varepsilon L/p)(de_s/dT)}\right]$$

From the Clausius-Clapeyron equation (2.31) and from $w \simeq \varepsilon e_s/p$

$$\frac{de_s}{dT} = \frac{m_v L 12}{R^* T^2}\frac{wp}{\varepsilon} = \frac{L}{R_v}\frac{wp}{T^2}$$

Thus,

$$\alpha_s = \frac{g}{cp}\left[\frac{1 + (L/R)w/T}{1 + (\varepsilon L^2/c_p R_v)w/T^2}\right]$$

Thus, α_s either as dT/dz or dT/dp, is expressed as α_{dry} times a factor and is always less than α_d. This result, of course, is only an approximation. For more exact treatment see Brunt (1939) or Haurwitz (1941).

In these relationships the specific heat at constant pressure is assumed to be that of dry air. For a mixture we should write

$$c_p^* = c_p\left[\frac{1 + (c_{pw}/c_p)w}{1 + (m/mw)w}\right] \simeq c_p\frac{1 + 1.95w}{1 + 1.609w} \qquad 2.35$$

i.e. $c_{pw} = 1.951$ and $c_p = 1.005$ J/g °C for dry air. If $w = 0.04$ then $c_p^* = c_p 1.0128 = 1.018$; hence, generally the error of putting $c_p = c_p^*$ is less than 2%.

Returning to equation 2.33 and comparing it with the differential form of Poisson's equation

$$c_p\frac{d\theta}{\theta} = c_p\frac{dT}{T} - R\frac{dp}{p}$$

shows that

$$-L\frac{dw}{T} = c_p\frac{d\theta}{\theta}$$

This shows that the potential temperature θ increases during the pseudo-adiabatic expansion, that is, dw is negative, because condensation water is precipitated and latent heat is released. The temperature decreases during the expansion but less rapidly than for dry air. If we write

$$\frac{dw}{T} \simeq d\left(\frac{w}{T}\right)$$

since $-wdT/T^2$ is small compared to dw/T, eqn 2.33 becomes

$$-L\,d\left(\frac{w}{T}\right) = c_p\frac{d\theta}{\theta}$$

$$-L\left(\frac{w}{T}\right) = c_p \ln\theta + K$$

The constant K may be evaluated for the temperature θ_o at which $w \simeq 0$ and the equation becomes

$$\theta = \theta_o e^{-Lw/c_p T} \qquad 2.36$$

which is known as the equivalent potential temperature θ_e.

2.1.6. Mixing of air masses.

For large air masses these adiabatic processes are a good approximation but for small parcels it needs to be realized that the air parcels do not move in isolation and that a certain amount of mixing occurs with the environmental air. Under normal conditions vertical mixing will tend to decrease the temperature and increase the moisture content in the upper portion of the mixed layer and vice versa in the lower layer. This will tend to decrease the relative humidity near the earth's surface and increase it in the upper portion of the mixed layer. An approximate picture of the ascent and mixing is obtained by considering a parcel of 1 kg of air at saturation, lifted in a pseudo-adiabatic process a small selected distance to a new T' and p, and mixed with M kg of environment air of temperature T and specific humidity H_s to make a mixture of (1 + M)kg. The temperature of the mixture could be approximated by the weighted mean

$$T_m = \frac{T' + MT}{1 + M} \qquad 2.37$$

and the specific humidity of the mixture by the weighted mean

$$H_{sm} = \frac{H_s' + MH_s}{1 + M} \qquad 2.38$$

where H_s' is the saturation specific humidity which the isolated parcel would have at T', p. Since the specific humidity, H_s, of the environmental air is less than the saturation value, the value for the mixture, H_{sm}, is less than saturation value and some of the cloud's water must evaporate. In order to evaporate enough of the cloud's water to keep the mixture saturated a certain amount of latent heat of evaporation has to be exchanged. Since no pressure change is involved the approximate form of the first law, eqn 2.33, yields $dT = -(L/c_p)dw$, and the evaporation of an amount $-dX$ of liquid water produces the amount of $+dw$ of vapour. Since $w \simeq H_s$, $dw \simeq dH_s$, the new temperature becomes

$$T'' = T_m + dT = T_m - \frac{L}{c_p} dH_s = T_m - \frac{L}{c_p}(H_s'' - H_{sm}) \qquad 2.39$$

where T'' and H_s'' are the actual values of temperature and specific humidity in the resulting cloud.

It is also necessary to determine whether or not there is enough liquid water in the cloud after lifting to accomplish saturation. The amount of water in liquid form in an isolated parcel going from point 1 to point 2 is given by

$$X_2 = X_1 + (H_{ss_1} - H_{ss_2}) \qquad 2.40$$

where H_{ss} is the saturation H_s and are known for given p and T of the points 1 and 2. A 100% entrainment is associated with a pressure change in the range of 100 to 500 mb.

2.2 Thermodynamic Diagrams.

The thermodynamic relationships in the preceding forms are very cumbersome to use, and graphical methods of solutions have been developed for daily application - the thermodynamic diagrams. The thermodynamic diagrams display graphically the major kinds of processes to which air may be subjected, i.e. *isobaric, isothermal, dry adiabatic* and *pseudo-adiabatic*. The main requirements for these diagrams are;
(1) The area enclosed by the lines representing any cyclic process should be pro-

portional to the change in energy or work done during the process.
(2) For ease of application as many as possible of the fundamental lines should be straight.
(3) The angle between the isotherms and the adiabats should be as large as possible, i.e. near 90°. Adiabats are θ = const lines. θ is a characteristic property of a parcel of air and is invariant during adiabatic processes.

The p versus v diagram is a thermodynamic diagram on which the element of work $dW = pdv$. Unfortunately the angle between the isotherms and adiabats on the p-v diagram is small (see Example 2.1) and hence, the intersection point is hard to define. The thermodynamic diagrams are designed to overcome this by the use of various transformations. For example, consider two variables A and B, each a function of one or more of the thermodynamic parameters, Fig. 2.7. The latter are determined by the state of the system so that it suffices to know p and v. Each closed curve on one plane corresponds to a closed curve on the other, but not necessarily of the same shape. We require that

$$-\oint p dv = \oint A dB$$

for any cyclic process, i.e.

$$\oint (p dv + A dB) = 0$$

For the closed line integral to be zero the integrand must be an exact differential, for example dC

$$p dv + A dB = dC$$

where $C = f(v, B)$. But

$$dC = \left(\frac{\partial C}{\partial v}\right)_B dv + \left(\frac{\partial C}{\partial B}\right)_v dB$$

Thus, the equal area transformation is satisfied by

$$p = \left(\frac{\partial C}{\partial v}\right)_B \quad \text{and} \quad A = \left(\frac{\partial C}{\partial B}\right)_v$$

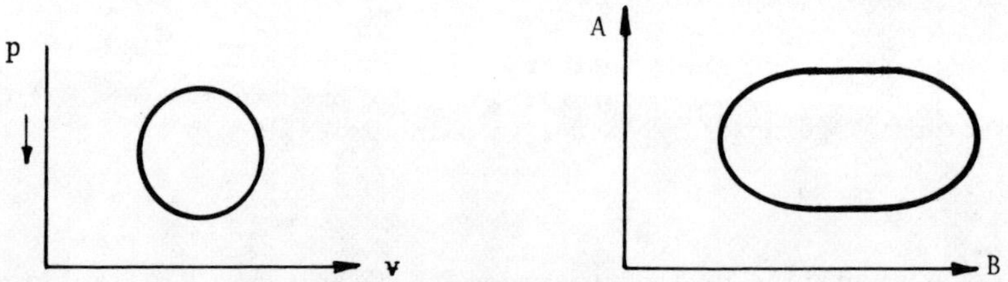

Fig. 2.7 Definition diagram for transformation of thermodynamic diagrams.

Further

$$\left(\frac{\partial p}{\partial B}\right)_v = \frac{\partial^2 C}{\partial v \partial B} \; ; \quad \left(\frac{\partial A}{\partial v}\right)_B = \frac{\partial^2 C}{\partial v \partial B}$$

whence,

$$\left(\frac{\partial A}{\partial v}\right)_B = \left(\frac{\partial p}{\partial B}\right)_v \qquad 2.41$$

and the areas are equal. Put $B \equiv T$, then

$$\frac{\partial A}{\partial v}\bigg|_T = \frac{\partial p}{\partial T}\bigg|_v = \frac{R}{v}$$

from $pv = RT$, and

$$\left(\frac{\partial A}{\partial v}\right)_T dv = R\,\frac{dv}{v}$$

or

$$A = R\,\ell n\,v + f(T)$$

Substituting from the logarithmic form of $pv = RT$ yields

$$A = -\,R\,\ell n\,p + [R\,\ell n\,R + R\,\ell n\,T + f(T)] \qquad 2.42$$

Since the terms in the brackets are constant or F(T) only the function can be chosen so that the bracket term becomes zero. Thus,

$$A = -\,R\,\ell n\,p; \quad B = T$$

This transformed graph is known as the *emagram* - energy per unit mass diagram. In this diagram the isobars and isotherms are straight and perpendicular, Fig. 2.8. From the logarithm of eqn 2.28 and θ = const

$$-\,\ell n\,p = -\,\frac{1}{\kappa}\,\ell n\,T + \text{const}$$

This shows that the dry air adiabats are logarithmic curves, becoming steeper with decreasing temperature. In the usual range of meteorological use they are nearly straight. The pseudo-adiabats

$$\theta_e = \theta_d\,\exp\left(\frac{Lw}{Tc_p}\right) \qquad 2.43$$

are markedly curved. These are in approximate form (in principle) plotted from eqn 2.33. The saturation mixing ratio lines are drawn from

$$w_s = M_r = \varepsilon\,\frac{e_s}{p - e_s} \qquad 2.44$$

A diagram of

$$A = c_p\,\ell n\,\theta, \; B = T$$

is obtained if the Poisson equation (2.28) is used instead of the equation of state. This yields the temperature - entropy plane - the *tephigram*, probably the most widely used thermodynamic diagram, Fig. 2.8. However, there are many more, each having a particular useful feature.

The thermodynamic diagrams provide a quick means of evaluation of the physical changes of air masses that move up or down. For example, Fig. 2.9, p = 1000 mb, T = 26 °C, relative humidity 60% and the chart shows saturation specific humidity of 22 g/kg. The air parcel would follow the dry adiabat up to *condensation* level which occurs at 0.60 x 22 = 13.2 g/kg and ∿ 1.1 km. Beyond that the movement would be along a pseudo-adiabat. Suppose we have a measured profile (by weather balloon) giving p, T and humidity. An air parcel at A when moved upwards (follows the dry adiabat) is colder than the local air and since it will then also be denser, would resist upward movement. However, if the parcel was forced past elevation B it would be warmer (follows along the pseudo-adiabat) and would tend to rise further. The two areas to either side of B express the amount of energy required to make

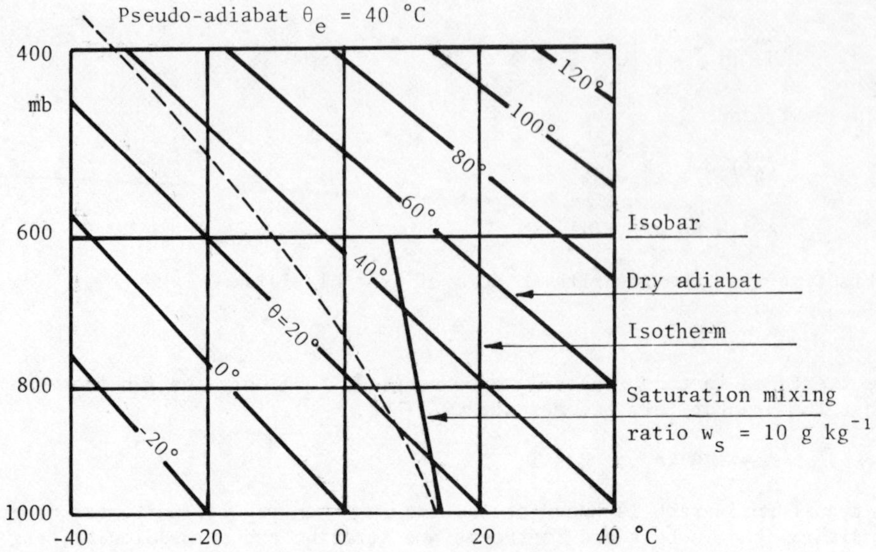

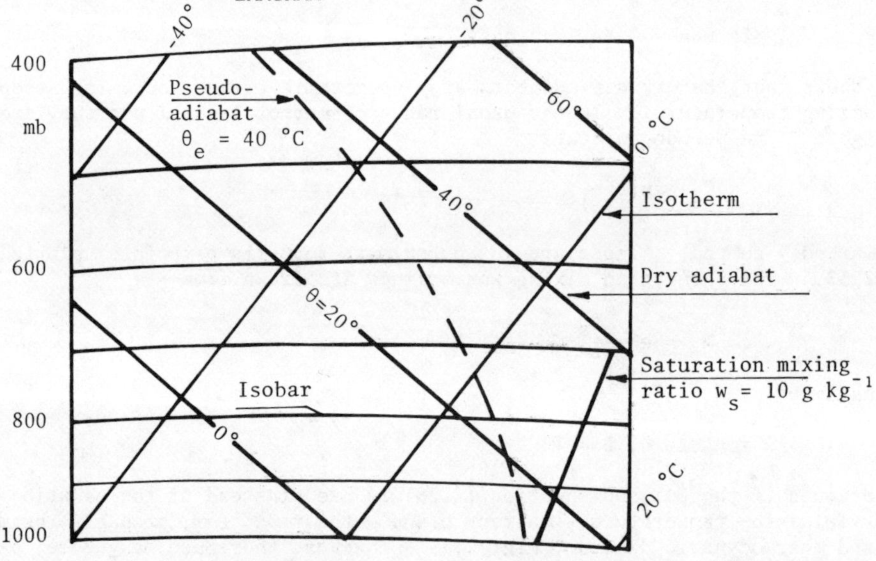

Fig. 2.8. Diagrammatic presentation of the *emagram* and *tephigram*.

(i) a unit mass of air at A to ascend to B (negative area - resists lifting) or
(ii) can be released by a unit mass at A if lifted beyond level B.

The difference between the two areas expresses the amount of *available energy*. If this difference is positive, energy will be available for the development of vertical currents. If the positive area is greater than the negative area the atmosphere has *latent instability*. This leads to the stability considerations of the air masses.

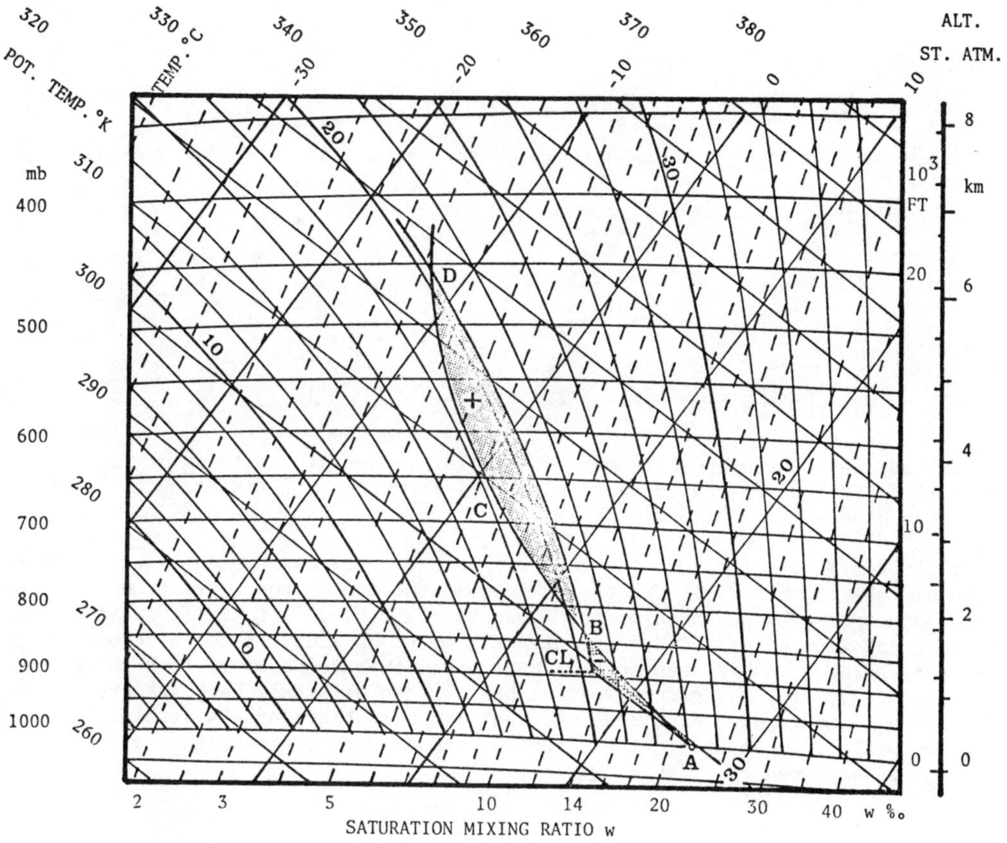

Fig. 2.9. A section from a tephigram. ABCD - measured profile, CL is condensation level or cloud base.

2.3 Atmospheric Stability.

We will consider first the atmospheric stability governed by

$$dp = -\rho g\,dz,\quad pv = RT,\quad \alpha_{dry} \text{ and } \alpha_s$$

Meteorologists use a variable ψ - the geopotential - instead of g, defined by one of the alternative expressions

$$d\psi = g\,dz = -dp/\rho = -v\,dp = -RT\,\frac{dp}{p}$$

and

$$\psi = \int_0^z g\,dz \qquad 2.45$$

is the potential energy acquired by a unit mass when lifted from z = 0 to z.

Special limiting types of atmosphere are
(i) *homogeneous atmosphere* is one of constant density. For this $dp = \rho R\,dT = -\rho g\,dz$ or

$$\frac{dT}{dz} = -\frac{g}{R} = -34.1\ °C/km \tag{2.46}$$

(ii) *isothermal atmosphere* described by

$$dp = -\frac{pg}{RT}dz \tag{2.47}$$

(iii) *constant lapse rate atmosphere* in which $T = T_o - \alpha z$, $\rho = p/RT$ and $dp = -\frac{pg}{RT}dz$. Hence,

$$\int_{p_o}^{p} \frac{dp}{p} = -\frac{g}{R}\int_{o}^{z} \frac{dz}{T_o - \alpha z}$$

and

$$\ln p/p_o = \frac{g}{R\alpha} \ln (T_o - \alpha z)/T_o$$

or

$$p = p_o\left(\frac{T_o - \alpha z}{T_o}\right)^{g/R\alpha} = p_o\left(\frac{T}{T_o}\right)^{g/R\alpha} \tag{2.48}$$

The dry adiabatic atmosphere is one of constant lapse rate, with $\alpha_d = g/c_p = 9.76$ °C/km (5.4 °F/1000 ft) and constant potential temperature θ (isentropic). For moist air T^* is used.

Stability could be defined as the condition where vertical movements in the atmosphere are absent or very restricted. Suppose the temperature variation in a layer of the atmosphere is as shwon by the lapse rate α_1, Fig. 2.10. If, for example, $\alpha_1 < \alpha_d$, and if a parcel of dry air at A was moved upwards, it cools according to the lapse rate α_d and is seen to be always at a lower temperature than the ambient air. Consequently, the parcel will be heavier and will fall back to level A, and vice versa if moved down from level A, that is this atmosphere is stable. The atmosphere would be unstable if the slopes α_1 and α_d were interchanged, $\alpha_1 > \alpha_d$. If $\alpha_1 = \alpha_d$, we have neutral stability.

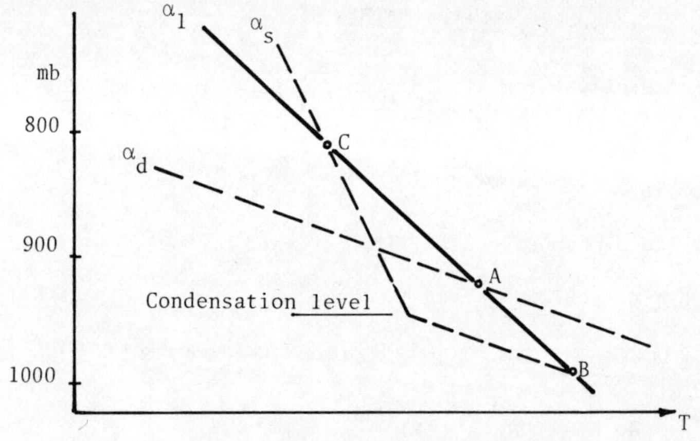

Fig. 2.10. Illustration of atmospheric stability.

The same argument applies to saturated air when α_s is used instead of α_d. Often the ascending air cools initially at the lapse rate α_d of a dry adiabatic atmosphere up to its condensation level, and thereafter at the lapse rate of the pseudo-adiabatic (saturated) atmosphere, α_s. A stable atmosphere can become unstable if, after the condensation level, $\alpha_s < \alpha$. For example, a parcel of air lifted from elevation B would first reach condensation level and then follow the lapse rate α_s. If lifted beyond level C the air will be warmer, and lighter, than ambient air. The atmosphere at this level has become unstable. The above are known as conditional instability. The atmosphere is stable for the ascent of dry air and unstable for the ascent of a parcel of saturated air. If $\alpha_s < \alpha_1$, the atmosphere will be absolutely stable, irrespective of whether it is dry or saturated. Beyond the condensation level, the stability can also be expressed in terms of the equivalent potential temperature (eqn 2.43). If $\partial\theta_e/\partial z > 0$, the saturated layer will be stable.

Convective instability. It may happen that an air column is stable but if the whole mass is lifted bodily it becomes unstable. Whether it becomes unstable or not depends on the moisture content. If the air at b, Fig. 2.11, is much drier than at a and the layer is lifted as a whole, point a will reach condensation level much sooner than b. After condensation level point a will cool at the rate of α_s. By the time the entire layer becomes saturated the temperature distribution is as shown by a'b', i.e. the temperature lapse rate α_1 is now less than α_s, and since the air is saturated it is *unstable*.

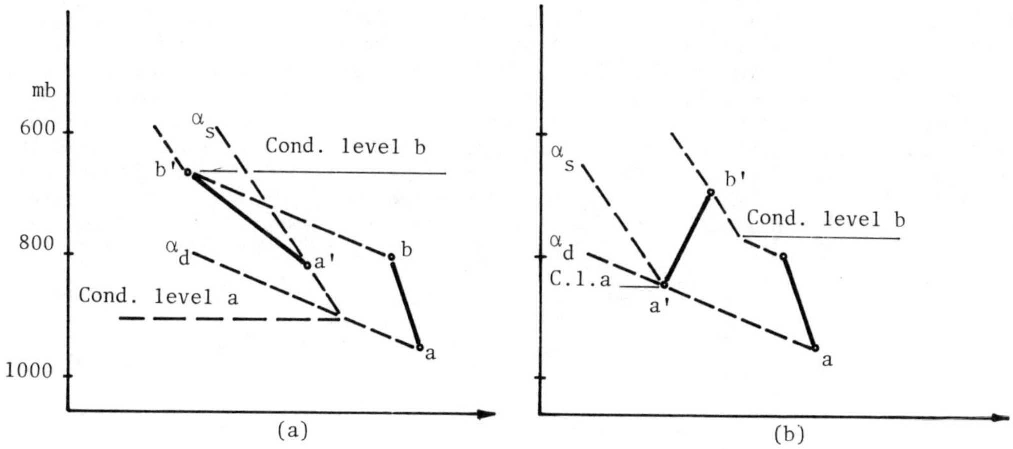

Fig. 2.11. Illustration of convective instability.

If the upper level has higher relative humidity than the lower portion the argument is reversed, and the air is seen to become more stable. This phenomenon is known as convective stability. Since the wet-bulb temperature varies along the moist adiabat in an adiabatic process, we can say that the air column is convectively unstable when the decrease in wet-bulb temperature with elevation exceeds the moist adiabatic lapse rate α_s.

2.4 Water Vapour Content of Air.

Fletcher (1962) wrote "Practically the whole of our usable water comes in the form of precipitation from the atmosphere, and few studies could be more important than those which lead to a complete understanding of how it is stored in the atmosphere and how it precipitates out". The physics of clouds, however, has been the subject of serious studies only over the last few decades.

Water undergoes three principal changes in the clouds: condensation, freezing and

deposition (sublimation). These phase changes have one common property, they do not begin in a continuous manner but require nucleation. Nucleation may be homogeneous or heterogeneous. In the former the embryos are a cluster of molecules, in the latter a foreign substance. For condensation to take place some degree of supersaturation is necessary. Any significant homogeneous nucleation requires more supersaturation than is usually found in the atmosphere. However, the atmosphere contains foreign nuclei in great abundance, and these offer surfaces upon which molecules can cluster with much less expenditure of energy than is required for homogeneous nucleation. Further important features are processes which prevent the cloud from either evaporatin or precipitating - cloud stability - and the process of precipitation itself. The mechanisms which can destroy the clouds stability are
(a) collision and coalescence of droplets after which the larger droplets can fall under action of gravity, and
(b) growth of ice crystals by transfer of vapour from neighbouring supercooled droplets.

Clouds may be characterized by their *liquid content, concentration of droplets,* and *drop size distribution.* Observations show little change in liquid water content from cloud to cloud and it is lower than predicted by theoretical models, based on the first law of thermodynamics. The implication is that the process is not adiabatic, and this is mainly due to the entrainment of dry surrounding air into the cloud. The droplets grow by condensation and coalescence ($T > 0$ °C) and attain a diameter of about 0.2 mm in order to precipitate against updraft in convective clouds (~ 1 ms^{-1}). For the droplets to grow, and precipitate, the cloud must have enough thickness. In a thin cloud a drop will never fall. It will grow in the upward motion and evaporate in the dry environment above the cloud. In a thick cloud the drop will grow until it starts falling. On its way down it will grow through coalescence much faster than on its way up. The raindrops may reach about 7 mm diameter and measured terminal fall velocities are of the order of 4.5 ms^{-1} for d = 1 mm and 9 ms^{-1} for d = 5 mm. Figure 2.12 illustrates drop size distributions as a function of intensity.

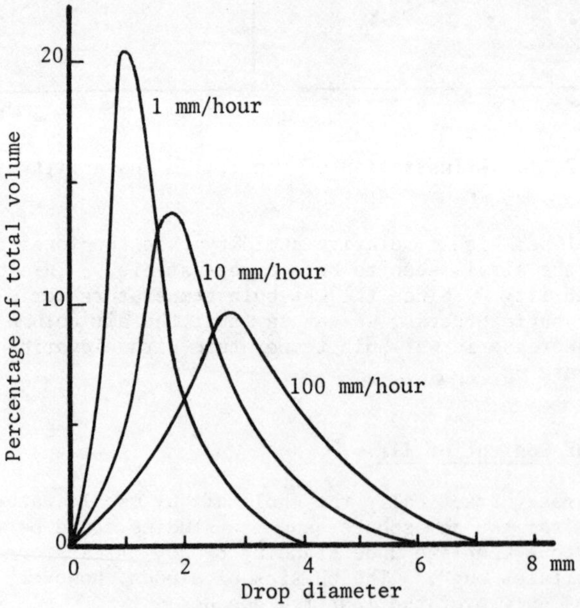

Fig. 2.12. Diagrammatic distribution of raindrop size.

In the case of cold clouds, as they can occur in continental air masses in temperate climates, and in particular in drought clouds, the droplets have so small a size that enormous vertical development of the clouds is necessary to produce rainfall by the condensation - coalescence mechanism. Here, ice nucleation at high altitudes is the most likely means of rainfall production. Snow forms when ice crystals grow by sublimation. We also see that in cloud seeding operations warm clouds should be seeded at their base with large condensation nuclei or with actual water droplets. Cold clouds with 0.5 to 1 g/m^3 of liquid water become unstable when the ice crystal concentration reaches about one per litre. Seeding should add these crystals at the top of cloud when the temperature there is not sufficiently low to create enough nuclei. Dry ice or silver iodide have been successfully used.

These, however, are all aspects outside the scope of treatment here. What the designer/planner would like to know is the total water vapour content of the atmosphere over a given place. For example, we could then calculate the maximum possible amount of rain that could fall. The total mass of water vapour within an air column of 1 cm^2 cross section is

$$W = \int_o^z \rho_v \, dz \qquad 2.49$$

where ρ_v is the vapour density or absolute humidity. Substituting from the hydrostatic relationship, $dp = -\rho g \, dz$

$$W = \int_p^{p_o} \frac{\rho_v}{\rho} \frac{1}{g} dp = \frac{1}{g} \int_p^{p_o} H_s dp$$

or

$$W_{(mm)} = 10 \int_p^{p_o} H_s dp_{(mb)} \qquad 2.50$$

where ρ is the total density of moist air and H_s is the specific humidity. If the relationship of H_s versus pressure can be established, then the integral is simply the area bounded by the curve H_s and the constant values of p and p_o. Alternatively, H_s could be replaced by vapour pressure, $H_s = 0.622 \, e/p$, which leads to

$$W = \frac{0.622}{g} \int_p^{p_o} e \frac{dp}{p} = \frac{0.622}{g} \int_p^{p_o} e \, d(\ell n \, p) \qquad 2.51$$

where W is in g/m^2, but since $\rho_w = 1$ and the area is one, $W = \rho V = AD = D$ the depth of precipitable water. Again, if a vapour pressure curve is plotted from sounding data on the adiabatic chart, with p on the logarithmic scale and T as the abscissa in natural scale (emagram), the area under this curve gives the water vapour content in the vertical column of air.

Unfortunately, vertical soundings are not always readily available, but order of magnitude calculations may be made using surface dew point observations (the temperature at which the moisture in air starts to condense). In a rain situation it can be assumed that the air mass is saturated and the vertical humidity distribution is represented by the dew point value at the surface, decreasing at the rate of α_s. The determination of precipitable water then depends on the estimate of the elevation of the top of the moist air masses.

The total precipitable water in saturated adiabatic atmospheres in terms of the wet-bulb potential temperature, °C, is shown in Table 2.3. Wet-bulb potential temperature for saturation condition is the same as the dew point at 1000 mb and is very close to the value given by the intersection of the wet adiabat α_s with the 1000 mb pressure line.

TABLE 2.3 Total Precipitable Water in Saturated Adiabatic Atmosphere

Wet-bulb pot. temp. °C	Precipitable water, mm	Wet-bulb pot. temp. °C	Precipitable water, mm
0	8.48	22	62.36
5	13.46	24	74.43
10	21.23	26	88.54
15	33.46	28	106.81
20	52.27	30	118.77

The depth of precipitable water in a column of air can be estimated from Fig. 2.13 or Table 2.4. The values are based on saturation at 1000 mb level and a pseudo-adiabatic lapse rate corresponding to surface wet-bulb temperature.

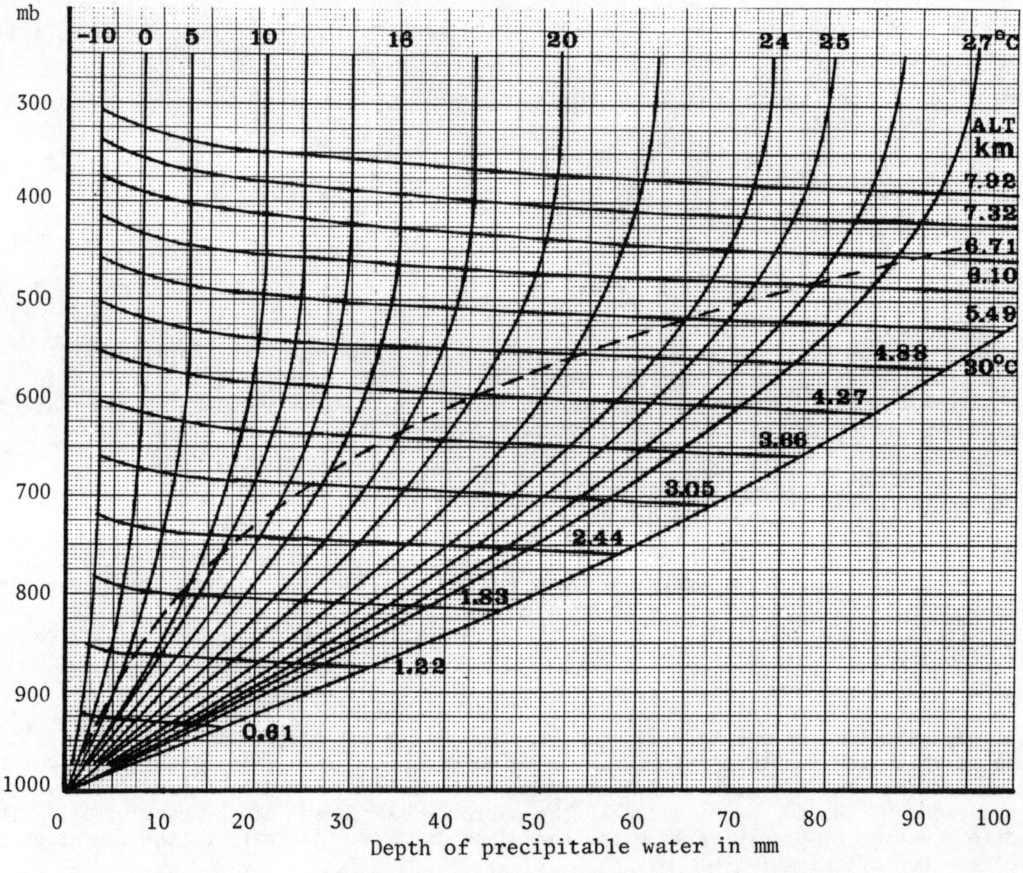

Fig. 2.13. Depth of precipitable water in a column of air of given height above 1000 mb based on pseudo-adiabatic lapse rate for the indicated surface temperature. Dashed line shows pressure at which 0 °C is attained by lifting of the air mass.

TABLE 2.4 Precipitable Water in mm in a Pseudo-Adiabatic Saturated Atmosphere between 1000 mb Surface and Indicated Pressure Level as a Function of Surface Wet-Bulb Potential Temperature °C

mb	Surface wet-bulb temperature °C																								
	0	2	4	6	8	10	11	12	13	14	15	16	17	18	19	20	21	22	23	24	25	26	27	28	29	30
990	0	0	0	1	1	1	1	1	1	1	1	1	1	1	1	1	1	2	2	2	2	2	2	2	2	3
980	1	1	1	1	1	1	2	2	2	2	2	2	2	2	3	3	3	3	3	4	4	4	4	5	5	5
970	1	1	1	2	2	2	2	3	3	3	3	3	4	4	4	4	5	5	5	5	6	6	7	7	7	8
960	1	2	2	2	3	3	3	3	4	4	4	4	5	5	5	6	6	6	7	7	8	8	9	9	10	11
950	2	2	2	3	3	4	4	4	4	5	5	6	6	6	7	7	8	8	9	9	10	10	11	12	12	13
940	2	2	3	3	4	4	5	5	5	6	6	7	7	7	8	9	9	10	10	11	12	12	13	14	15	16
930	2	3	3	4	4	5	5	6	6	7	7	8	8	9	9	10	11	11	12	13	14	14	15	16	17	18
920	3	3	4	4	5	6	6	7	7	8	8	9	9	10	10	11	12	13	14	14	15	16	17	19	20	21
910	3	3	4	5	5	6	7	7	8	8	9	10	10	11	12	13	13	14	15	16	17	18	20	21	22	23
900	3	4	4	5	6	7	7	8	9	9	10	11	11	12	13	14	15	16	17	18	19	20	22	23	24	26
890	4	4	5	6	7	8	8	9	9	10	11	12	12	13	14	15	16	17	18	20	21	22	24	25	27	28
880	4	4	5	6	7	8	9	9	10	11	12	12	13	14	15	16	17	19	20	21	23	24	26	27	29	31
870	4	5	6	7	8	9	9	10	11	12	13	13	14	15	16	18	19	20	21	23	24	26	28	29	31	33
860	4	5	6	7	8	9	10	11	12	12	13	14	15	16	18	19	20	21	23	24	26	28	30	32	34	36
850	5	5	6	7	9	10	11	11	12	13	14	15	16	18	19	20	21	23	24	26	28	30	32	34	36	38
840	5	6	7	8	9	10	11	12	13	14	15	16	17	19	20	21	23	24	26	28	30	32	34	36	38	40
830	5	6	7	8	9	11	12	13	14	15	16	17	18	19	21	22	24	26	27	29	31	33	35	38	40	43
820	5	6	7	8	10	11	12	13	14	15	17	18	19	20	22	24	25	27	29	31	33	35	37	40	42	45
810	5	6	8	9	10	12	13	14	15	16	17	19	20	21	23	25	26	28	30	32	34	37	39	42	44	47
800	6	7	8	9	11	12	13	15	16	17	18	19	21	22	24	26	28	29	32	34	36	38	41	44	46	49
790	6	7	8	9	11	13	14	15	16	17	19	20	22	23	25	27	29	31	33	35	38	40	43	46	49	52
780	6	7	8	10	11	13	14	16	17	18	19	21	23	24	26	28	30	32	34	37	39	42	45	48	51	54
770	6	7	9	10	12	14	15	16	17	19	20	22	23	25	27	29	31	33	35	38	41	43	46	49	53	56
760	6	7	9	10	12	14	15	17	18	19	21	22	24	26	28	30	32	34	37	39	42	45	48	51	55	58
750	6	8	9	11	13	15	16	17	18	20	21	23	25	27	29	31	33	35	38	41	44	47	50	53	57	60
740	7	8	9	11	13	15	16	18	19	20	22	24	26	28	30	32	34	37	39	42	45	48	51	55	59	62
730	7	8	9	11	13	15	17	18	20	21	23	24	26	28	30	33	35	38	40	43	46	50	53	57	60	64
720	7	8	10	11	13	16	17	18	20	22	23	25	27	29	31	34	36	39	42	45	48	51	55	58	62	66
710	7	8	10	12	14	16	17	19	20	22	24	26	28	30	32	35	37	40	43	46	49	53	56	60	64	68
700	7	8	10	12	14	16	18	19	21	23	24	26	28	31	33	35	38	41	44	47	50	54	58	62	66	70
690	7	9	10	12	14	17	18	20	21	23	25	27	29	31	34	36	39	42	45	48	52	55	59	63	68	72
680	7	9	10	12	15	17	19	20	22	24	25	27	30	32	34	37	40	43	46	49	53	57	61	65	69	74
670	7	9	11	12	15	17	19	20	22	24	26	28	30	33	35	38	41	44	47	51	54	58	62	67	71	76
660	8	9	11	13	15	18	19	21	23	24	26	29	31	33	36	39	42	45	48	52	55	60	64	68	73	78
650	8	9	11	13	15	18	19	21	23	25	27	29	31	34	37	39	42	46	49	53	57	61	65	70	75	80
640	8	9	11	13	15	18	20	21	23	25	27	29	32	35	37	40	43	46	50	54	58	62	67	71	76	81
630	8	9	11	13	16	18	20	22	24	26	28	30	32	35	38	41	44	47	51	55	59	63	68	73	78	83
620	8	9	11	13	16	19	20	22	24	26	28	30	33	36	38	42	45	48	52	56	60	65	69	74	79	85
610	8	9	11	13	16	19	20	22	24	26	28	31	33	36	39	42	45	49	53	57	61	66	71	76	81	87
600	8	9	11	13	16	19	21	23	25	27	29	31	34	37	40	43	46	50	54	58	62	67	72	77	82	88
590	8	10	11	14	16	19	21	23	25	27	29	32	34	37	40	43	47	51	55	59	63	68	73	78	84	90
580	8	10	11	14	16	19	21	23	25	27	30	32	35	38	41	44	48	51	55	60	64	69	74	80	85	91
570	8	10	12	14	16	20	21	23	25	27	30	32	35	38	41	45	48	52	56	61	65	70	75	81	87	93
560	8	10	12	14	17	20	21	23	26	28	30	33	36	39	42	45	49	53	57	61	66	71	77	82	88	94
550	8	10	12	14	17	20	22	24	26	28	30	33	36	39	42	46	49	53	58	62	67	72	78	83	90	96

cont.

mb	Surface wet-bulb temperature °C 0	2	4	6	8	10	11	12	13	14	15	16	17	18	19	20	21	22	23	24	25	26	27	28	29	30
540	8	10	12	14	17	20	22	24	26	28	31	33	36	39	43	46	50	54	58	63	68	73	79	85	91	97
530	8	10	12	14	17	20	22	24	26	28	31	34	37	40	43	47	50	55	59	64	69	74	80	86	92	99
520	8	10	12	14	17	20	22	24	26	29	31	34	37	40	43	47	51	55	60	64	70	75	81	87	93	100
510	8	10	12	14	17	20	22	24	26	29	31	34	37	40	44	48	51	56	60	65	70	76	82	88	95	102
500	8	10	12	14	17	20	22	24	27	29	32	34	37	41	44	48	52	56	61	66	71	77	83	89	96	103
490	8	10	12	14	17	21	22	25	27	29	32	35	38	41	45	48	52	57	61	66	72	78	84	90	97	104
480	8	10	12	14	17	21	23	25	27	29	32	35	38	41	45	49	53	57	62	67	73	78	85	91	98	105
470	8	10	12	14	17	21	23	25	27	29	32	35	38	42	45	49	53	58	62	68	73	79	85	92	99	106
460	8	10	12	14	17	21	23	25	27	30	32	35	38	42	45	49	54	58	63	68	74	80	86	93	100	108
450	8	10	12	14	17	21	23	25	27	30	32	35	39	42	46	50	54	58	63	69	74	81	87	94	101	109
440	8	10	12	15	17	21	23	25	27	30	33	35	39	42	46	50	54	59	64	69	75	81	88	95	101	110
430	8	10	12	15	17	21	23	25	27	30	33	36	39	42	46	50	55	59	64	70	76	82	88	96	103	111
420	8	10	12	15	18	21	23	25	27	30	33	36	39	43	46	50	55	60	65	70	76	82	89	96	104	112
410	8	10	12	15	18	21	23	25	27	30	33	36	39	43	47	51	55	60	65	71	77	83	90	97	105	113
400	8	10	12	15	18	21	23	25	28	30	33	36	39	43	47	51	55	60	65	71	77	84	90	98	105	114
390	8	10	12	15	18	21	23	25	28	30	33	36	39	43	47	51	56	60	66	71	77	84	91	98	106	115
380	8	10	12	15	18	21	23	25	28	30	33	36	39	43	47	51	56	61	66	72	78	85	92	99	107	115
370	8	10	12	15	18	21	23	25	28	30	33	36	40	43	47	51	56	61	66	72	78	85	92	100	108	116
360	8	10	12	15	18	21	23	25	28	30	33	36	40	43	47	51	56	61	66	72	79	85	93	100	108	117
350	8	10	12	15	18	21	23	25	28	30	33	36	40	43	47	52	56	61	67	73	79	86	93	101	109	118
340	8	10	12	15	18	21	23	25	28	30	33	36	40	43	47	52	56	61	67	73	79	86	93	101	109	118
330	8	10	12	15	18	21	23	25	28	30	33	36	40	43	47	52	56	61	67	73	79	86	94	102	110	119
320	8	10	12	15	18	21	23	25	28	30	33	36	40	44	48	52	57	62	67	73	80	87	94	102	111	120
310	8	10	12	15	18	21	23	25	28	30	33	36	40	44	48	52	57	62	67	73	80	87	94	102	111	120
300	8	10	12	15	18	21	23	25	28	30	33	36	40	44	48	52	57	62	67	74	80	87	95	103	111	121
290	8	10	12	15	18	21	23	25	28	30	33	36	40	44	48	52	57	62	68	74	80	87	95	103	112	121
280	8	10	12	15	18	21	23	25	28	30	33	36	40	44	48	52	57	62	68	74	80	88	95	103	112	121
270	8	10	12	15	18	21	23	25	28	30	33	36	40	44	48	52	57	62	68	74	81	88	95	104	112	122
260	8	10	12	15	18	21	23	25	28	30	33	36	40	44	48	52	57	62	68	74	81	88	96	104	113	122
250	8	10	12	15	18	21	23	25	28	30	33	36	40	44	48	52	57	62	68	74	81	88	96	104	113	122
240	8	10	12	15	18	21	23	25	28	30	33	36	40	44	48	52	57	62	68	74	81	88	96	104	113	123
230	8	10	12	15	18	21	23	25	28	30	33	36	40	44	48	52	57	62	68	74	81	88	96	104	113	123
220	8	10	12	15	18	21	23	25	28	30	33	36	40	44	48	52	57	62	68	74	81	88	96	104	113	123
210	8	10	12	15	18	21	23	25	28	30	33	36	40	44	48	52	57	62	68	74	81	88	96	105	114	123
200	8	10	12	15	18	21	23	25	28	30	33	36	40	44	48	52	57	62	68	74	81	88	96	105	114	123

From WMO - No. 233.TP.126, 1969. For the International Standard Atmosphere pressure, p, in mb is related to height, z, in m through

$$z = 44\ 308\ [1 - (p/1013.2)^{0.19023}]$$

The average annual values of the moisture content vary over the surface of the earth. There is a steady increase in the mean meridional value (average over the latitude around the world) from the poles, where mean depth $\bar{W}$ is just a few mm, to the equator with about 43 mm. The large desert areas of Tibet, Central Mexico, the high areas of Western United States, the Himalayas, Central Africa, etc. show low values of $\bar{W}$, whereas oceanic tropics, equatorial regions of West Africa and South America (Starr et al. 1965) have high values of $\bar{W}$.

Example 2.3 Saturated air at a ground temperature of 20 °C and pressure of 1000 mb flows over a region. Assume a constant lapse rate

$$\alpha_s = (g/c_p)\{[1 + (L/R)w/T]/[1 + (\varepsilon L^2/c_p R_v)w/T^2]\}$$

as given by conditions on the ground, and calculate the mixing ratios for the lower 5 km of the atmosphere and its water content. In the above equation $w = \rho_v/\rho_a$ is the mixing ratio, L is the latent heat of evaporation, $L = 2501 - 2.37(T - 273)$ J/g, $R = 0.28704$ J/g °C, $c_p \simeq 1.003 + 0.80w$ (J/g °K) per unit mass of moist air, $R_v = 1.609R$ and $\varepsilon = 0.622$.

The energy equation

$$\Delta Q = c_p dT - v dp$$

can be written for a saturated atmosphere as

$$- L dw_s = c_p dT - \frac{RT}{p - e_s} d(p - e_s)$$

$$L \frac{dw_s}{T} + c_p \frac{dT}{T} - R \frac{dp}{p} \simeq 0$$

Here $\frac{dw}{T} \simeq d(\frac{w}{T})$ since $- w \frac{dT}{T^2} << \frac{dw}{T}$. Hence,

$$L\, d(\frac{w_s}{T}) + c_p \frac{dT}{T} - R \frac{dp}{p} = 0$$

$$L(\frac{w_s}{T}) - L(\frac{w_s}{T})_o + c_p \ln \frac{T}{T_o} - R \ln \frac{p}{p_o} = 0$$

or

$$L \frac{w_s}{T} + c_p \ln \frac{T}{T_o} = L(\frac{w_s}{T})_o + R \ln \frac{p}{p_o}$$

In order to evaluate α_s the mixing ratio $w_s = 0.622\, e_s/(p - e_s)$ has to be calculated. The saturation vapour pressure can be read from tables or calculated from

$$e_s = 6.11 \times 10^{7.5\, t/(t + 237.3)}$$

where t is temperature in °C. Thus, at t = 20 °C, e_s = 23.389 mb (23.373 Table 2.2) $w_s = 14.886 \times 10^{-3}$ or 14.9 g/kg, c_p = 1.0149 J/g °K = 1014.9 J/kg °K, and

$$\alpha_s = \frac{9.81}{1014.9} \left[\frac{1 + (2459.6/0.28704)(0.01490/293)}{1 + (0.622 \times 2459.6^2/1.0149 \times 1.609 \times 0.28704)(0.01490/293^2)}\right]$$

$$= 5.80 \text{ °C/km}$$

For a constant lapse rate atmosphere $T = T_0 - \alpha z$. Then $dp = (pg/RT)dz$, $\ln(p/p_0) = (g/RT) \ln(T_0 - \alpha z)/T_0$ or

$$p = p_0[(T_0 - \alpha z)/T_0]^{g/R\alpha} = p_0(1 - \frac{0.0058}{293} z)^{5.89}$$

where $g/R\alpha = 9.81/287.04 \times 0.0058 = 5.89$. Thus,

z km	0	1	2	3	4	5
p mb	1000	899	788	697	615	541
$R \ln(p/p_0)$	0	-33.77	-68.39	-103.61	-139.54	-176.34

<u>At z = 0</u>: p_0 = 1000 mb, t = 20 °C or T_0 = 293 °K, $(e_s)_0$ = 23.4 mb, $(w_s)_0 = 0.622 \times e_s/(p - e_s) = 14.9 \times 10^{-3}$ and $L(w_s/T)_0$ = 124.7 (J/kg °C).
<u>At z = 1 km</u>: $L(w_s/T)_0$ = 124.7, $R \ln(p/p_0)$ = -33.8, $c_p \simeq$ 1012 J/kg °C, $L(w_s/T) + c_p \ln(T/T_0)$ = 90.9.

T	e_s	w	L	$L\frac{w}{T}$	$c_p \ell n \frac{T}{T_o}$	$L\frac{w}{T} + c_p \ell n \frac{T}{T_o}$
288	17.044	12.158×10^{-3}	2465.45	104.06	-17.46	86.60
289	18.173	12.980×10^{-3}	2463.08	110.63	-13.95	96.68

90.94 - 86.60 = 4.34, 96.68 - 86.60 = 10.08, 3.34/10.08 = 0.43, <u>$T \simeq 288.4$ °K, $e_s \simeq$ 17.529 mb, $w_s \simeq 12.51$ g/kg.</u>

<u>At z = 2 km</u>: $L(w_s/T) + c_p\ \ell n(T/T_o) = 124.7 - 68.4 = 56.3$, $c_p \simeq 1010$

283	12.272	9.84×10^{-3}	2477.30	86.14	-35.07	51.07
284	13.119	10.53×10^{-3}	2474.93	91.77	-31.51	60.26

<u>$T \simeq 283.6$ °K, $e_s \simeq 12.755$ mb, $w_s = 10.23$ g/kg.</u>

<u>At z = 3 km</u>: $L(w_s/T) + c_p\ \ell n(T/T_o) = 124.7 - 103.6 = 21.1$, $c_p \simeq 1009$

278	8.7192	7.88×10^{-3}	2489.15	70.55	-53.02	17.53
279	9.3465	8.45×10^{-3}	2486.78	75.35	-49.40	25.95

<u>$T \simeq 278.4$ °K, $e_s \simeq 8.98$ mb, $w_s \simeq 8.12$ g/kg.</u>

<u>At z = 4 km</u>: $L(w_s/T) + c_p\ \ell n(T/T_o) = 124.7 - 139.5 = -14.8$, $c_p \simeq 1008$

273	6.1078	6.239×10^{-3}	2501.00	57.16	-71.27	-14.11
272	5.623	5.739×10^{-3}	2503.37	52.82	-74.97	-22.15

<u>$T \simeq 272.9$ °K, $e_s \simeq 6.064$ mb, $w_s \simeq 6.198$ g/kg.</u>

<u>At z = 5 km</u>: $L(w_s/T) + c_p\ \ell n(T/T_o) = 124.7 - 176.3 = -51.60$, $c_p \simeq 1006$

266	3.379	3.909×10^{-3}	2517.59	37.00	-97.26	-60.26
267	3.685	4.266×10^{-3}	2515.22	40.18	-93.48	-53.30

<u>$T \simeq 267.2$ °K, $e_s \simeq 3.76$ mb, $w_s \simeq 4.35$ g/kg.</u>

This trial and error solution can be avoided by the use of thermodynamic diagrams (the tephigram).

The total water mass within an air column is

$$W = \int_0^z \rho_v\, dz = \int_p^{p_o} \frac{\rho_v}{\rho} \frac{1}{g} \int_p^z H_s dp$$

The specific humidity H_s could be replaced by vapour pressure from $H_s \simeq 0.622 \frac{e}{p}$

$$W = \frac{0.622}{g} \int_p^{p_o} e \frac{dp}{p} = \frac{0.622}{g} \int_p^{p_o} e\ d(\ell n\ p)$$

Dimensionally $W = 0.622/9.81(s^2 m^{-1})100(kg\ m\ s^{-2} m^{-2}) = 6.34\ kg/m^2$

p	ℓn p	Δℓn p	mean e_s	e Δ(ℓn p)
1000	6.91			
		0.12	½(23.37 + 17.53)	2.454
889	6.79			
		0.12	½(17.53 + 12.76)	1.817
788	6.67			
		0.12	½(12.76 + 9.98)	1.364
697	6.55			
		0.13	½(9.98 + 6.06)	1.043
615	6.42			
		0.13	½(6.06 + 3.76)	0.638
514	6.29			
				7.316

$W = 6.34 \times 7.316 = 46.385\ kg/m^2$ = <u>46.385 mm</u>

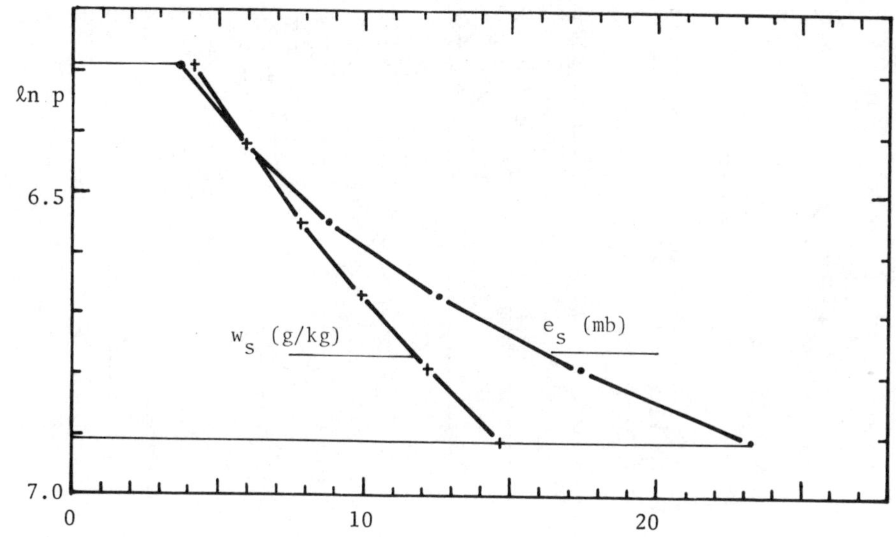

Ex. 2.3. Mixing ratio and vapour pressure in lower atmosphere.

2.5 Atmospheric Circulation

The non-uniform distribution of net radiation over the surface of the earth gives rise to energy gradients which lead to redistribution of energy through oceanic and atmospheric circulations. The atmosphere is heated at its base and cooled at the top and this leads to convective motions. The atmospheric motions are composed of a mean circulation and a highly variable system of smaller movements or eddies. The mean circulation determines the climate and the smaller transient movements the weather. In the oceans the heating and cooling is predominantly at the surface and thermal convection is not of primary importance. Indeed, the oceans are only indirectly involved in the hydrological processes. The primary oceanic currents are maintained by the predominant surface wind pattern. The currents, in turn, are modified by the boundaries formed by the land masses and by the Coriolis acceleration. A general pattern is shown in Fig. 2.14. Most of the convective currents in the oceans result from density differences of surface waters when cool surface water is blown on to warmer water or more saline water onto a less saline one. Salinity can be increased by increased evaporation as is the case at trade wind latitudes where high sea temperatures and continuous dry winds create evaporation peaks, or by surface freezing. Figure 2.15 illustrates the pattern of temperature, salinity and dissolved oxygen variation with depth in ocean at middle and low latitudes. For details of these oceanic features the reader is referred to texts on Oceanography.

Space will permit only an outline discussion of the atmospheric circulation. Atmospheric phenomena, however, are so dominant in hydrology, in the narrow sense, that some knowledge of these is essential. Atmospheric movements caused by the uneven input of energy on a global scale lead in turn to the redistribution of that energy, i.e. of heat and moisture. The atmospheric movements are further complicated by the rotation of the earth. Although, due to the complexities of the problem, the complete analytical weather prediction model has eluded formulation, a great deal has been learned about the general features from simplified models and from analysis of measured parameters. These show well established patterns for long term average pressure, wind velocity, direction, and many other parameters.

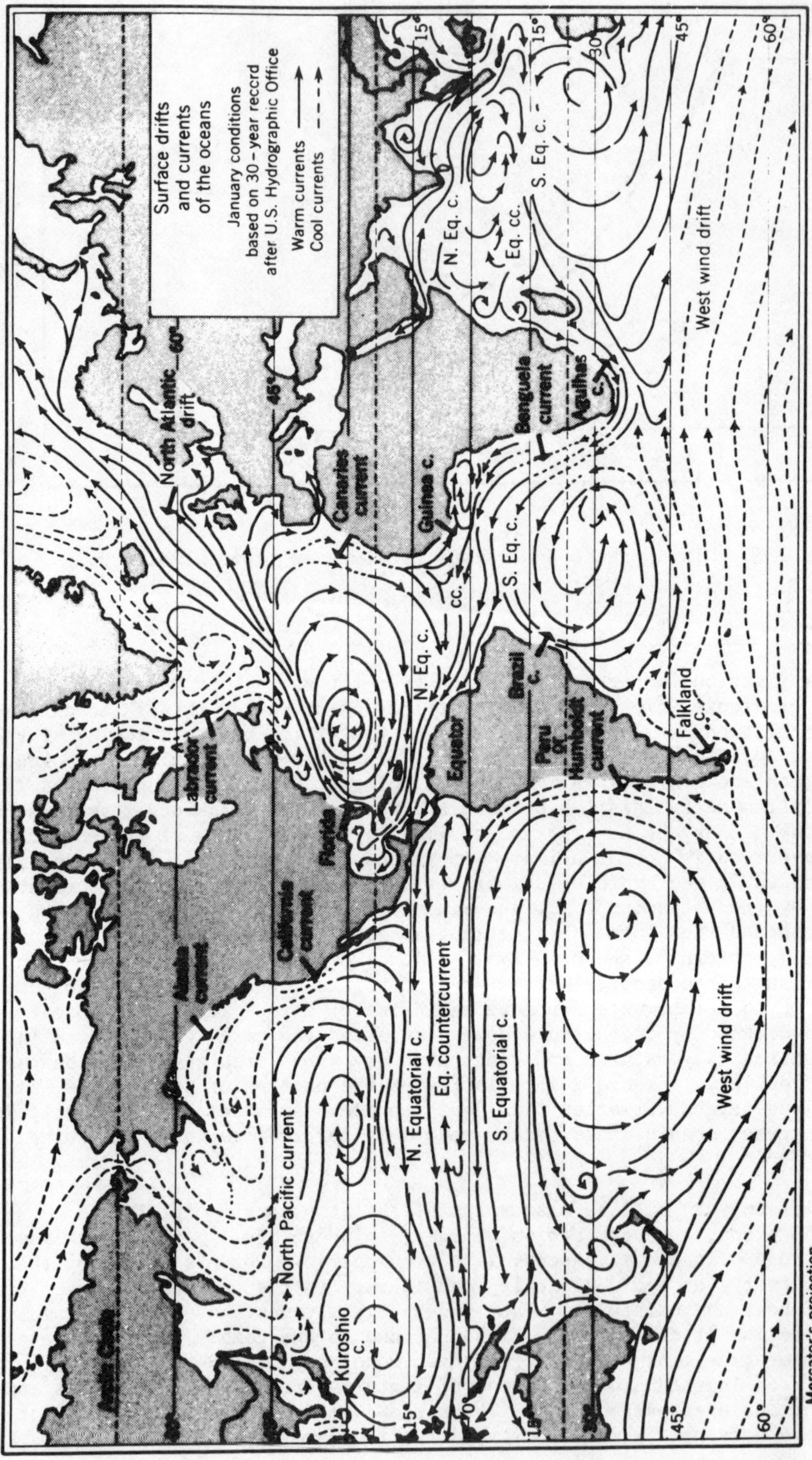

Fig. 2.14. Surface drifts and currents of the oceans in January. (Data from U.S. Navy Oceanographic Office), according to Strahler and Strahler (1974).

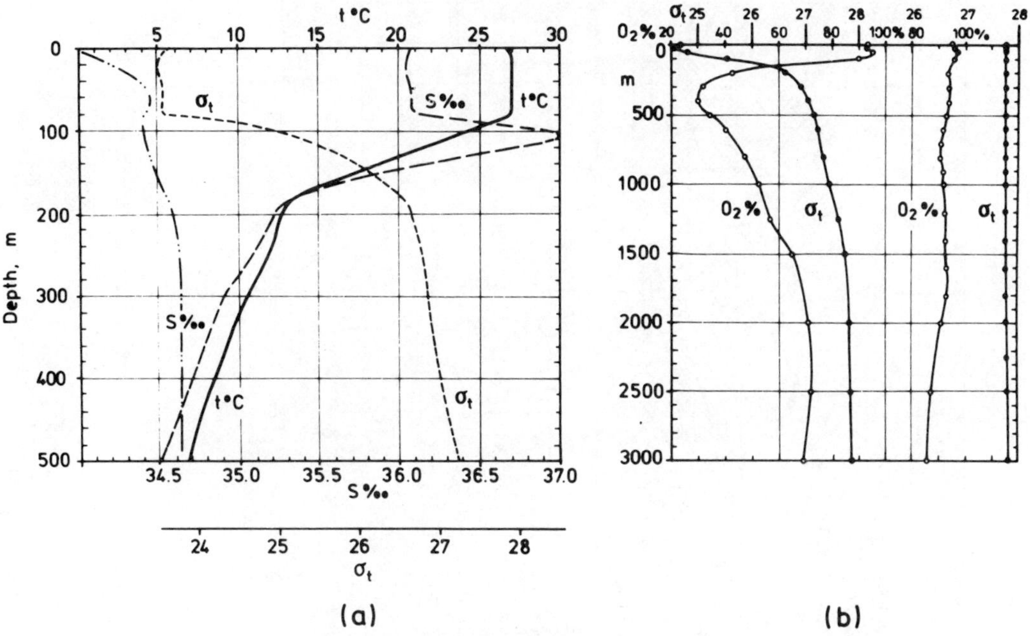

Fig. 2.15. (a) Vertical temperature, salinity σ_t and density curves at "Meteor" Stn 256 (2.4 °S, 39.3 °W). The chain-dotted line is for salinity in Antarctic ("Will. Scoreby" 554, 63°20'S, 17°23'W), $\sigma_t = (\rho - 1) \times 1000$. Temperature below 1000 m is fairly constant at a few degrees C. In tropics salinity has a minimum at about 800 m, but in depth greater than about 1500 m salinity is approximately constant at 34-35‰.
(b) Vertical distribution of oxygen and density at 10 °S (left) and at 55°3'N, 44°46'W.

For example, the winter and summer average pressure patterns are shown in Fig. 2.16. Note, the cellular pattern around the polar regions. In addition the South Pole is a pressure high for both seasons and has a number of low pressure centres around it at abour 70° latitude. The corresponding meridional distributions are plotted in Fig. 2.17 and show the effect of the land masses of the Northern Hemisphere.

Figure 2.18 shows the pattern of the mean summer and winter wind directions. Notice how the large anticyclonic centres of diverging flow coincide with the centres of high pressure in Fig. 2.16, and the less dominant converging cyclone pattern with the low pressure centres, for example, the January low over Australia and corresponding convergent pattern of wind directions. Prominent are also the lines of convergence in the equatorial trough of low pressures, Fig. 2.18.

There are many more features of wind that could be plotted in terms of some average, but reference is made only to the geographical distribution of the meridional or the average zonal component of wind. This, according to Mintz and Dean (1952), is shown in Fig. 2.19 and shows the tropical easterly trade wind belts and the westerly belts

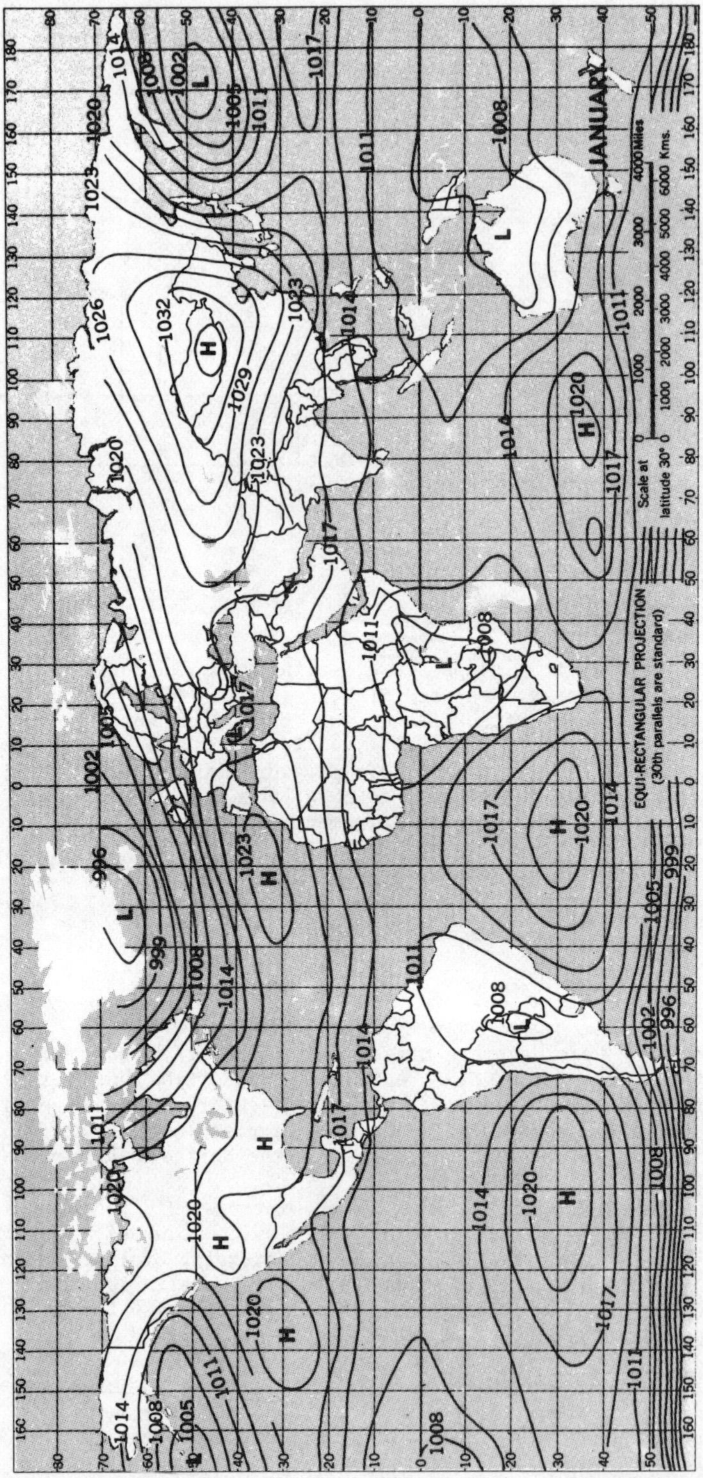

January

Fig. 2.16a. Distribution of mean sea-level pressure intensity in mb in January, after Mintz and Dean (1952) from Eagleson (1970).

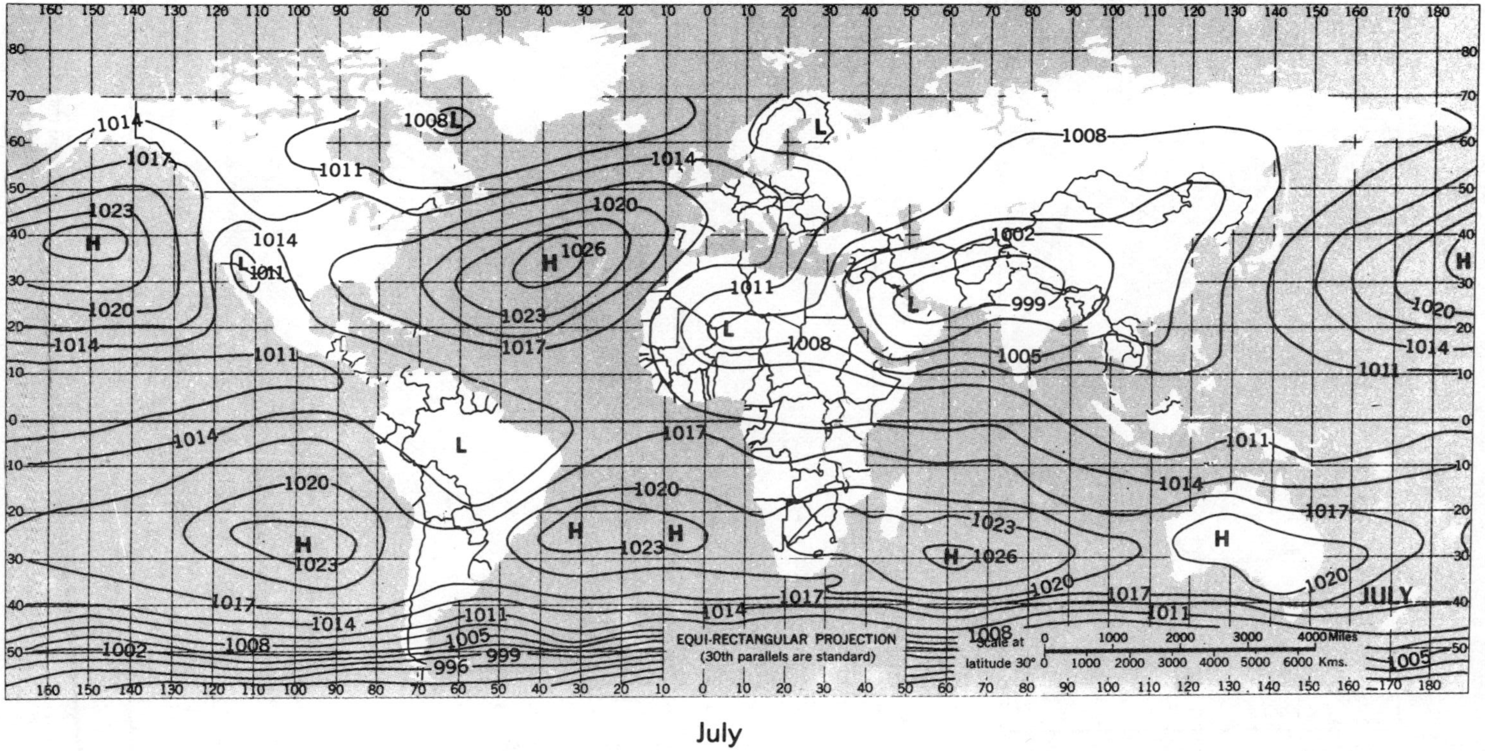

Fig. 2.16b. Distribution of mean sea-level pressure intensity in mb in July, after Mintz and Dean (1952) from Eagleson (1970).

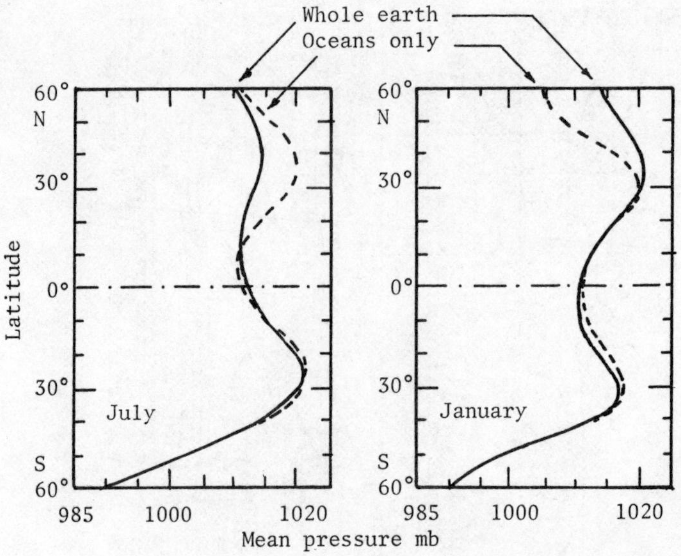

Fig. 2.17. Meridional distribution of average sea-level pressure, after Mintz and Dean (1952).

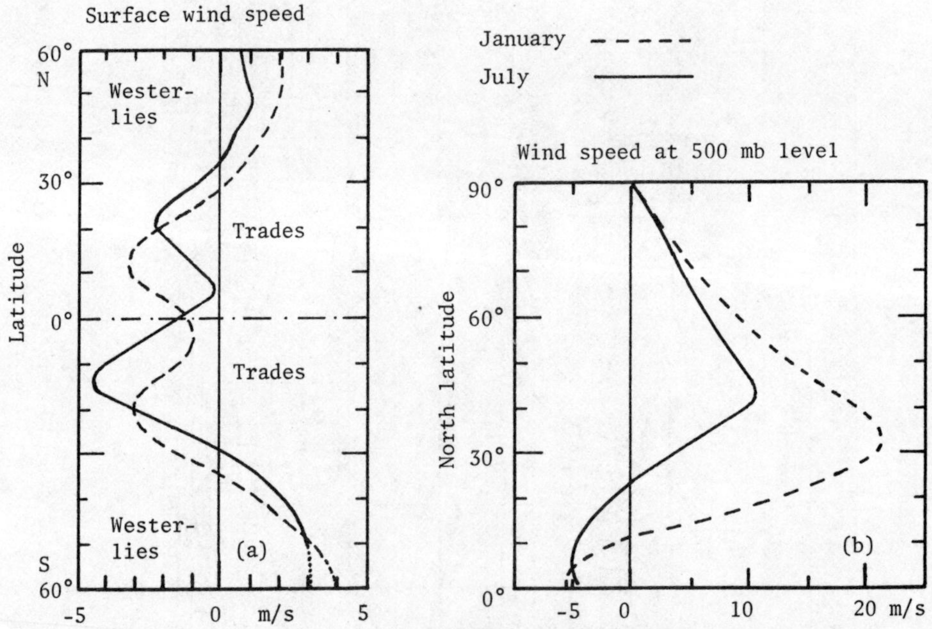

Fig. 2.19. Average zonal component of the observed mean wind, (a) surface, (b) mb level, after Mintz and Dean (1952).

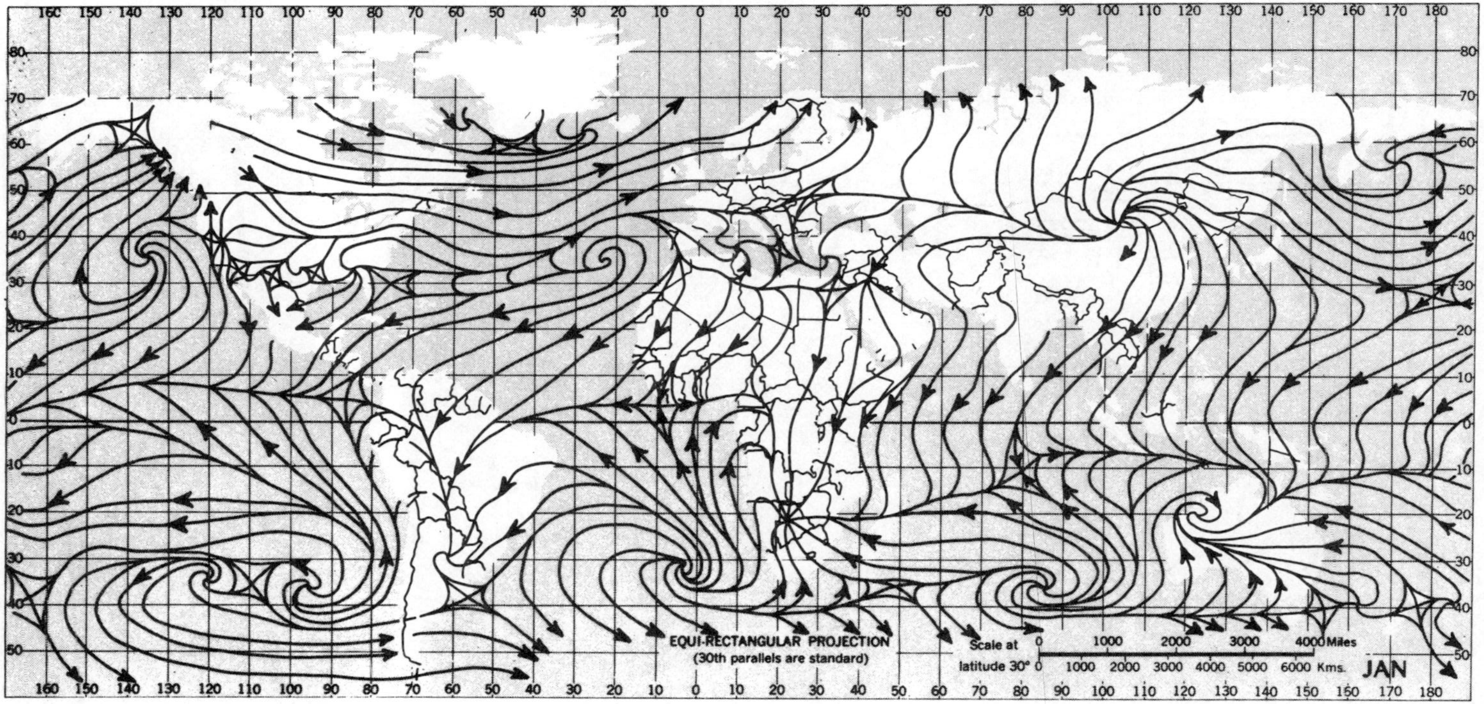

Fig. 2.18a. Mean direction of surface wind in January, after Mintz and Dean (1952) from Eagleson (1970). Over the oceans and over the U.S.A. the streamlines show the direction of the mean wind vector, over the remaining lands the mean wind direction. Over the oceans the mean wind is based on the Greenwich mean noon shipboard observations (McDonald, 1938).

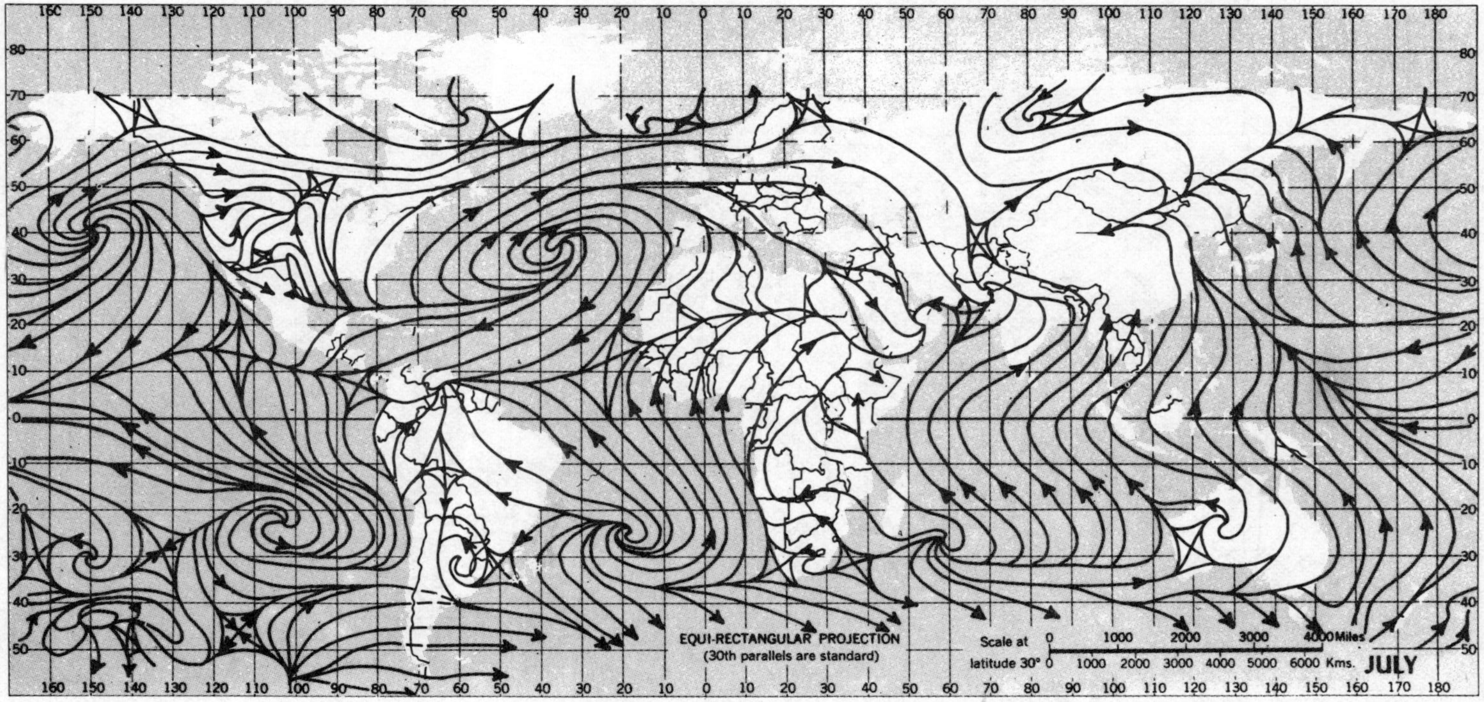

Fig. 2.18b. Mean direction of surface wind in July, after Mintz and Dean (1952) from Eagleson (1970). Over the oceans and over the U.S.A. the streamlines show the direction of the mean wind vector, over the remaining lands the mean wind direction. Over the oceans the mean wind is based on the Greenwich mean noon shipboard observations (McDonald, 1938).

of the middle latitudes. In addition, the distribution of the average magnitude of the zonal component at the surface and at the 500 mb level is shown. The 500 mb level is approximately in the middle of the troposphere both with respect to elevation and mass of the atmosphere. The distribution of the mean zonal winds is shown in Fig. 2.20. The zones of maximum velocity are called the jet stream. The jet streams vary in magnitude and location with seasons. Note, the associated steep velocity gradients at altitudes well outside the influence of the earth's boundary layer. When the streamlines and isovels at the 500 mb level are plotted as a contour map with the Poles in the centre, the lines near the Poles indicate a wobbly flow around the axis of rotation, Fig. 2.21, but the streamlines are smoothed out as the latitudes of the jet stream are approached. The transition from the westerlies to the lower latitude easterlies is through a number of anticyclonic cells. The undulations in the westerly flow of the upper air are known as Rossby waves. The development of these waves gives rise to the high and low pressure centres seen in Fig. 2.16.

The classical theory of the atmospheric circulation goes back to Hadley (1735). The basic argument is that since the equatorial regions receive more heat than the poles, the air will be rising in the tropics and falling at the polar regions. Superimposed on this is the rotation of the earth, which would set the air masses in motion about the axis, and the Coriolis acceleration. The angular momentum, along with mass, energy and water, must be conserved, although the last two may change their form. For such a large scale circulation to exist the conservation of angular momentum would require wind speeds in the upper atmosphere which are very much greater than observed. This model was modified by Bergeron and later by Rossby, on the basis that the poleward moving air cools at 1-2 °C per day and as a result the upper current sinks at about 30° latitude. This is supposed to lead to the three cell pattern, Fig. 2.22, which appears when zonal wind records (averaged over all longitudes) are plotted. If we look at Fig. 2.22 we see that in the zone of trade winds, the wind has an easterly component which through surface resistance will exert an eastward torque $T(\phi)$ on the atmosphere, that is momentum flows from the earth to the atmosphere. At higher latitudes, where the winds are westerly, momentum is extracted from the atmosphere. Hence, there must be a poleward flux, $M(\phi)$, of the eastward angular momentum. Detailed studies indicate, however, that the transport of angular momentum is mainly by the transient horizontal eddies while that by the meridional cells is quite small. Generally, the momentum flux is given by the Reynolds stress $\tau = \rho\varepsilon\, \partial\bar{u}/\partial y$ which indicates transport in the direction of decreasing mean velocity. The $\partial(\bar{U} \cos \phi)/\partial z$ term is positive from 0 to about 40° latitude so that for poleward momentum transport the momentum exchange coefficient must be negative. Correspondingly, in this zone the transient horizontal eddies supply energy to the mean meridional velocity and support it against dissipation by turbulence and viscous shear. It is therefore concluded that the primary thermal driving mechanism of the atmosphere operates through the generation of individual disturbances and not through the maintenance of mean meridional circulations (Starr and White, 1954).

A further important deduction from the direction of momentum flux $M(\phi)$ is that the zonal average of the correlation between meridional and longitudinal velocities, u and v, must be positive. For this the atmospheric streamlines must have waves and these waves must have a meridional asymmetry at any elevation. For poleward momentum flux the streamline picture is, according to Starr (1948), as illustrated in Fig. 2.23.

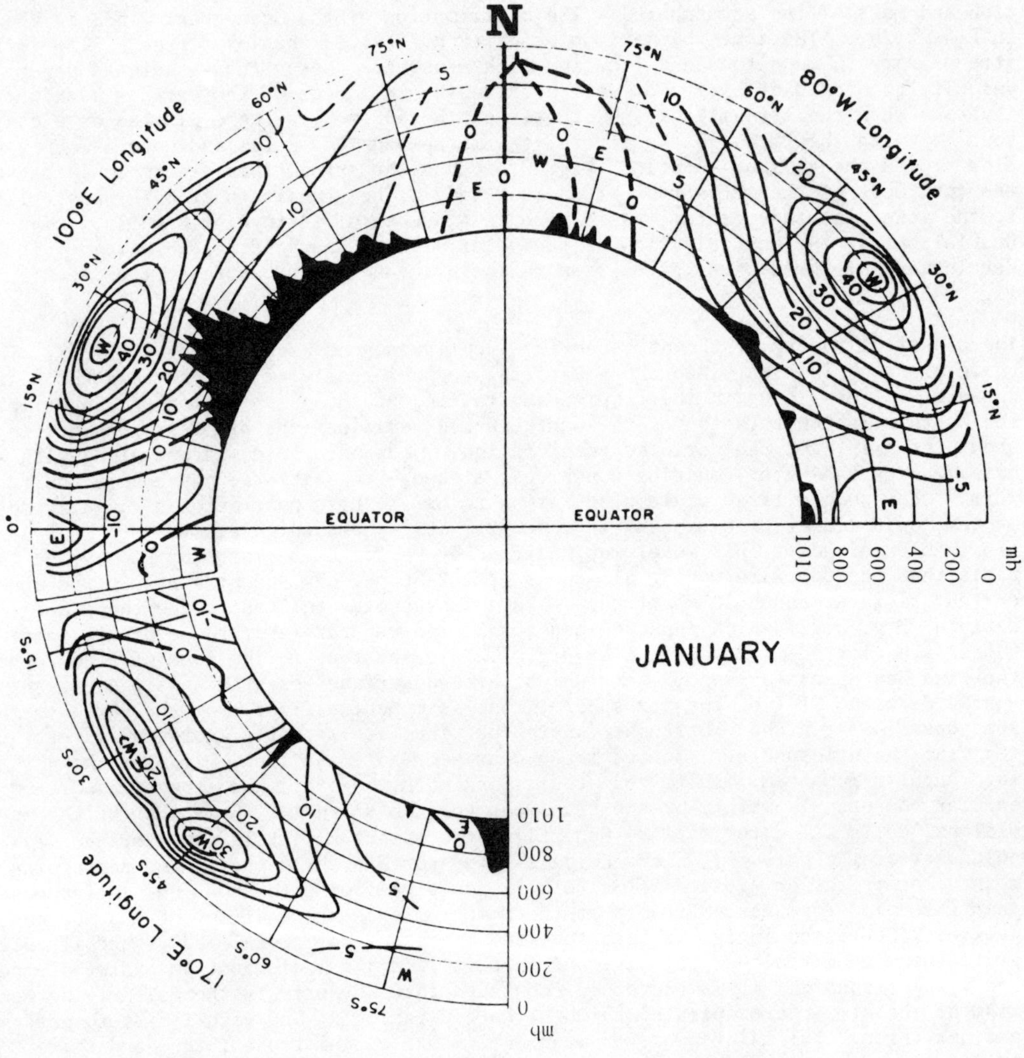

Fig. 2.20a. Meridional cross sections of the mean zonal wind in January, in metres per second. Note that because of the linear scale of pressure the atmosphere is greatly compressed (Mintz and Dean, 1952).

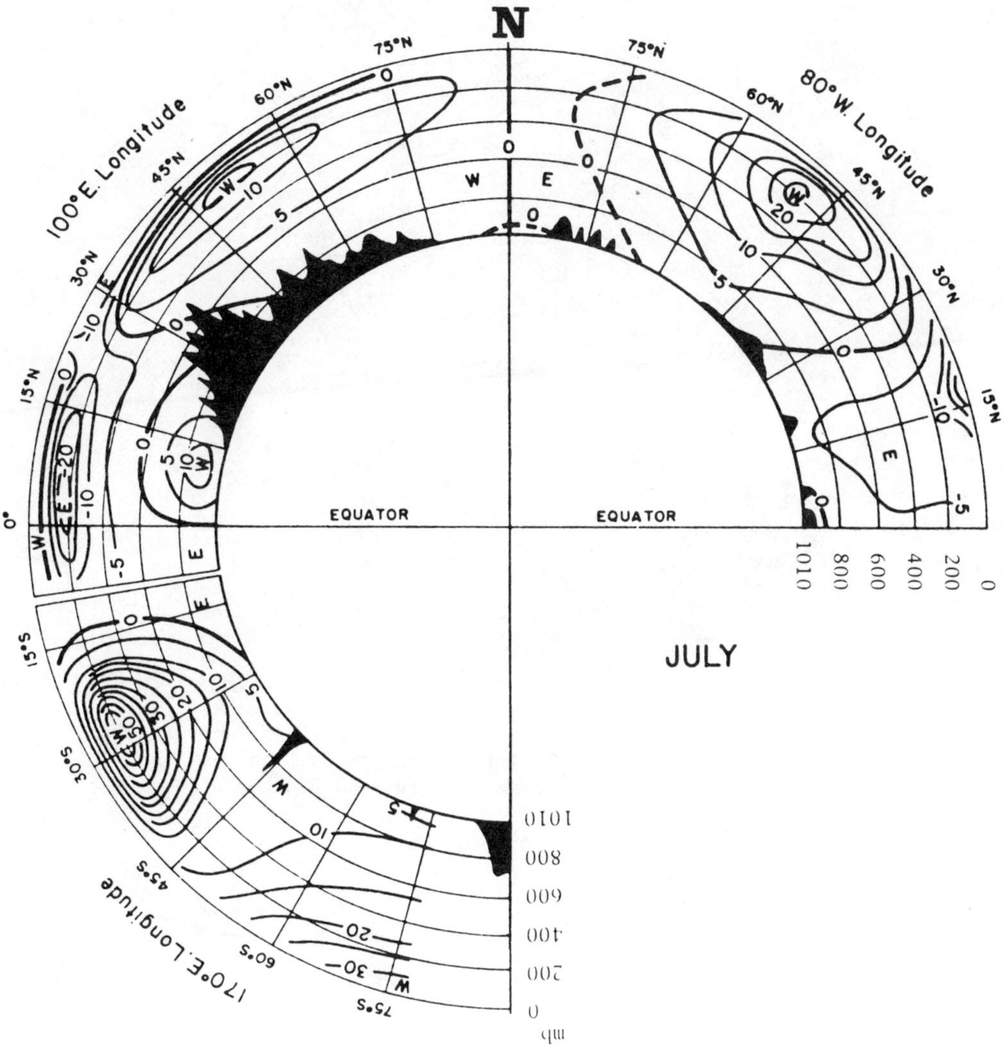

Fig. 2.20b. Meridional cross sections of the mean zonal wind in July, in metres per second. Note that because of the linear scale of pressure the atmosphere is greatly compressed (Mintz and Dean, 1952).

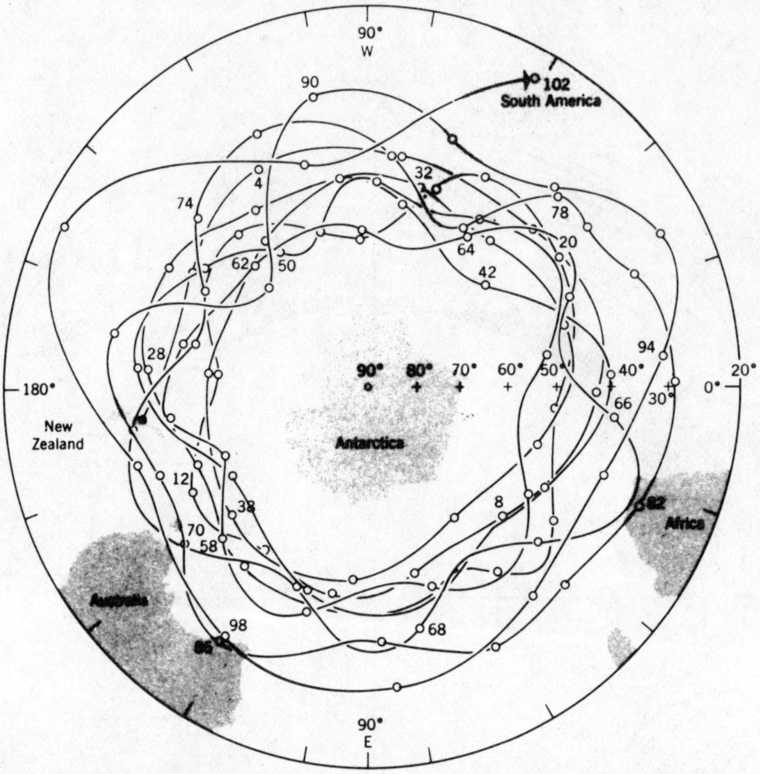

Fig. 2.21. Complete flight trajectory for balloon No.79R, launched from Christchurch, New Zealand. Flight level 200 mb (12 km), Lally and Lichfield, (1969).

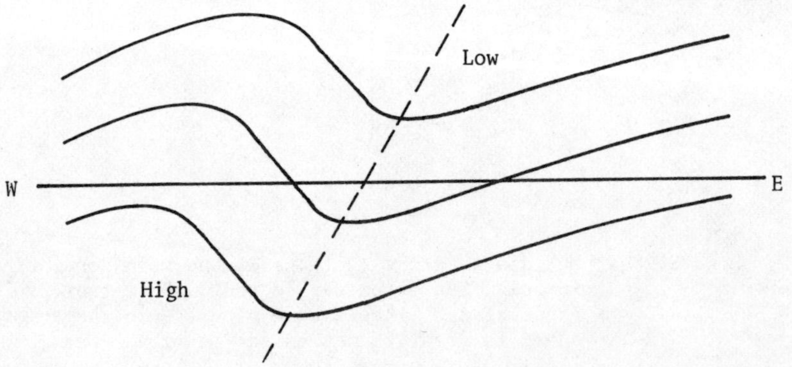

Fig. 2.23. Diagrammatic illustration of the asymmetric waves in the horizontal atmospheric streamlines associated with poleward momentum flux.

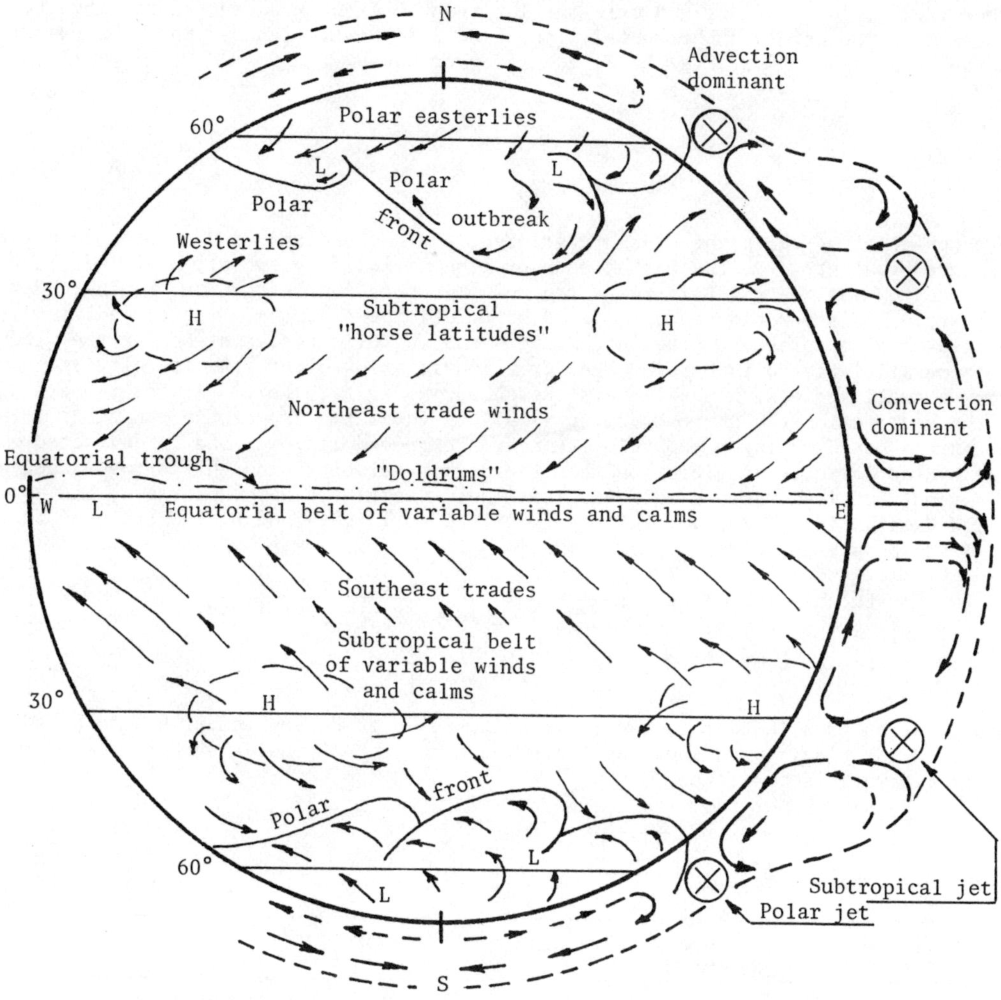

Fig. 2.22. Idealized diagram of global surface winds.

2.5.1. The winds.

The total acceleration for any atmospheric motion is the sum of

$$\underset{\sim}{a} = \underset{\sim}{a}\text{ (pressure gradient)} + \underset{\sim}{g} + \underset{\sim}{a}\text{ (coriolis)} + \underset{\sim}{a}\text{ (friction)}$$

and the equations of motion are

$$Du/dt + 2\omega(w \cos\phi - v \sin\phi) = -(1/\rho)\partial p/\partial x + F_x$$

$$Dv/Dt + 2\omega\, u \sin\phi = -(1/\rho)\partial p/\partial y + F_y$$

$$Dw/Dt + 2\omega\, u \cos\phi = -(1/\rho)\partial p/\partial z - g + F_z \qquad 2.52$$

where F_x, F_y and F_z are components of friction force, ω is the earth's rotational speed (7.29×10^{-5} rad/sec) and ϕ is the latitude. If only $\underset{\sim}{a}_p$ and $\underset{\sim}{a}_c$ are involved (g has no component in a horizontal plane) and friction is ignored then the result is the *geostrophic wind*, Fig. 2.24, and is described by

$$-\frac{1}{\rho}\frac{\partial p}{\partial x} = -2v\,\omega\sin\phi$$

$$-\frac{1}{\rho}\frac{\partial p}{\partial y} = 2u\,\omega\sin\phi \qquad 2.53$$

On a non-rotating earth the pressure gradient would cause the wind to blow down the pressure gradient. On the earth, however, the Coriolis acceleration will deflect the wind towards the right in the Northern Hemisphere and left in the Southern Hemisphere. If air motion is assumed to be frictionless a steady state condition is reached when the pressure gradient balances the Coriolis acceleration and the wind blows parallel to the isobars. When friction is allowed for, the velocity vector makes a small angle with the isobars, as shown in Fig. 2.24(b). If the isobars are in the x-direction the angle is given simply by the ratio v/u of the eqn 2.53. In the eart's boundary layer the velocity near the ground is lower and is deflected less by the Coriolis acceleration than the higher velocity winds higher up. Hence, an assembly of wind vectors on a vertical would describe a warped surface, the so-called Ekman spiral (1905). The geostrophic wind is governed by the ratio of the Euler and Rossby numbers

$$\text{Euler number} = \text{Eu} = \frac{V}{(2\Delta p/\rho)^{\frac{1}{2}}} = \frac{\text{inertial force}}{\text{pressure gradient force}}$$

$$\text{Rossby number} = \text{Ro} = \frac{V}{\omega L} = \frac{\text{inertial force}}{\text{Coriolis force}}$$

When friction is allowed for, the ratio of Reynolds to Rossby number has to be added. This ratio is also known as the Ekman number. (Reynolds number = Re = VL/ν = inertial force/viscous force).

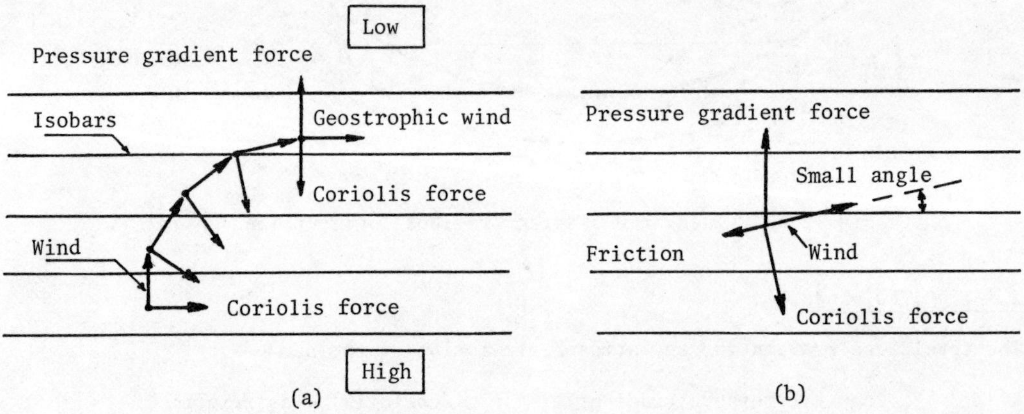

Fig. 2.24. Illustration of geostropic wind, (a) frictionless, (b) with friction.

Gradient wind. When the isobars are curved the wind will be following a curved path and the centripetal acceleration will enter the problem. Such a motion is best analysed in polar coordinates. For a frictionless flow in a plane the equations of motion are

$$v_r \frac{\partial v_r}{\partial r} + v_\theta \frac{\partial v_r}{r\partial\theta} - \frac{v_\theta^2}{r} = v_\theta f - \frac{1}{\rho}\frac{\partial p}{\partial r}$$
$$v_r \frac{\partial v_\theta}{\partial r} + v_\theta \frac{\partial v_\theta}{r\partial\theta} + \frac{v_\theta v_r}{r} = - v_r f - \frac{1}{\rho}\frac{\partial p}{r\partial\theta} \qquad 2.54$$

where $f = 2\omega \sin \phi$. For the case of concentric circular isobars and symmetrical velocity distribution $\partial p/\partial\theta = 0$, $\partial v_\theta/\partial\theta = \partial v_r/\partial\theta = 0$, $v_r = 0$ and the equations reduce to

$$\frac{v_\theta^2}{r} + fv_\theta = \frac{1}{\rho}\frac{\partial p}{\partial r} \qquad 2.55$$

The wind v_θ is called the gradient wind

$$v_\theta = - \frac{fr}{2} \pm [(\frac{fr}{2})^2 + \frac{r}{\rho}\frac{\partial p}{\partial r}] \qquad 2.56$$

where r is positive outwards and θ is measured counter-clockwise. Rotation about a low pressure centre is called *cyclonic motion* and is counter-clockwise in the Northern Hemisphere and clockwise in the Southern Hemisphere. The anticyclonic motion is about the high pressure centre and is anticlockwise south of the equator.

When the scale of the cyclonic motion is small it becomes a problem of balance between the pressure gradient and the centrifugal force

$$\frac{v_\theta^2}{r} = \frac{1}{\rho}\frac{\partial p}{\partial r} \qquad 2.57$$

Tornados, hurricanes, water-spouts, etc. are described by this relationship. The associated winds are referred to as *cyclostrophic*.

The hydrostatic relationship, eqn 2.8, the gas law, eqn 2.9, and eqn 2.53 for the geostrophic wind lead after a small amount of manipulation to

$$\frac{\partial u}{\partial z} = - \frac{g}{fT}\frac{\partial T}{\partial y} + \frac{u}{T}\frac{\partial T}{\partial z}$$
$$\frac{\partial v}{\partial z} = \frac{g}{fT}\frac{\partial T}{\partial x} + \frac{v}{T}\frac{\partial T}{\partial z} \qquad 2.58$$

These are the differential equations for the *thermal wind*.

When at any elevation $p = \rho RT$ is satisfied then $T = T(p)$ and we have a *barotropic atmosphere* where

$$\frac{\partial T}{\partial x} = \frac{dT}{dp}\frac{\partial p}{\partial x}, \quad \frac{\partial T}{\partial y} = \frac{dT}{dp}\frac{\partial p}{\partial y} \quad \text{and} \quad \frac{\partial T}{\partial z} = \frac{dT}{dp}\frac{\partial p}{\partial z} \qquad 2.59$$

In a barotropic atmosphere the constant pressure surfaces, constant density surfaces and constant temperature surfaces are all parallel. Using eqn 2.53 (with $F = 0$) and eqns 2.59 and 2.58, we can show that

$$\frac{\partial u}{\partial z} = \frac{\partial v}{\partial z} = 0$$

or that geostrophic wind in a barotropic atmosphere does not vary with height.

An atmosphere where these simple relationships are not satisfied is called *baroclinic*. Generally, the atmosphere is strongly baroclinic. The vertical temperatu-

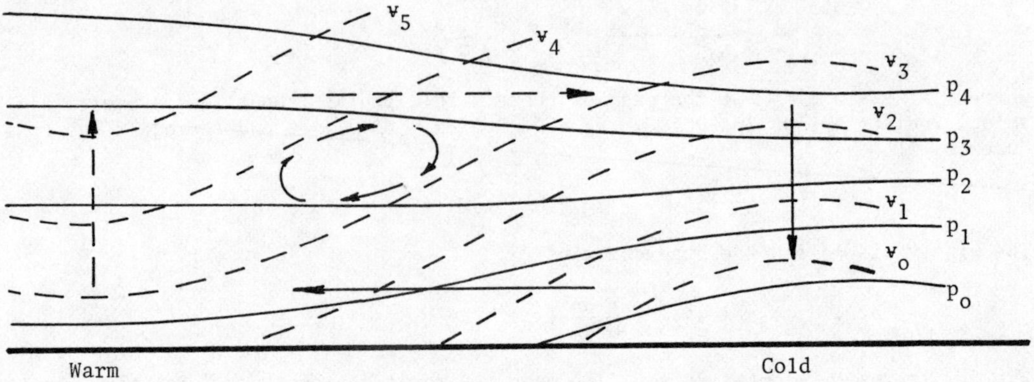

Fig. 2.25. Vertical cross section of a baroclinic atmosphere, pressure p decreases and specific volume v increases with elevation.

re gradient terms may be neglected in comparison with the horizontal gradients. This leads to

$$\frac{\partial u}{\partial z} \simeq -\frac{g}{fT}\frac{\partial T}{\partial y}$$
$$\frac{\partial v}{\partial z} \simeq \frac{g}{fT}\frac{\partial T}{\partial x} \qquad 2.60$$

which show that when there is a horizontal temperature gradient the geostrophic wind increases with elevation. In the atmosphere pressure decreases and specific volume increases ($v = 1/\rho$) with height and at a given pressure the specific volume increases with temperature. The pressure decreases more rapidly with elevation in cold than in warm air at the same levels. These conditions are illustrated in Fig. 2.25. The tubes formed by the p and v surfaces are called *solenoids*, which are characteristic of a baroclinic atmosphere. A barotropic atmosphere is one without solenoids.

A monsoon (from the arabic word mawsim for season) is a thermal wind smaller in scale than the cyclone and anticyclone. During the summer the land is warmer than the ocean, the air rises and this leads to an inflow of warm moist air from the ocean. In winter the direction of this circulation reverses.

Drainage of the cold air mass from a high plateau into the valley below is known as katabatic wind. The foehn wind is also a flow of air down the slope but it does not arise from drainage of air. It occurs when the prevailing winds in warm, moist air are directed against a mountain. The forced ascent on the windward side may cause clouds to form and leads to precipitation, but during most of this ascent the air is cooled at a moist adiabatic rate. By the time the air passes over the mountain it has lost much of its moisture and as it descends the lee slopes it warms at the dry adiabatic rate and is warmer than at the same elevation on the windward side.

2.5.2. Fronts

An air mass which stays over a region for a time adjusts through radiative and convective energy exchange to the temperature and humidity of the underlying area and becomes essentially homogeneous at any given elevation. Large areas, such as tropical oceans, the plains of arctic North America and Asia, the Sahara, and others, are good sources for formation of large fairly homogeneous air masses. These are classified according to their sources as polar or tropical, maritime or continental, and accordingly cP refers to continental polar air mass, mP, mT, etc. Eventually temperature gradients or pressure gradients will cause the air mass to move away from its source area. After it has left the source region an air mass is further characterized as *cold* or *warm*, depending on whether it is colder or warmer than the underlying surface.

A sharp contrast of temperature and humidity may exist across the boundary that separates two air masses of differing properties. Such a boundary is called a frontal zone or surface, or just a *front*. When a cold air mass moves into a region occupied by warm air, and forces the latter up, we have a *cold front*. When warm air is the active mass, which pushes the cold local air, the meteorologist calls it a *warm front*. A front is basically a discontinuity in temperature and is a sloping boundary, Fig. 2.26.

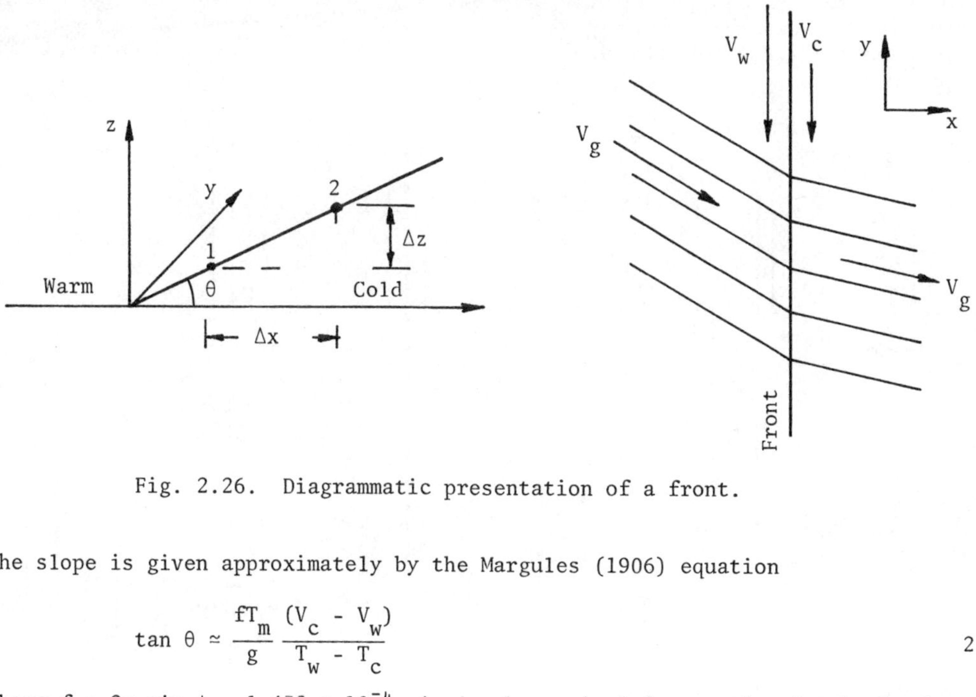

Fig. 2.26. Diagrammatic presentation of a front.

The slope is given approximately by the Margules (1906) equation

$$\tan\theta \simeq \frac{fT_m}{g}\,\frac{(V_c - V_w)}{T_w - T_c} \qquad 2.61$$

where $f = 2\omega \sin\phi = 1.458 \times 10^{-4} \sin\phi$, where $\sin\phi$ is negative in the Southern Hemisphere, $T_m = \frac{1}{2}(T_W + T_C)$, r is the radius of curvature of the front in plan and V_C and V_W are the wind velocities along the front in the cold and warm air, respectively. The range of slopes is from 1:50 for cold fronts to 1:250 for some warm fronts with 1:100 as a typical value.

For continuity in pressure at any elevation across the frontal surface

$$\left(\frac{\partial p}{\partial y}\right)_c = \left(\frac{\partial p}{\partial y}\right)_w$$

For two points along the frontal surface the hydrostatic relationship is

$$p_1 - \left(\frac{\partial p}{\partial x}\right)_c \Delta x - \gamma_c \Delta z = p_2 = p_1 - \gamma_w \Delta z - \left(\frac{\partial p}{\partial x}\right)_w \Delta x$$

or

$$\frac{dz}{dx} = - \frac{(\partial p/\partial x)_c - (\partial p/\partial x)_w}{\gamma_c - \gamma_w}$$

which shows that the gradient normal to the front must be larger in the cold air

$$\left(\frac{\partial p}{\partial x}\right)_c > \left(\frac{\partial p}{\partial x}\right)_w$$

and that there must be a discontinuity at the interface. Likewise, there must be a discontinuity in the component of the gradient wind V_g along the front. Hence, if $V_c > V_w$, Fig. 2.27(a), $(V_c - V_w)$ is negative and from eqn 2.61 tan θ is negative in the Northern and positive in the Southern Hemisphere (since f is negative). In the Northern Hemisphere the front would slope westward, which is an impossible condition because cold air would have to be above warm air. In the Southern Hemisphere the front would slope eastward as required, the wind is clockwise and the low pressure is to the north. In Fig. 2.27(c) $(V_c - V_w)$ is positive and it is a possible situation in the Northern Hemisphere. In a similar manner, all combinations can be analysed.

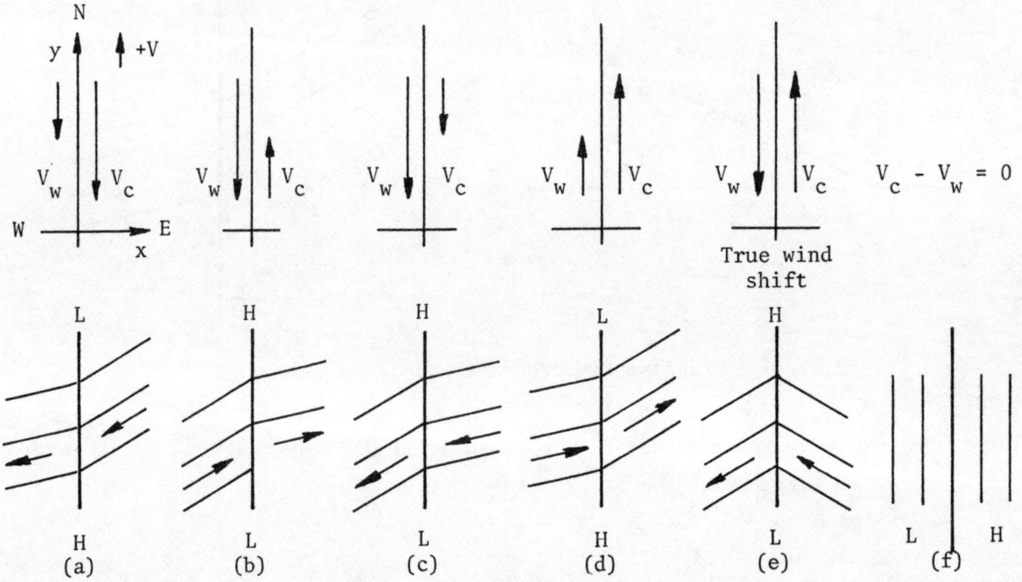

Fig. 2.27. Illustration of possible and impossible fronts.

It is seen that fronts are associated with wind shear or wind shift. Isobars drawn across a front with wind discontinuity make an angle of less than 180° toward low pressure, or along the front the kink in isobars points towards the high pressure. Figure 2.28 shows an idealized cross section through a cold and a warm front. It was discussed earlier, and weather maps for conditions near the earth's surface show that the isobars in the middle latitudes (30-60°) are not straight lines around the globe. Instead there are centres of low and high pressue at about 1500 km apart. These centres are known as cyclones and anticyclones, respectively, and usually travel from west to east. The most important day to day changes in weather are connected with these centres. Figure 2.29 shows a plan and cross section of a polar front cyclone. Figure 2.30 illustrates the life cycle of a cyclone. Figure 2.31 illustrates the warm and cold occluded fronts. For further details on weather disturbances the reader is referred to texts on meteorology, e.g. Riel (1972, 1954).

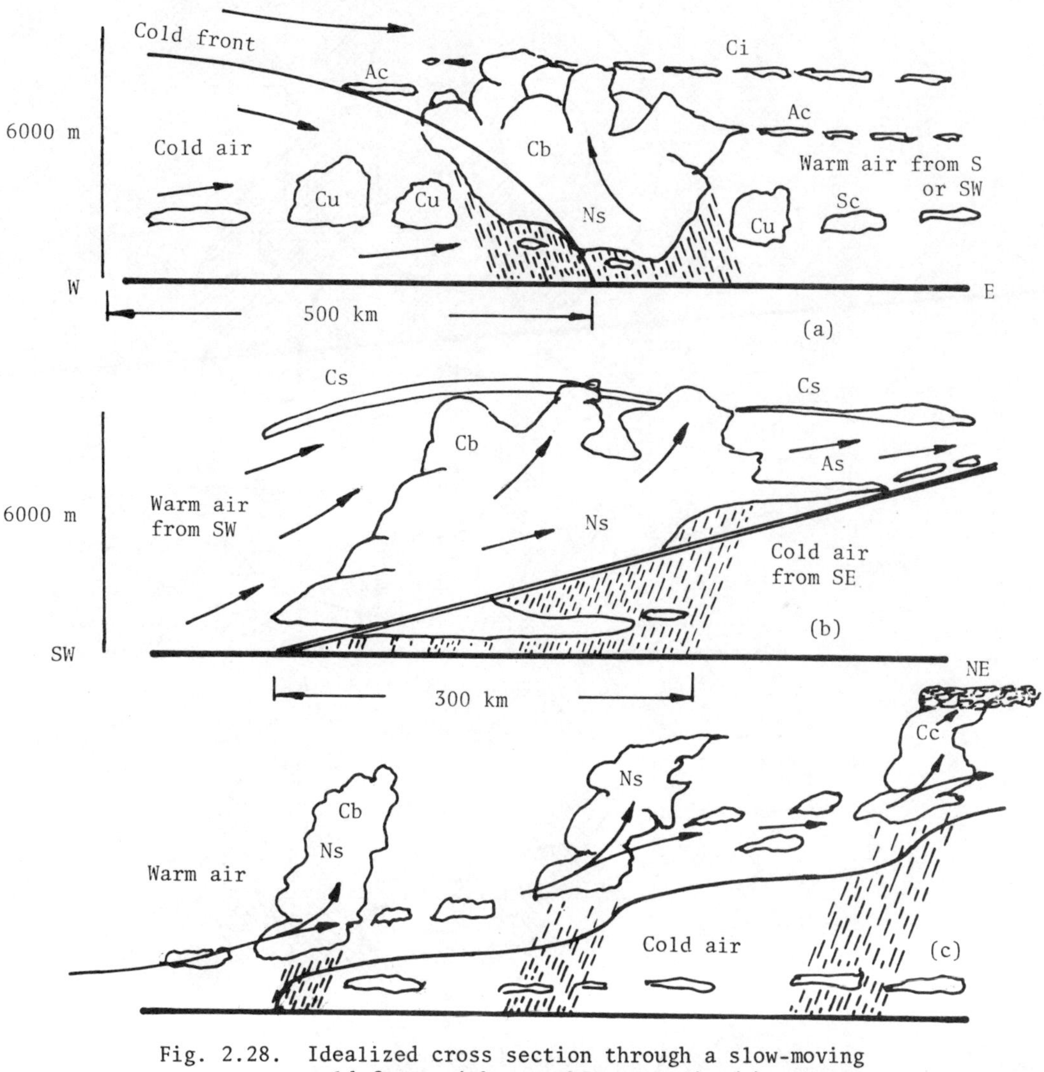

Fig. 2.28. Idealized cross section through a slow-moving cold front with unstable warm air (a), a warm front (b) and wavy warm front (c) in a mature cyclone. With stable warm air there is no mushrooming on the top of the cloud. Cloud designations: Ac - altocumulus, As - altostratus, Cc - cirrocumulus, Ci - cirrus, Cu - cumulus, Cb - cumolonimbus, Cs - cirrostratus, Ns - nimbostratus, Sc - stratocumulus. Not shown is stratus (St), a low uniform layer cloud resembling fog.

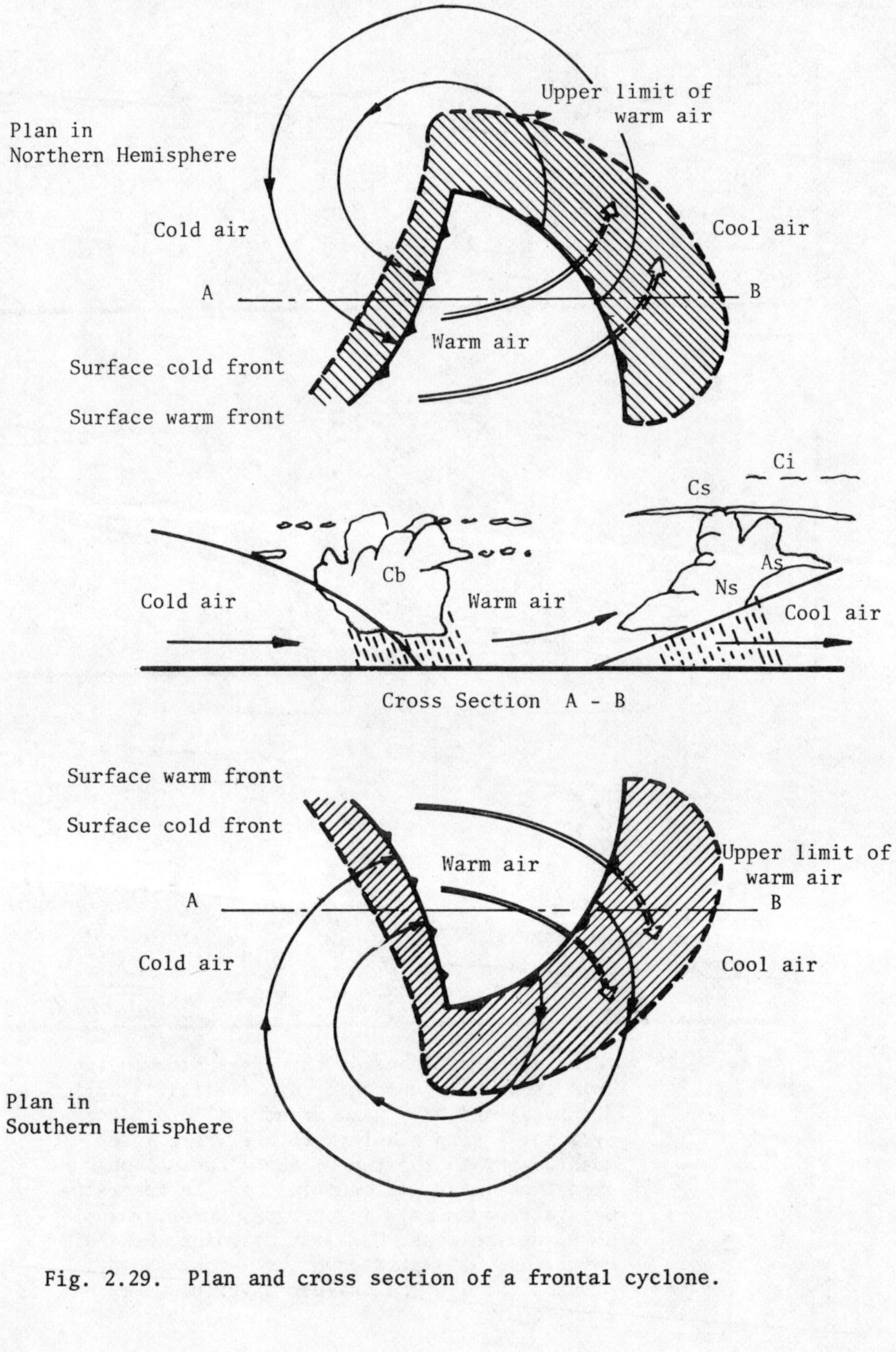

Fig. 2.29. Plan and cross section of a frontal cyclone.

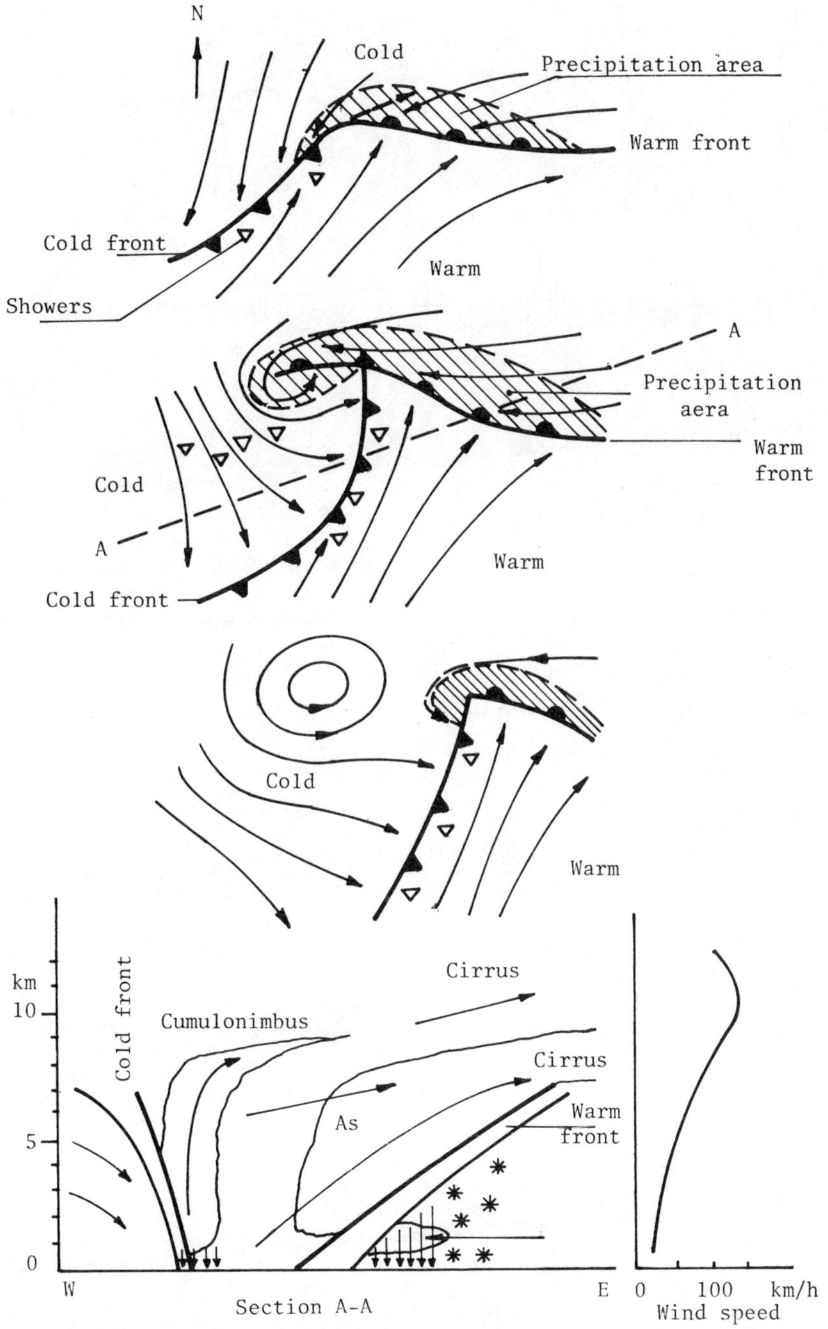

Fig. 2.30. Three stages in the life cycle of a cyclone in Northern Hemisphere and a vertical cross section along the line A-A. The vertical scale is 50 to 100 times the horizontal scale.

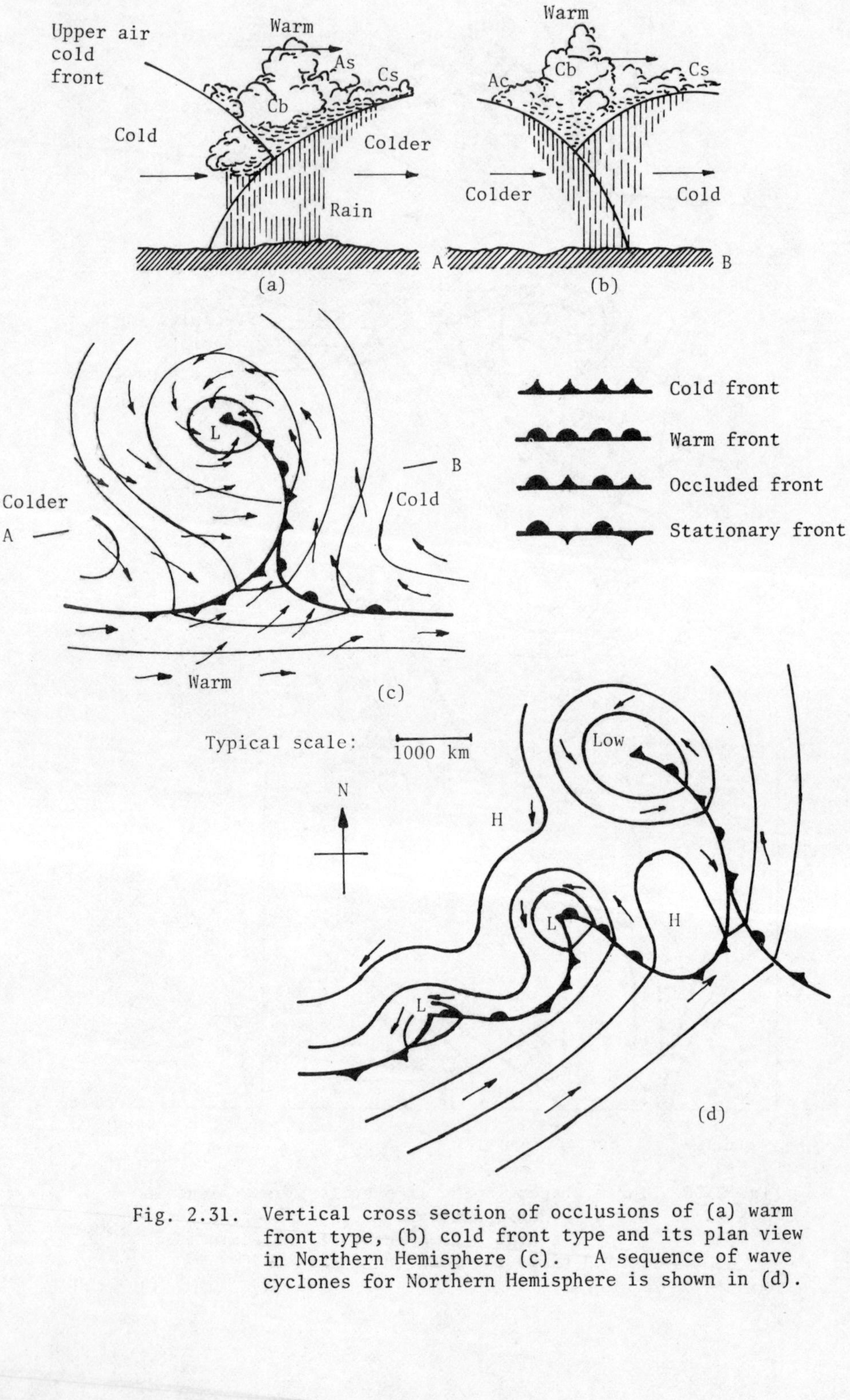

Fig. 2.31. Vertical cross section of occlusions of (a) warm front type, (b) cold front type and its plan view in Northern Hemisphere (c). A sequence of wave cyclones for Northern Hemisphere is shown in (d).

2.6 Standard Atmospheres

International Standard Atmosphere (ICAN)

1. Dry air of constant composition at all altitudes.
2. $g = 980.62$ cm/sec^2.
3. Temperature at mean sea level 15 °C and pressure 1013.2 mb.
4. Lapse rate gives temperature at any altitude z as $t = 15 - 0.0065z$ °C.
5. Above 11 000 m $t = -56.5$ °C = const.
6. Pressure p and elevation z

$$p = 1013.2\ (1 - \frac{0.0065}{288}\ z)^{5.2568}\ \text{mb}$$

$$z = 44\ 318\ [1 - (\frac{p}{1013.2})^{0.19023}]\ \text{m}$$

US Standard Atmosphere

Pressure at MSL = 760 mm Hg = 1013.25 mb, $g = 980.665$ cm/sec^2, $t = 15$ °C = 288 °K. Lapse rate 0.65 °C per 100 m up to 10 769 m (35 332 ft) where p = 234 mb, above that $t = -55$ °C = 218 °K = const.

Selection of Atmospheric Data

Mean molecular weight of dry air is m = 28.966 g/mol, mR = R*, R* = 8.31443 J/mol °C, $c_p = 1.0048$ J/g °C, $c_v = 0.7159$ J/g °C. The ratio of $c_p/c_v = 1.403$ at 0 °C and 1.0133 bar and 1.401 at 100 °C and 1.0133 bar, $c_p - c_v = R$, $R = 0.28704$ J/g °C, $\kappa = R/c_p = 0.286$, $R_v = 1.609$ R. Density of air $\rho = 1.2928$ kg/m^3 at 0 °C and 1.0133 bar. For standard atmospheres $\kappa = 0.286$.

TABLE 2.5 Density of Air in kg/m^3

Pressure	Virtual temperature °C											
mb	-70	-60	-50	-40	-30	-20	-10	0	10	20	30	40
100	0.172	0.164	0.156	0.150	0.143	0.138	0.132	0.128	0.123	0.119	0.115	0.111
200	0.343	0.327	0.312	0.229	0.287	0.275	0.265	0.255	0.246	0.238	0.230	0.223
300	0.514	0.491	0.468	0.449	0.430	0.413	0.397	0.383	0.369	0.357	0.345	0.334
400	0.686	0.654	0.625	0.598	0.573	0.550	0.530	0.510	0.492	0.475	0.460	0.446
500	0.858	0.818	0.781	0.748	0.717	0.689	0.662	0.648	0.615	0.594	0.575	0.556
600	1.030	0.981	0.937	0.897	0.860	0.826	0.795	0.766	0.738	0.713	0.689	0.668
700	1.202	1.146	1.095	1.047	1.004	0.965	0.927	0.894	0.862	0.833	0.805	0.779
800	1.374	1.310	1.250	1.197	1.146	1.102	1.059	1.020	0.986	0.952	0.920	0.891
900	1.544	1.472	1.406	1.345	1.290	1.239	1.192	1.148	1.108	1.071	1.035	1.002
1000	1.715	1.635	1.562	1.495	1.434	1.376	1.325	1.276	1.230	1.190	1.150	1.113
1100	1.887	1.801	1.720	1.645	1.578	1.515	1.459	1.405	1.354	1.308	1.265	1.225

Specific heat capacities of water vapour at low vapour pressures

$$c_{vv} = 3R_v = 1.3858\ \text{J/g °C}$$

$$c_{pv} = 4R_v = 1.8464\ \text{J/g °C}$$

Specific heats per unit mass of humid air are

$$c_p = \frac{c_{pa} + wc_{pv}}{1 + w} \qquad c_v = \frac{c_{va} + wc_{vv}}{1 + w}$$

where c_{pa} and c_{pv} are the specific heats at constant pressure of dry air and water vapour, respectively, and c_{va} and c_{vv} are those at constant volume. Per unit mass of dry air these reduce to

$$c_p = c_{pa} + wc_{pv} \qquad c_v = c_{va} + wc_{vv}$$

Saturation vapour pressure over water in terms of temperature T in °K is

$$\frac{de_s}{dT} = \frac{e_s}{T^2}(6790.5 - 5.02808T + 4916.8\text{x}10^{-0.304T}T^2 + 174209\text{x}10^{-1302.88/T})$$

over ice

$$\frac{de_s}{dT} = \frac{e_{si}}{T^2}(5721.9 + 3.56654T - 0.0073908T^2)$$

and over salt water

$$e_{ss} = e_s\ (1 - 0.000537s)$$

where s is salinity in parts per thousand.

Saturation vapour pressure for -40 °C < t < 40 °C can be expressed as

$$e_s = 6.11 \times 10^{at/(t+b)}$$

where over water a = 7.5, b = 237.3 and over ice a = 9.5, b = 265.5.

TABLE 2.6 Water Vapour Pressure and Specific Volume

°C	e_s mb	v ideal gas / v observed
-10	2.86	1.0003
0	6.11	1.0005
10	12.27	1.0008
20	23.27	1.0012
30	42.43	1.0018
40	73.78	1.0027

Summary of More Common Meteorological Temperatures.
1. Actual temperature T in °K
2. Virtual temperature $[(1 + w/\varepsilon)/(1 + w)]T$
3. Saturation temperatures, Fig. 2.32:

(a) Dew point temperature T_d is the temperature at which vapour pressure e equals saturation vapour pressure e_s when air is cooled at constant pressure and constant mixing ratio w, i.e. p, w and e are constant, $T_d < T$.

(b) Condensation-level temperature T_c is the temperature at saturation of an ascending parcel of air, i.e. p and e decrease, w is constant, $T_c < T_d < T$, $T_c = T(e_c/e)^{\kappa}$.

(c) Wet-bulb temperature T_w is the temperature shown by the thermometer whose bulb is covered by a moist wick when an air stream at constant pressure, temperature and mixing ratio flows past it. The bulb is cooled by evaporation from the wet wick. Equilibrium conditions occur when $L(w' - w) = (c_p + wc_p')(T - T_w)$, where w' is the mixing ratio of air leaving the bulb and c_p is the specific heat at constant pressure of the water vapour. When the air leaving the wick is saturated $w' = \varepsilon e_s/(p - e_s)$, $T_w > T_d$. There is also a pseudo-wet-bulb temperature T_{sw} which differs from T_w significantly only when $T - T_w$ is large and humidity is low.

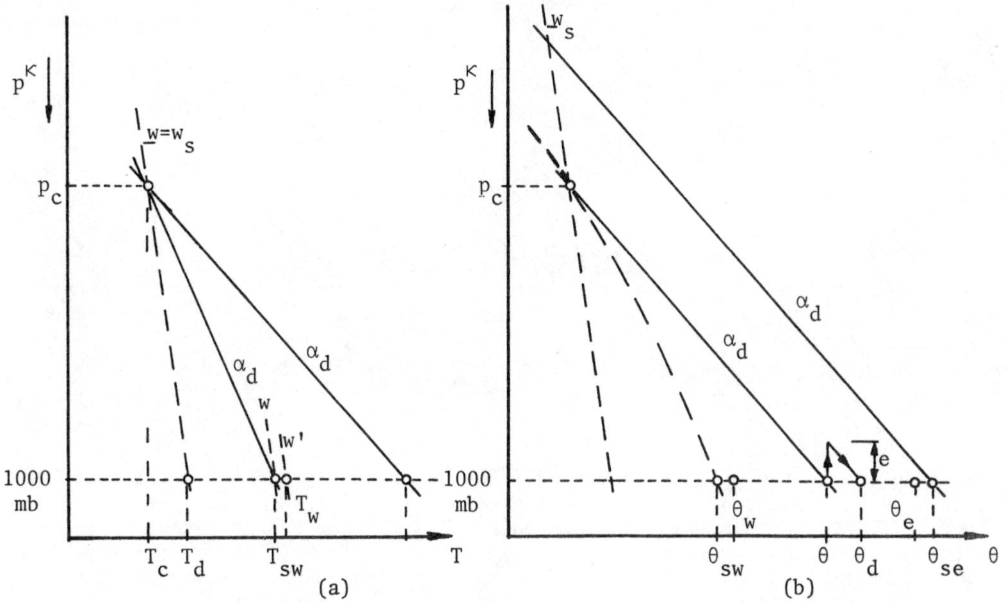

Fig. 2.32. Definition diagram of saturation temperatures (a) and of potential temperatures (b).

4. Potential temperatures, Fig. 2.32:

(a) Potential temperature $\theta = T(1000/p)^\kappa$.

(b) Partial potential temperature

$$\theta_d = T[1000/(p - e)]^\kappa = \theta[(1 + e/(p\ e)]^\kappa \simeq \theta(1 + 0.461\ w).$$

(c) Pseudo-wet-bulb potential temperature θ_{sw} is approximately defined by

$$c_p\ \ell n\ T_1/\theta_{sw} - R_d\ \ell n(p_1 - e_{s1})/(1000 - e_{s2}) + (w_{s1}L_1)/T_1 - (w_{s2}L_2/\theta_{sw} = 0$$

where subscript 1 refers to a location on the wet adiabatic and 2 at 1000 mb.

(d) Wet-bulb potential temperature $\theta_w \simeq \theta_{sw}$.

(e) Pseudo-equivalent potential temperature $\theta_{se} = T_0(1000/p_o)^\kappa \simeq \theta_d \exp[(L_w/T_c c_p)]$ where T_0 and p_0 are the temperature and pressure at the elevation where $w = 0$. The equivalent potential temperature θ_e refers to the value of θ that the humid air would attain when all the water is condensed out at constant pressure and the heat of evaporation is used to warm the air.

Chapter 3

PRECIPITATION

Precipitation is the primary source of fresh water supply and its records are the basis of most studies dealing with water supply in all its forms, floods, and droughts. Atmospheric moisture is obtained through evaporation and transpiration. The major source of supply of vapour is tropical oceans. Cooling of the moist air masses leads to precipitation. There are four major processes which cause cooling:
1. *Orographic lifting*. Where a mountain range intercepts the moist air flow, the air is forced to ascend, and as it cools it will precipitate some or all of its water content, as rain, snow or both.
2. *Convection*. Differential heating or advection can lead to the air becoming locally more buoyant. The air mass may then rise to levels where it becomes saturated, forms clouds and precipitates. Thunderstorms are examples of the convective process.
3. *Convergence*. Wind fields may converge and force air to rise.
4. *Fronts*. The low pressure areas of temperate and polar regions usually have frontal systems. These are the interfaces between large warm and cool air masses in motion.

Any of these processes may act alone or in combination with any of the others. The amount of precipitation depends on the availability of moisture in the atmosphere. Orographic barriers can lead to very large amounts of precipitation per annum, whereas the greatest intensities are associated with short term convective storms.

The amount of precipitation can be defiend as an accumulated total volume for any selected period. Precipitation, as a function of time, is a highly variable function. This variability is present within a storm record as well as in the record of annual totals, Fig. 3.1. The record looks noisy, i.e. there is a random variation, which obscures the mean value as well as any long term trends and periodicities which may exist, and it shows the stochastic nature of the record.

Likewise, the distribution of presipitation in space - the areal distribution - is uneven. From a number of point measurements we can plot the equal depth of precipitation contours for a given storm - the isohyetal map, Fig. 3.2. The meteorological system, which gave cause to this exceptional rainfall, was associated with warm moist air from the north-east meeting cooler air from the south-east, the ascent of the warm air being accentuated by the continuous range of mountains west of Hawke's Bay. The track of the storm touched New Zealand only across the east coast promontory, and everywhere (except on the East Cape) the ground winds were the cool winds from the south-east. The distribution of rainfall was similar on each

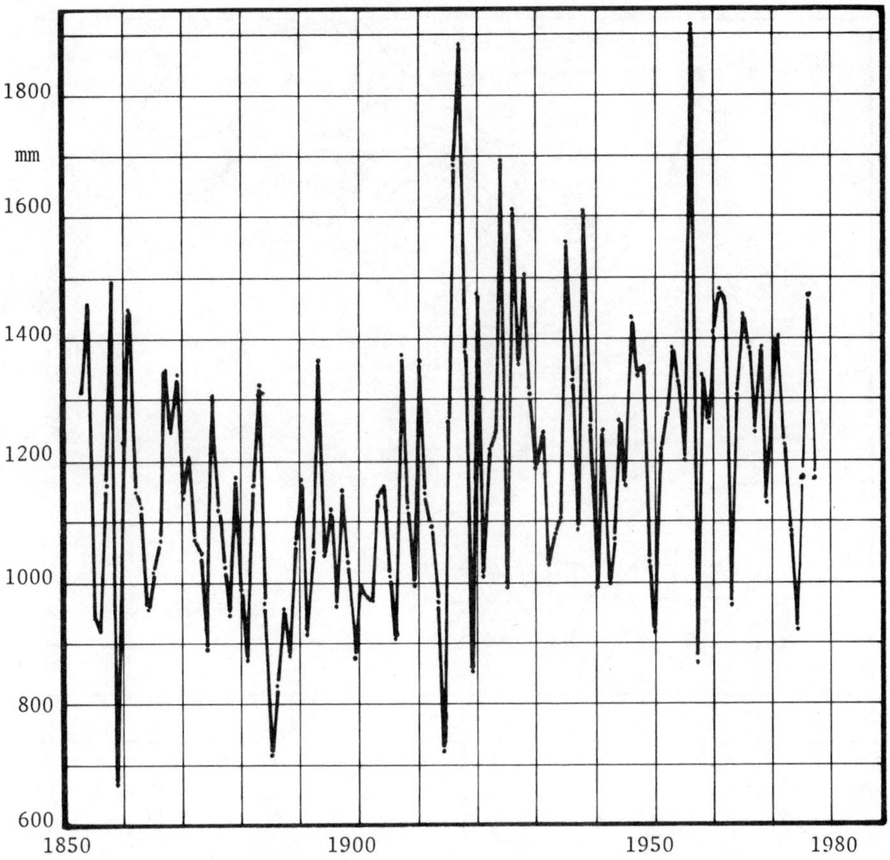

Fig. 3.1. Annual precipitation at Auckland, N.Z., over 125 years.

day despite the movement of the storm, thus pointing to the dominant influence of the topography. The area of very heavy rainfall on each day was a well-defined area about 16 km wide, parallel to the coast of Hawke's Bay and extending from Kotemaori down to a point somewhere in the Tutaekuri catchment, a distance of 55-56 km. In the same area in March, 1924, a similar storm caused 511.6 mm and 419.1 mm of rain in 10 hours at Rissington and Eskdale, respectively, and 241.8 mm in 9 hours at Napier.

Generally, the structure of clouds which yield the precipitation is not uniform. As an oversimplified picture, the cloud could be imagined to be a random conglomeration of "cells" of random size, which move about in a random manner relative to the mean motion in a lower intensity "background" cloud. Hence, if we imagine a set of identical (statistically) clouds, then a rain gauge at a given point would yield a measurement, which depends on the number and size of the "blobs" passing over it.

There are many models for storm structures. The deterministic features are present primarily in the temporal and spatial distributions of precipitations of the same scale. The meteorologists refer to the

1. *microscale* or *convective scale*, which is the smallest and consits predominantly

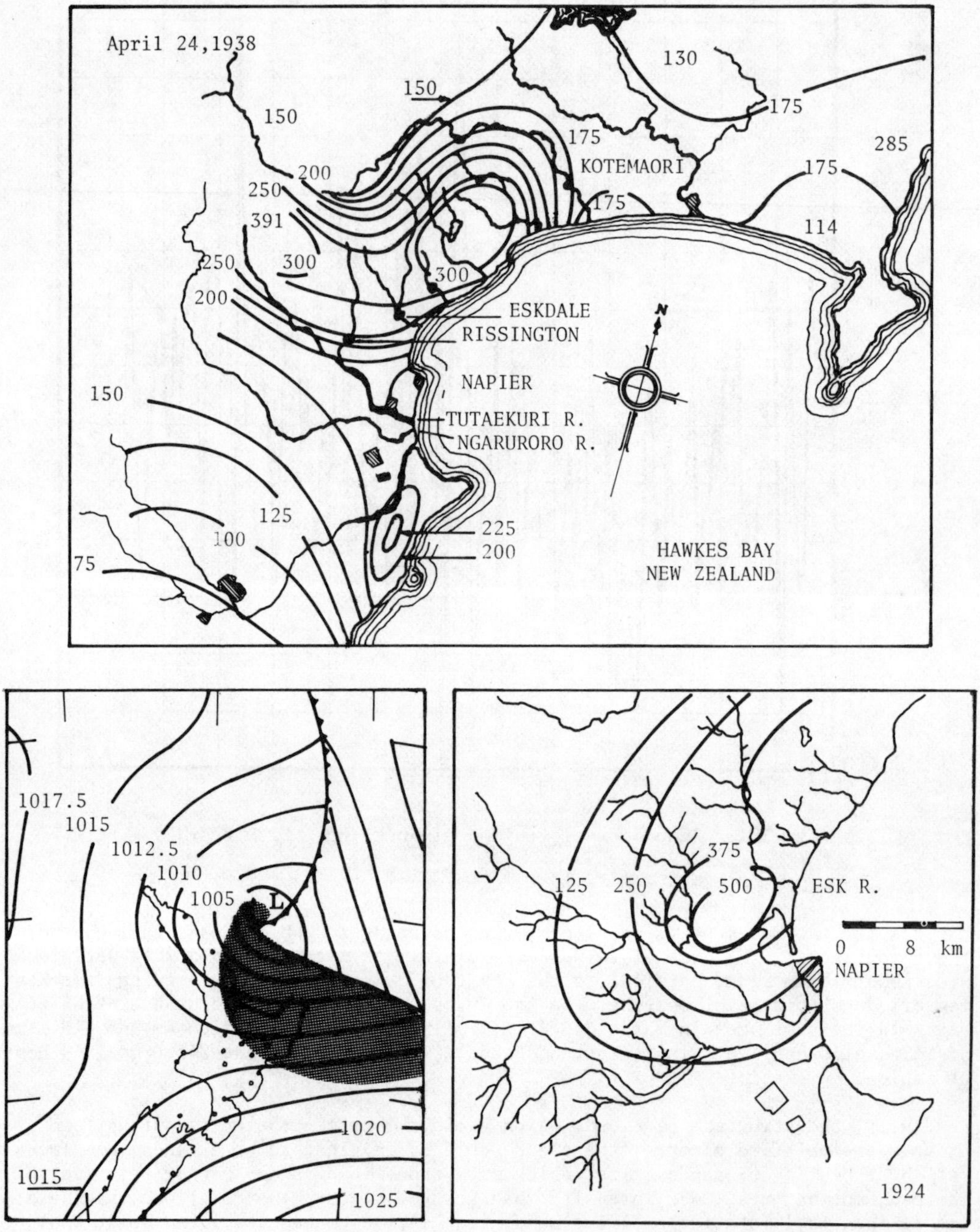

Fig. 3.2. Isohyetal and weather maps for a storm rainfall, April 24, 1938, together with the isohyetal map for another storm on the same area 14 years earlier. Isohyets in mm.

of convective cells. The cells measure a few kilometers in size and contain vigorous up and down drafts.
2. *mesoscale*. The convective cells are produced for some time. They grow, rise and decay. The agglomeration of these cells into one at a given time leads to the mesoscale, Fig. 3.3, which is of the order of 5-50 km in size. Thunderstorms belong to this scale.
3. *synoptic scale*, which describes storm systems associated with fronts and low pressure centres and is hundreds of kilometers in size.

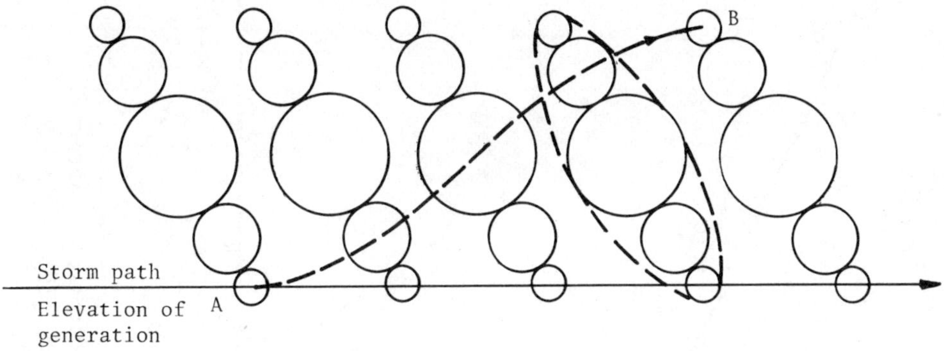

Fig. 3.3. Diagrammatic illustration of propagation and growth of cells in a convective storm. The dashed line, A-B, indicates the path of a cell from generation to decay. The group of cells at a given instant, within the dotted envelope, defines the mesoscale.

Most regions have maps, which show the mean annual precipitation isohyets, as well as maps showing maximum daily recorded rainfall, etc. The rainfall intensities are highly variable. Figure 3.4 shows a plot of the world's greatest observed point rainfalls, according to Jennings (1950), to which a few additional points have been added. The points from La Réunion are for a gauge location with extreme orographic effects, both a rapid change of elevation and a funnelling effect.

Snow. In many parts of the world a substantial proportion of the precipitation is in the form of snow. Snowfall over an area tends to be more uniform than rainfall because of the lightness of snow and the mixing effect during the fall, but, except within forests, the snow cover tends to be very uneven. Within mountains the snow cover varies widely because of wind effects, snow drifts, snow slides, etc. Over tundra and prairies the cover is more uniform, although local detail can vary. Shelter belts, buildings, and gullies, etc, may cause massive accumulation of snow.

Snow depth information alone provides little information on the equivalent water yield. For the latter the density of the snowpack has to be known. The snowpack on the ground is subject to consolidation, wind, and heat transfers, and its structure changes, i.e. the snow ages. Ageing in turn changes its density. The density of freshly fallen snow is highly variable and is related to the temperature at ground level. Observed density values range from 4-340 kg/m^3. Below are a few average values. The bulk of the data would be within the ± 30 kg/m^3 band.

°C	-10	-6	-2	0	4
kg/m^3	40	60	100	120	200

Compaction and structural changes in freshly fallen snow start immediately and snow

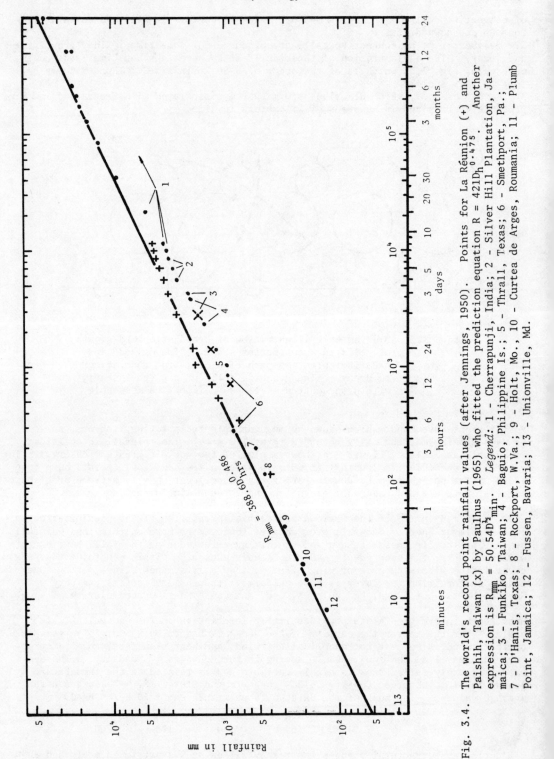

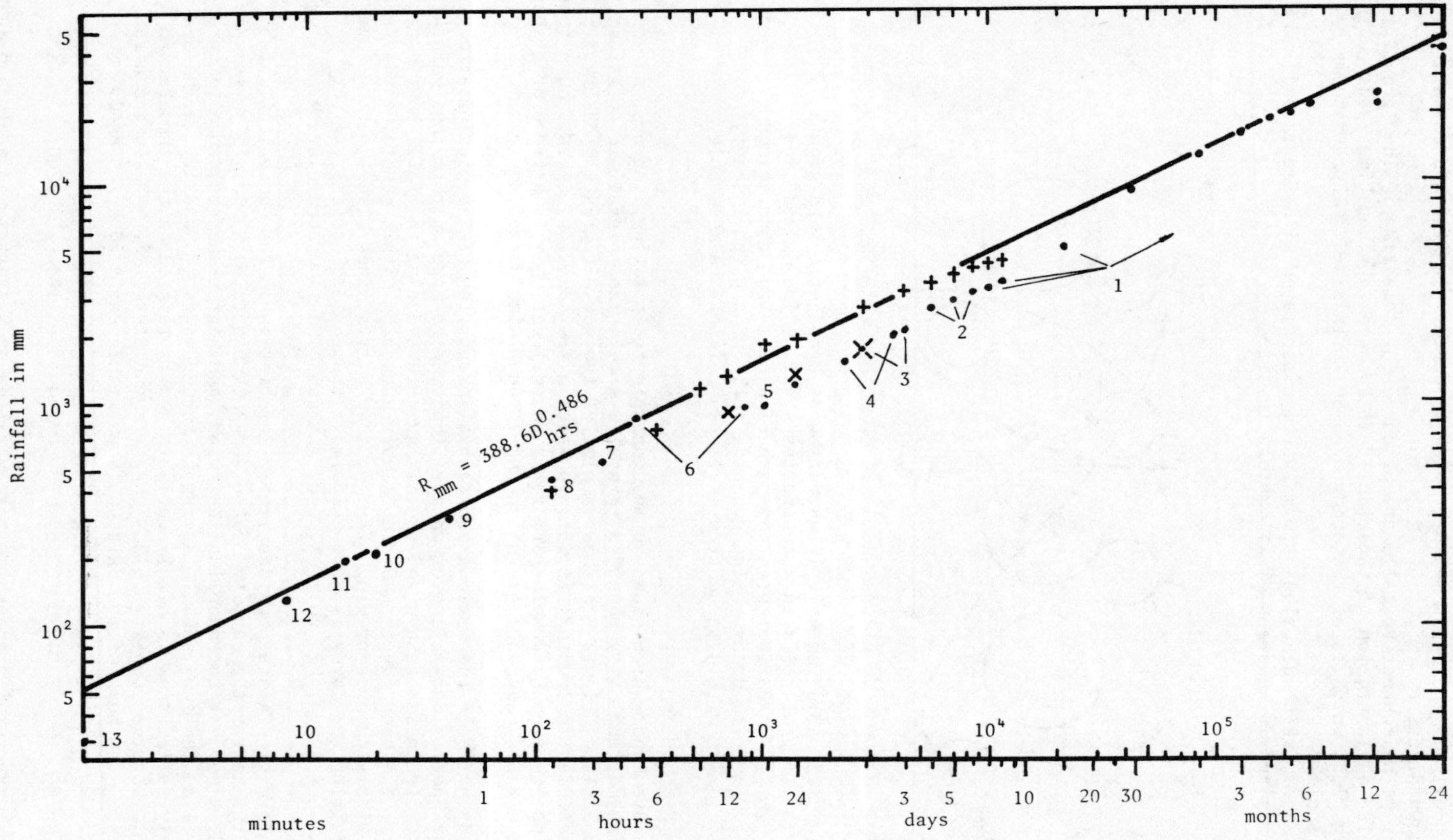

Fig. 3.4. The world's record point rainfall values (after Jennings, 1950). Points for La Réunion (+) and Paishih, Taiwan (x) by Paulhus (1965) who fitted the prediction equation $R = 421D_h^{0.475}$. Another expression is $R_{mm} = 50.54D^{\frac{1}{2}}_{min}$. *Legend:* 1 - Cherrapunji, India; 2 - Silver Hill Plantation, Jamaica; 3 - Funkiko, Taiwan; 4 - Baguio, Philippine Is.; 5 - Thrall, Texas; 6 - Smethport, Pa.; 7 - D'Hanis, Texas; 8 - Rockport, W.Va.; 9 - Holt, Mo., 10 - Curtea de Arges, Roumania; 11 - Plumb Point, Jamaica; 12 - Fussen, Bavaria; 13 - Unionville, Md.

with a density of 40 kg/m^3 may become snow with a density of 160 kg/m^3 in 24 hours.

3.1 Measurement of Precipitation

Measurement of precipitation is a specialized topic which is covered by meteorological texts. Apart from a few brief comments, it is not intended to enter here into a discussion of the various instruments and measuring techniques.

In general, the network of measuring instruments can be subdivided into national and local networks. The former are extensive in coverage whereas the latter usually refer to much more closely spaced instrumentation of specific catchments. The fall on the catchment as measured by the gauges depends on the density of the network. There is an upper saturation limit for the number of gauges beyond which the mean of the measured values will not vary with the addition of further gauges. At a lesser number of gauges the confidence limits of the measurement are a function of the number of gauges, Neff (1965). In catchment practice it is usual to use one gauge for 0-10 ha and up to 3 gauges for 100 ha. The density of the network decreases with increasing area. The number of gauges necessary to achieve a measurement of the rainfall to within selected confidence limits can be determined by assuming that the area is meteorologically homogeneous, the distribution is normal and that there are no systematic sampling errors. Then the value of precipitation, P, must lie between $\bar{P} \pm (S/\sqrt{N})t$, where $S = [\Sigma(P_i - \bar{P})^2/(N - 1)]^{\frac{1}{2}}$, summed from 1 to N, is the sample standard deviation and t is the Students t. For the mean to lie within $\pm \Delta P$ the condition $(S/\sqrt{N})t \leq \Delta P$ has to be stisfied. Using past observations to evaluate S we can arrive at an estimate for N. Gauge spacing also has an effect. A uniform grid network would be convenient, but the grid spacing should be reduced across zones of intense climatic variation, for example, up a steep slope. Most of the existing gauge networks are irregular. A detailed design method for rainfall networks was proposed by Rodriguez-Iturbe and Mejia (1974). They consider spatial and time correlation, the number of stations and the network geometry. For further reference see Eagleson (1967), Rodda et al. (1969) and Stol (1972).

There are two principal types of rain gauges: storage, in which the total precipitation is determined by direct measurement of the amount of water at 12 or 24 hour intervals, and the recording gauges, which make or transmit an accumulative record as the rain falls. The recording gauges employ either a float linked to a pen in a storage chamber, which is equipped with an automatic siphoning of the chamber when it is full, or two collecting chambers under a funnel. The chamber is balanced so, that it tips over and brings the other chamber under the funnel each time an amount corresponding to, for example, 0.2 mm of rain has been collected. Each changeover makes or breaks an electrical circuit for recording purposes. The variety of snow measuring devices is far greater but it is very difficult to achieve measurements which compare in accuracy with rain gauging reults.

3.2 Analysis of Precipitation Data

Of all hydrological data, data on precipitation are most readily available and have been collected for the longest periods. Likewise, the network of precipitation recorders is much finer than, for example, that of stream flow gauging stations. However, prior to analysis all data should be checked. The records should be homogeneous and complete. Over the years the instruments, their location, and observers change. These changes may cause instrumental and observational errors. Frequently records, particularly older records, indicate trends which are not due to meteorological causes at all.

The *double-mass curve* technique is used to check the precipitation data for consistency. The technique is based on the observation that the mean accumulated precipitation for a number of gauges is not very sensitive to changes at individual stations because many of the errors compensate, whereas the cumulative curve for a single gauge is immediately affected by changes at that station. Thus, if the total accumulated precipitation for all the other gauges is plotted against that for the gauge under scrutiny, the result should be an unbroken straight line, provided the entire record for the single gauge has been observed at the same site under the same conditions, Fig. 3.5. If there is a change in slope at a given point in time, it is usually possible to find the cause, for example, the station was shifted, gauge type changed, etc. The earlier records can then be adjusted by multiplying by the ratio of the slopes. In areas with marked seasonal variation in precipitation comparisons are best done on a seasonal basis.

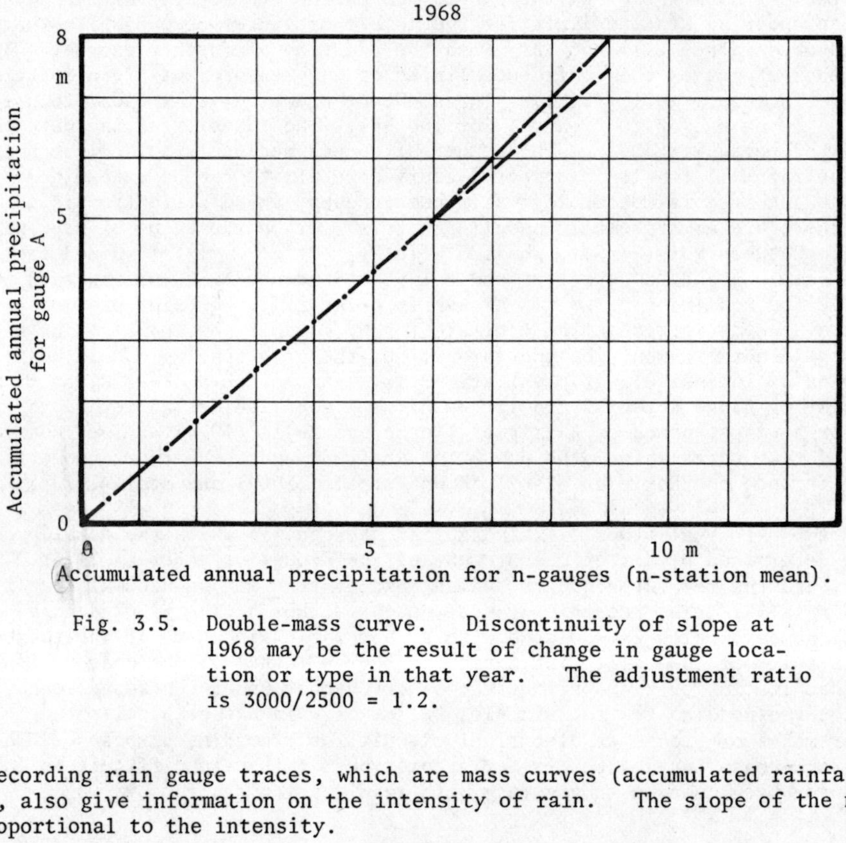

Fig. 3.5. Double-mass curve. Discontinuity of slope at 1968 may be the result of change in gauge location or type in that year. The adjustment ratio is 3000/2500 = 1.2.

The recording rain gauge traces, which are mass curves (accumulated rainfall against time), also give information on the intensity of rain. The slope of the mass curve is proportional to the intensity.

The records may be analysed for distribution of intensities and frequencies, including extremes, as well as for areal distribution and for physical relationships.

The records of individual gauges provide point values which allow plotting of isohyetal maps and estimation of storm yield. Statistical analysis of each record also gives estimates for maximum recorded rainfalls, annual rainfalls, and 10 minute rainfalls, etc., for selected return periods (probability, e.g. a 50 year return period means a 1 in 50 chance of the event occurring in any one year). From analysis of several point records one could plot isohyetal maps for given return period rainfalls.

3.2.1. Intensity - duration - frequency relationships.

For each record we can calculate (see Chapter 12) the probability levels for given intensities and durations. These, when plotted, will give the intensity - duration - frequency curves, Fig. 3.6.

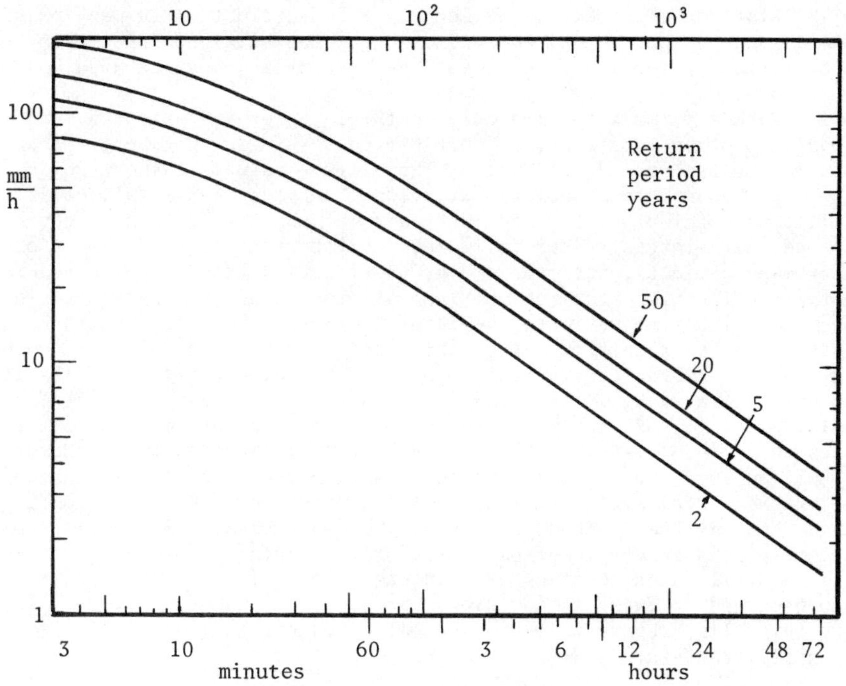

Fig. 3.6. Diagrammatic intensity - duration- frequency curves.

Instead of intensity total depth is frequently used, in which case the curves will slope upwards. The depth (or intensity) - duration - frequency curves can be extended from one gauge location to regional relationships by averaging data from gauges within a homogeneous region. The station grouping may be based on the analysis of variance. For numerical work on computers these curves can be expressed for i in mm/h or mm/min as

$$i = \frac{kT^x}{(t + c)^n} \qquad 3.1$$

where T is the return period in years, t is the duration of rain in hours or minutes, respectively and k, x, c, n are constants determined from the curves. Writing the expression for intensity in logarithmic form yields

$$\log i = (\log k + x \log T) - n \log (t + c)$$

which is an equation of the form y = b + mx. Thus, n = the slope of the line, x = the spacing of the curves for the various recurrence intervals T, k determines the vertical position of the lines as a set, i.e. the set of lines is shifted vertically by changing k only. It is often found that if a constant value (c) is added to all the values of t, then the set of lightly curved lines plots approximately straight on a log-log plot. If the lines do plot sufficiently straight to start with, the value of c is zero. The term kT^x is at times replaced by a constant D, which is interpreted as the instantaneous rainfall depth for a given return period. In the USSR the expression for i in mm/min is used with c = 1, i.e. $i = D/(t + 1)^n$, where $n \simeq 0.5$ to 0.70 with the lower value in the mountainous regions (Maksimov, 1964). It is seen that k = i when T and (t + c) are equal to 1.0, x is the change in log i per log-cycle of T, and n is the negative slope of the lines, i.e. $n = (\Delta \log i)/[\Delta \log(t + c)]$ = change in log i per log-cycle t. In many cases the slope n varies to such an extent that approximation by one straight line is not sufficient, and approximation by two straight lines (making a break) must be used.

In general, such a formula applies only to the location investigated, and is rarely transferable to other locations. Short-period rainfalls, however, appear to be almost free of geographical dependence. The extreme values of short-period rainfalls of up to 2 to 3 hours are associated with local convective rainfall cells, which have similar physical properties in most parts of the world. The frequencies of short-period, high-intensity rainfalls form an important part of the information needed for many projects, but records for short-period rainfalls are generally scarce because their collection requires continuously recording rain gauges. The general requirement is either to estimate the rainfall depth, when the duration and return period are given, or to estimate the return period, when the rainfall depth and duration are given. From the various studies of high-intensity rainfalls it appears that the data, if at all, is only marginally dependent on geographical location. Hershfield, et al. (1955) analysed the fairly extensive data available for the United States and pointed out its potential value in application to other regions. Reich (1963) applied these relationships to Africa and Bell (1969) combined the data from the United States including Hawaii, Alaska and Puerto Rico with data from Australia and South Africa. The U.S. Weather Bureau (Hershfield, 1961) recommends an empirical relationship derived from short-duration data, according to which the depth of a t_{min} rainfall has a constant ratio to the depth of a 1 hour rainfall of the same return period. These ratios are 0.29, 0.45, 0.57 and 0.79 for 5, 10, 15 and 30 minute rainfalls, respectively. The estimated errors in the data from throughout the United States are 5-8%.

Bell (1969) proposed an empirical relationship for the high-intensity, short-period rainfall depth and frequency as

$$\frac{P_T^t}{P_{10}^t} = 0.21 \ln T + 0.52, \quad 2 \leq T \leq 100 \qquad 3.2$$

where P_T^t is the depth of a t-minute, T-year return period rainfall and P_{10}^t is the depth of a t-minute, 10-year return period rainfall. The depth-frequency relationship was combined with depth-duration data by first expressing the ratio of P_T^t to the 60-minute depth, P_T^{60}, as

$$\frac{P_T^t}{P_T^{60}} = 0.54\, t^{0.25} - 0.50, \quad 5 \leq t \leq 120 \text{ min} \qquad 3.3$$

and then combining equations 3.2 and 3.3 to give

$$P_T^t = (0.21 \ln T + 0.52)(0.54\, t^{0.25} - 0.50)P_{10}^{60} \qquad 3.4$$

for $2 \leq T \leq 100$ years, $5 \leq t \leq$ minutes. For locations where the records are inadequate to determine the 1 h - 10 year rainfall, the following empirical relationships are proposed

$$P_2^{60} = 6.69 \times 10^{-3} M N^{0.33}, \qquad 0 < M \leq 2.0, \quad 1 < N \leq 80$$

$$P_2^{60} = 8.27 \times 10^{-3} M^{0.67} N^{0.33}, \qquad 2.0 < M \leq 4.5, \quad 1 < N \leq 80 \qquad 3.5$$

where P_2^{60} is the 60-minute, 2-year return period rainfall depth in mm, M is the mean maximum annual daily precipitation in mm, and N is the mean annual number of thunderstorm days. The 2-year, 1-hour rainfall may be used to estimate values for other durations and frequencies, using.

$$P_T^t = (0.35 \ln T + 0.76)(0.54\, t^{0.25} - 0.50) P_2^{60}$$

$$\text{for } 2 \leq T \leq 100 \text{ years}, \quad 5 \leq t \leq \text{minutes} \qquad 3.6$$

Bell claimed that this equation fits the curves by Hershfield and Wilson (1957) and Reich (1963) with deviations less than 10%.

Results like those in Fig. 3.6 are obtained, for example, from extreme value analysis, which consider the annual maxima of rainfalls of given duration and return period. These maxima come from different storms. Within a given duration, for a storm of a given frequency, the within storm depths are usually less than the corresponding yields for the extreme values, that is, the 3 hour depth in a 6 hour 50 year return period storm is less than the 3 hour depth in a 3 hour 50 year storm. Figure 3.7 indicates adjustment factors used by the U.S. Weather Bureau.

3.2.2 Depth - area relationships

The catchment rainfall may be estimated by averaging the rainfall measured at the gauge locations or from the isohyetal map. For averaging of the rainfall the Thiessen polygon method is convenient, particularly for repeated calculations. The concept of the method is illustrated in Fig. 3.8. The bisectors of the lines connecting the gauges subdivide the catchment. The area thus allocated to each station is measured and expressed as a percentage of the total area. These percentages are then used as "weights", e.g.:

Station	Areal weight	Rainfall mm	Proportion mm
1	0.35	15.2	5.32
2	0.25	10.0	2.50
3	0.30	14.0	4.20
4	0.10	17.5	1.75
	1.00	Mean rainfall =	13.77

The method assumes linear variation between stations and makes no allowance for topographical effects. Whitmore et al. (1960) modified the method to take into account the station altitudes. The automation of the method was discussed by Diskin (1969, 1970) with the computer determination of the coefficients from spatial coordinates for the rain gauges.

The summation of the rainfall between the isohyets over the catchment is still one of the best methods. An experienced analyst can make allowances for topographical effects in drawing the isohyetal map. Salter (1972) extended this method to computer analysis using the computer plotting facility. In addition to these three methods, more complex mathematical techniques have been introduced, such as the trend surfaces described by polynomials (Mandeville and Rodda, 1970), quadratic and cubic surfaces, double Fourier surfaces, bi-cubic splines and multiquadric analysis (Shaw and Lynn, 1972, and Lee et al., 1974). The multiquadric analysis is well suited for computer use.

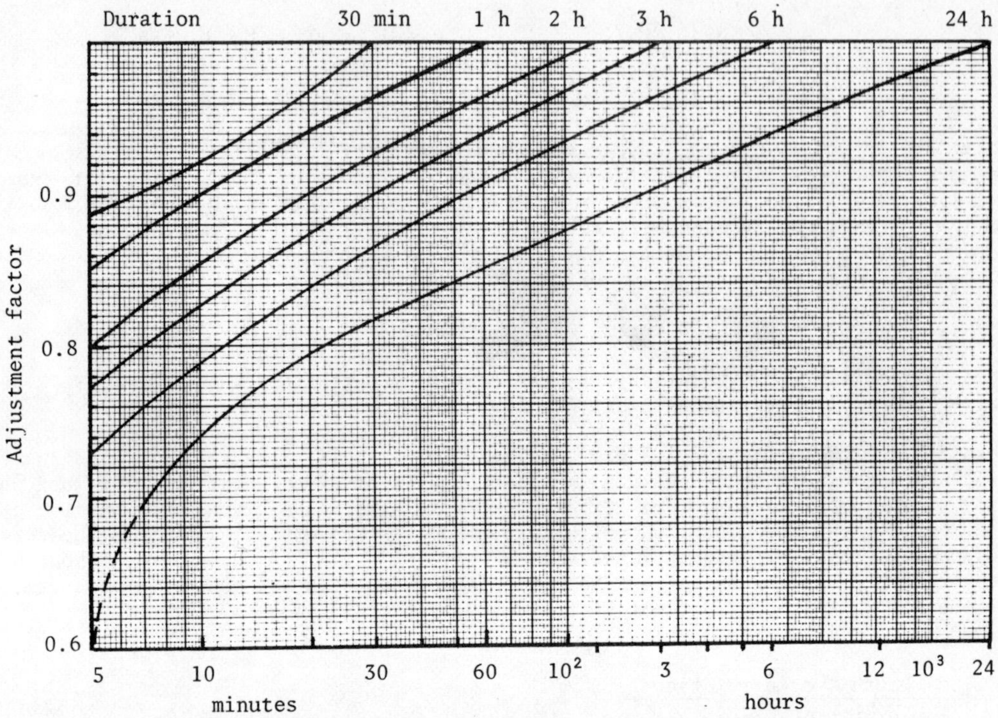

Fig. 3.7. Within storm adjustment factors used by the U.S. Weather Bureau. This shows, e.g., that the maximum 1 hour rainfall depth in a 24 hour storm is only 85% of that in a 1 hour storm of the same frequency.

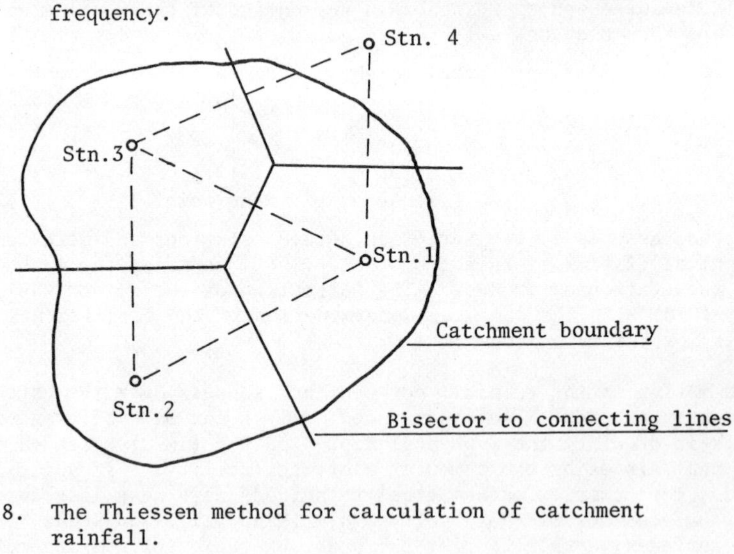

Fig. 3.8. The Thiessen method for calculation of catchment rainfall.

Depth-area relationships may be either *storm-centred* or related to *fixed-locations*, Fig. 3.9. The storm-centred relationships are used in studies of probable maximum precipitation. The fixed-location relationships are used for frequency studies and are based on different parts of the storm, giving lower values than the storm-centred relationships. The rainfall recorded at a gauge is a good sample of the amount falling on a small area surrounding the gauge. As the area increases, the correlation between the point rainfall and the mean rainfall over the area diminishes. The rate at which the areal mean departs from the point rainfall depends on the characteristics of the rainfall. The mean from short high-intensity rainfalls is very sensitive to the size of the area, whereas persistent rainfalls associated with warm fronts give uniform precipitation over a large area.

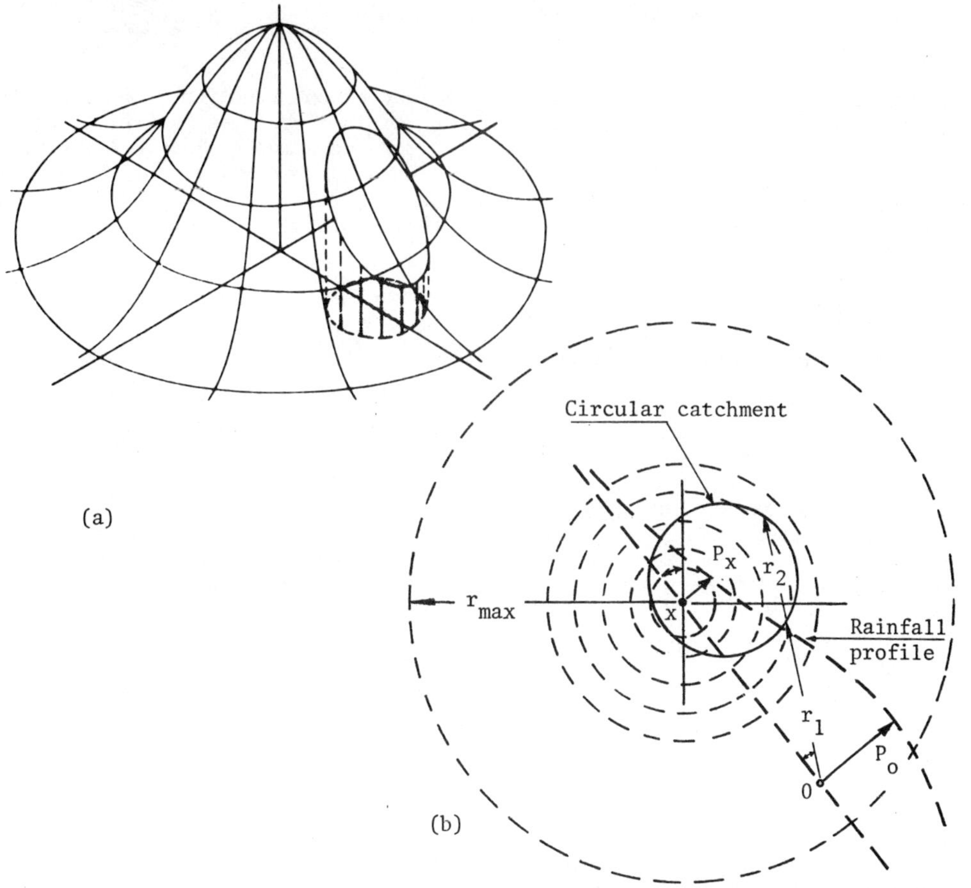

Rain gauge at x, storm focal point at 0.

Fig. 3.9. (a) Isometric sketch of rainfall depth distribution over a storm-centred catchment (the base circle area) and over a fixed location area (shaded circular area). Note, that the amount of rain on the fixed catchment will depend on the location and direction of storm movement. (b) Diagrammatic illustr. of areal reduction of areal mean rainfall depth.

Hershfield and Wilson (1960) found from storm-centred data on tropical and non-tropical storms that the ratio of the average rainfall over the whole area to that on 10 sq. mi (25 km^2) was as shown in Table 3.1. They also concluded that "the shape of the area-depth curve is not a function of the type of the storm".

TABLE 3.1 Ratio of average rainfall to that over 25 km^2

Time hrs	Area km^2 250	500	1 250	2 500	12 500
6	0.85	0.80	0.73	0.65	0.46
12	0.89	0.85	0.79	0.72	0.51
24	0.92	0.88	0.82	0.76	0.58
48	0.93	0.90	0.84	0.80	0.63

The depth-area relationship depends on the method of construction. By using the isohyetal maps, it is assumed that the maximum rainfall observed in a given storm is approximately equal to the maximum rainfall that occurred. The Thiessen polygon method yields a maximum observed value, which is a mean value over the area element, as defined by the polygon for the maximum station, and is less than the maximum that occurred. Frequently, the mean depth of rainfall $\bar{P}$ varies almost linearly with log A, for example

$$\bar{P} = P_1 - c\log(A/A_1)$$

or

$$A/A_1 = \exp[2.3(\bar{P} - P_1)/c] \qquad 3.7$$

where A_1 is the value of the area, A, obtained by extrapolating the line to a rainfall depth, P_1, and c is the slope of the line. Alexander (1963) showed that the difference between the mean depth, $\bar{P}$, and the minimum isohyetal depth, P^*, is constant and equal to c/2.3. The difference is hence independent of the area A. For a given storm and return period this difference can be determined from the corresponding depth-area curve. If $A = A_1$ is the total area covered by the storm, then $P^* = 0$ and $P_1 = \bar{P}$. The lines through the point A_1 represent a family of lines given by $f(P) = e^{P/m}$, where the mean depth, m, is measured above a zero defined at the point of truncation A_1. Comparison with the equation above shows that $m = c/2.3$, and that the lines representing P^* and $\bar{P}$ are parallel lines a distance d_1 or c/2.3 apart. Thus, by plotting the logarithm of the area against the minimum depth, P^*, of the isohyet enclosing the area, the mean depth $\bar{P}$ is obtained using $\bar{P} = P^* + P_1$.

The plot of the isohyetal values against the equivalent radius of a circular area equal to that enclosed by the isohyets is known as the *rain profile*.

Numerous formulae have been developed for the relationship between the average depth of rainfall and area. These express the mean depth $\bar{P}$ for the given area centred around P_{max}. Frühling (1894) proposed that for sudden downpours the ratio of P at a distance x from the storm centre to P_{max} at the centre is proportional to $\sqrt{x}$, i.e. for x in km

$$P = P_{max}(1 - \sqrt{x/12}), \qquad x < 12 \text{ km} \qquad 3.8$$

By integration the average depth of rainfall for a circular area of radius x is

$$\bar{P} = P_{max}\{1 - [2/(5\sqrt{3})]\sqrt{x}\} \simeq P_{max}(1 - 0.17\, A^{\frac{1}{4}}) \qquad 3.9$$

and $\bar{P}/P_{max} = 0.2$ for the entire area of 12 km radius. Horton (1924) proposed a relationship of the form

$$\bar{P} = P_{max} \exp(-kA^n) \qquad 3.10$$

and Boyer (1957)

$$P = P_{max} \exp(-bx) = P_{max} \exp(-bkA^{\frac{1}{2}}) \qquad 3.11$$

where $k = \pi^{-\frac{1}{2}}$ for circular isohyets and increases as the shape of the area departs from circularity. The average

$$\bar{P} = 2P_{max} b^{-2}x^{-2}\{1 - (1 + bx)\exp(-bx)\}$$

$$= \frac{5.2P_{max}}{b^2x^2}\{1 - (1 + 0.62bx)\exp(-0.62bx)\} \qquad 3.12$$

where $b = 0.0235$ for $A > 260$ km^2. Court (1961) introduced the Gaussian bivariate distribution for which the isohyets are elliptical. With the x-axis along the major axis of the ellipses

$$P_{x,y} = P_{max} \exp(-a^2b^2 - b^2y^2) \qquad 3.13$$

where the parameters a and b define the scale and ellipticity (a = b gives circular isohyets). Any elliptical isohyet passing through points $x,y = (\pm c/a, 0), (0, \pm c/b)$ is given by $a^2x^2 + b^2y^2 = c^2$ and the average precipitation inside this isohyet is

$$\bar{P} = (P_{max}/c^2)\{1 - \exp(-c^2)\} \qquad 3.14$$

The $P_{max}/2$ isohyet enclosed half of the storm rainfall and from the corresponding distances x and y, along the major and minor axis of the $P_{max}/2$ isohyet, the values of a and b can be estimated

$$a = 0.832/x \qquad b = 0.832/y \qquad 3.15$$

The various formulae predict vastly different rainfall depth gradients from the centre of the storm outwards. Short-duration storms tend to have steeper precipitation gradients than those of longer duration which usually cover larger areas. Court concluded in his discussion and comparison of the various area-depth relationships that "if such an expression can be found, its parameter(s) will depend on the area and duration". This implies a *depth-area-duration relationship*.

A theoretical treatment of the problem of conversion of a point rainfall to an areal rainfall was attempted by Roche (1963). Roche sought to answer the question: "Given the point rainfall for a certain level of probability at an arbitrary point on the area what is the average rainfall over the area for the same level of probability?" Rodriguez-Iturbe and Mejia (1974a) discuss the solution by Roche and propose a simpler method. They deduce that the rainfall reduction factor $\bar{P}_A/P$ depends entirely on the correlation coefficient between the point rainfalls at two randomly chosen points in the area under consideration and is equal to $K = \sqrt{r}$, where r represents the expected correlation coefficient between two randomly chosen points in the area A. Two different spatial correlation structures are identified as being important for the areal process, the exponentially decaying structure

$$r_1(s) = e^{-|hs|} \qquad 3.16$$

and the Bessel type correlation structure

$$r_1(s) = sbK_1(sb) \qquad 3.17$$

where h and b are parameters, s is the distance between the two points, and $K_1(sb)$

is the modified Bessel function of the second kind. Figure 3.10 shows the reduction factor K as a function of Ah^2 and Ab^2, respectively. The characteristic distance depends on the size and shape of the area. Matern (1960) computed the mean distance between two randomly chosen points for the following shapes of unit area:

circle 0.5108, hexagon 0.5126, square 0.5214, equilateral triangle 0.5544, rectangle (with side ratio 2) 0.5691, (with side ratio 4) 0.7137 and (with side ratio 6) 1.3426.

Thus, a unit rectangle with side ratio 2 has a diagonal of 1.58 and a similar catchment with a diagonal of 200 km would have a characteristic distance 0.5691(200/1.58)= 72 km. Equations 3.16 and 3.17 yield h and b. The square of these multiplied by $A(km^2)$ gives Ah^2 and Ab^2 and Fig. 3.10 gives the correction factor K.

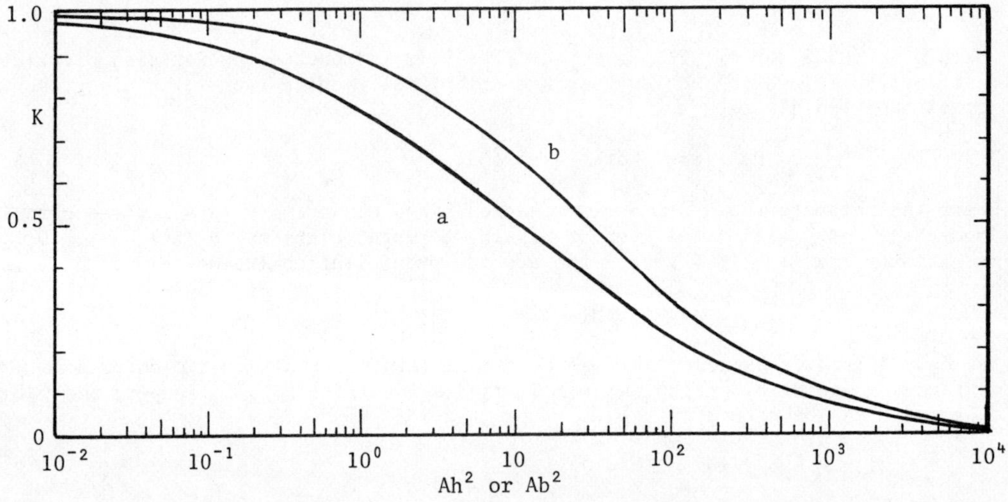

Fig. 3.10. Spatial reduction of point rainfall intensity for the case of exponential correlation structure (a) and Bessel-type correlation structure (b), according to Rodriguez-Iturbe and Mejia (1974a).

3.2.3. Depth - area - duration relationships

The depth-area-duration (DAD) relationships provide the designer with important information on temporal and spatial variation of rainfall for a given area and the DAD also provide one of the simplest methods of transposing of the storm data. For a given storm with one centre the depth-area relationship is derived using the isohyets as boundaries of individual areas, working from the centre outwards. For each individual storm the rainfall data can be tabulated from the isohyetal maps as a matrix where each column refers to rainfall totals of a particular duration and the rows give the area involved, starting with station maximum over 25 km^2 and proceeding, for example, in steps to 100, 200, etc. km^2. The size of area increments used will depend on the type of the storm. For widespread storms the point data from the rain gauge is usually assumed to apply to 25 km^2. For the majority of hydrologic requirements 6 hour time increments have been found satisfactory. Longer increments may be warranted with extensive storms on very large catchments and shorter with intense convective storms over a small area. In principle, this means that isohyetal maps are required for each 6 hour period. Instead of tabulation the data could be plotted as average precipitation against area.

Storms with multiple centres are analysed by dividing the area into zones. After the depth-area data for each zone has been calculated the results are combined within each isohyet by adding their respective accumulated areas and accumulated volumes. The average depth at each step is then obtained by dividing the volume by the respective areas. Details of the computational procedure are described in WMO - No. 237.TP.129 (1969) manual, together with a worked example. The end result may then be tabulated in the matrix form or plotted as average precipitation against area with duration as a parameter, Fig. 3.11. For each duration the line is drawn as an envelope curve.

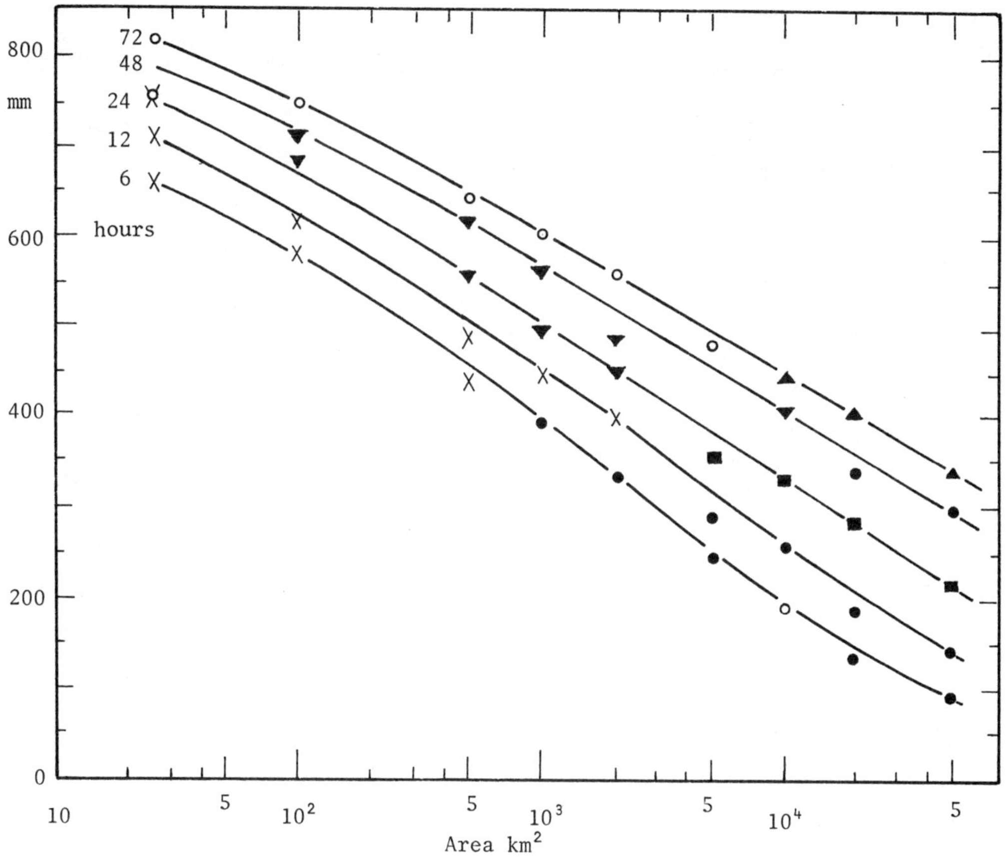

Fig. 3.11. Diagrammatic presentation of maximum depth-area-duration curves for a catchment. Symbols indicate separate storms. Note the enveloping of data points.

The information from a single rain gauge can be applied to the catchment with the aid of regional depth-area-duration data. Where this data is not available the point value of precipitation P(T,t) has to be multiplied with a suitable factor which converts the point value of the t-hour rainfall of T-years return period to an estimate for the whole area to which it is to be applied. Relatively little is known in a quantitative sense about this factor, mainly because of the scarcity of sizeable catchments with dense rain-gauge networks and long records. A fairly widely used relationship is the one shown in Fig. 3.12 from U.S. Weather Bureau

Tech. Paper No. 29, Pt. 3, which was derived from a study of seven dense networks of gauges in eastern U.S.A. and relates to major storms. The networks covered areas from 250 to 1000 km^2 and the records varied between 7 and 15 years. Within the general region the relationship appeared to be independent of geographical location. However, each of the curves in Fig. 3.12 was fitted to a set of points with considerable scatter. The ratio of the areal precipitation depth $\bar{P}_A$ to the precipitation depth P at a point was expressed by Leclerc and Schaake (1972) as

$$K = \frac{\bar{P}_A}{P} = 1 - \exp(-1.1\, t_r^{\frac{1}{4}}) + \exp(-1.1\, t_r^{\frac{1}{4}} - 0.01\, A) \qquad 3.18$$

where t_r is the duration in hours and A is the area in square miles. Figure 3.12 also shows the reduction factors developed by the U.S. Weather Bureau (1960) for the western United States.

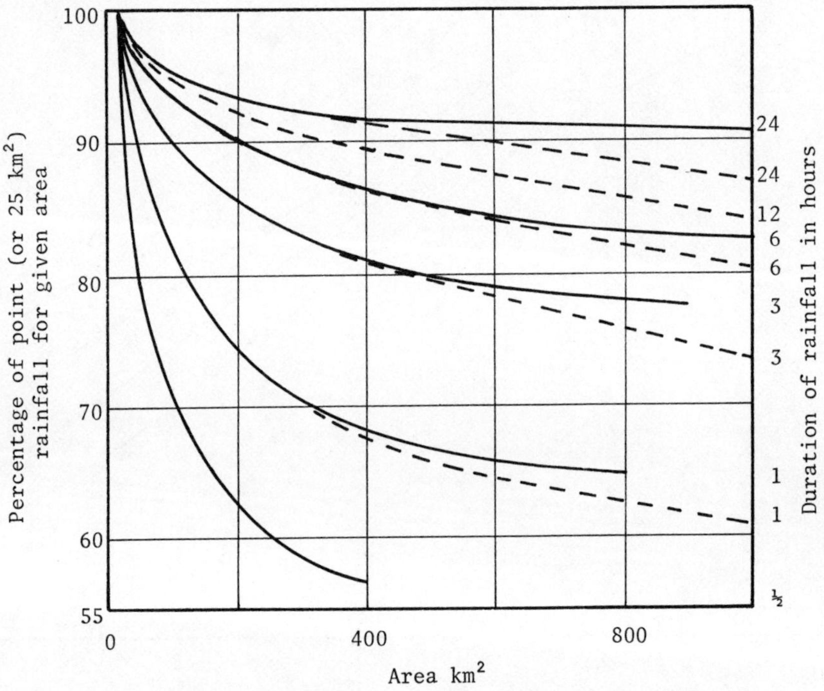

Fig. 3.12. Areal reduction of point rainfall, according to U.S. Weather Bureau. Full lines are for eastern and dotted for western United States.

3.2.4. Maximization of storms

Under this heading come the techniques of *moisture maximization, sequential maximization* and *spatial maximization* of storms.

Moisture maximization of storms in place is based on the concept of moisture content of air. A vertical air column has a moisture content given by

$$W = \frac{1}{g} \int_{P_1}^{P_2} H_s\, dp \qquad 3.19$$

where W is the mass per unit area (g/cm^2) and H_s is the specific humidity. Provided the distribution of H_s and the limits of the integral are known, the calculation of W gives the total water content of the air column and hence the maximum amount of water that could precipitate.

The most intense precipitations on drainage areas up to a few hundred km^2 are from thunderstorms. These are common in areas with warm temperatures and in tropical regions. The thunderstorm, or a related group of thunderstorms, is characterized by an inflow at a low level of very moist air, which condenses soon after it starts to rise. The yield of precipitable water varies with the wet-bulb potential temperature θ_w, which is the same as the 1000 mb dew point temperature. A measure of this variation is the specific humidity difference along a saturation adiabat between the low level inflow and the level of the top of the cloud. The variation of the specific humidity, H_s, is usually expressed as the difference between the 900 mb (1 km) and 200 mb (11-13 km) values. The moisture maximization ratio is then expressed as

$$r_m = \frac{(H_{s\ 900} - H_{s\ 200})_m}{(H_{s\ 900} - H_{s\ 200})_s} = \frac{(\Delta H_s)_m}{(\Delta H_s)_s} \qquad 3.20$$

where m refers to the condition of maximum wet-bulb potential temperature for the season and region, and s to that for the storm. The difference in specific humidity for tall clouds and for moderate clouds is shown in Fig. 3.13.

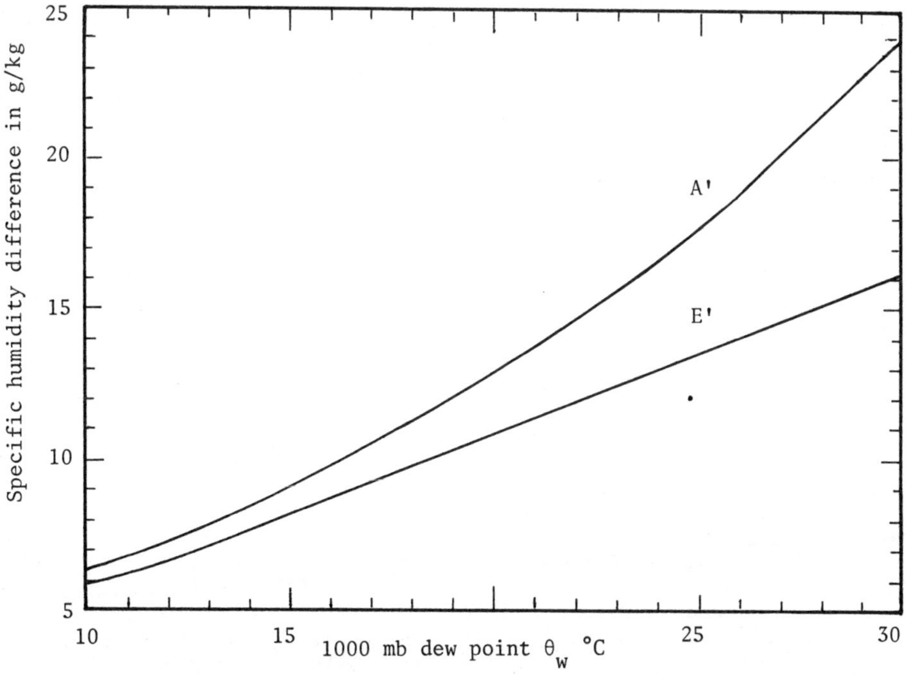

Fig. 3.13. Specific humidity differences along saturation adiabats. The A' curve refers to tall cloud specific humidity difference $(H_{s\ 900} - H_{s\ 200})_{\theta_w}$, and E' to that for clouds of moderate height $(H_{s\ 900} - H_{s\ 400})_{\theta_w}$. (WMO, 1969).

Since H_s and the mixing ratio w are numerically very nearly equal [H_s(g/kg) = w/(1 - w/1000)], the difference may also be read off the tephigram. However, $H_{s\ 200}$ is a low value and eqn 3.20 can be approximated by

$$r_m = \frac{(H_{s\ 900})_m}{(H_{s\ 900})_s} \simeq \frac{w_m}{w_s} \qquad 3.21$$

The thunderstorm clouds reach great heights. The cloud height varies appreciably with temperature. The cloud height variation is almost linear from 300 mb (9000 m) at θ_w = 10 °C to 190 mb (12 000 m) at θ_w = 20 °C with a ceiling at about 100 mb (16 000 m) corresponding to θ_w = 26 °C or more. An adjustment for the cloud height is obtained by multiplying r_m by the ratio $(mb)_m/(mb)_s$ of the pressures at the top of the respective clouds.

On large drainage areas extreme quantities of water are derived from cyclonic storms covering thousands of km^2. The clouds of cyclonic storms are generally not as thick as those of thunderstorms and seldom exceed the 400 mb level. The moisture maximization ratio in terms of ($H_{s\ 900}$ - $H_{s\ 400}$) or ($H_{s\ 1000}$ - $H_{s\ 400}$) is used in maximizing storm yields from cyclonic storms, although the thunderstorm model is not readily adaptable to these situations.

In order to apply the maximization process, two saturation adiabats have to be identified; one for the storm to be maximized (usually at the time and place of the heaviest rainfall) and the other for the warmest condition that could be expected at the same place and season. The dew point value must relate to the inflowing air masses and not to extraordinary evaporation on a hot day from local lakes and swamps. To counteract errors in dew point measurement the maximum dew point value should be based on several consecutive measurements. The *highest persisting 12-hour dew point* is one criterion in use. It is the highest value equalled or exceeded in a 12 hour period. For example, the following 3-hourly measurements in °C, 18, 20, 22, 24, 22, 23, 20, give 22 °C as the highest persisting 12-hour dew point value. Alternatively the average, say, over 12 hours could be used. The storm dew point should relate to the warmest air mass.

The moisture maximization estimates the total moisture content of the air mass at a given location, temperature and humidity. It is an upper estimate of the possible depth of precipitation. The actual precipitation is always less because not all atmospheric moisture is precipitated.

The above concepts gave rise to the terms "maximum precipitation" and "probable maximum precipitation". Both were aimed at giving an estimate of the maximum or upper limit of flow for design of major hydraulic structures but both terms have been controversial. The theoretician does not accept that there is an upper limit for the yield which may result from the interaction of all the factors involved in producing rainfall. Yevjevich (1968) wrote that "the thickness of an air mass that produces the maximum precipitation can always be 1 metre greater than assumed". At the same time if the planet Earth is assumed to be a constant mass system, then there must be an upper limit to the amount of water that can precipitate. Rather than to define an upper limit, the statistician prefers to define the probability of a certain value being exceeded. The term "maximum possible precipitation" was subject to severe criticism on account of the limited data available on extremes and the limited understanding of the processes leading to such an event. As a compromise, the even more controversial and contradictory term "the probable maximum precipitation" (PMP) was introduced. The PMP method was intended somehow to yield a limiting value so that the level of risk need not be considered. It is the aspect of the risk involved that is essentially at the centre of most of the controversy. As Benson (1964) wrote:"A maximum has no probability associated with it, except a limiting exceedance probability of zero".

The extreme events are an area where more research is needed into the probabilities of occurrence of the various factors that combine to produce these exceptional rainfalls. The studies should include the interactions of these factors and the resultant probabilities of these rainfalls and runoffs. The runoffs and their probabilities should include the effects of the catchment characteristics. One of the most promising lines of statistical studies is the analysis of multi-station data in both time and space. In this way all the information is used. Work along these lines was started by Alexander (1963) and will be referred to later. For other than extreme events, frequency analysis of existing records is the principal method in use for estimation of rainfall yields and return periods.

The moisture maximization is associated with *wind maximization* in orographic regions when it is observed that rainfalls over a mountain range vary in proportion to the speed of the moist winds blowing against the range.

Sequential maximization refers to the rearrangement of observed storms, or parts of these, into a sequence which has minimum duration intervals between the rainfalls. This involves the study of storm types, associated with heavy rainfalls in or near the project area, movements of surface and upper-air, lows and highs, etc. From the study of sequences of storms in and near the area of interest the minimum time interval between storms which produce heavy rain are estimated. This interval may be days for large areas or hours between downpours over small catchments. The actual storms used in the sequence may have occurred years apart, but the sequence must be compatible with meteorological requirements.

The *time sequences* of rainfalls provide important information to the designer, in addition to the magnitude of the design storm, and can be estimated from past records. By analysing past records one can establish the probability transition matrices for the various storm sequences. For example, one can calculate for a storm yielding x mm of rainfall the probabilities of it being followed by x + 1 mm, x + 2 mm, ... and x - 1 mm, x - 2 mm, ... of rainfall. Similarly, one could calculate the probability of this storm being followed by an interval of given duration before the next rainfall is likely to occur. However, for such analysis detailed records are necessary and extraction of the information is a laborious task.

For more detailed studies of the probability distributions of storm characteristics, the time series of point rainfall observations is separated into individual events. Sariahmed and Kisiel (1968) used for this purpose the rank correlation coefficient combined with a significance test. Alternatively, one may aim at simulation of the record with its internal dependencies as was done by Raudkivi and Lawgun (1974) using a basic time unit of 10 min. The durations of the rainfalls were modelled by an autoregressive scheme which also took into account the skewness of the distribution of the random component. The yields within the rainfalls were obtained with a first order Markov model. The time intervals between the rainfalls were found in a large number of cases to be independently distributed in time, and were simulated by the Monte Carlo process with the aid of the cumulative distribution of the historic data. Many distributions have been fitted to describe the time intervals between rainfalls but almost all of these have failed to describe the short interval end of the distribution.

Under climatic conditions where storms are primarily associated with fronts the within rainfall distribution of yield is approximately random, where the rainfall is defined as continuous rain. However, taken over the period of the storm (the time it takes for the front to pass) the average distribution of yields shows that about a third of the total falls within an hour or so. Similarly, convective (thunder) storms usually have very pronounced high intensity peaks.

Spatial maximization refers to transposition of storms that have occurred in or near the area of interst to the location in the catchment which leads to the maximum run-off. The concept is based on the argument that within certain climatic boundaries the location of storms is determined by chance alone. Hence, such a storm could also have passed over the catchment in question. This transportable set of storms in a homogeneous region includes, as a sub-set, those storms which produced the local observed record. For such a homogeneous region of area A, the transposition probability P_t of a storm centring over a catchment of size a is $P_t = a/A$ and this may be combined with the probability of a certain rainfall depth. In addition, the probability of occurrence of a storm of rank r in the period of record N is $P_r = r/N$. A further complication is that the depth from the storm decreases with increasing area. Therefore, in order to rank storms the depths have to be associated with area.

The first step in transposing a storm is to identify the time and location of the heaviest rainfall and the meteorological causes of it. Next, with the aid of past weather maps the region is outlined in which this type of storm is common. Tracks of tropical and extratropical cyclones are usually shown in published records and climatological charts. The transpositons must be compatible with the weather patterns and meteorological requirements. The third step is to delineate topographical limits, such as mountain ranges and coastlines. Coastal storms usually extend only a limited distance inland.

The methods of transposition of a storm from place of occurrence to a particular catchment could be subdivided as
(a) transposition of isohyetal patterns without modification,
(b) transposition of the depth-area-duration relationships developed from isohyetal patterns, and
(c) transposition of maximized depth-area-duration relationships.

The transposition of isohyetal pattern from a storm that has occurred in an adjacent area is only justified in regions with small differences in relief. Both the original location and the area to which the storm is applied should be free of orographic influences on precipitation, or have equal influence. The orientation of the storm must be in keeping with meteorological experience in the study area which usually is smaller than the area covered by the isohyetal pattern.

In regions for which depth-area-duration relationships have been developed these provide the simplest method of storm transposition. For a particular storm the graph or table gives for the given area the depths for selected durations. The data can be rearranged so as to give the shape of the hyetograph expected for the locality or several sequences may be studied to obtain the peak discharge. If there are large discrepancies between the boundaries and the shape of the isohyetal lines the depth-area- duration method leads to over estimation, particularly when the area is rather elongated.

In a number of countries estimates of the "probable maximum precipitation" in the form of deph-area-duration relationships have been developed, either as graphs or maps, and these provide the designer with quick estimates.

Adjustments to moisture content (precipitable water) are made with the aid of dew point data at 1000 mb. The adjustment factor is the ratio of the amount of precipitable water for the dew point condition at the transposed location to that where the storm occurred.

A barrier adjustment is applied when the storm passes over a range or high country (up to 1000 m in height, otherwise they become climatic barriers), which blocks off some of the inflow of moist air. The adjustment factor is the ratio of the amount of precipitable water in an air column at the crest of the range to that at the foot of the range on the windward side.

A further step in the maximization procedure is the maximized and transposed rainfall data. The rainfall depths from all such storms are plotted against duration and a curve is drawn through the maximum values. These maxima for different durations may come from different storms. Similarly, the rainfall depths from these storms are plotted against area and again the envelope is drawn. From these two plots one can prepare a plot of probable maximum precipitation depth against area with duration as a parameter, i.e. one line for each duration. Not all of the data are of equal accuracy or reliability and some storms may satisfy the transposition criteria better than others. It is therefore essential to use a measure of judgement in drawing the lines. Details of these procedures are given in WMO (1973), Manual for Estimation of Probable Maximum Precipitation.

Finally, it should be noted that the methods of statistical frequency analysis of point rainfall data are a form of maximization. The point data can be used with the area adjustment factors.

Example 3.1. Measurements of specific humidity in the lower 5 km of the atmosphere yielded the following data:

mb:	990	950	910	850	830	820	815	800	750	700	650	600	560
g/kg:	13	10.5	10	10.2	10	9	8	7	6.1	5.6	5	4.1	2.8

Calculate the depth of precipitable water in this atmosphere.

For a column of unit horizontal cross-sectional area the mass of water vapour is $dm_v = \rho_v dh$ or

$$m_v = \int_0^h \rho_v dh$$

Substituting for $dp = -\rho g\, dh$ leads to

$$m_v = \frac{1}{g}\int_p^{p_o} (\rho_v/\rho)dp = \frac{1}{g}\int_p^{p_o} H_s\, dp$$

since g is essentially constant and $H_s = \rho_v/\rho$, where $\rho = \rho_a + \rho_v$. This is eqn 2.50 or 3.19. Since 1 kg of water is equivalent to 1 mm of water per m^2 and 1 mb = 100 Pa = 100(kg m/s^2)/m^2, $m_v = (100/9.81)\int H_s p_{(mb)}$ or when H_s is expressed in g/kg (using the numerical value of grams) then m_v is in mm of precipitable water

$$D(mm) = 0.01 \int_p^{p_o} H_s\, P_{mb}$$

The depth is now readily determined from the areas of the plot of pressure versus specific humidity. The scale used is equivalent to 1 cm^2 ≡ 50 units of the integral or 0.5 mm of water.

p mb	Δp mb	H_s	$\bar{H}_s$	ΔD mm
990		13.0		
950	40	10.5	11.75	4.700
910	40	10.0	10.25	4.100
850	60	10.2	10.1	6.060
830	20	10.2	10.1	2.020
820	10	9.0	9.5	0.950
815	5	8.0	8.5	0.425
800	15	7.0	7.5	1.125

p mb	Δp mb	H_s	$\bar{H}_s$	ΔD mm
750	50	6.1	6.55	3.275
700	50	5.6	5.85	2.925
650	50	5.0	5.3	2.650
600	50	4.1	4.55	2.275
560	40	2.8	3.45	1.380
				31.890
500	60	0.0	1.4	0.840

Thus, the total water content of the lower 5 km of the atmosphere is approximately 31.9 mm and probably less than 1 mm above the 560 mb level.

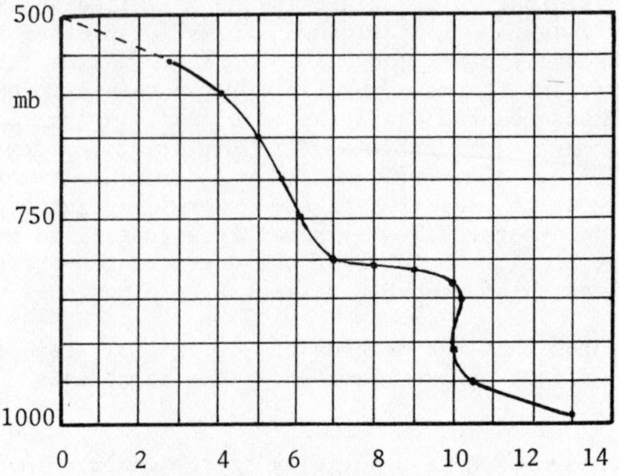

Plot of atmospheric pressure in the lower atmosphere versus specific humidity for Example 3.1.

Example 3.2. An observed rainstorm over an area of 2000 km^2 yielded a depth of 200 mm in 6 hours. It occurred over a barrier of 300 m and had a 1000 mb dew point value of 20 °C. This storm is to be transposed to a similar sized catchment in the same meteorologically homogeneous zone with a moisture barrier of 500 m in height. The highest 1000 mb dew point during this season at the transposed site was 24 °C and 25 °C at the actual location.

For the observed storm:

The moisture content W from 1000 to 200 mb at 20 °C: 52 mm

From $dp = -\rho g\, dz$ and $p = \rho RT$,

$$\frac{dp}{p} = -\frac{g}{RT}\,dz = -\frac{g\,dz}{R(T_o - \alpha z)} \qquad \log\frac{p}{p_o} = \frac{g}{R\alpha}\log\frac{T_o - \alpha z}{T_o}$$

or

$$p = p_o\left(\frac{T_o - \alpha z}{T_o}\right)^{g/R\alpha}$$

For the international standard atmosphere the lapse rate α = 0.0065 °C/m and $g/R\alpha \simeq 5.26$. The dry air lapse rate $dT/dz = -g/c_p \simeq 9.8$ °C/km. The saturated air lapse rate is given by eqn 2.34 and is always slightly less than that for dry air. For the given conditions the mixing ratio $w \simeq 0.015$ and the bracketed term of eqn 2.34 becomes ~ 0.6 or $\alpha_s \simeq 0.0058$ °C/m. For first estimates

$$p = p_o\left(\frac{T_o - 0.0065\,z}{T_o}\right)^{5.26}$$

is adequate. For order of magnitude estimates 3% reduction per 100 m may be used. Thus, for T_o = 293 and z = 300 m, p ≃ 965 mb and Table 2.4 gives W = 5 mm

Hence, residual moisture content above 300 m level is 47 mm

For the project area:

Total W from 1000 to 200 mb at 24 °C	74 mm
Reduction of W from 1000 mb to 944 mb (500 m) at 24 °C	10 mm
Residual moisture content above 500 m	64 mm

$$\text{Adjustment factor} = \frac{\text{Residual W of project area}}{\text{Residual W of observed storm}} = \frac{64}{47} = 1.36$$

The adjusted depth for the 6 hour storm at project area is 1.36 x 200 ≃ 272 mm

Alternatively:

(1) Observed storm: $\frac{\text{W from 300 m to 200 mb at 25 °C}}{\text{W from 300 m to 200 mb at 20 °C}} = \frac{74}{47}$

(2) Project site: $\frac{\text{W from 300 m to 200 mb at 24 °C}}{\text{W from 300 m to 200 mb at 25 °C}} = \frac{68}{74}$

(3) Height difference: $\frac{\text{W from 500 m to 200 mb at 24 °C}}{\text{W from 300 m to 200 mb at 24 °C}} = \frac{64}{68}$

$$\text{Total adjustment} = \frac{74}{47} \times \frac{68}{74} \times \frac{64}{68} = \frac{64}{47} = \frac{\text{W from 500 m to 200 mb at 24 °C}}{\text{W from 300 m to 200 mb at 20 °C}} = 1.36$$

It should be noted, however, that it is common to neglect elevation adjustment when large-area storms are transposed from adjacent flat regions into hilly country of less than 1000 m elevation. It is assumed that the foothills stimulate convection and increase rainfall which will offset the decrease of moisture with higher elevation. A discussion of transposition and maximization may be found in WMO Tech. Note No. 98, 1960.

3.2.5. General information on precipitation

The general information on precipitation may be presented in many ways. For quick reference most countries have equal precipitation contour maps (isohyetal maps) for periods of, for example, 10 min, 1 hour, 6 hour, 12 hour, 24 hour precipitation as well as for mean annual values, such as Fig. 3.14. For given locations prediction formulae or graphs are available. The local historical information, when analysed and plotted, gives yield versus return period data for various durations, either in graphical form, such as Fig. 3.6, or as a prediction formula.

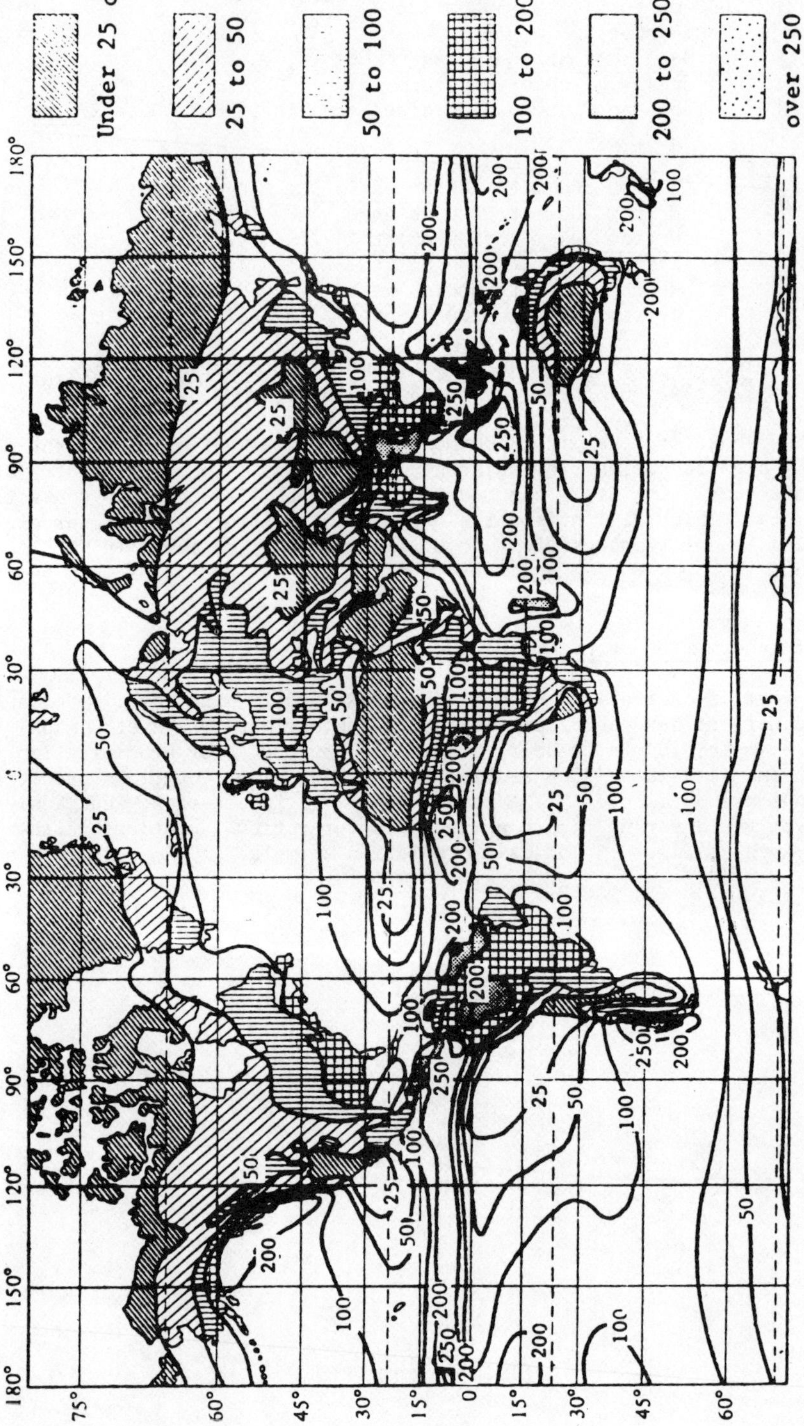

Fig. 3.14. Mean annual precipitation in cm.

Chapter 4

EVAPORATION AND TRANSPIRATION

Since about 70% of the annual precipitation on the land surface of the earth is returned to the atmosphere by evaporation and transpiration, it is clear that evaporation and transpiration are important elements of the hydrological cycle. They also play a major part in all water balance studies and in design of all water resources projects. The local rate of evaporation depends on the availability of water and of energy, and on a multitude of factors describing the nature of the surface, vegetation, humidity, wind, etc. The energy supply is primarily from solar radiation and it varies with latitude, season (Fig. 1.6), weather, exposure of the surface and its reflective properties (albedo). For transport of moisture away from the evaporating surfaces, wind and its turbulence have the most important role.

In calm conditions the air just above ground or over water tends to become saturated with water vapour and the rate of evaporation decreases rapidly. However, wind by convection can take the moist air away and bring in drier air or remove the moist air by turbulent dispersion, or both, and thus allows an unhindered evaporation. I general, the higher the wind speed the more effective is the removal of moisture. The rate of evaporation is also strongly affected by absolute humidity deficit ($e_s - e$). Moist air can absorb less additional water vapour than dry air. In the evaporation process the liquid changes to the gaseous phase. For this change the water molecules must acquire enough energy to break through the water surface into the atmosphere. The water molecules in the atmosphere are also in continuous motion and some return to the liquid mass. An equilibrium state is reached when the number of water molecules escaping equals that returning. At this stage the vapour pressure or concentration gradient vanishes and the rate of diffusion (vapour transport) goes to zero. However, if the air mass is replaced by a drier one the process will continue indefinitely. The energy needed for the evaporation process is known as the latent heat of vaporisation. With increasing temperature the energy content of water and the rate of evaporation, as well as the apparent capacity of air to "absorb" water vapour, increase. Air, of course, is only incidental. The saturation vapour pressure is a function of temperature only, i.e. that of a plane surface of pure water and no air need be involved. The evaporation from open water is up to about 4500 mm per annum in arid zones. From land surfaces in temperate climate the evaporation is of the order of 450 mm per annum and about 75-100 mm in arid climates, the latter depends entirely on the availability of water.

Only a small fraction of the water needed by a plant is retained in the plant structure. Most of it is transpired from the leafy parts of the plant into the atmosphere. Transpiration proceeds almost entirely by day under the influence of solar radiation. At night the pores, or stomata, of plants close up and very little moisture leaves the plant. Here, the cactus and pineapple are the more notable exceptions. Their stomata open only at night. Under field conditions it is practically impossible to differentiate between evaporation and transpiration and the two are lumped together as *evapotranspiration*. The total amount of evapotranspiration depends on precipitation, temperature, humidity, etc., as well as on the type of cultivation and vegetation.

Distinction must be made between potential evapotranspiration and what actually takes place. The potential evapotranspiration refers to rates which are not limited by availability of water to evaporate or to transpire. The actual rates are governed by the availability of water and, for example, under desert conditions may by very low.

4.1 Evaluation of Evaporation

The methods used to estimate the amount of evaporation, or evapotranspiration may be grouped as follows:

- measurement from evaporation pans, etc.
- empirical formulae
- water budget methods
- mass transfer methods
- energy budget methods

4.1.1. Measurement of evaporation

Evaporation pan measurements are relatively simple and the results are in a reasonably constant ratio to evaporation from large open water surfaces and from region to region. The pan measurements, although not limited by the availability of water, do not give the potential evaporation because the moisture transport away from the pan is not confined and does not represent the areal conditions. There are a number of types of pan: sunken, raised and floating. The *sunken pan* is placed below ground level, which reduces the effects of heat exchange through the side walls, but they collect rubbish, are more difficult to maintain and to install, and there is an unknown amount of heat transfer from pan to soil. The *raised pan* is the standard instrument in many parts of the world. The class A pan used in U.S.A., Canada and a number of other countries is a 1219 mm (4 ft) diameter (7000 cm^2 surface area), 254 mm deep tray set on timber supports, so that the bottom is about 15 cm above ground. The water level is maintained within 50-75 mm from the top. Pans with 3000 cm^2 surface area, and less, are also in use. The potential evaporation is related to the pan measurement through a coefficient C. For the raised pan C is of the order of 0.69, and for the sunken pan $C \simeq 0.79$. However, considerable variation is possible when comparing results from a pan to lake, due firstly, to differences in the amounts of heat flow across the pan, and secondly, to large variations in advected heat flow across the lake. If the atmosphere is stable then the vertical vapour pressure gradient reduces with distance downwind over the lake. It has been found experimentally that the evaporation from a lake surface decreases proportional to the 2/3-power of the distance downwind. Kohler et al. (1955) suggested an empirical equation for evaluation of evaporation from shallow and deep lakes. Wiesner (1970) gave for the change of evaporation rate with distance downwind

$$\frac{\partial E}{\partial x} = \text{const } u^{0.78} x_o^{-0.11}$$

where x_0 is the strip length. The vapour transfer process depends, in addition to the wind velocity and distance downwind, also on the wave height, that is, on the surface roughness. The wave height for a given wind speed is a function of the distance downwind and depends on whether the water is flowing or stationary. On flowing water the wave heights for the same wind speed and location are substantially reduced, even though the change in relative velocity is usually quite small.

In temperate climates the ratio of evaporation from bare wet soil to potential evaporation is about 0.9 and from grass to potential evaporation 0.5 to 0.8, depending on type and length of grass. One simple method of evaluation of the pan coefficients is by comparison with values from pans larger than 20 m^2. Although the *floating pan* in the lake surface would give the best results for lake evaporation, their use is curtailed by operational difficulties, wave action, accessibility, etc.

4.1.2. Empirical evaporation formulae

Most of the empirical formulae are based on the aerodynamic relationships of the form

$$E = N\, f(u)(e_s - e) \qquad 4.1$$

where f(u) is a function in terms of wind speed, e_s is the saturation vapour pressure at surface temperature, and e the actual vapour pressure at a defined height above the surface, N is a constant. If f(u) is equal to the wind velocity in ms^{-1}, measured 2 m above water surface in the middle of the lake, water surface temperature is measured at mid-lake and e in mb is measured on the shore outside the lake's vapour blanket (a few km away from the lake), then according to Harbeck (1962) the evaporation from the lake in cm per day can be evaluated with

$$N = 0.0291/A^{0.05} \qquad 4.2$$

where A is the area of water surface in m^2. (Harbeck measured E in in./day, u in mph, e in mb and $N = 0.00338/A^{0.05}$, where A is in acres).

Thornthwaite (1948) presented a method for estimation of evaporation from climatic data. The formula is

$$E = C\, T_m^a \qquad 4.3$$

where E is evaporation or potential evapotranspiration in cm, T_m is the monthly mean temperature, °C, a is an exponent and C is a coefficient. The exponent a is evaluated in terms of the annual heat index I as

$$a = 6.75 \times 10^{-7} I^3 - 7.71 \times 10^{-5} I^2 + 0.0179\, I + 0.492 \qquad 4.4$$

where

$$I = \sum_{1}^{12} \left(\frac{T_m}{5}\right)^{1.514} \qquad 4.5$$

For 30 days a month and 12 hours of sunshine per day

$$E = 1.62 \left(\frac{10\, T_m}{I}\right)^a \qquad 4.6$$

or the actual potential evaporation

$$E_a = E\, \frac{Dh}{360} \qquad 4.7$$

where D is the number of days in month and h is the average number of hours of sunshine.

However, any empirical formula is dependent on local climatic conditions, and even then can only give an indication of the average values of evaporation.

4.1.3. Water budget and mass transfer methods

The *water budget* method for evaluation of evapotranspiration is essentially an application of the conservation of mass principle

$$\text{precip} + \text{Evaporation} = -\text{Inflow} + \text{Outflow} + \Delta\ \text{Storage}$$

or

$$\text{Precipit.} + \text{ET} + \text{Surface flow} + \text{Groundw. flow} + \Delta\ \text{Storage} = 0$$

The difficulties of estimation of storage changes within the catchment, in lakes, in groundwater aquifers, etc., render the idea of limited value at other than special sites and conditions.

The *mass transfer* approach was originated by Dalton, who recognized that the relationship between evaporation and vapour pressure could be expressed as transport down the concentration gradient

$$E = b(e_s - e) \tag{4.8}$$

where b is an empirical constant. This can be modified for the turbulent diffusion condition by writing

$$E = \frac{Nf(u)(e_s - e)}{f(k)} \tag{4.9}$$

where f(k) is a surface roughness parameter. By combining the boundary layer and turbulent mixing concepts, various expressions can be obtained. Widely quoted are expressions by Sverdrup (1946)

$$E = \frac{0.623\ \rho\kappa^2\ u_8(e_s - e_8)}{p(\ell n\ 800/k)^2} \tag{4.10}$$

and by Thornthwaite and Holzman (1939)

$$E = \frac{0.623\ \rho\kappa^2(u_8 - u_2)(e_2 - e_8)}{p(\ell n\ 800/200)^2}$$

where E is evaporation in g $cm^{-2}s^{-1}$ or cm s^{-1} (instead of kg $m^{-2}s^{-1}$ or mm m^{-2}), ρ is density of air in g/cm^3, $\kappa = 0.4$ is the von Karman constant, u_2, u_8, are wind speeds in cm/s at 2 m and 8 m above the surface, respectively, e_s is saturation vapour pressure at surface temperature in mb and e_2, e_8 those at 2 and 8 m elevation, p is atmospheric pressure in mb and k is the roughness height in cm.

The mass transfer equations appear to give reasonable estimates for periods longer than a day and require only a limited amount of field data but at two levels.

4.1.4. Energy budget method

The *energy budget method* is the most rigorous and detailed. Its main function is to give a good understanding of the processes involved. Only in special cases is there enough data from field measurements to enable numerical application. The method is based on the principle of conservation of energy and the amount of evaporation is calculated from the energy available. Consider a soil or water column which extends from the surface to a depth where vertical heat exchange is negligible, Fig. 4.1.

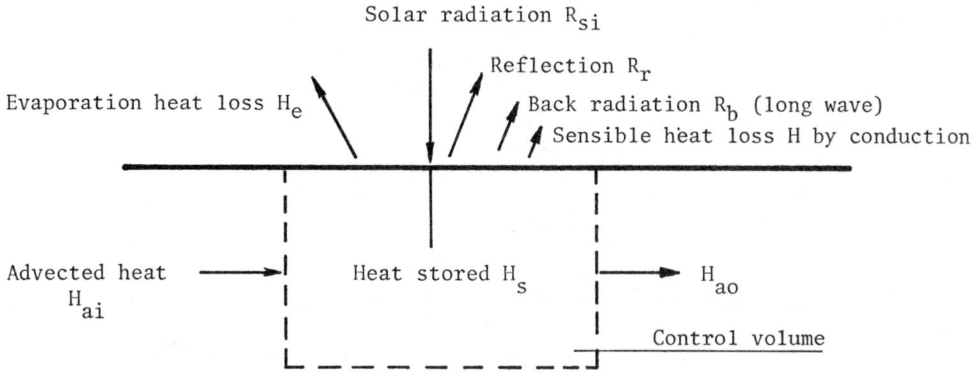

Fig. 4.1. Illustration of energy budget applied to a control volume.

For balance

$$R_{si} - R_r - R_b - H - H_e + H_{ai} - H_{ao} = H_s \qquad 4.12$$

Here $R_{si} - R_r - R_b = R_n$, is the net radiation in W m^{-2} received; H is the heat loss from the control element by conduction; H_e = LE is the latent heat flux due to evaporation (negative) or condensation (positive), where L is latent heat (kJ kg^{-1}) and E is the rate of evaporation (kg $m^{-2}s^{-1}$); H_{ai} and H_{ao} refer to rates of heat flow into and out of the volume due to advection. For a soil column $H_{ai} - H_{ao} \simeq 0$, but for a control volume in water the amount of lateral heat transport, by currents, for example, may be very significant. The sum of the right hand terms must equal the heat stored in the control volume H_s.

For land surface

$$R_n + H_s + H + H_e \simeq 0 \qquad 4.13$$

This equation is valid for any time period but it does ignore a few minor terms of energy, such as from photosynthesis, respiration, precipitation, and melting of snow.

The partitioning of the net radiation energy R_n at the surface of the earth is strongly influenced by the availability of water. *On totally dry* surfaces, Fig. 4.2, all of R_n goes into sensible heat in soil and air.

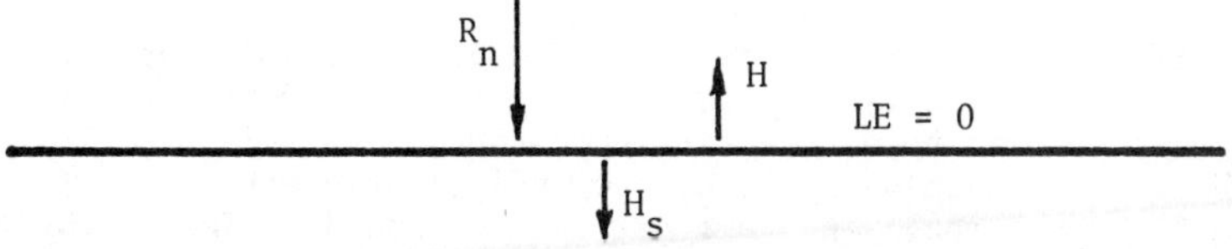

Fig. 4.2. Partitioning of the net radiation energy at a dry surface.

Surface temperatures exceed air temperature during most of the daylight period. The air near the surface is hotter and lighter, and the lowest few hundred metres of the atmosphere are unstable. Measurement of R_n and H_s allows deduction of heat transfer H. Figure 4.3 is an example of such measurements.

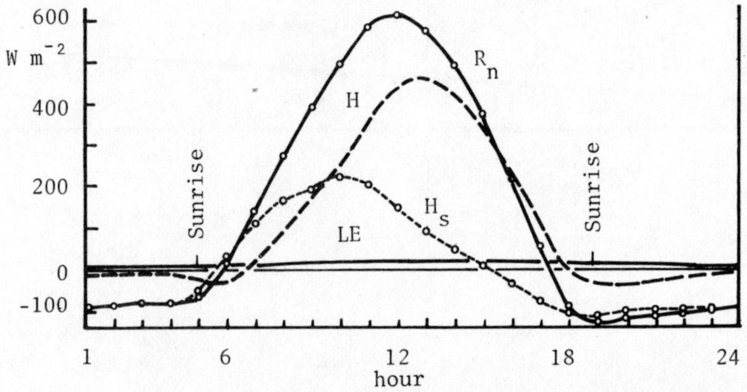

Fig. 4.3. Average diurnal variation of the components of the surface energy balance over bare soil at El Mirage, California, dry lake bed, June 9-11, 1950, (after Vehrencamp, 1953).

Conversely, if water completely covers the surface, Fig. 4.4, as a thin film, or the surface is covered completely by vegetation, with negligible stomatal resistance, one might think that the net radiant energy goes entirely to evaporation and soil heat, and that heat conduction and convection, H, would be negligible.

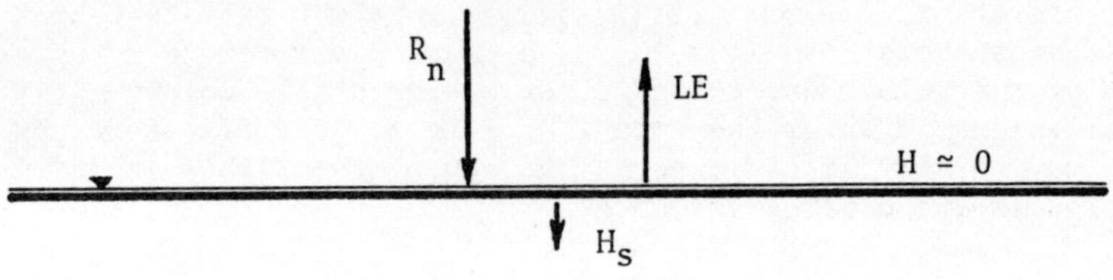

Fig. 4.4. Model for partitioning of the net radiation energy at a surface covered by a film of water.

Surface temperatures should remain at or near air temperature, and neutral stability conditions should prevail during the daytime period. Experiments show, however, that then the ratio of LE/R_n can be any value greater than about 2/3. Occurrence of $LE/R_n > 1.0$ are attributed to advection - horizontal convection of heat by air masses. $LE/R_n <$ about 2/3 arises where the plant cover of the ground is not complete and the stomatal resistance of plants is not negligible.

The climatic effects are illustrated by Fig. 4.5. The data for Davis covers 3½ years and is for a well-irrigated perennial ryegrass. During the spring months some energy goes to heating of the air above the ground and LE is less than R_n. From June onwards the mean daily LE corresponds closely with R_n. In the figure, data of some 100 dry strong wind days has been excluded. The LE curve for Sept.-Nov. is a little higher when data of the days with strong advected heat is not excluded. In contrast, the data from Denmark, for short clover-grass mixture, not only shows a lower net radiation but also that a much lower fraction of R_n goes

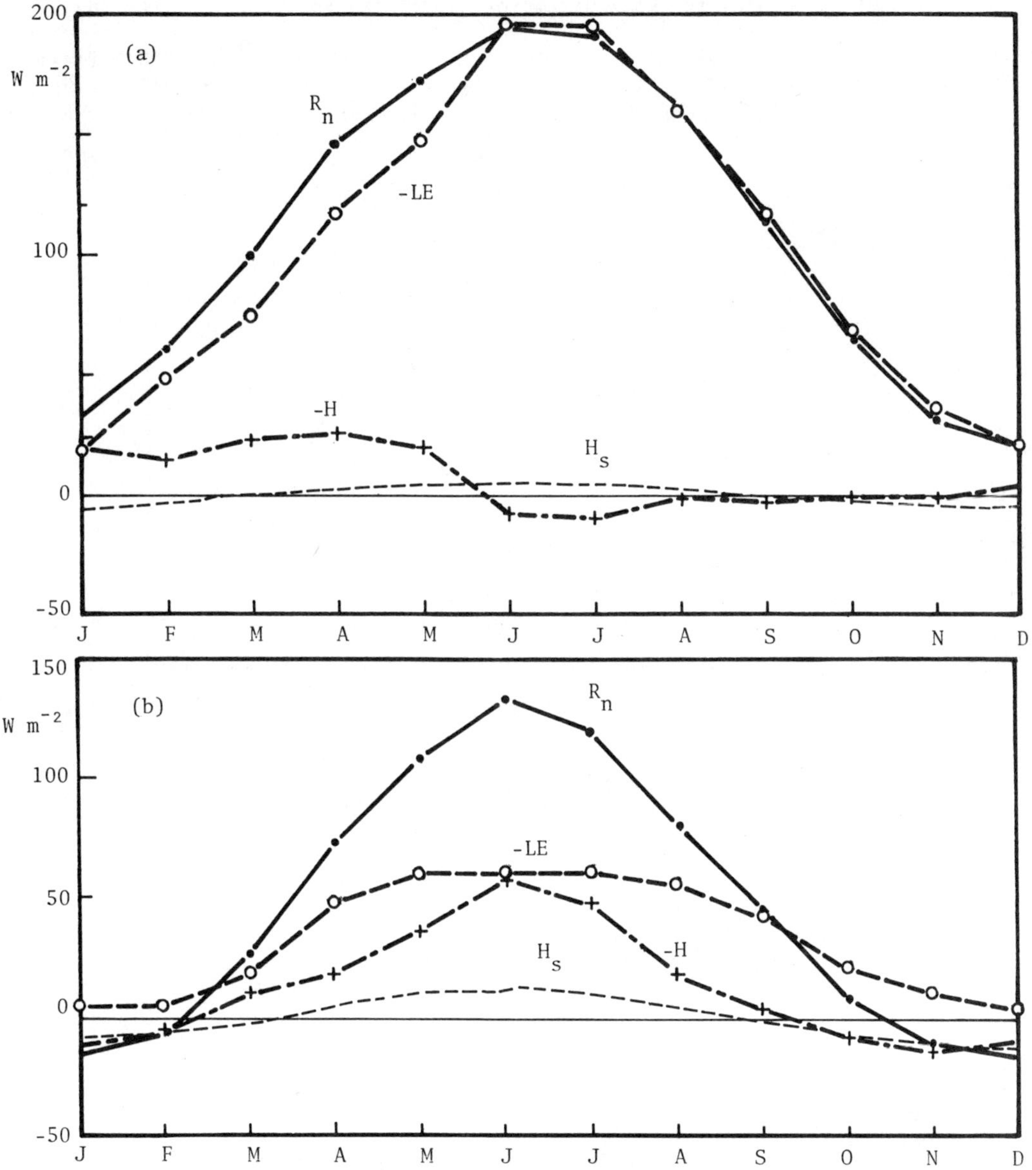

Fig. 4.5. (a) Energy balance for a well-watered, perennial ryegrass cover at Davis, California, 1960-1963. Approximately 100 days of dry, strong-wind days excluded from the 3-½ years of record. (After Pruitt, 1971).
(b) Energy balance for 10 years for a clover-grass mixture at Copenhagen, Denmark, 1955-64. R_n, LE and H_S were measured with H calculated from the energy balance. (After Aslyng & Jensen, 1965).

into evaporation. For winter months R_n goes negative and the energy for evaporation comes from the warmer air masses moving in from the sea. During the summer about half of R_n is used to heat the cooler air masses arriving from the sea.

At the other extreme to Copenhagen is the data by van Bavel and Fritschen (1964) for conditions in Arizona, Fig. 4.6. Apart from the much greater value of R_n, it is seen that LE is always greater than R_n. This is due to consistent strong advection of heat from nearby desert areas. In humid warm climates LE is less during the daytime than R_n because the vapour pressure difference between the surface and air is less than under arid conditions.

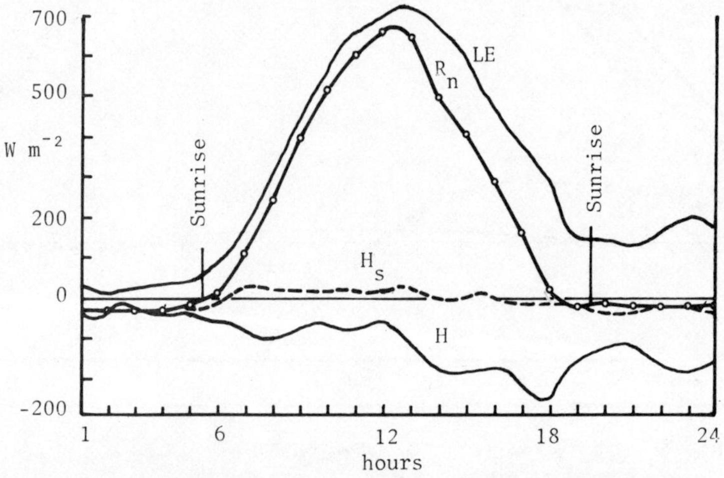

Fig. 4.6. Average diurnal variation of the components of surface energy balance over Sudan-grass at Tempe, Arizona, July 20, 1962. (After van Bavel and Fritschen, 1964).

The next step is to establish methods for evaluation of the individual terms in eqn 4.13.

4.1.4.1. Net radiation

The net amount of energy received on the earth's surface from solar radiation could be evaluated by direct measurement (using a pyrheliometer) or by indirect evaluation using parameters which can be measured more readily and cheaply than the direct net radiation. A procedure for estimation makes use of tabulated radiation values, I_o, received at the outer surface of the earth's atmosphere, Table 4.1. This value of I_o must be adjusted to account for losses due to absorption, reflection and scattering. The actual sun and sky radiation received at the earth's surface may be expressed as

$$R_s = I_o(a + b\frac{n}{D}) \tag{4.14}$$

where n is the actual number of hours of bright sunshine and D is the possible maximum number of hours of sunshine, and a and b are constants for a particular region, Table 4.2. A substantial amount of this energy may be reflected back to the space. The amount depends on the coefficient of reflection r, of the surface involved, the

TABLE 4.1 Solar Radiation Flux I_0 at the Outer Limit of the Atmosphere W m^{-2} as a Daily Average

Latitude	Jan. 13	Feb. 4	Feb. 26	Mar. 21	Apr. 13	May 6	May 29	June 22	July 15	Aug. 8	Aug. 31	Sept. 23	Oct. 16	Nov. 8	Nov. 30	Dec. 22
90°	-	-	-	-	205.0	374.1	484.1	522.0	481.7	370.7	202.6	-	-	-	-	-
80°	-	-	3.4	75.1	205.0	368.3	476.8	513.7	474.9	365.4	202.6	74.1	3.4	-	-	-
70°	-	11.6	63.5	148.8	254.4	363.0	455.0	490.4	452.6	359.6	251.5	146.8	62.5	11.6	-	-
60°	35.4	70.7	133.7	216.6	307.7	392.0	452.6	474.4	450.2	388.2	304.8	214.2	132.3	70.7	34.9	23.7
50°	99.3	140.0	203.0	278.6	354.7	420.1	464.2	479.3	462.3	416.3	351.3	275.2	200.6	138.6	98.9	85.3
40°	169.6	210.3	268.0	332.4	391.1	441.0	471.0	480.2	468.6	436.6	386.7	328.1	264.1	207.9	168.6	153.6
30°	239.4	275.2	324.7	375.6	419.2	450.2	468.6	472.5	465.2	446.3	414.8	370.7	321.3	273.3	238.4	225.8
20°	305.3	334.8	372.6	407.5	433.2	447.3	453.1	453.1	450.7	443.9	428.4	402.7	368.3	331.9	303.8	293.2
10°	364.4	385.2	409.5	427.4	434.7	432.7	426.9	423.0	425.0	429.3	429.8	422.1	404.6	382.3	362.5	355.2
0°	414.3	425.5	434.2	433.7	423.0	405.6	389.6	382.8	387.7	402.2	418.2	428.9	429.3	421.6	412.4	408.5
-10°	453.6	453.6	446.3	427.4	399.0	368.3	342.6	332.9	341.1	364.9	394.5	422.1	441.0	449.2	451.1	452.1
-20°	481.2	469.1	444.8	407.5	363.4	319.8	287.4	274.8	285.9	316.9	359.1	402.7	439.5	464.7	478.8	484.1
-30°	496.7	471.5	430.3	375.6	316.9	263.1	225.3	211.3	224.4	260.7	313.0	370.7	425.0	467.1	494.3	504.5
-40°	500.1	461.8	401.2	332.4	260.7	200.1	159.4	143.9	158.9	198.2	258.3	328.1	396.9	457.4	497.7	513.2
-50°	493.3	440.5	364.4	278.6	197.7	133.7	93.5	80.0	93.0	132.8	195.8	275.2	360.0	436.6	491.4	511.7
-60°	480.7	410.4	315.9	216.6	130.4	67.8	33.0	22.8	33.0	67.4	128.9	214.2	312.1	407.1	478.3	506.9
-70°	483.6	380.4	261.2	148.8	61.5	11.5	-	-	-	11.1	61.1	14.8	257.8	377.0	481.8	523.8
-80°	506.9	385.7	210.3	75.1	3.4	-	-	-	-	-	3.4	74.1	207.9	382.8	504.5	548.5
-90°	514.6	392.0	210.3	-	-	-	-	-	-	-	-	-	207.9	388.2	511.7	556.8
Longitude of the sun	292.5°	315°	337.5°	0°	22.5°	45°	67.5°	90°	112.5°	135°	157.5°	180°	202.5°	225°	247.5°	270°

TABLE 4.2 Values of Coefficients in Eqn 4.14

Location	a	b	Ref.
Southern England	0.18	0.55	Penman (1948)
Virginia, U.S.A.	0.22	0.54	Kimball (1914)
Canberra, Australia	0.25	0.54	Prescott (1940)
Southern Sask. Canada	0.25 May, Aug.	0.60	McKay, after
	0.34 Sept., Oct.	0.52	Gray (1970)
New Zealand	0.25	0.54	DeLisle (1966)

albedo. The average tabulated values for r, Table 1.4, have to be used where no actual measured values are available. The reflection will vary with solar altitude, with cloudiness and of course with the type of ground and the species of vegetation. Thus, from the incoming solar radiation only $R_S(1 - r)$ is retained.

A substantial fraction of this absorbed energy is reradiated by the earth as longwave radiation, particularly at night when the sky is clearer and the air drier than at daytime. Most bodies also absorb longwave radiation much more readily than shortwave radiation, e.g. r for longwave radiation from snow is of the order of only 0.005 and 0.11 from sand. Of the longwave radiation from the surface of the earth about 94% is absorbed by the atmosphere, and 6% escapes direct to space, Fig. 1.8. Another 10% escapes from the atmosphere and about 84% of the original back radiation returns to the surface from the atmosphere (some authors quote as low as 82%). In arid regions the net longwave loss is greater at night than in humid regions, but so is the gain of shortwave radiation during the day. The ratio of the longwave radiation down from the atmosphere, R_{LA}, to that upwards from the earth, R_{LE}, is the emissivity, ε, and can be expressed by

$$\frac{R_{LA}}{R_{LE}} = \varepsilon = A + B\sqrt{e_2} \qquad 4.15$$

where e is the vapour pressure in mb at 2 m elevation. Published values of A and B range from 0.43-0.74 for A and 0.029-0.081 for B. In temperate regions A is smaller and B is larger than in arid regions. Over snow fields the vapour pressure remains approximately constant (between 3 to 9 mb) and the clear sky value of $\varepsilon \simeq 0.757$. Under clear skies the net longwave radiation is

$$R_{LN} = R_{LA} - R_{LE} = R_{LE}(\varepsilon - 1) \qquad 4.16$$

The back radiation of earth is calculated as for a black body, eqn 1.8,

$$R_b = R_{LE} = \sigma T^4 \qquad 4.17$$

where $\sigma = 5.67 \times 10^{-8}$ W m^{-2} $°K^{-4}$, and

$$R_{LN} = \sigma T^4(\varepsilon - 1) \qquad 4.18$$

An additional modification is for cloudiness in the form of a factor $(\alpha + \beta n/D)$ for which Penman gave $(0.1 + 0.9\ n/D)$. The effect of cloud cover on longwave radiation could be approximated by

$$\frac{R_{LN}}{R_{LNC}} = 1 - kC \qquad 4.19$$

where R_{LNC} is the clear sky value, C is the fraction of the sky covered by clouds.

An empirical value for the coefficient k for snow surfaces is

$$k = 1 - 0.0787\ z \tag{4.20}$$

where z is the elevation of cloud base in km.

Correction of the Stefan-Boltzmann constant σ according to the nature of the radiating surface may also be necessary. An average coefficient for water surface is 0.97σ, giving

$$R_{LN} = 0.97\ \sigma T^4(\varepsilon - 1)(0.1 + 0.9\ n/D) \tag{4.21}$$

Thus, the total net radiation received is

$$R_n = R_s(1 - r) - R_{LN} \tag{4.22}$$

4.1.4.2. Assembly of the evaporation equation

The energy input in the form of the total net radiation received may be obtained from direct measurement or, for example, with the aid of eqn 4.22. In order to evaluate eqn 4.12 the other terms involved have to be expressed in functional form.

The *heat storage* term H_s can be evaluated when the temperature distribution throughout the control volume is known.

The *net advected energy* term is far more difficult to evaluate. It is necessary to measure the temperatures and volumes of all the components over the selected time interval, the surface flows, rainfall, seepage, currents, etc.

The measurement of the *heat transfer* term, H, resulting from conduction, is most difficult. Therefore, H is taken to be in a given ratio to the heat involved in the evaporation/condensation process, H_e.

The *heat transport* is given (see eqn 1.1) by

$$H = -K_H \frac{d}{dz}(c_p\rho T) = -\rho c_p K_H \frac{\Delta^\circ T}{\Delta z} \tag{4.23}$$

where $c_p\rho T$ is the enthalpy ($h = e_i + pv$, where e_i is the internal energy and v is the specific volume, $1/\rho$); $p = \rho RT$; $c_p = c_v + R$ and c_p and c_v are the specific heats at constant pressure and constant volume, respectively, and R is the gas constant. The term K_H is a transfer coefficient known as the thermal diffusivity (m^2s^{-1}) and is equal to the thermal conductivity $W\ m^{-1}K^{-1}$ divided by density ρ times the specific heat capacity at constant pressure. Equation 4.23 is analogous to $\tau = -\mu du/dy = -\nu d(\rho u)/dy$, where the derivative is the gradient of momentum (in laminar flow) at the wall. The gradient in eqn 4.23 is that of enthalpy. In still air eqn 4.23 describes heat conduction and K_H is dependent on the properties of the gas, likewise in laminar flow. In turbulent flow K_H like the momentum transfer coefficient ε (eddy viscosity) becomes a function of elevation, i.e. distance from the boundary. If changes of elevation are involved the heat loss due to expansion of gases can be separated from that due to diffusion by inclusion of the dry adiabatic lapse rate in eqn 4.23 as

$$H = -\rho c_p K_H \frac{\partial}{\partial z}(T + \alpha_d z)$$

Similarly, the *transport of latent heat* is given by another diffusion relationship as

$$H_e = LE = -L\rho K_v \frac{\Delta q}{\Delta z} \qquad 4.24$$

where L is the latent heat (negative for evaporation, positive for condensation) of water per unit mass

$$L \simeq 2500.78 - 2.37(T - 273) \text{ kJ kg}^{-1} \qquad 4.25$$

K_v is the vapour diffusivity (m^2s^{-1}) and is a function of elevation, $K_v = f(z)$, i.e. distance from the surface, q is the specific humidity, defined as the ratio of the mass of water vapour to the mass of moist air containing this vapour, i.e.

$$q = \frac{\rho_v}{\rho_a + \rho_v} \simeq 0.622 \frac{e}{p} \qquad 4.26$$

The ratio H/LE is known as the Bowen ratio. Substituting form eqn 4.26 it becomes

$$\frac{H}{LE} = \beta = \frac{c_p K_H}{LK_v} \frac{\Delta T}{\Delta q} = \frac{c_p K_H}{LK_v \varepsilon_m} p \frac{\Delta T}{\Delta e} = C_B p \frac{T_s - T}{e_s - e} \qquad 4.27$$

where T_s and e_s are the saturation temperature and vapour pressure, respectively at the surface, and C_B is the Bowen constant

$$C_B \simeq \frac{c_p}{0.622L} \frac{K_H}{K_v} \simeq 6.4 \times 10^{-4} K_H/K_v \simeq 6.1 \times 10^{-4} [°C^{-1}] \qquad 4.28$$

for average wind conditions which implies $K_H/K_v \simeq 0.95$. The product $C_B p = \gamma$ is a psychrometric constant. An empirical relationship for the water-air system, which is in agreement with theory, gives for restricted range of air velocities $(e_s - e)/(T - T_s) = 0.000660(1 + 0.00115T_s)p$, where $T_s = T_w$ is the wet-bulb temperature and p is in mb. For dry air $c_p = 1.005$ kJ/(kg °C). For moist air $c_p^* = c_p[1 + (c_{pw}/c_p)w]/[1 + (m/m_w)w] = c_p(1 + 1.95w)/(1 + 1.609w)$.

From equations 4.24 and 4.26 the evaporation rate is

$$E = -0.622 K_v \frac{\rho}{p} \frac{\Delta e}{\Delta z} \text{ (kg m}^{-2}\text{s}^{-1}) \qquad 4.29$$

and is independent of elevation because of the conservation of mass requirement. In general, the rate of evaporation is a function of pressure, temperature and salinity of water. The variation with temperature is expressed as vapour pressure versus temperature at constant pressure with the relative humidity as a parameter, Table 2.2.

The water temperature relative to air is also important. If the water of a lake, for example, is warmer than air, it will have a higher vapour pressure than the saturation value at air temperature and the vapour pressure gradient encourages intense evaporation. This is the reason why mist forms over lakes and swamps during cool calm nights.

In nature temperature, pressure and wind effects vary with elevation and it is difficult to isolate the effects of a single variable. Field plots of evaporation rate versus elevation (on mountain sides) show a linear decrease up to about 3000-3500 m, above which the evaporation rate remains constant. The plot of temperature with elevation is essentially of the same shape.

The actual rate of evaporation from water or a moist surface depends very strongly on the transport of vapour away from this surface. In the absence of such transport the air becomes saturated, vapour concentration gradient and evaporation rate go to zero. Wind is the most important transporter of moisture. By convection and turbulent diffusion large quantities of moist air at the surface are replaced by drier air and the process of evaporation can continue unhindered. Advection by horizontal wind alone is effective only over a limited length of surface because the air becomes saturated. It is the turbulent diffusion which will maintain the upward transport of moisture and the turbulent transfer coefficient is orders of magnitude greater than the coefficient of diffusion in still air. The transfer process is that of momentum and is characterized by shear stress

$$\tau = \mu \frac{du}{dz} + \rho \ell^2 \left|\frac{du}{dz}\right| \frac{du}{dz} \simeq \rho \varepsilon \frac{du}{dz}$$

where $\rho\varepsilon \, du/dz$ signifies the turbulent transfer of momentum (and mass) across a plane at elevation z and is known as the Reynolds stress $\tau = -\rho\overline{u'w'}$. Thus, we have to introduce an aerodynamic relationship.

For vertical momentum flux, independent of elevation, we have

$$\tau_z = \tau_o = \rho u_*^2 = \rho\varepsilon \frac{\partial u}{\partial z}$$

or

$$\varepsilon = \frac{u_*^2}{du/dz} \qquad 4.30$$

where u_* is the shear velocity. Multiplying and dividing eqn 4.29 by eqn 4.30 yields

$$E = 0.622 \frac{K_v}{\varepsilon} \frac{\rho u_*^2}{p} \frac{e_1 - e_2}{u_2 - u_1} \qquad 4.31$$

where Δe and Δu are differences across the elevation Δz, with the higher elevation denoted by subscript 2.

The velocity distribution over a rough boundary is of the form

$$\frac{\bar{u}}{u_*} = C_1 \,\ell n \frac{z}{k} + C_2 \qquad 4.32$$

where $C_1 = 1/\kappa = 2.5$ and $C_2 = 8.5$. Introducing this into eqn 4.31 yields

$$E = 0.622 \frac{K_v}{\varepsilon} \frac{\rho}{pC_1^2} \frac{(\bar{u}_2 - \bar{u}_1)(e_1 - e_2)}{\ell n^2 (z_2/z_1)} \qquad 4.33$$

where

$$\bar{u}_2 = u_*(C_1 \,\ell n\, z_2/k + C_2)$$

$$\bar{u}_1 = u_*(C_1 \,\ell n\, z_1/k + C_2)$$

$$\bar{u}_2 - \bar{u}_1 = u_* C_1 \,\ell n\, z_2/z_1$$

and

$$u_* = (\bar{u}_2 - \bar{u}_1)/C_1 \,\ell n\, z_2/z_1$$

If we assume z_1 to be in the vapour saturated film where $\bar{u}_1 = 0$, $z_1 = k$ the roughness height and $e_1 = e_s$, then

$$E = 0.622 \frac{K_V}{\varepsilon} \frac{\rho}{pC_1^2} \frac{\bar{u}_2(e_{os} - e_2)}{\ell n^2 \; z/k}$$

or

$$E = (\frac{K_V}{\varepsilon} \frac{\rho\bar{u}}{p}) \; f(\ell n \frac{z}{k})(e_{os} - e_2)$$

or

$$E = (a + b\bar{u})(e_{os} - e_2) \qquad 4.34$$

where a has been added to allow for the fact that $E \neq 0$ when $\bar{u} = 0$. Equation 4.34 is, of course, of the same form as eqn 4.1, but the derivation has been based on physical concepts.

Penman (1948) combined the energy balance and aerodynamic considerations which had been introduced into literature (c.f., for example, eqn 4.10) into a coherent treatment. Equation 4.33 can be written as

$$LE = LB(e_1 - e_2) \qquad 4.35$$

where

$$B = (0.622 \frac{K_V}{\varepsilon} \frac{\rho}{pC_1^2}) \frac{\bar{u}_2}{\ell n^2 \; z_2/k}$$

$$k = z_1 \quad \text{and} \quad \bar{u}_1 = 0$$

The bracketed term can be evaluated for given pressure, temperature and humidity conditions using $C_1 = 2.5$ and an estimated value of the Schmidt number ε/K_V, as a first approximation unity is used. Values of the aerodynamic roughness associated with various types of vegetation are shown in Table 4.3 for wind speeds $\bar{u}_2 \sim 2 \; ms^{-1}$. Several authors have expressed the relationship between the roughness height, k, and the height of vegetation, h, by

$$\log k = a + b \log h$$

where the a values range from -0.883 to -1.385 and the values of b from 0.997 to 1.417, Tanner and Pelton (1960), Sellers (1965), i.e. $k \sim h/10$.

From eqn 4.27, using eqn 4.35 and assuming saturation conditions at the surface, we obtain

$$H = \gamma LE \frac{T_s - T_2}{e_{os} - e_2} = \gamma LB(T_s - T_2) \qquad 4.36$$

Penman introduced

$$\Delta = \frac{e_{os} - e_{2s}}{T_s - T_2} \qquad 4.37$$

where e_{os} is the saturation vapour pressure at the surface temperature T_s and e_{2s} is the saturation vapour pressure at elevation 2 where $T = T_2$. Substitution of Δ in eqn 4.36 yields

$$H = \frac{\gamma}{\Delta} LB(e_{os} - e_{2s})$$

$$= \frac{\gamma}{\Delta} LB(e_{os} - e_2) - \frac{\gamma}{\Delta} LB(e_{2s} - e_2) = \frac{\gamma}{\Delta} LE - \frac{\gamma}{\Delta} LB(e_{2s} - e_2)$$

TABLE 4.3 Aerodynamic Roughness Heights

Type of surface	Height of vegetation h mm	Roughness height k mm	Author
Open water, u_2 = 2.1 m/s		0.01	van Bavel (1966)
Smooth mud flats		0.01	Deacon (1953)
Tarmac		0.02	Rider et al. (1963)
Dry lake bed		0.03	Vehrencamp (1951)
Smooth snow		0.05	Priestly (1959)
Wet soil u_2 = 1.8 m/s		0.2	van Bavel (1966)
Flat desert		0.3	Deacon (1953)
Snow on flat land		1	Priestly (1959)
Short even mowed grass	15	2	Priestly (1959)
Grass	20-30	3.2	Rider (1954)
	40	1.4	Rider et al. (1963)
	50-60	7.5	Covey et al. (1958)
u_2 = 1.48 m/s	600-700	154	Deacon (1953)
u_2 = 3.43 m/s		114	Deacon (1953)
u_2 = 6.22 m/s		80	Deacon (1953)
Alfalfa u_2 = 2.60 m/s	152	27.2	Tanner & Pelton (1960)
u_2 = 6.25 m/s		24.5	Tanner & Pelton (1960)
Wheat $u_{1.7}$ = 1.90 m/s	600	233	Penman & Long (1960)
$u_{1.7}$ = 3.84 m/s		220	Penman & Long (1960)
Corn	900	20	From Eagleson (1970)
	1750	95	
u_4 = 0.29	2200	845	Wright & Lemon (1962)
u_4 = 2.12		742	Wright & Lemon (1962)
$u_{5.2}$ = 0.35	3000	1270	Wright & Lemon (1962)
$u_{5.2}$ = 1.98		715	Wright & Lemon (1962)
Sugar cane	1000	40	From Eagleson (1970)
	4000	90	From Eagleson (1970)
Brushwood	1500	140	Tanner & Pelton (1960)
Citrus orchard	3350	2830*	Kepner et al. (1942)
Orchards	3500	500	Tanner & Pelton (1960)
Conifer forests	5550	2830*	Baumgartner (1956)
	500	650	From Eagleson (1970)
	27000	3000	From Eagleson (1970)
Deciduous forest	17000	2700	From Eagleson (1970)
Large city (Tokyo)		1650	Yamamoto and Shimanuki (1964)

* Large; unstable atmosph. boundary layers?

From energy balance, eqn 4.13,

$$R_n + H_s - LE - H = 0$$

where the evaporation energy H_e = - LE and transfer of heat to air, H, with a negative sign, have been inserted. Substitution of H from above leads to

$$LE(1 + \frac{\gamma}{\Delta}) = (R_n + H_s) + \frac{\gamma}{\Delta} LB(e_{2s} - e_2)$$

$$LE = \frac{(\Delta/\gamma)(R_n + H_s) + LB(e_{2s} - e_2)}{1 + \Delta/\gamma} \qquad 4.38$$

This is the modified Penman equation for potential evaporation. In it L and γ are constants, Δ is the known slope of the e_s versus T curve taken at the air temperature, $(e_{2s} - e_2)$ is the vapour pressure deficit and B is a factor which depends on the surface roughness and wind profile. The slope Δ of the saturation vapour pressure curve can be estimated from Table 2.2. The dimensionless values of Δ/γ at T_2, according to van Bavel (1966) are presented (with the addition of sub-zero temperature values) as a function of air temperature in Table 4.4. The Δ/γ values are evaluated at T_2, which is easier to measure than T_s and has been found to cause only a small error. Comparison of the calculated potential evaporation with measured values by van Bavel (1966) shows very good agreement.

TABLE 4.4 Δ/γ Versus T in °C at 1000 mb Over Ice and Water

T	Ice Δ/γ	Water Δ/γ	T	Δ/γ	T	Δ/γ	T	Δ/γ	T	Δ/γ	T	Δ/γ
0	0.762		0	0.67	10	1.23	20	2.14	30	3.57	40	5.70
-1	0.708	0.631	1	0.72	11	1.30	21	2.26	31	3.75	41	5.96
-2	0.657	0.592	2	0.76	12	1.38	22	2.38	32	3.93	42	6.23
-3	0.609	0.555	3	0.81	13	1.46	23	2.51	33	4.12	43	6.51
-4	0.565	0.520	4	0.86	14	1.55	24	2.64	34	4.32	44	6.80
-5	0.523	0.489	5	0.92	15	1.64	25	2.78	35	4.53	45	7.10
-6	0.484	0.455	6	0.97	16	1.73	26	2.92	36	4.75	46	7.41
-7	0.448	0.426	7	1.03	17	1.82	27	3.08	37	4.97	47	7.73
-8	0.414	0.398	8	1.10	18	1.93	28	3.23	38	5.20	48	8.09
-9	0.383	0.372	9	1.16	19	2.03	29	3.40	39	5.45	49	8.42
-10	0.354	0.347	10	1.23	20	2.14	30	3.57	40	5.70	50	8.77

Calculated using $\gamma = 0.000660(1 + 0.00115T_s)p$ and the expressions for de_s/dT as given in Section 2.6. Values at pressures other than 1000 mb are obtained by multiplying the values in the table by (1000/p).

Example 4.1. Estimate the rate of potential evaporation from a pasture at latitude 40° S on a clear day in January when the air temperature is 25 °C, pressure is 1000 mb, relative humidity is 50% and the wind speed at 2 m above ground is 6 m/s. Assume that $K_v = \varepsilon$ and that the temperatures in the ground do not change.

By eqn 4.22 the net radiation is $R_n = R_s(1 - r) - R_{Ln}$, in which by eqn 4.14 the $R_s = I_0(a + bn/D)$ where for clear sky $n/D = 1$ and from Table 4.2 the value of $a = 0.25$ and $b = 0.54$. From Table 1.4 the coefficient of reflection is 0.15-0.25, assume 0.20. Table 4.1 gives the value of I_0 as 500.1 W/m². Hence,

$$R_s = 500.1(0.25 + 0.54) = 395.08 \text{ W/m}^2$$

The net longwave radiation is given by eqn 4.18 and clear sky value of emissivity ε

$$R_{Ln} = 5.67 \times 10^{-8} \times 298^4 (0.757 - 1) = -108.66 \text{ W/m}^2$$

Thus,

$$R_n = 395.08 \times 0.8 - 108.66 = 207.40 \text{ W/m}^2$$

LE is given by eqn 4.38

$$LE = \frac{(\Delta/\gamma)(R_n + H_s) + LB(e_{2s} - e_2)}{1 + \Delta/\gamma}$$

in which $\Delta/\gamma \simeq 2.78$ from Table 4.4, L is given by eqn 4.25 as

$$L = 2500.78 - 2.37 \times 25 = 2441.53 \text{ J/g}$$

Table 2.2 gives e_{2s} = 31.671 mb and accordingly $e_2 = 0.5e_{2s}$ = 15.84 mb, $e_{2s} - e_2$ = 15.84 mb. B is given by eqn 4.35, in which k ≃ 0.32 cm from Table 4.3. At 50% relative humidity the mixing ratio is w = 0.622 x 15.84/(1000 - 15.84) = 0.010011 and the virtual temperature is T* = 298[(1 + 0.01/0.622)/(1 + 0.01)] = 299.8 °K. The density of air is then 1.163 kg/m³ (from Table 2.5) and

$$B = 0.622 \frac{1.163(\text{kg/m}^3)}{1000(\text{mb}) \times 2.5^2} \frac{6(\text{m/s})}{\ell n^2(2/0.0032)} = 0.01675 \times 10^{-3}\left(\frac{\text{kg}}{\text{m}^2\text{mb s}}\right)$$

$LB(e_{2s} - e_2) = 2441.53 \times 10^3 (\text{J/kg}) \times 0.016756 \times 10^{-3} (\text{kg m}^{-2}\text{mb}^{-1}\text{s}^{-1}) \times 15.84$ (mb) = 648.02(J $m^{-2}s^{-1}$). Introduction of these values in eqn 4.38 yields

$$LE = \frac{2.78 \times 207.40 + 648.02}{3.78} = 323.97 \ (\text{W/m}^2)$$

and

$$E = \frac{323.97(\text{J m}^{-2}\text{s}^{-1})}{2441.53\text{x}10^3(\text{J kg}^{-1})} = 0.132691 \times 10^{-3}(\text{kg m}^{-2}\text{s}^{-1}).$$

$$= 0.133 \times 10^{-3} \text{ mm/s} = 0.47 \text{ mm/hour}$$

Example 4.2. A thermal power station is to take its cooling water from a river at 19 °C and return it through a diffuser, laid across the river, at 28 °C. It has been calculated that at 5 km downstream the river temperature will be uniform at 20 °C and the hot surface layer has been eliminated. The width of the river is 250 m and the flow rate is 150 m^3s^{-1}. Estimate the amount of heat that goes into evaporation, into radiation and is carried away by convection under clear sky conditions. The air temperature at 2 m is 15 °C, relative humidity at 2 m is 60%, atmospheric pressure is 1000 mb, wind speed at 2 m above water surface is 2.5 m s^{-1} and zero at 0.02 mm, density of air is 1.2 kg m^{-3}, the Bowen constant C_B = 6.1 x 10^{-4} °K^{-1}, and the Schmidt number ε/K_V = 0.7. Neglect the heat loss into the ground.

The energy balance in the river before the addition of the cooling water is given by

$$(R_{sn} - R_{Ln} - LE - H)_o = 0$$

and after the addition of cooling water by

$$(R_{sn} + \Delta Q - R_{Ln} - LE - H)_1 = 0$$

in which ΔQ is the heat added by the cooling water and other symbols have their previously defined meanings. The advected heat input must be balanced by heat lost from the warm water. The loss occurs through evaporation, through longwave radiation, convection of heat by the flow of the river, and convection of sensible heat to the atmosphere from the water.

For calculation of evaporation the stretch of the river may be considered to be a 5 km long pond. The value of net shortwave radiation received, R_{sn}, is not changed by addition of the cooling water. The energy balance for before and after is then

$$\Delta Q - R_{Ln1} - LE_1 - H_1 = - R_{Lno} - LE_o - H_o$$

$$\Delta Q = R_{Ln1} - R_{Lno} + LE_1 - LE_o + H_1 - H_o = \Delta R_{Ln} + \Delta LE + \Delta H$$

$$= (R_{Ln1} - R_{Lno}) + (1 + \beta)(LE_1 - LE_o)$$

The Bowen ratio β = H/LE is

$$\beta = C_B p \frac{T_s - T}{e_s - e} = \frac{6.1 \times 10^{-4}\ {}^\circ K^{-4} \times 10^5 Pa(24 - 15)\ {}^\circ K}{(29.83 - 0.6 \times 17.04)10^2 Pa} = 0.280$$

where 24 °C is the mean water surface temperature and the vapour pressures are from Table 2.2. The corresponding value for maximum temperature of 28 °C is $\beta = 0.288$. By eqn 4.31 evaporation is

$$E = 0.622 \frac{K_v}{\varepsilon} \frac{\rho u_*^2}{p} \frac{e_1 - e_2}{\bar{u}_2 - \bar{u}_1}$$

in which $u_* = \sqrt{\tau/\rho} = \kappa y\ du/dy$ and $u = (u_*/0.40)\ell n\ y/y' = 5.75 u_* \log 2/2 \times 10^{-5}$. With the values given $\bar{u}_2 - \bar{u}_1 = 2.5$ m/s and $u_* = 2.5/5.75 \times 5 = 0.087$ m/s and $0.622(1/0.7)1.2 \times 0.0870^2 \times 10^{-5} = 8.0707 \times 10^{-8}$.

For 19 and 15 °C respect. $(e_1 - e_2)_o = 21.96 - 0.6 \times 17.04 = 11.74$ mb $= 1174$ Pa

For 28 and 15 °C respect. $(e_1 - e_2)_1 = 37.80 - 0.6 \times 17.04 = 27.52$ mb $= 2752$ Pa

For 24 and 15 °C respect. $\overline{(e_1 - e_2)}_1 = 29.83 - 0.6 \times 17.04 = 19.61$ mb $= 1961$ Pa

Hence,

$$E_o = 8.0707 \times 10^{-8}(1174/2.5) = 3.79 \times 10^{-5} kg\ m^{-2}s^{-1}$$

$$E_1 = 8.0707 \times 10^{-8}(2752/2.5) = 8.88 \times 10^{-5} kg\ m^{-2}s^{-1}$$

$$\bar{E}_1 = 8.0707 \times 10^{-8}(1961/2.5) = 6.33 \times 10^{-5} kg\ m^{-2}s^{-1}$$

Latent heat of evaporation $L = 2500.78 - 2.37 \times 15 = 2.465$ MJ/kg for average conditions and 2.434 MJ/kg for 28 °C. The surface area $A = 5 \times 10^3 \times 250 = 1.25 \times 10^6 m^2$, and with these values

$$(\Delta LE)_{max} = 2.465 \times 10^6 \times 1.25 \times 10^6 (E_1 - E_o) = 156.85 \text{ MW}$$

$$(\Delta LE)_{aver.} = 2.465 \times 10^6 \times 1.25 \times 10^6 (E_1 - E_o) = 78.27 \text{ MW}$$

The longwave radiation is given by eqn 4.21 for clear sky as $R_{Ln} = 0.97\ \sigma T^4(\varepsilon - 1)A$, whence

$$\Delta R_{Ln} = 0.97\sigma(4T^3)\Delta T\ A = 0.97 x 5.67 x 10^{-8} x 4 x 297^3 x 5 x 1.25 x 10^6 = 36.02 \text{ MW}$$

in which the average temperature of the reach of 24 °C and the corresponding increment of 5 °C have been used. Likewise, $(\Delta R_{Ln})_{max} = 67.49$ MW. Thus,

$$\Delta Q_{max} = (1 + \beta)\Delta LE + \Delta R_{Ln} = 1.288 \times 156.85 + 67.49 = 270 \text{ MW}$$

$$\Delta Q_{aver.} = (1 + \beta)\Delta LE + \Delta R_{Ln} = 1.280 \times 78.27 + 36.02 = 136 \text{ MW}$$

The increase in evaporation is given by $(E_1 - E_o)$, or

$$E_{max} = (8.88 - 3.79)10^{-5} kg\ m^{-2}s^{-1} = 5.09 x 10^{-8} m^3 m^{-2} s^{-1} = 0.18 \text{ mm/hour}$$

$$E_{aver.} = (6.33 - 3.79)10^{-5} kg\ m^{-2}s^{-1} = 2.54 x 10^{-8} m^3 m^{-2} s^{-1} = 0.09 \text{ mm/hour}$$

The heat carried away by the flow of the river (convection) is

$$150 \times 10^3 (kg/s) \times 1(^\circ C) \times 4186.8\ (J/kg\ ^\circ C) = 628 \text{ MW}$$

4.1.4.3. Extensions to the energy balance method.

Considerable divergence of opinion on the effect of wind on LE is evident in literature. This is mainly because, unlike for water, the evaporation from a land surface may increase or decrease with wind speed. The following discussion follows closely that by Slatyer and McIlroy (1961). They expressed the Bowen ratio as

$$\beta = \frac{c_p}{L}\frac{K_H}{K_v}\frac{\Delta T}{\Delta q} = \delta\frac{\Delta T}{\Delta q} \qquad 4.39$$

On occasions of strong thermal convection the ratio K_H/K_v may vary appreciably. However, when working close to the ground, above the laminar layer and within the layer of strong turbulence generated by the steep wind velocity gradient, K_H is approximately equal to K_V. This layer also accommodates important heat exchanges near the ground. For example, heat transported downwards by wind could partly compensate for the heat loss through radiation during the night.

Equation 4.13 can be written (with H_s as a loss) in the form

$$E = \frac{R_n - H_s}{L}\left(\frac{LE}{H + LE}\right) = \left(\frac{R - H_s}{L}\right)\left(\frac{1}{1+\beta}\right) \qquad 4.40$$

for evaluation of β. For a given accuracy in E the accuracy required in β diminishes as β becomes smaller as under relatively moist conditions. Slatyer and McIlroy expressed this relationship in terms of the wet-bulb temperature.

When a stream of air at p, T and w (mixing ratio) flows past a thermometer bulb, which is covered with a wet wick, water will evaporate from the wick and the bulb temperature will decrease. The energy equation for the equilibrium condition, when the air stream leaving the wick is saturated, is

$$(T - T_w)(c_p + wc_{pv}) = (w_s - w)L_{ev} \qquad 4.41$$

where T and T_w are the temperatures of the approaching and leaving (wet-bulb temp.) air, respectively, w and w_s are mixing ratios of the approach and leaving (saturated) air, c_p is the specific heat of dry air and c_{pv} that of water vapour (eqn 2.35). The value of w_s is determined by T_w and may be calculated from

$$w_s = \frac{0.622\, e_s}{p - e_s} \qquad 4.42$$

where e_s is the saturation vapour pressure at T_w. The simultaneous solution of eqns 4.41 and 4.42 is made easy by the Kiefer's (1941) hygrometric chart or by tables in terms of $(T - T_w)$.

When the change in heat content of water vapour is neglected ($wc_{pv} \ll c_p$), eqn 4.41 reduces to

$$w = w_s - \frac{c_p}{L_{ev}}(T - T_w) \simeq \varepsilon\frac{e}{p}$$

or

$$e = e_s - Ap(T - T_w) \qquad 4.43$$

where $A = c_p/\varepsilon L_{ev}$ and e_s is the saturation vapour pressure at the wet-bulb temperature T_w. Writing $k = Ap$ and $D = T - T_w$, eqn 4.43 becomes

$$e = e_s - kD$$

or

$$\Delta e = e_2 - e_1 = e_{2s} - e_{1s} - Ap[T_2 - T_{2w} - (T_1 - T_{1w})]$$
$$= S(T_{2w} - T_{1w}) - k(D_2 - D_1) = S\Delta T_w - k\Delta D$$
$$= (S + k)\Delta T_w - k\Delta T = k[(\frac{S + k}{k})\Delta T_w - \Delta T] \quad 4.44$$

where

$$S(T_{2w} - T_{1w}) = (e_{2s} - e_{1s})_{T_w}$$

and is approximately equal to the slope of the saturation vapour curve for water at small differences.

Slatyer and McIlroy approximated the specific humidity difference by

$$\Delta q = s\Delta T_w - \delta\Delta D = (s + \delta)\Delta T_w - \delta\Delta T \quad 4.45$$

where

$$s \simeq \varepsilon S/p , \qquad \delta = \frac{0.622\ A}{1 - 0.378\ e/p} \simeq \frac{0.63\ k}{p} = 0.63\ A =$$

0.42 mg/g per °C and $(s + \delta)/\delta \simeq (S + k)/k$. Equation 4.45 is within 1% for vapour pressures up to 25 mb (saturation at 21 °C) and within 4% up to 60 mb (saturation at 36 °C). From eqns 4.39 and 4.45

$$\frac{1}{1 + \beta} = \frac{\Delta q}{\Delta q + \delta\Delta T} = 1 - \frac{\delta}{s + \delta}\frac{\Delta T}{\Delta T_w}$$

and eqn 4.40 becomes

$$E = (\frac{R_n - H_s}{L})(1 - \frac{\delta}{s + \delta}\frac{\Delta T}{\Delta T_w}) \quad 4.46$$

When over an extensive uniformly moist surface, the air close to the ground is saturated (or nearly saturated) the wet-bulb temperature profile will concide with the dry-bulb profile or $\Delta T/\Delta T_w \rightarrow 1.0$ and LE may be approximated by

$$LE = (R_n - H_s)(1 - \frac{\delta}{s + \delta}) = (R_n - H_s)(\frac{s}{s + \delta}) \quad 4.47$$

This condition is described by McIlroy as the *equilibrium evaporation* and could have been derived directly from eqn 4.38. The air and the ground moisture are adjusted to each other, but even if the air just above the surface were essentially saturated, a strong lapse rate in the dry-bulb temperature profile (under significant radiation levels) would still allow a strong specific humidity gradient upwards, and a significant vapour flux. For a wet surface, this is the "Arctic sea smoke" situation.

The weighting factor $s/(s + \delta)$ is a function of the mean surface temperature (moist surface) and of the wet-bulb temperature of the air at a reference level above the surface. The values of $s/(s + \delta)$ range form 0.4 at $\bar{T} \simeq 0$ °C to about 0.8 at $\bar{T} \simeq$ 30 °C. Thus, in very humid conditions with the air nearly saturated and for a wet surface the ratio of $LE/(R_n + H_s) \simeq 0.8$ in tropical areas and 0.4-0.6 in cool climates.

When the air above the ground is not saturated it will take up more moisture than indicated by eqn 4.47 and less heat by convection. Under severe conditions the air may even supply heat to meet the latent heat requirements.

By analogy to eqns 4.23 and 4.24 the transfer through a layer of air from z = 0 to z could be expressed as

$$H = c_p \rho \bar{K} \frac{T_o - T_z}{z} = h \, \Delta T \qquad 4.48$$

and

$$LE = L \rho \bar{K} \frac{q_o - q_z}{z} = \frac{L}{c_p} h \, \Delta q \qquad 4.49$$

where

$$h = \frac{c_p \rho \bar{K}}{z}$$

represents an overall heat conductance for the whole layer and $\bar{K}$ is a weighted average transfer coefficient over the height 0 to z. From the definitions of A and δ, $\delta \simeq c_p/L_{ev}$ and using eqn 4.45, eqn 4.49 can be expressed as

$$Le = \frac{h}{\delta}(s \, \Delta T_w + \Delta D) = \frac{hs}{\delta} \Delta T_w - h \, \Delta D \qquad 4.50$$

and eqn 4.49 as

$$H = h(\Delta T_w + \Delta D) \qquad 4.51$$

With these substitutions eqn 4.13 becomes

$$R_n - H_s = H + LE = h \, \Delta T_w (1 - \frac{s}{\delta})$$

or

$$h \, \Delta T_w = (\frac{\delta}{\delta + s})(R_n - H_s)$$

Substituting this in eqn 4.50 yields

$$LE = (\frac{s}{s + \delta})(R_n - H_s) - h(D - D_o) \qquad 4.52$$

The factor h represents a heat transfer coefficient which depends primarily on wind speed but would also vary with crop or surface roughness and stability conditions. Pruitt (1971) expressed h as

$$h = 0.002(1 + u)$$

where u is the wind speed in m/s. For effectively moist conditions, or for vegetation with low stomatal resistance, $D_o \simeq 0$ and eqn 4.52 follows directly from eqn 4.38, since h is uniquely related to B. The first term in eqn 4.52 depends on the time of day and will be a maximum at mid-day. The second term, expressed as $0.002(1 + u)D$, is seen to increase linearly with the wind velocity. When D_o is not negligible the slope will further depend on air temperature and wet-bulb depression; the higher the air temperature and wet-bulb depression the steeper the slope. The expression for h should also include the effect of aerodynamic roughness of the surface or crop. At low wind speeds and high humidities (low values of D) the effect of roughness should be much less than at high wind speeds and lower humidities.

4.2 Evaporation from Snow

Snow surfaces have peculiar properties which have to be recognized in the evaporation studies. Snow surfaces have high and variable reflectivity coefficients which depend on the age of snow, its water content, depth, etc. The evaporation involves solid- liquid-gas states. Whereas the heat of fusion is approximately 335 kJ/kg, about 2826-2847 kJ/kg are required to sublimate snow. Therefore, the mass of water melted is usually substantially greater than the amount evaporated. The limiting upper boundary temperature is 0 °C which corresponds to saturated vapour pressure of 6.11 mb over the snow surface. However, for evaporation to occur there must be a vapour pressure gradient from snow to air. Since saturation vapour pressure increases with temperature warm air must have a lower relative humidity than cold air when at the same vapour pressure. Thus, if the relative humidity is 99.9% at 0 °C, at 5 °C it must be less than 70%, at 10 °C less than 49.7%, at 15 °C less than 35.7%, etc. As a consequence, the rising air temperature may transfer more heat to snow, but if the relative humidity remains the same the evaporation potential decreases and more heat is available for melting.

The energy balance may be applied to evaporation from snow in an analogous manner to evaporation from water or wet surfaces. The evaporation energy is again determined with the aid of the Bowen ratio, eqn 4.27, as

$$\frac{H}{LE} = \beta \simeq 0.61 \frac{T_s - T_a}{e_s - e_a} \frac{p}{1000}$$

where T_s, e_s and T_a, e_a refer to temperature in °C and vapour pressure in mb of snow and air, respectively, and p is the atmospheric pressure in mb. The net radiation and heat conduction into soil have to be measured or calculated in order to determine H.

In polar latitudes where snow cover prevails during winter months the net shortwave radiation received is small and seldom even balances the longwave back radiation. Thus, the major input of energy is by turbulent transfer from warm air flowing over the snow surface. Order of magnitude estimates of evaporation can be obtained from eqn 3.10 with $\kappa = 0.40$ and $k = 2.5$ mm. Two other equations widely used are the Kuzmin equation (USSR)

$$E = (0.18 + 0.098u_{10})(e_{os} - e_2) \qquad 4.53$$

where E is evaporation from snow in mm/day, u_{10} is average daily wind speed in m/s at 10 m height, and the Central Sierra Snow Laboratory equation (Snow Hydrology, U.S. Dept. of Commerce, 1956)

$$E = 0.0063(z_a z_b)^{-1/6}(e_{os} - e_a)u_b \qquad 4.54$$

where E is in in./day, u_b is wind speed in mph at z_b ft above ground, and e_a is vapour pressure in mb at z_a ft above ground.

In spring the net shortwave radiation increases rapidly and becomes important in both evaporation and snow melt, but turbulent heat transfer by the warm wind still remains the dominant factor.

4.3 Evapotranspiration

Transpiration itself is a complex biological process which depends on the plant, soil and atmosphere. The rate at which moisture can be taken up by the plant depends on the root system and on the soil. As the soil loses moisture the pore pressure deficiency or capillary potential increases at different rates for different soils. The moisture absorbed by the roots is passed through the plant to the leaves. The nature of the leaf system controls the rate of transpiration. The energy for the transpiration process is derived from the osmotic potential between the soil water and sap solution, and capillary potential in the leaves. The theoretical approach to the transpiration problem is again based on the energy budget. The discussion of this theory is beyond the scope here. The reader is referred first to the short note in Oxford Biology Readers series by Rutter (1972), and further to Briggs (1967), Kozlowski (1964), Kramer (1969) and Slatyer (1967).

The planner of an irrigation project, for example, is primarily interested in the consumptive use of a given crop, that is, how much water has to be provided and at what rate to satisfy the needs of evapotranspiration for production of a good crop.

Again the standard is the potential evapotranspiration, which is the quantity of water used by evaporation and transpiration, if not limited by availability of water. The determination of the actual evapotranspiration is not easy. Special lysimeters enable the determination of exact water and heat balance but this method is restricted to relatively small samples. Indication of the average condition in small sample plots may also be obtained by measurement of soil moisture depletion. An illustration of a soil moisture deficiency diagram is shown in Fig. 4.8. The soil is the moisture reservoir for the plants. The quantity of water held by the soil between field capacity and the wilting point is a function of the soil type. The rate at which soil moisture is available to the plants depends on the root surface area in contact with the soil and the rate at which moisture can move through the soil.

A variety of energy and water balance methods are applied to larger areas. Some are related to pan evaporation, others to the net radiation energy. Models of transpiration from vegetation depend on meteorological parameters and on surface parameters such as the aerodynamic and stomatal resistance to diffusion of water vapour from the plants. The resistance to molecular and turbulent diffusion of water vapour between leaf surfaces and air above at a reference height is expressed as

$$r_a = \frac{c_s - c}{Q} \qquad 4.55$$

where c_s and c are the vapour concentration (mg/cm^3) at leaf surface and reference level respectively and Q is the flux of vapour per unit horizontal area (g/cm^2s). The inverse of r_a is the transport velocity. The concept of resistance may be more apparent from the following argument. Consider a leaf at a uniform temperature T_o. The vapour pressure at its surface is e_o, and that of ambient air e. The transpiration rate is then proportional to $(e_o - e)/r_a$ which is an analogous expression to Ohm's law; a potential difference maintaining a current or flux of water vapour across an external diffusion resistance in the air surrounding the leaf, i.e.

$$EL = \frac{\rho c_p}{\gamma} \frac{e_o - e}{r_a} \qquad 4.56$$

The same value of r_a may be related to diffusion of heat.

The rate of diffusion within the leaf may be expressed in the same way as proportional to $[e_s(T_o) - e_o]/r_\ell$, where r_ℓ is the internal resistnace of the leaf. This internal resistance can be treated as a free water surface if $(e_o - e)$ is replaced by $[e_s(T_o) - e]/(1 + r_\ell/r_a)$ or the constant γ in the evaporation equation is expressed as $\gamma^* = \gamma(1 + r_\ell/r_a)$.

The wind velocity at reference level in near-neutral conditions of temperature gradient is given by

$$u(z) = \frac{u_*}{\kappa} \ln \frac{z - d}{k} \qquad 4.57$$

where d is the zero displacement, i.e. the height which makes wind speed proportional to $\ln(z - d)$. Penman and Long (1960) related r_a from measurements to wind speed and roughness height k as

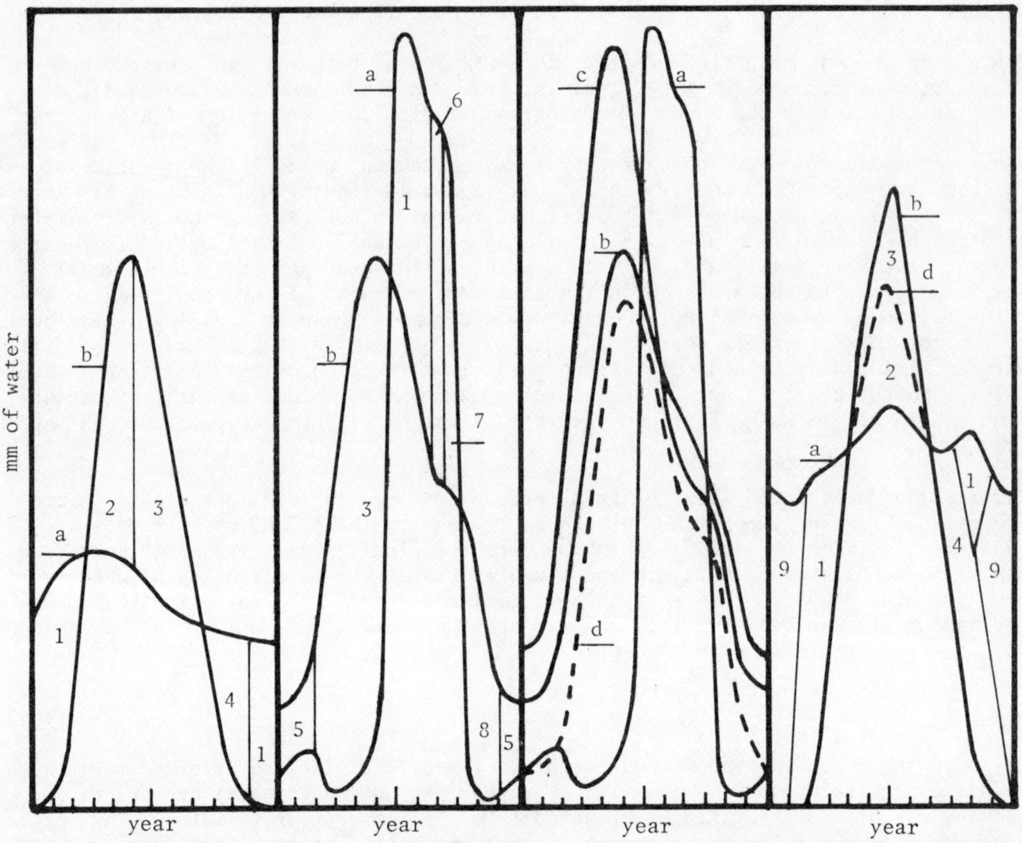

Fig. 4.8. Illustration of annual soil moisture deficiency diagram. The first panel is for a limited growth season. It shows water deficiency, both rate and total, despite fairly uniform rainfall throughout the year. Hence, little or no runoff or deep percolation can be expected from arable land. The second and third panels are for a year-round growing season, and show a large water deficiency before the rainy season. The fourth panel is similar to the first and shows a condition where some of the precipitation is in the form of snow (9). Legend: (a) rainfall, (b) P.E.T. curve, (c) class A evaporation pan curve, (d) P.E.T. curve for design purposes.
(1) runoff, (2) soil moisture utilization, (3) moisture deficiency, (4) soil moisture accretion, (5) soil moisture utilization from deep soils (water deficiency in shallow soil), (6) soil moisture recharge-shallow soils, (7) soil moisture recharge-deep soils (6 & 7 indicate proportion of rainfall excess not a period of time), (8) soil moisture utilization-shallow soils.

$$r_a = \frac{\rho c_p}{\gamma L}\,\frac{e_o - e}{E} = \frac{[\ell n(z - d)/k]^2}{\kappa^2 u} = \frac{1}{\kappa u_*}\,\ell n(\frac{z - d}{k}) \qquad 4.58$$

in which they used $\kappa = 0.41$. The wind speed u and vapour pressure e are measured at height z above the crop surface.

Chamberlain (1966) wrote eqn 4.58 for r_a with k' instead of k, where $k' = ke^{-\kappa B^{-1}}$ and B^{-1} is a dimensionless parameter which increases with the wind velocity. Chamberlain gives a few measured values for B^{-1}, over short and tall grass, which range from 7 to 12 for u_* from 0.135 to 1.76 m/s. The resistance value r_a is made dimensionless by multiplying it with u_*, i.e. $r_+(z) = u_* r_a$. It is analogous to a dimensionless resistance for momentum

$$f_+(z) = \frac{u_* \rho u(z)}{\tau} = \frac{u(z)}{u_*} = \frac{1}{\kappa}\,\ell n\,\frac{z}{k} \qquad 4.59$$

and is determined by the geometry of the boundary surface alone.

The single leaf idea is extended to the surface resistance, r_s, of a crop, pasture or forest cover by writing, in analogy to r_ℓ

$$r_s = \frac{\rho c_p}{\gamma L}\,[\frac{e_s(T_o) - e_o}{E}] \qquad 4.60$$

The determination of T_o is not easy but when the wind, humidity and temperature profiles above the crop are of the same shape then, plotting of T against u at a number of heights gives a straight line from which $T = T_o$ at u = o.

The maximum rates of transpiration are related to weather conditions through

$$\frac{LE}{H} = \frac{\Delta/\gamma + r_i/r_a}{\Delta/\gamma + (r_s + r_a)/r_a} \qquad 4.61$$

where

$$r_i = \rho c_p\,\frac{e_s(T) - e}{\gamma H} \qquad 4.62$$

When $r_i = r_s + r_a$ the heat supply goes all to evaporation, LE = H. Table 4.5 shows some values given by Monteith (1965).

TABLE 4.5 Crop Parameters and Annual Transpiration

	Reflectivity %	k mm	r_i $s\ m^{-1}$	r_a $s\ m^{-1}$	r_s $s\ m^{-1}$	$\frac{\Delta}{\gamma}$	E $\frac{mm}{yr}$	$\frac{LE}{H}$ %
				Thames Valley				
Short grass	25	1	80	110	50	1.3	470	76
Tall crop	25	50	80	110	50	1.3	580	95
Pine forest	15	2500	70	2.5	100	1.3	480	69
				Sacramento Valley				
Short grass	25	1	110	110	50	2.4	1480	96
Tall crop	25	50	110	36	50	2.4	1870	117
Pine forest	15	2500	90	2.5	100	2.4	1820	100

Stewart and Thom (1973) expanded the treatment of the resistance problem and applied it to a pine forest. They found that the Chamberlain $B^{-1} \simeq 2$ to 3 at $u_* \simeq 0.75$ m s^{-1} and that for given weather conditions intercepted water will evaporate at about five times the corresponding transpiration rate. This is quite different from the behaviour of short crops and grasses which transpire at about the same rate, when soil water deficits are small, as the evaporation of intercepted water from them under the same weather conditions.

Gash and Stewart (1975) rearranged the Monteith-Penman equation

$$LE = \frac{\Delta Q + \rho c_p [e_s(T_o) - e]/r_a}{\Delta + \gamma(1 + r_i/r_a)} \qquad 4.63$$

(Monteith, 1965) where Q is the rate of energy supply ($Q = R_n - H_S - H - P$, where P is net rate of energy absorption by photosynthesis and respiration) for surface resistance r_S as

$$r_s = (\frac{\Delta L}{c_p} \beta r_H - r_v) + (1 + \beta) r_I \qquad 4.64$$

where Δ is the slope of the saturated specific humidity curve at the mean of the surface and air temperature, β is the Bowen ratio, r_H and r_v are the aerodynamic resistances to transfer of heat and water vapour respectively, and r_I is the climatological resistance and is given by eqn 4.62 when H is replaced by Q.

For routine application, available data on evapotranspiration rates are correlated with monthly mean temperature, T_m, percentage of daylight hours, n, length of growing period and precipitation. For example, Blaney and Criddle (1950) related these factors and a monthly crop coefficient k as

$$cu = \frac{kT_m n}{100} = kf \qquad 4.65$$

where cu is the consumptive use and $T_m n/100 = f$ is the monthly consumptive use factor. The seasonal consumptive use is obtained by summing the monthly values over the growing season as

$$CU = \Sigma\, kf = K\, \Sigma f = KF$$

values of k and K for various crops are given in specialized texts.

Every crop has its own optimum water requirement which depends on soil texture, fertility and climate. Too much as well as too little water will reduce the yield. Figure 4.9 illustrates this yield-moisture relationship. Some crops are more sensitive to moisture conditions than others. The summer water need of pastures in temperate climates is of the order of 55% of the raised pan evaporation rate. However, the water demand is also a function of the growing season as illustrated in Fig. 4.10.

Most crops have a low water demand in early stages of growth. The demand reaches a maximum when the crop approaches maturity. In meadows the demand drops suddenly to a low value after haymaking but the demand increases rapidly as the growth recommences. Generally, short vegetation types, such as grass, transpire freely when moisture is available at about the same rate as intercepted water would evaporate from then under the same over head conditions. However, as referred to above transpiration from pine forest (and probably other forests) have been found to be only one fifth of the evaporation rate of the intercepted water under the same atmospheric conditions (Stewart and Thom. 1973). An additional feature of forests is that

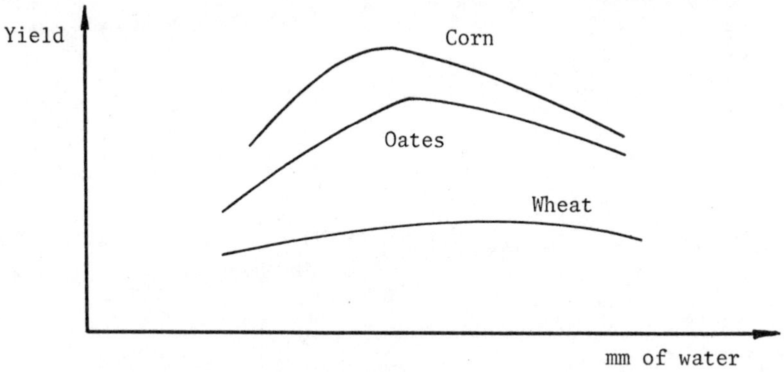

Fig. 4.9. Diagrammatic illustration of yield-moisture relationship for crops.

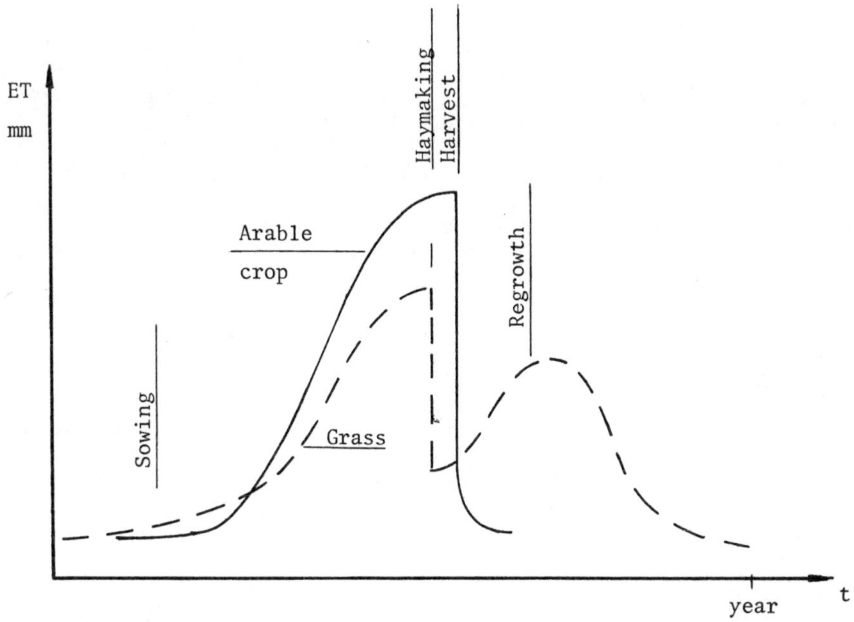

Fig. 4.10. Diagrammatic illustration of water demand as a function of growing season and crop.

through evapotranspiration they extract large quantities of water from the soil to depths approaching 30 m. The associated soil pressure deficiency may significantly contribute to slope stability. Felling of the forest allows the soil to become saturated and this may lead to slipping of the hill sides.

Chapter 5

INTERCEPTION

Some of the precipitation is intercepted by vegetation (and roofs) before reaching the ground. The volume of water caught is referred to as interception. The portion of intercepted water which is retained in storage on the vegetal cover and evaporates is called the interception loss. The interception loss may be a very significant factor in water balance studies where its influence depends on the type, density, etc. of the vegetation. The density ranges from desert plants to tropical rain forests. In humid forested regions about 25% of the annual precipitation may become interception loss. Helvey and Patric (1965) suggest that this loss is about 250 mm. Pierce et al. (1970) observed an increase of stream flow of 346 mm for the year after the clearing of the catchment of all woody vegetation. The total annual precipitation was 1304 mm. Similar results are reported by Swank and Helvey (1970). In contrast, interception is usually neglected altogether in studies of major storm effects and floods.

The studies of interception may be subdivided into field measurements and theoretical modelling. The modelling of interception on a catchment generally starts from the idealized leaf of the particular varieties of plants present. In a light uniform rain water droplets collect on the surface of each leaf. These droplets grow and coalesce. Eventually the size of the individual droplets becomes larger than can be held by the surface tension forces. The drop will then move across the leaf, merge with others on the way, and on pendant leaves will fall off the tip of the leaf. Where the leaves point upwards, the drop moves towards the stem and may flow down the stem of the plant. The cumulative volume of drops from a given point on the leaf is shown in Fig. 5.1. A similar picture is obtained for the cumulative volume of drops from N points, except that the step heights will vary widely. Continuity requires that for the catchment the total precipitation equals the interception storage plus the through-fall or net precipitation on the ground, assuming that during the precipitation the evaporation losses are negligible. Thus, the cumulative through-fall function must become asymptotic to the unit gradient as the storage becomes filled, Fig. 5.1. The cumulative curve must also pass through the origin. The rate of net precipitation or through-fall is the gradient of the cumulative curve, that is, differentiated with respect to time for uniform rate of rainfall, Fig. 5.2. The area above the rate of through-fall curve, and below the rate of precipitation asymptote, gives the volume of the interception storage.

For application, the rate of precipitation P_o and the rate of through-fall to the ground P_g, for a given condition of vegetation, are usually related by simple linear regression

$$P_g = aP_o - b$$

where $a < 1$ because of evaporation. For a given instant, continuity requires that

$$P_o = P_g + \Delta S + E$$

where ΔS is change of storage in the preceding interval and E is the evaporation during that interval.

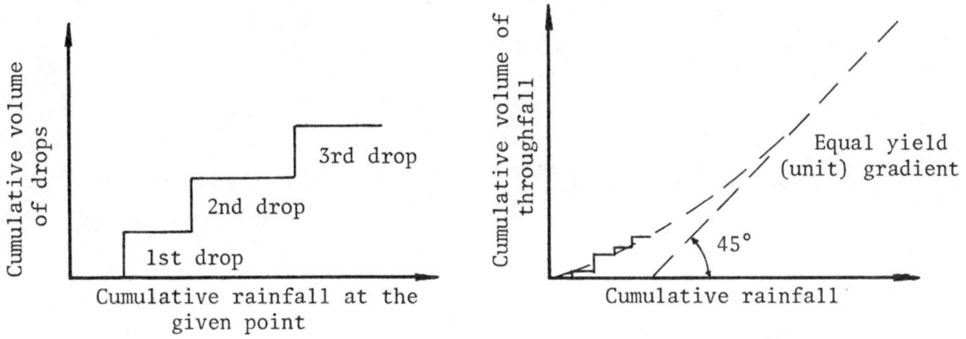

Fig. 5.1. Illustration of cumulative throughfall

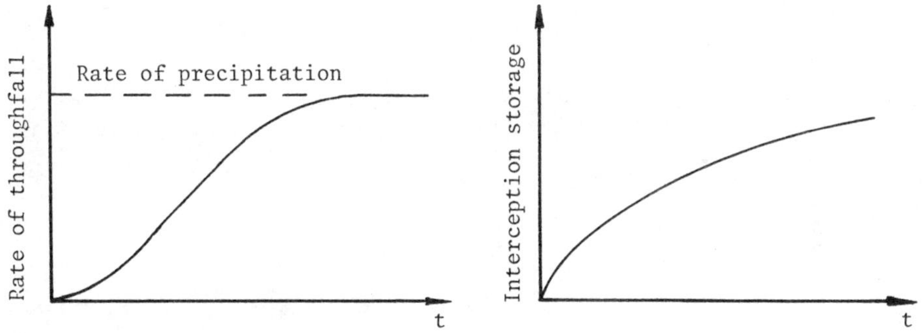

Fig. 5.2. Diagrammatic presentation of the rate of throughfall and interception storage.

The energy balance models (Waggoner and Reifsnyder, 1968; Murphy and Knoerr, 1972, 1975) assume a homogeneous vegetation with a closed canopy of large extent. For this one-dimensional model the upper boundary is assumed to be the upper portion of the aerodynamic boundary layer and the lower boundary the soil surface. This lower boundary excludes the soil with its large thermal lag. Within these boundaries the forest is divided into four regions: aerodynamic boundary layer above the forest, the canopy air space, the leaf surfaces and the litter surface. Murphy and Knoerr (1975) showed by such a simulation model that a forest canopy wetted by rainfall will partition more of the absorbed radiant energy into latent heat exchange than an unwetted canopy in the same environment. This is at the expense of sensible heat transfer from the canopy to the atmosphere and a smaller decrease in longwave radiation from the canopy.

Field studies with grass (McMillan and Burgy, 1960) showed no appreciable difference between evapotranspiration from wetted and from unwetted vigorous grass covers. They concluded that "the data indicate that evaporation from wetted leaf surfaces may replace all or part of normal transpiration and that the entire plant-soil system should be considered in evaluating interception loss".

5.1 Measurement of Interception

The interception loss is usually determined by measuring the total precipitation P_o over the area, the throughfall of precipitation, P_g, to the ground under the vegetation and the amount of water reaching the ground as flow down the stems or trunks S_f. Then

$$I = P_o - P_g - S_f$$

Horton (1919) expressed the total interception loss I for the projected area of the canopy by

$$I = S_v + RE\ t$$

where S_v is the storage capacity of the vegetation over that area, R is the ratio of the surface area of vegetation to the projected area, E is the evaporation rate from the vegetal surfaces and t is the duration of rainfall. This relationship assumes that the total storage capacity S_v is available at the beginning of the rainfall and is filled by it. The initial storage capacity depends on prior precipitation. In addition the rate and amount of rainfall will have bearing on the interception loss. Many sets of experimental interception loss, I, have been fitted by an equation of the form

$$I = a + bP_o^n$$

where a, b and n are constants and n is often equal to unity. Values of a and b are frequently related to the height of plants as $a = \alpha h$ (Horton, 1919; Gray, 1970).

Meriam (1960) modified the Horton equation to

$$I = S_v(1 - e^{-P_o/S_v}) + RE\ t$$

which appears to fit the observations fairly well. The last term is difficult to estimate in terms of the individual values and is, therefore, expressed as KP where $K = RE\ t/P$ is assumed to remain constant. This implies a constant relationship between the evaporation and precipitation, which is difficult to justify theoretically. If P becomes large this equation reduces to the Horton equation. Seasonal interception losses are fitted by relationships of the form

$$\Sigma I = a\Sigma P_o + bn$$

where n is the number of storms per season. Similar expressions can be written for throughfall ΣP_g, stem flow ΣS_f, and interception loss ΣL.

Evaporation of intercepted water and evapotranspiration have in the past generally been treated as separate terms in the hydrologic cycle because they were believed to be independent of each other. However, this concept of mutual independence has been questioned because both are energy dependent processes and the amount of energy available at a given time is limited. It has been suggested (Leyton and Carlisle, 1959; Penman, 1963) that energy used to evaporate intercepted water must come from that which would have been used for evapotranspiration. The hypothesis is that the energy available for evaporation under a given set of conditions is a constant fraction of solar radiation, represented by the potential evaporation, and energy used to evaporate intercepted water reduces that available for evapotranspiration or evaporation from the ground. Accordingly, interception loss would not be an additional loss but rather a compensation for transpiration that would have occurred. Experimental results, however, have shown that evaporation from wetted forest trees is more than evapotranspiration from dry trees under similar conditions (Rutter, 1959; Frankenberger, 1960). Lysimeter experiments by Waggoner et al. (1969) showed that the total evaporation rate from corn crop wetted by spray irrigation was much greater than before wetting. This experimental evidence lead to the hypothesis that although the process is energy dependent, more energy is avaialble for evaporation from wet plants than dry plants under the same conditions. The added energy is assumed to come from different proportioning of the radiant energy input.

The measurement of the total precipitation over the vegetal cover follows standard rain gauging procedures. The rain gauge sites should be large and open near the study area and have the same slope and aspect. Where these are not available small clearings are cut to provide at least a 45° unobstructed sky view angle. In areas with strong winds even larger openings are required. Placing of gauges at treetop level is also being studied. The number of gauges required to give the desired accuracy is also determined according to the standard rain gauging techniques. Two to four gauges are usually adequate.

The measurement of throughfall is a more difficult task and depends to some extent on the type of the vegetal cover. Standard rain gauges can be used under a tall stand of vegetation (Helvey and Patric, 1965). Standard rain gauges give results which may be compared directly with measurements of the total precipitation. The throughfall is highly variable over the area and, therefore, a much larger number of gauges is required (about ten times as many as for P_O). The effects of the variation can be further reduced by periodically moving the gauges to new randomly chosen sites. Under low shrubs non-standard gauges are used at ground level, such as a network of sheet-metal or plastic troughs, collection trays, etc. The existence of a secondary vegetal cover may further complicate the problem. A dense undergrowth may require a network of troughs with or without rain gauges at a level above the undergrowth. The interception loss of litter cover may be calculated by determining its moisture content immediately after the rainfall event and periodically thereafter.

The stem flow of trees is relatively easily measured. A narrow funnel type collar is sealed to the tree trunk and the stem flow is collected into a container. The collar needs to be fairly narrow so as not to collect much of the throughfall and the container should prevent evaporation of the collected water. The amount of stem flow varies with the species and may be 1 to 15% of the precipitation, P_O, for smooth bark trees and about 2-3% for rough bark species.

5.2 Interception of Snow

Interception of snow, Fig. 5.3, is not as well understood as of rain. The approach of the hydrologists has been influenced by statements like that by Kittredge (1948), "Snow is subject to interception just as is rain, and it has been shown that the magnitudes of interception of snow are not very different from those of rain". However, interception of snow is primarily a storage or delay effect and very little of it becomes interception loss. Satterlund and Haupt (1970) found that losses returning to atmosphere from intercepted snow were only about 5% of total snowfall. They found that 46% of intercepted snow was washed off the tree by rain, 25% fell off as large lumps of snow and the remainder by drip and small releases of snow.

They also concluded that meteorological factors were more important than the morphological differences between trees.

Fig. 5.3. Interception of snow. Photo by Mr E. Wengi.

The conventional methods for determining snow interception are modelled after those for rain. The snowfall is measured at a number of points under the forest and compared with measurements in a nearby opening "free from forest effects". The selection of the open site is a difficult task. A completely open exposed area is not suitable because of drifting snow and the effects of wind in reducing the snow catch in the gauges. At the same time experiments have shown that the size of the openings in the forest has a pronounced effect upon the snow catch. Costin et al. (1961) observed snow accumulation in open areas, areas forested with varieties of evergreen eucalyptus trees, and in forest openings of various size. Least snow was found in open areas, more under the trees and most in forest openings. Openings up to 10 tree heights in width increased snow catch on sheltered slopes. On exposed slopes openings of smaller size were more effective. The differences were attributed to redistribution by wind. Indeed, redistribution by wind appears to be one of the most important features. Generally, there appears to be a strong increase in snow near the centre of the opening which is at least partly offset by a decrease in snow below the trees downwind, Fig. 5.4.

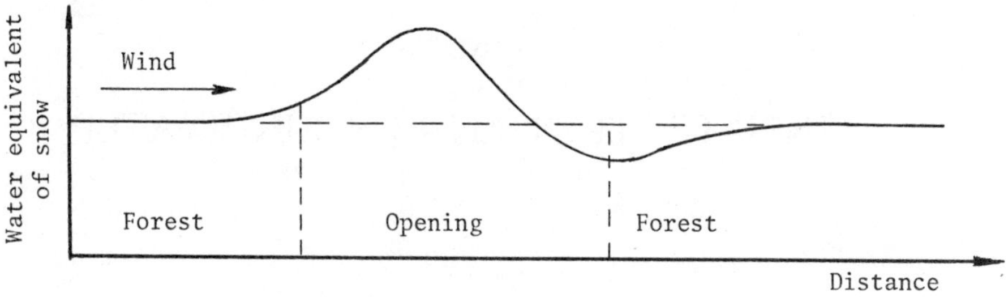

Fig. 5.4. Illustration of snow catch in a forest clearing.

There are many reports on measurement of snow accumulation in various sized open areas. It is common to find that the differences between forest and the open field are explained by more rapid evaporation from snow held on tree crowns than from snow on the ground. There seems to be a consensus of opinion that there are great differences in the energy and vapour balance between the intercepted snow and snow on open ground but these are unlikely to explain why there is more snow in the forest openings. This accumulation of snow is believed to be the result of redistribution of snow. Loss of intercepted snow must occur through evaporation and sublimation and these processes require considerably more energy than does the melting process. Hence, although the forested areas receive a larger amount of energy, due to the much higher absorptivity of the trees than the open snow field, the bulk of it would go into melting. The melting water either drips down or converts the snow into ice.

A different aspect of snow interception by trees is the breakage it can cause. Shidei (1954) reported laboratory and field measurements of actual snow loads on trees and related these to snow properties, air temperatures, sunshine, and wind velocities. The snow loading of trees was found to be strongly temperature dependent, increasing from a low value at -6 °C to a maximum at -1 °C and decreasing rapidly with further increase in temperature. The most favourable conditions for maximum accumulation of snow on trees are slight winds, falling air temperature and no sunshine. There are several reasons for the temperature effect: the cohesion of snow is greatest at just below freezing temperature, temperatures falling below °C may be associated with initial wetting of foliage and branches, and the angle of repose of snow increases with temperature.

Chapter 6

INFILTRATION AND GROUNDWATER

Precipitation which does not evaporate becomes either runoff or infiltrates into the ground or both. Infiltration is one of the most difficult elements of the hydrologic cycle to quantify. Experimental studies have shown great differences in the infiltration capacity of soils of different texture and structure, and of the same soil under various types of vegetation. In addition, most soils are strongly anisotropic with properties varying in both horizontal and vertical directions. Further, the anisotropy of the soil profile acquires time dependent properties through growth of vegetation, biotic activity, wetting and drying, freezing and thawing, tillage, etc., and consequently the value of the permeability of the surface layers is seldom a constant.

Groundwater is all interstitial water below the water table. The water table is the upper surface of the completely saturated ground. Directly above the water table is the capillary fringe held by capillary forces. The height of this capillary rise depends on the pore size and can be from less than a cm to several metres. The entire zone from the water table to the surface is known as the zone of aeration or zone of suspended water. This zone contains water in the form of attached films on the surfaces of the particles, as capillary water suspended in the pores by menisci, Fig. 6.1, and water seeping downwards under the action of gravity. The water content of the soil, after all the water moving under the action of gravity has drained away, is called the field capacity. Most of the active root development is within the top 3 m of the zone of aeration. The zone of aeration is also where the destructive chemical actions and weathering of rocks occurs.

A water molecule, which escapes from a horizontal free surface to vapour, has to overcome the attraction of the molecules that surround it. This requires the input of latent heat of evaporation. In partially saturated soils the forces to be overcome are increased by the curvature of the interface between the adsorbed film of water and the air-water vapour mixture, and the adsorptive forces of the solids. Therefore the vapour pressure in the atmosphere within the voids is always less than over a free horizontal water surface at the same temperature and humidity conditions. Figure 6.2 illustrates typical variation of vapour pressure and relative humidity in the soil with moisture content. Below the relative humidity of about 93% (in this case) large changes in relative humidity are associated with small changes in moisture content and adsorptive forces dominate. Above this value large changes in moisture content have only a small effect on relative humidity and capillary forces dominate. The boundary between these two regions coincides approximately with the wilting point. The attached films of water are held by electro-chemical forces.

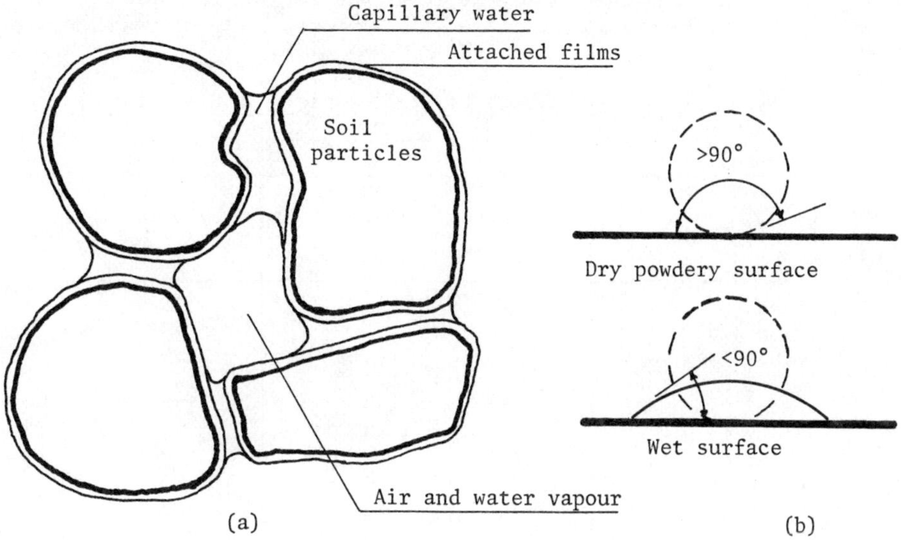

Fig. 6.1. (a) Illustration of forms of water storage in soil. (b) A drop of water on dry powdery surface and on wet surface.

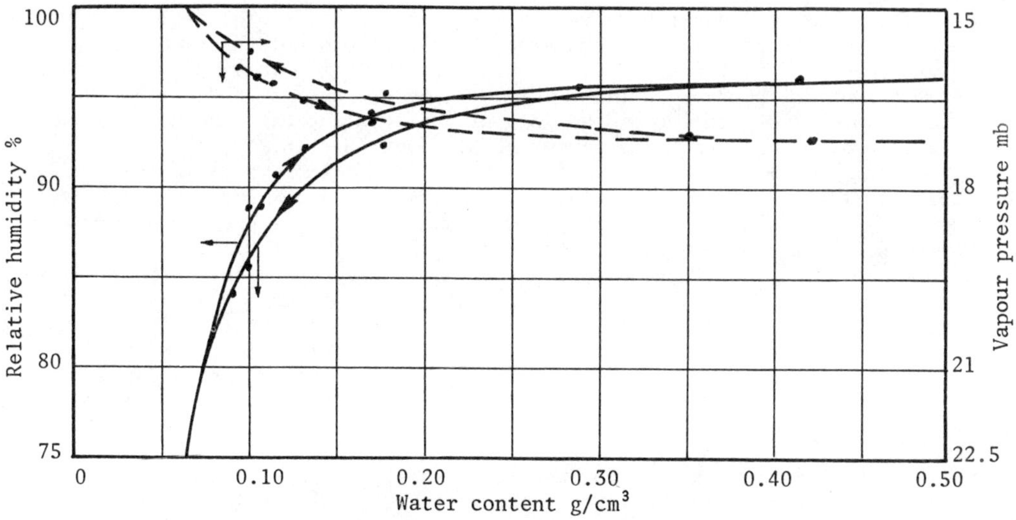

Fig. 6.2. Relationship between relative humidity and moisture content in Greywacke loam (Nguyen, 1974).

The pore pressure deficiency is a measure of the depletion of these films. The pore pressure deficiency is zero after the gravity water has drained away and before evaporation commences, i.e., at field capacity. The pore pressure deficiency increases with the depletion of the adsorbed films by evapotranspiration and can reach about 100 MPa (1000 bar) in a dried soil. The suction of plant roots is limited to about 700-800 kPa (7-8 bar). The wilting point for potatoes and soil bacteria is at about -1 bar, whereas that for wheat is at -40 bar and for some soil fungi at -50 bar. Table 6.1 shows some average values of field capacity and moisture content at permanent wilting point.

TABLE 6.1 Average Values of Field Capacity and Moisture at Wilting Point

Soil type	Field capacity mm	Permanent wilting point mm	Soil moisture available at field capacity mm
Sand	100	25	75
Fine sand	120	30	90
Sandy loam	160	50	110
Fine sandy loam	220	70	150
Loam	270	100	170
Silty loam	280	110	170
Light clay loam	300	130	170
Clay loam	320	150	170
Heavy clay loam	330	170	160
Clay	330	210	120

(After "Snow Hydrology" U.S. Army Corps of Engineers, North Pacific Division, Portland, Oregon, 1956).

Capillary water is held in pores by balanced film forces. The capillary force depends on the angle of contact between the media in contact. The angle is a measure of the relative strength of the attractions of the liquid molecules to each other and to the molecules of the solid. For minerals the angle is very small but for oily surfaces or dry powdery surfaces the angle of contact is greater than 90°. The powdery surface can influence the initial infiltration rate because drops of water remain in drop form until the powder becomes wet (Fig. 6.1).

When water is applied to a dry soil no water moves beyond the surface layer before the adsorbed films have been restored. Only after the films have been completed will water percolate through the wet soil under the action of gravity. Since the driving force is the sum of the pore pressure deficiency and gravity the infiltration starts at a high rate. The rate decreases as the films are satisfied and finally settles down to a lower constant steady-state rate of infiltration if the supply of water continues at a constant rate, Fig. 6.3. This rate is referred to as the infiltration capacity, f_c, and is the rate of flow through the soil under gravity. Note that the rate of infiltration is strongly dependent on the initial moisture content of the soil.

The rate at which water enters into the ground is also strongly influenced by the structure of the surface of the soil. Water entering through large pores, Fig. 6.4a & b, spreads both vertically and horizontally under the action of pore pressure deficiency and gravity. The pressure in the open pore is governed by the local piezometric head and is greater than atmospheric, whereas that in the soil is less. The pressure in a closed cavity is approximately atmospheric and the cavity remains as an air bubble, it does not fill. Hence, large pores isolated from the free water will not contribute to infiltration because the piezometric head increases in the direction of such isolated voids. In fact, the air pressure increases in such an air bubble to a value greater than atmospheric, depending on the rate of infiltration. In general, for water to infiltrate soil air must be displaced. Wilson and Luthin (1963) observed soil air pressures up to 14 cm of H_2O in homogeneous soil

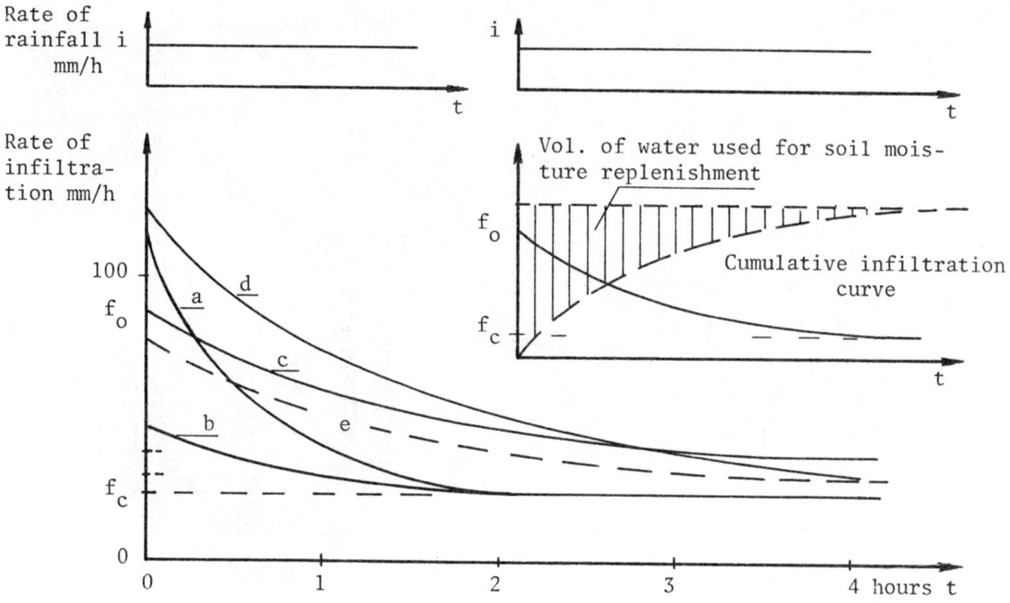

Fig. 6.3. Diagrammatic illustration of variation of infiltration rate with time when precipitation rate exceeds infiltration rate. (a) and (b) are for a grassland on fine sandy loam where (a) is for initially dry and (b) for initially wet condition. (c) is for moist sandy soil, (d) and (e) illustrate infiltration rates of the same soil where (d) is for good cover of vegetation and (e) is for bare soil.

columns vented to the atmosphere. Thus, only surface-connected pores are positive head flow tubes and increase the rate of infiltration. Dixon (1966) and Dixon and Peterson (1971) showed that large open pores, which in themselves represent a negligible fraction of the total surface area, contribute greatly to infiltration. This large contribution to infiltration is to be expected because, according to the Poiseuille equation, the flow rate increases with the 4th power of the tube diameter, that is, at 1 mm diameter tube could carry 10 000 times as much water as a tube of 0.1 mm diameter. At the same time, it requires only about 3 cm of H_2O of internal air pressure by the displaced air to eliminate the large infiltration contribution of the 1 mm diameter pores. A rise in soil air pressure equivalent to 3 cm of H_2O can occur when 3 mm of water infiltrates a 10 m deep soil which contains initially 10% of air. This shows that such a rise in soil air pressure could be expected during wetting.

Dixon demonstrated the effect of large pores with the aid of six idealized forms, Fig. 6.4c, which include the influence of small surface features. The diagram (A) shows a channel, for example, an insect burrow, from a small depression to a small rise in the soil surface. The soil surface is covered by vegetation. During rainfall, water collects in the hollow over the opening and flows down the tube while the air is readily exhausted from the high end of the hole which is not submerged. From the tube itself the water readily flows into the soil by capillary suction. The important features of this diagram are (i) the surface cover which prevents compaction and closure of the inlet orifice, (ii) the small depression and (iii) the ex-

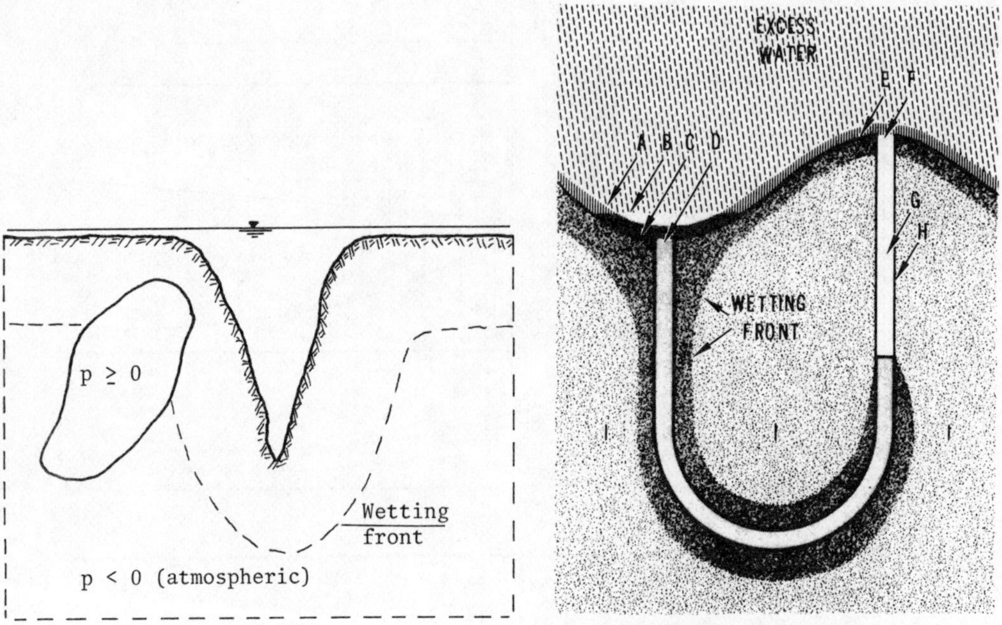

(a) (b)

Fig. 6.4 a & b. Illustration of the effect of large openings. (a) Water filled openings assist spreading but soil exhaust pressure (displaced air) and surface capillary tension prevent water form entering an isolated cavity. (b) The channel concept of Dixon and Peterson (1971), where A is soil surface cover, B is free water surface, C is a microdepression, D is a water intake opening, E is an air exhaust opening, F is a subsurface channel and H is the porous wall of a channel in soil I.

haust at the crest of a small surface feature. Diagrams (B) and (C) illustrate the same geometrical situation as (A) except for the protection by vegetation. Compaction of the surface by rain drops has led to partial (B) and total (C) closure of the tube and strong reduction of infiltration. Diagram (D) differs from (A) only by surface roughness. There is no optimal location for exhaust. The thin layer of water over the surface seals off both holes. The ports receive too little water for rapid water intake and too much for exhaustion of displaced air under low pressure. There will be an intermittent taking in of water and exhaustion of air. Diagrams (E) and (F) show the effect of sealing of the surface.

These diagrams also illustrate the effect that compaction and inwash has in general. The pounding of large drops and the tendency of fine particles to wash into the voids can close off the larger voids and cause a significant drop in the rate of infiltration. Vegetation can absorb most of the kinetic energy of raindrops and protect the soil from compaction, as well as from erosion. Wischmeier and Smith (1958) estimated the kinetic energy of rain striking the ground at terminal velocity to be

$$KE = 118.8 + 87.21 \log_{10} X$$

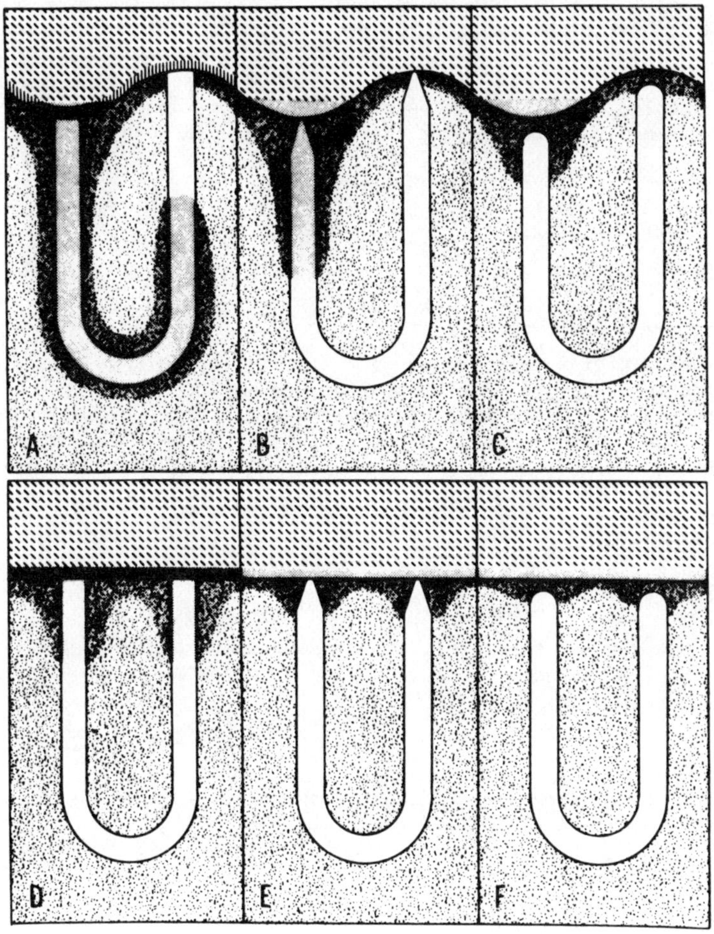

Fig. 6.4c. Six idealized channel systems which differ only at soil surface. Channel systems A, B and C have rough surfaces with open, constricted and closed water intake and air exhaust orifices, whereas channel systems D, E and F have smooth surfaces with open, constricted and closed orifices, respectively. (Dixon, R.M. and Peterson, A.E., 1971).

where KE is in Nm/m^2 per cm of rain and X is rainfall intensity in mm/hour. ($Y = 916 + 331 \log_{10} X$, where Y is in ton-ft/acre-inch and X is in in./h). Compaction due to animals and any other traffic has a similar effect.

Dense vegetal cover promotes high rates of infiltration. The layer of organic debris forms a spongelike surface which retains moisture and prevents compaction and closure of large pores and channels. Under such a cover there is usually an abundance of burrowing animals and insects. In addition root systems provide ingress for water into the soil. Roots of growing plants shrink whenever evapotranspiration exceeds water entry to the roots. Together with the shrinkage of the soil this creates air vents and hydraulic conduits. In the course of a day roots may shrink to about 60% of their maximum diameter. Cracks which form due to shrinkage in some soils may drastically alter the rate of infiltration.

The rate of infiltration is influenced by a number of additional variables. Temperature, for example, affects the rate of infiltration through variation of viscosity. The lower the temperature the higher the viscosity, and hence, a lower rate of infiltration. Entrapped air reduces the permeability of the soil. Leaching out of soluble soils can also affect permeability. Many soils swell when wetted and this introduces a time dependent variation of permeability. Depth of water over the surface also affects the rate of infiltration.

6.1 Calculation of Infiltration Rate

At low intensities of rainfall the water goes into replenishment of soil moisture. Any excess percolates to groundwater storage and may eventually become groundwater runoff. With high intensity rainfalls more water arrives than can infiltrate at the ground surface. The rainfall excess, P_e (mm), becomes surface runoff and can be expressed as

$$P_e = P - S_{sm} - f_c t \qquad 6.1$$

where P is rainfall (mm), S_{sm} is the replenishment of soil moisture and f_c is the infiltration capacity.

The extreme complexity of the infiltration process has excluded exact solutions. Of the semi-empirical methods the Horton (1940) equation (Gardner and Widstoe, 1921) is one of the best known

$$f = f_c + (f_o - f_c)e^{-kt} = a + be^{-kt} \qquad 6.2$$

where f is the infiltration rate, f_c and f_o are the infiltration capacity and the initial value respectively, and exp(-kt) gives the decay of infiltration rate with time. The value of k can be determined by plotting observed data as $\log(f - f_c)$ against time, Fig. 6.5.

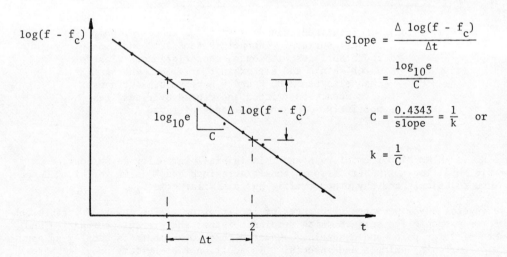

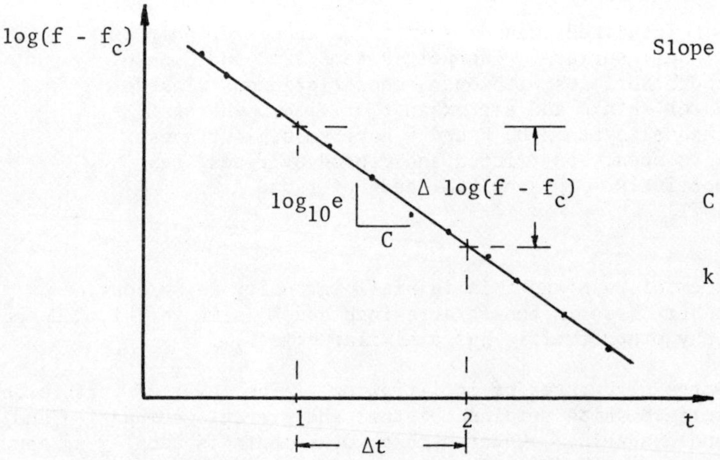

$$\text{Slope} = \frac{\Delta \log(f - f_c)}{\Delta t} = \frac{\log_{10} e}{C}$$

$$C = \frac{0.4343}{\text{slope}} = \frac{1}{k} \quad \text{or}$$

$$k = \frac{1}{C}$$

Fig. 6.5. Evaluation of the constant k in eqn 6.2.

The volume of water that has infiltrated from t = 0 to t is obtained by integration of eqn 6.2

$$F = \int_0^t f dt = f_c t + \frac{1}{k}(f_o - f_c)(1 - e^{-kt}) \qquad 6.3$$

The Horton equation requires knowledge of the initial rate of infiltration and of the rate of reduction with time. Errors in the estimate of these values can lead to serious errors in the calculation of the amounts of infiltration, particularly for longer periods of time. Philip (1957) developed an equation with time as the dependent variable but found it very unwieldy.

Holtan (1961) related the infiltration to the depletion of the water storage capacity of the soil as

$$f = aF_p^n + f_c \qquad 6.4$$

where F_p = S - F is the potential amount of infiltration up to the time when the rate of infiltration becomes constant (in mm of water), S is water storage capacity of the soil at the beginning of the rainfall and is the difference in water content at saturation and at permanent wilting point (mm of water) in the surface layer (the A horizon in the agricultural soils), and F is the accumulated infiltration (mm) from t = 0 to t = t_c when f = f_c. The extent to which S is utilized appears to depend on the type of vegetation present. Cropping and depletion of organic matter reduce soil porosity by the order of 10%. The reduction in the porosity in the form of the larger non-capillary channels may be more than 50%. These non-capillary channels are effective in distributing water and enable the filling of the remaining pores by the slower capillary processes. The reduction of these channels leads to a steady state infiltration condition where the water bypasses, through these channels, lumps of soil which over the duration of the rainfall will not become saturated. Holtan found from field measurements on test areas that F_p could be as low as 30% of the available pore space and up to 100% for good grass cover, Fig. 6.6.

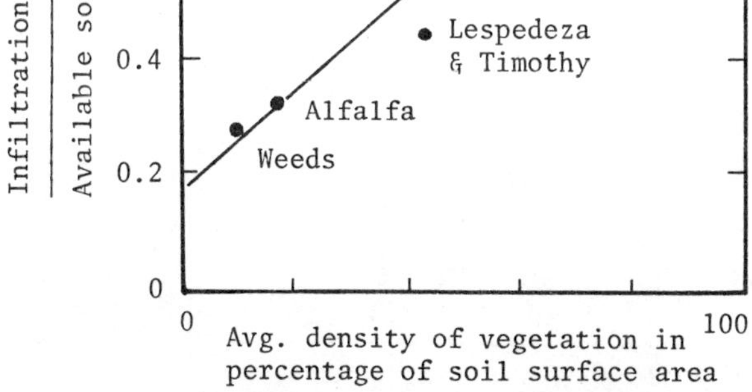

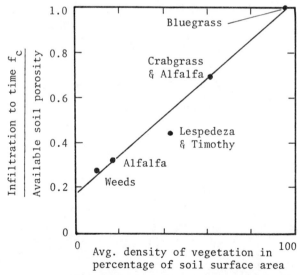

Fig. 6.6 The portion of soil porosity exhausted before infiltration becomes constant as a function of vegetative density (Holtan, 1961).

The percentage depends on the amount of ground cover by roots or stems. For his experiments Holtan found (in inch units) $n = 1.387$, $0.25 < a < 0.82$, with $\bar{a} = 0.62$.

If the available pore space (storage potential) is subdivided into the volume of capillary water and gravitational seepage flow volume, the recovery of infiltration between periods of rain may be estimated. The rate of removal of gravity water into groundwater storage can be estimated either by Darcy's law or by assuming that it takes place at f_c. The depletion of capillary and adsorbed water could be estimated using rates appropriate to evapotranspiration. Equation 6.4 also shows that if $F = S$, the rate of infiltration is equal to the infiltration capacity f_c through the control layer. Holtan (1971) modified eqn 6.4 to

$$f = GI\ ka\ S^n + f_c \qquad 6.5$$

where GI is a growth index equal to the ratio of evapotranspiration to potential evapotranspiration, and allows for the season; a is the infiltration capacity in inches or cm per hour per unit of pore space available for storage and is assumed to vary with the type of vegetation as discussed above; $k = 1$ if a is measured in in./h; S is the pore space available for storage of water in the A-horizon; and n is a function of soil texture ($n \simeq 1.4$ for silt loams). As above, S can be partitioned into gravity drainage and soil moisture available to plants. The term $ka\ S^n$ has to be determined empirically. Some values according to Holtan are shown in Table 6.2. The growth index is illustrated in Fig. 6.7.

TABLE 6.2 Estimates of Vegetative Parameter a in Eqn 6.4.

Land use or cover	Condition and rating of soil surface*			
Fallow	After row crop	0.10	After sod	0.30
Row crops	Poor	0.10	Good	0.20
Small grains	"	0.20	"	0.30
Hay (legumes)	"	0.20	"	0.40
Pasture (bunch grass)	"	0.20	"	0.40
Hay (sod)	"	0.40	"	0.60
Temporary pasture (sod)	"	0.40	"	0.60
Permanent pasture (sod)	"	0.80	"	1.00
Woods and forests	"	0.80	"	1.00

*Adjustments needed for "weeds" and "grazing".

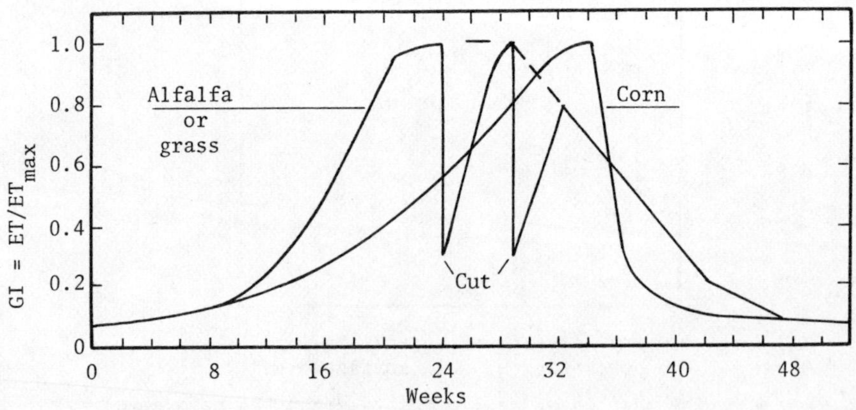

Fig. 6.7. Growth index for irrigated corn and hay, Coshocton, Ohio (Holtan, 1971).

The infiltration volume F was expressed by Kirkham and Fang (1949) as

$$F = ct^{\frac{1}{2}} + a$$

and Philip (1957) expressed F in the form of a series, which converges for all except very large values of time, as

$$F = At^{\frac{1}{2}} + Bt + Ct^{3/2} \cdots$$

where the constants are integrals and functions of θ.

6.2 Infiltration into Frozen Soils

Although a substantial fraction of the total land surface is covered by snow and is frozen over part of the year, the amount of data available on infiltration in frozen soils is very limited. The process of infiltration, in this case, depends on many additional factors. The behaviour of the infiltration rates into frozen soils is illustrated in Fig. 6.8.

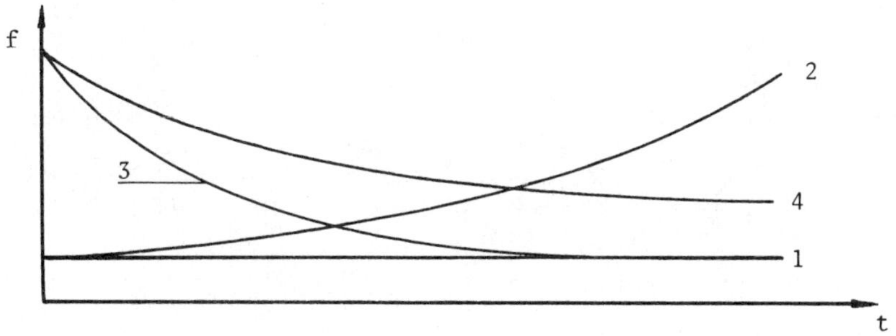

Fig. 6.8. Illustration of trends in infiltration rates into frozen soils.

Curve 1 characterizes infiltration into a soil which has been frozen in saturated condition or when an impervious ice layer has developed during a thaw and freeze cycle. The rate of infiltration is very low and any snow melt or rain has to become runoff. Curve 2 illustrates conditions when the soil has been frozen at less than field capacity. Some of the meltwater is now able to penetrate into the soil carrying heat into the pores. Slowly, as the soil warms and ice melts from the pores, the infiltration rate increases. Increases during the melting period of up to 8 times have been reported. In time the rate of infiltration will start to decrease when the high moisture content of the soil becomes the controlling factor. Curve 3 characterizes the infiltration rate into a soil which has been frozen at a low moisture content, but at the time of snowmelt the soil temperature is well below freezing. The water freezes into the pores and reduces the rate of infiltration to a very low value. Finally, curve 4 refers to conditions where the soil has been frozen at low moisture content and is at the time of snowmelt near or above freezing temperature. This condition is similar to infiltration into unfrozen soil. The ice in the pores gradually melts as the water penetrates the soil.

This multitude of functional forms of the infiltration rate implies that the yield of runoff or infiltration from a snow field cannot be described completely in terms of climatic or meteorological conditions at the time of snowmelt.

6.3 Analysis of Moisture Movement in Partially Saturated Soil

Soil contains water in liquid and vapour form. Under *isothermal* conditions vapour concentration gradients are absent and any moisture transfer is primarily in the form of a viscous flow under the capillary potential difference. Of other forces the osmotic potential may be significant, particularly in connection with plant intake of soil water. Percolation through a partially saturated soil is described by Darcy's law

$$\underset{\sim}{q} = - K \text{ grad } h = - K\nabla h \qquad 6.6$$

where q is the bulk velocity vector; $K = k\gamma/\mu$ is the hydraulic conductivity or transmission coefficient, or coefficient of permeability [L/T] and is a function of the moisture content, $\gamma = \rho g$ where ρ is the mass per unit volume and g is the gravitational acceleration, μ is the dynamic or absolute viscosity of fluid; k is the intrinsic permeability or permeability [L^2]; and grad h is the piezometric gradient. The intrinsic permeability $k = cd^2f(n)$ where c is a coefficient with values in the range of 1/150 to 1/300 with 1/180 being conventionally used, d is the grain diameter and f(n) is a function of the porosity n

$$f(n) = (n - n_o)^3/(1 - n)^2 \qquad 6.7$$

where n_o is the irreducible or ineffective porosity. The piezometric head $h = p_c/\gamma + z = \psi + z$, where $p_c/\gamma = \psi$ is the capillary potential and frequently dominates in unsaturated soils. (There are other potential contributions, e.g. osmotic , but their overall effect is small). Capillary potential varies with moisture content of the soil and the shape and size of the pores in the soil. The dependence of capillary suction on moisture content is not a single-valued function. Figure 6.9 illustrates a typical hysteresis loop.

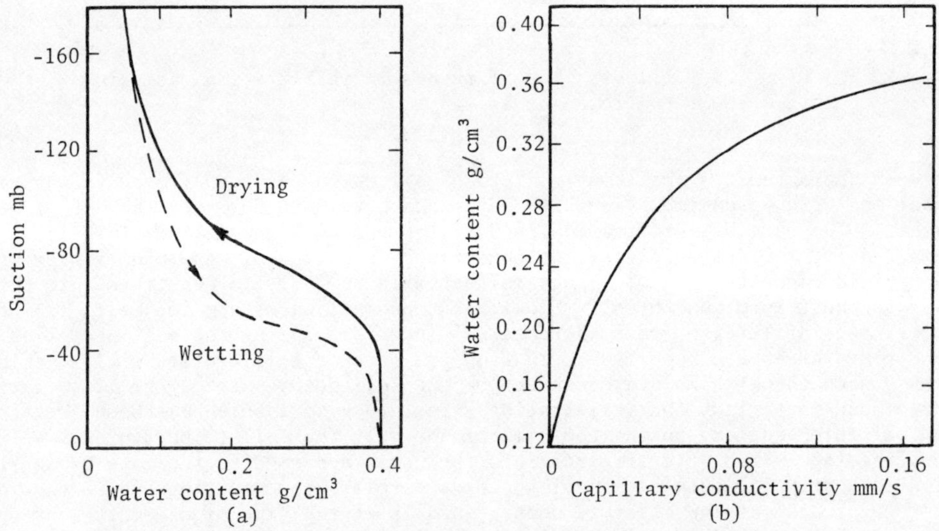

Fig. 6.9. Capillary suction and capillary conductivity as functions of moisture content. (a) Hysteresis effects on moisture-suction characteristics of a 50-500 µm sand sample, after Jackson, Reginato and van Bavel (1965). (b) After Rogers and Klute (1971).

The hysteresis is the result of instability of the air-water interface as larger pores are filled during the wetting. A secondary hysteresis arises from filling of different pores on different occasions when the moisture content is the same. A detailed discussion is given by Childs (1969). For flow, the entrapped air bubbles are part of the impervious soil matrix. With decrease of air in the soil, more pores are opened to the flow and K increases, until the K value at saturation, K_s, is reached. Experimental values by Brooks and Corey (1966) in Fig. 6.10 illustrate the variation of the intrinsic relative permeability of the wetting phase k_{rw} and of the non-wetting phase k_{rnw} (escaping air), k_r is the ratio of k at a given moisture content to that at saturation. Saturation is defined as the ratio of the volume of water in the voids to the total volume of voids, s = θ/n, where θ is the moisture content (volume of liquid per bulk volume of soil) and n is the porosity.

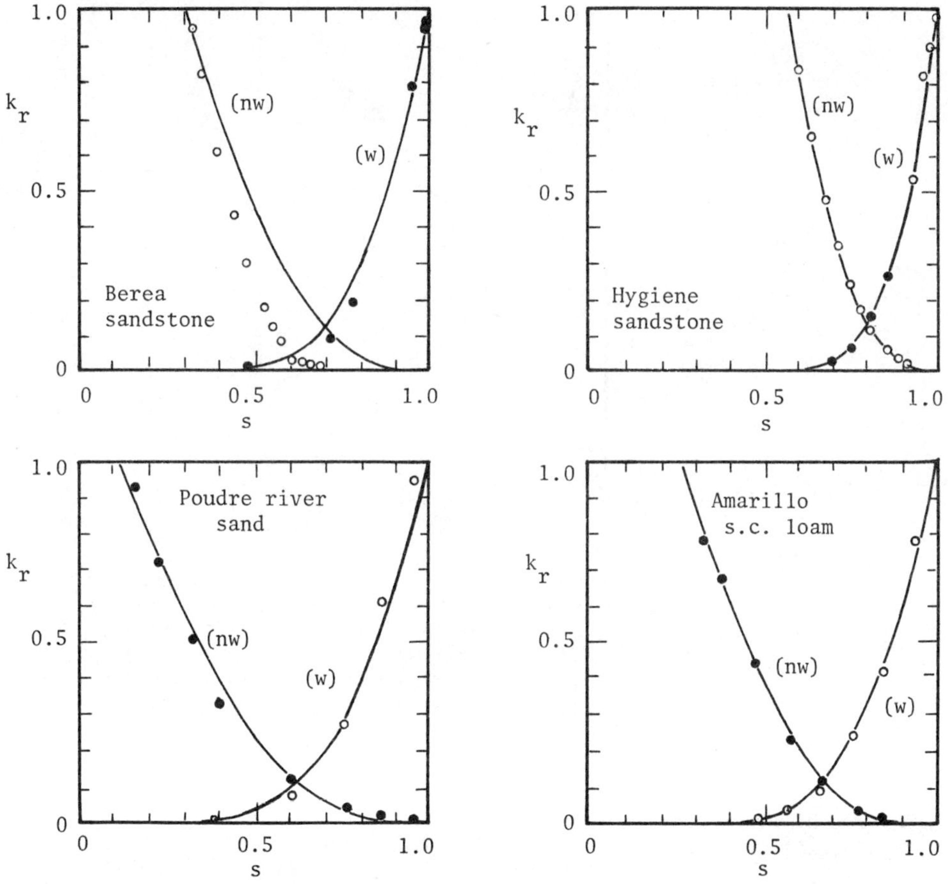

Fig. 6.10. Theoretical curves and experimental data for relative permeabilities of the non-wetting (nw) and wetting (w) phases as a function of saturation for two consolidated and two unconsolidated porous media, Brooks and Corey (1966).

The theoretical function for the wetting phase in Fig. 6.10 was expressed as

$$k_{rw} = s_e^{(2+3\lambda)/\lambda} = (p_b/p_c)^{2+3\lambda} \qquad 6.8$$

where $s_e = (s - s_0)/(1 - s_0)$, s_0 is the saturation corresponding to the residual moisture content θ_0, p_b is the bubbling pressure and is the pressure at which the first continuous air-filled passage is formed as the suction on a water filled sample is increased, p_c is the capillary pressure at a given moisture content, and λ is a parameter which, like s_0 and p_c, depends on the type of the soil and has different values for wetting and drying cycles. Figure 6.11 illustrates the variation of ψ and K with moisture content θ.

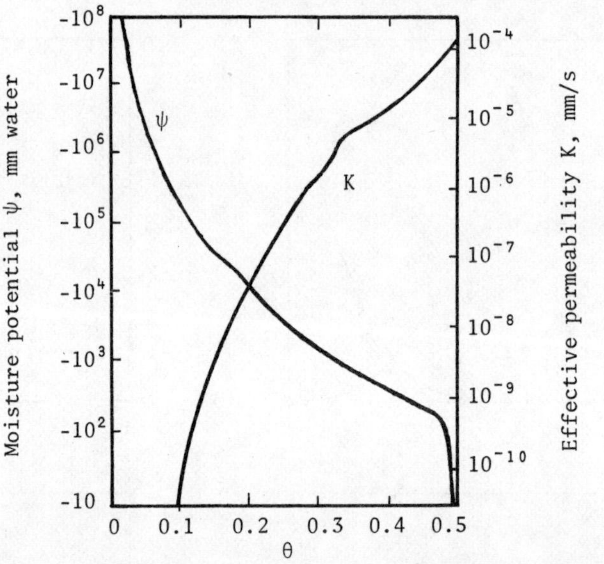

Fig. 6.11. Moisture potential, ψ, and hydraulic conductivity, K, plotted against volumetric moisture content, θ, after data by Moore (1939) for Yolo light clay ($\theta > 0.22$). ψ extended on basis of other data, K by modification of method of computation of Childs and Collis-George (1950), according to Philip (1957a).

The permeability K can be expressed as a function of ψ by eliminating θ, but because of hysteresis in both ψ and K, a single-valued function can be obtained, in general, only through approximation. Plots of $K(\psi)/K_s$ as ordinate versus ψ are usually half bell-shaped. The function $K(\psi)/K_s = 1$ when $\psi = 0$ and is asymptotic to the abscissa $[K(\psi)/K_s = 0]$ at large values of $-\psi$. The relationship between K, ψ and θ has to be obtained from tests with particular soil samples.

Irmay (1954) showed that when $K^{1/3}$ is plotted against saturation, s, experimental data define a straight line

$$(K/K_o)^{1/3} = (s - s_0)/(1 - s_0)$$

where s_0 is the ineffective saturation. The value of K(s) is almost unaffected by hysteresis and is the same for all drying and wetting branches.

Assuming that the soil matrix does not change, the density of water ρ is constant, and $p = p_c$ for the partially saturated soil, then the continuity requirement is

$$\frac{\partial u}{\partial x} + \frac{\partial v}{\partial y} + \frac{\partial w}{\partial z} = - \frac{\partial \theta}{\partial t}$$

or

$$- \nabla \cdot \underset{\sim}{q} = \frac{\partial \theta}{\partial t} \qquad 6.9$$

In swelling soils the constant soil matrix assumption could be unacceptable. Then the continuity equation has to incorporate the matrix change as a function of moisture content. Substituting of $\underset{\sim}{q}$ from eqn 6.6 yields

$$\frac{\partial \theta}{\partial t} = \nabla \cdot \{K \nabla(\psi + z)\}$$

$$= \nabla \cdot \{K \nabla(\psi) + K\underset{\sim}{k}\}$$

$$= \nabla \cdot \{K \nabla(\psi)\} + \frac{\partial K}{\partial z}$$

where

$$\nabla = \underset{\sim}{i} \frac{\partial}{\partial x} + \underset{\sim}{j} \frac{\partial}{\partial y} + \underset{\sim}{k} \frac{\partial}{\partial z}$$

This equation can be converted to an equation in θ by using $K = K(\theta)$ and $\psi = \psi(\theta)$ or to an equation in ψ by using $K = K(\psi)$ and $\theta = \theta(\psi)$, the inverse of $\psi = \psi(\theta)$. As pointed out earlier, these are not in general single-valued functions because of hysteresis, and in application an average relationship or the correct branch of the loop has to be used.

In terms of the moisture content θ

$$\frac{\partial \theta}{\partial t} = \nabla \cdot \{K(\theta) \nabla[\psi(\theta)]\} + \frac{\partial}{\partial z} [K(\theta)]$$

$$= \nabla \cdot \{K(\theta) \frac{d\psi(\theta)}{d\theta} \nabla(\theta)\} + \frac{\partial}{\partial z} [K(\theta)]$$

$$= \nabla \cdot \{D(\theta) \nabla(\theta)\} + \frac{\partial}{\partial z} [K(\theta)] \qquad 6.10$$

where

$$D(\theta) = K(\theta) \frac{d\psi(\theta)}{d\theta} \qquad 6.11$$

was introduced by Buckingham (1907) and is called the soil water diffusivity with dimensions $[L^2/T]$.

In terms of soil moisture potential ψ

$$\frac{\partial \theta}{\partial t} = \frac{d\theta(\psi)}{d\psi} \frac{\partial \psi}{\partial t}$$

and

$$\frac{d\theta}{d\psi} \frac{\partial \psi}{\partial t} = \nabla \cdot \{K(\psi) \nabla\psi\} + \frac{\partial}{\partial z} [K(\psi)] \qquad 6.12$$

Equations 6.10 and 6.12 have to be solved numerically. Closed solutions are available for simplified forms only. The parameters $K(\psi)$, $D(\theta)$ and $d\theta/d\psi$ vary markedly in most soils with water content or matrix suction head. The first simplifying assumption is to replace the hysteresis loop with a unique function. The θ-based equation is preferred for numerical solutions in unsaturated flow because changes in θ and D are two to three orders of magnitude smaller than the corresponding changes in ψ and $d\theta/d\psi$, and the round off errors are correspondingly smaller. As θ approaches saturation ($s = 1$, $\theta = n$), however, the driving potential becomes independent of θ, $D(\theta)$ tends to infinity and the numerical calculations diverge. Therefore, solutions which have to go through to saturated state have to use eqn 6.12, which for saturated conditions reduces to the Laplace equation with K constant and $d\theta/d\psi$ zero. This equation was first discussed by Richards (1931).

For one-dimensional infiltration in the vertical direction

$$\frac{\partial \theta}{\partial t} = \frac{\partial}{\partial z}\left[D(\theta)\frac{\partial \theta}{\partial z} + K(\theta)\right] \qquad 6.13$$

Figure 6.12a illustrates the zones of infiltration. If gravitational effects are negligible or K is approximately constant, the K term can be dropped. Then

$$\frac{\partial \theta}{\partial t} = \frac{\partial}{\partial z}\left[D(\theta)\frac{\partial \theta}{\partial z}\right] \qquad 6.14$$

These conditions are met approximately in the early stages of infiltration or late stages of exfiltration when the moisture content is low. The results obtained agree reasonably well with observation, except that hysteresis effects are underestimated.

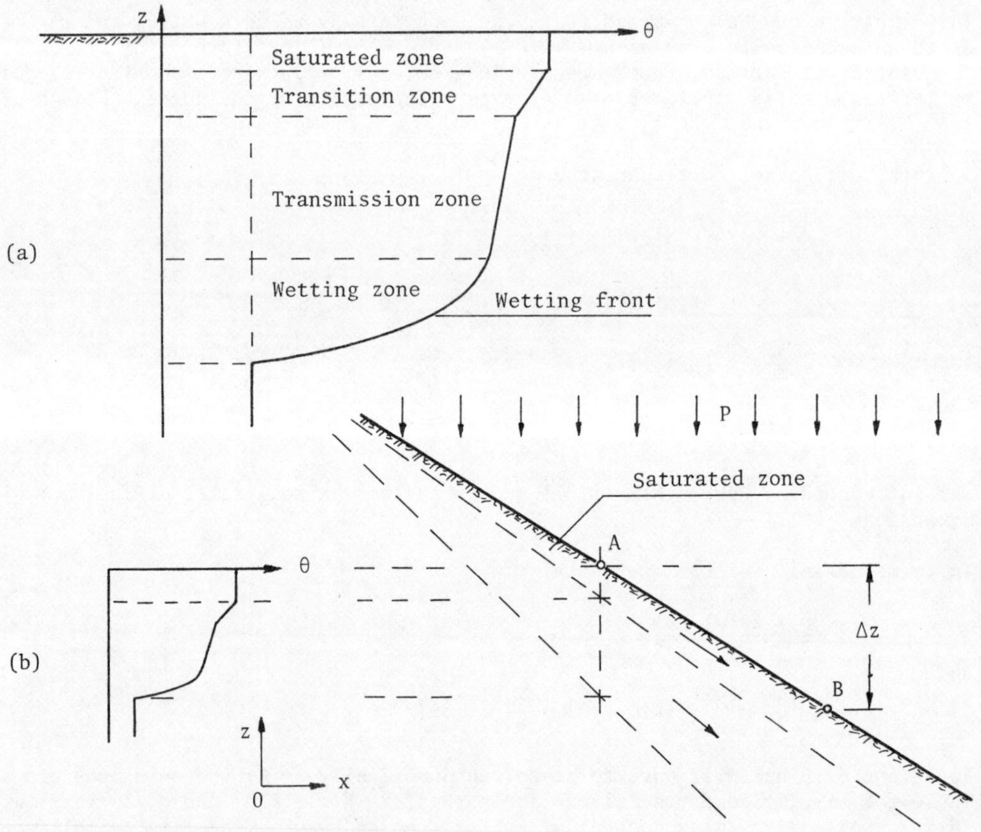

Fig. 6.12. (a) Diagrammatic illustration of the zones of infiltration. (b) Infiltration into sloping ground.

Equation 6.14 reduces to a simple diffusion equation

$$\frac{\partial\theta}{\partial t} = D\frac{\partial^2\theta}{\partial z^2} \qquad 6.15$$

when $D(\theta)$ can be replaced by a constant average value of diffusivity. This equation is treated at some length by Carslaw and Jaeger (1959) and Crank (1956).

The solution of eqn 6.15 depends on boundary conditions. If the rate of rainfall exceeds the infiltration capacity, the soil will be saturated at the surface and for a semi-infinite slab of soil the boundary conditions are

$$\theta = \begin{cases} \theta_i & z \leq 0, \quad t = 0 \\ \theta_o & z = 0, \quad t > 0 \end{cases} \qquad 6.16$$

where θ_i is the initial moisture content of the entire soil field and θ_o (= n, the porosity) is the constant moisture content at the surface during infiltration. Equation 6.15 and these boundary conditions are satisfied by

$$\frac{\theta_o - \theta}{\theta_o - \theta_i} = \mathrm{erf}\left[\frac{|z|}{2\sqrt{Dt}}\right] \qquad 6.16a$$

At a depth $|z| = 4\sqrt{Dt}$, $\theta - \theta_i$ is negligible and this length is a measure of the influence of events at the surface. The total volume of infiltration through a unit area after time t when gravity is neglected is

$$V_f = 2(\theta_o - \theta_i)\sqrt{Dt/\pi} \qquad 6.16b$$

and the rate of infiltration is given by the rate of diffusion at z = 0, namely

$$f_c = - D\frac{\partial\theta}{\partial z}\bigg|_{z=0} \qquad 6.16c$$

or

$$f_c = \left| - D\frac{\partial\theta}{\partial z} - K\right|_{z=0} \qquad 6.16d$$

if gradients of K or gravitational effects or both cannot be neglected. Hence,

$$f_c = - (\theta_o - \theta_i)\sqrt{D/\pi t} \qquad 6.16e$$

or

$$f_c = - (\theta_o - \theta_i)\sqrt{D/\pi t} - K_o \qquad 6.16f$$

where $K = K_o$ at $t = z = 0$. Elimination of t between eqns 6.16b and c yields a convenient expression for estimation of f_c when a known volume of infiltration or evaporation has occurred. Equations 6.16e and f show that the rate of infiltration decreases with time, and this is in accordance with experience. However, at t = 0 they give $f_c = \infty$. This situation can be improved by solving eqn 6.15 by separation of variables (the product method) and finding the velocity as a function of z and t. The result is

$$w = f = \Sigma_{m=0}^{m=\infty} (A \cos mz + B \sin mz)e^{-D^2mt}$$

If

$$f = f_c, \quad z = 0, \quad t = \infty$$
$$f = f_i, \quad z = 0, \quad t = 0$$

then

$$f = f_c + (f_i - f_c)e^{-D^2mt}$$

which is the Horton equation.

When infiltration or evaporation occurs at a constant rate, f_o, as when the rainfall intensity $i < f_c$ or when the potential evaporation rate PE is smaller than the rate of exfiltration from the soil, then the moisture content is

$$\theta = \frac{2f_o}{D}\,[\sqrt{(Dt/\pi)}\,\exp(-z^2/4Dt) - \tfrac{1}{2}|z|\,\mathrm{erfc}\,|z|/(2\sqrt{Dt})] \qquad 6.17$$

(Carslaw and Jaeger, 1959).

Equation 6.14 is more difficult to handle but it can be converted into an ordinary differential equation with the aid of the Boltzmann transformation

$$\eta = zt^{-\frac{1}{2}} \text{ which is here written as } \eta = \tfrac{1}{2}z(D_ot)^{-\frac{1}{2}}$$

where D_o is the diffusivity at the surface moisture content $\theta = \theta_o$, as follows:

$$\frac{\partial\theta}{\partial z} = \frac{1}{2\sqrt{D_ot}}\,\frac{d\theta}{d\eta}$$

$$\frac{\partial}{\partial z}\,[D(\theta)\,\frac{\partial\theta}{\partial z}] = \frac{1}{4D_ot}\,\frac{d}{d\eta}\,[D(\theta)\,\frac{d\theta}{d\eta}]$$

$$\frac{\partial\theta}{\partial t} = -\,\frac{z}{4\sqrt{D_ot^3}}\,\frac{d\theta}{d\eta}$$

$$-\,2\eta\,\frac{d\theta}{d\eta} = \frac{d}{d\eta}\,[\frac{D(\theta)}{D_o}\,\frac{d\theta}{d\eta}] \qquad 6.18$$

The boundary conditions of eqn 6.16 now become

$$\theta = \begin{cases} \theta_i & \eta \to -\infty \\ \theta_o & \eta = 0 \end{cases} \qquad 6.18a$$

and eqn 6.16c becomes

$$f_c = \tfrac{1}{2}\,(\frac{D_o}{t})^{\frac{1}{2}}(\theta_i - \theta_o)\frac{d\Theta}{d\eta}\Big|_{\eta=0} \qquad 6.18b$$

where $\Theta = (\theta - \theta_o)/(\theta_i - \theta_o)$ = relative water content. If gravity is significant then again a -K term has to be added on the right hand side. When a weighted mean diffusivity, $\bar{D} = (\pi/4)(d\Theta/d\eta)^2_{\eta=0}$, is introduced comparison of eqn 6.16f with eqn 6.18b (with the -K term) leads to

$$f_c = (\theta_i - \theta_o)\sqrt{D_o\bar{D}/\pi t} - K$$

Crank (1956) using numerical experiments expressed the best weighting as

$$\bar{D} = \frac{5}{3(\theta_o - \theta_i)^{5/3}}\int_{\theta_i}^{\theta_o}(\theta - \theta_i)^{2/3}D(\theta)\,d\theta$$

for infiltration and as

$$\bar{D} = \frac{1.85}{(\theta_i - \theta_o)^{1.85}} \int_{\theta_o}^{\theta_i} (\theta - \theta_o)^{0.85} D(\theta)d\theta$$

for desorption. It has been found, however, that except for extreme water contents, an expression for $D(\theta)$ of the form

$$\frac{D(\theta)}{D_o} = \exp[\beta(\theta - \theta_o)] \qquad 6.18c$$

can be used, where β is a parameter which depends on soil type. Both Crank (1956) and Gardner (1959) solved eqns 6.18 and 6.18a numerically using eqn 6.18c for a range of D_i/D_o values, where D_i is the diffusivity at the initial moisture content θ_i. The results are shown in Fig. 6.13a & b. The line $D_i/D_o = 1$ is the solution of eqn 6.16a. Substitution of eqn 6.18c into the expressions for $\bar{D}$ leads to solutions for $\bar{D}/D_o$ in terms of D_i/D_o, shown in Fig. 6.13c. Figure 6.13c shows that for any initial corresponding boundary condition $\bar{D}$ is greater for infiltration than for exfiltration, i.e., water enters soil more readily than it escapes (capillary hysteresis).

Hanks and Bowers (1962) solved eqn 6.14 numerically. They wrote it as

$$\frac{\partial\theta}{\partial t} = \frac{\partial}{\partial x} (K \frac{\partial h}{\partial x})$$

where K was obtained from eqn 6.11 and $h = \psi + z$. They used $\eta = zt^{-\frac{1}{2}}$ and converted eqn 6.18 to

$$-\frac{\eta}{2} = \frac{d}{d\theta} (D \frac{d\theta}{d\eta})$$

by multiplying through with $d\eta/d\theta$. Figure 6.14 summarizes their results. Note, that infiltration is governed by the least permeable soil layer once the wetting front has reached this layer.

In practice the surface water content does not remain constant. It increases during the rain until saturation condition is reached and decreases during evaporation. With heavy rain or high rates of evaporation the transient stage is of short duration and its effect on the results is small. If the transient stage is long incremental changes of the boundary conditions become necessary. The infinite medium assumption leads to no error as long as the penetration depth satisfies

$$4\sqrt{D_o T} << Z$$

where T is the duration of the rainfall or evaporation and Z is the total depth of the soil.

The solution of eqn 6.13 is discussed by Philip (1960, 1957b part 1 & 6). It leads to the series solution referred to earlier. Further refinements to the solution were made by Parlange (1971).

Although it is implied in eqn 6.10 that the infiltration type of flow is not limited to vertical movement of moisture, the point is emphasized here; some interesting flows in partially saturated porous media are approximately two-dimensional in vertical planes. One such flow is infiltration when the surface of the ground is sloping, as sketched in Fig. 6.12b. At any location, there is a wetting front,

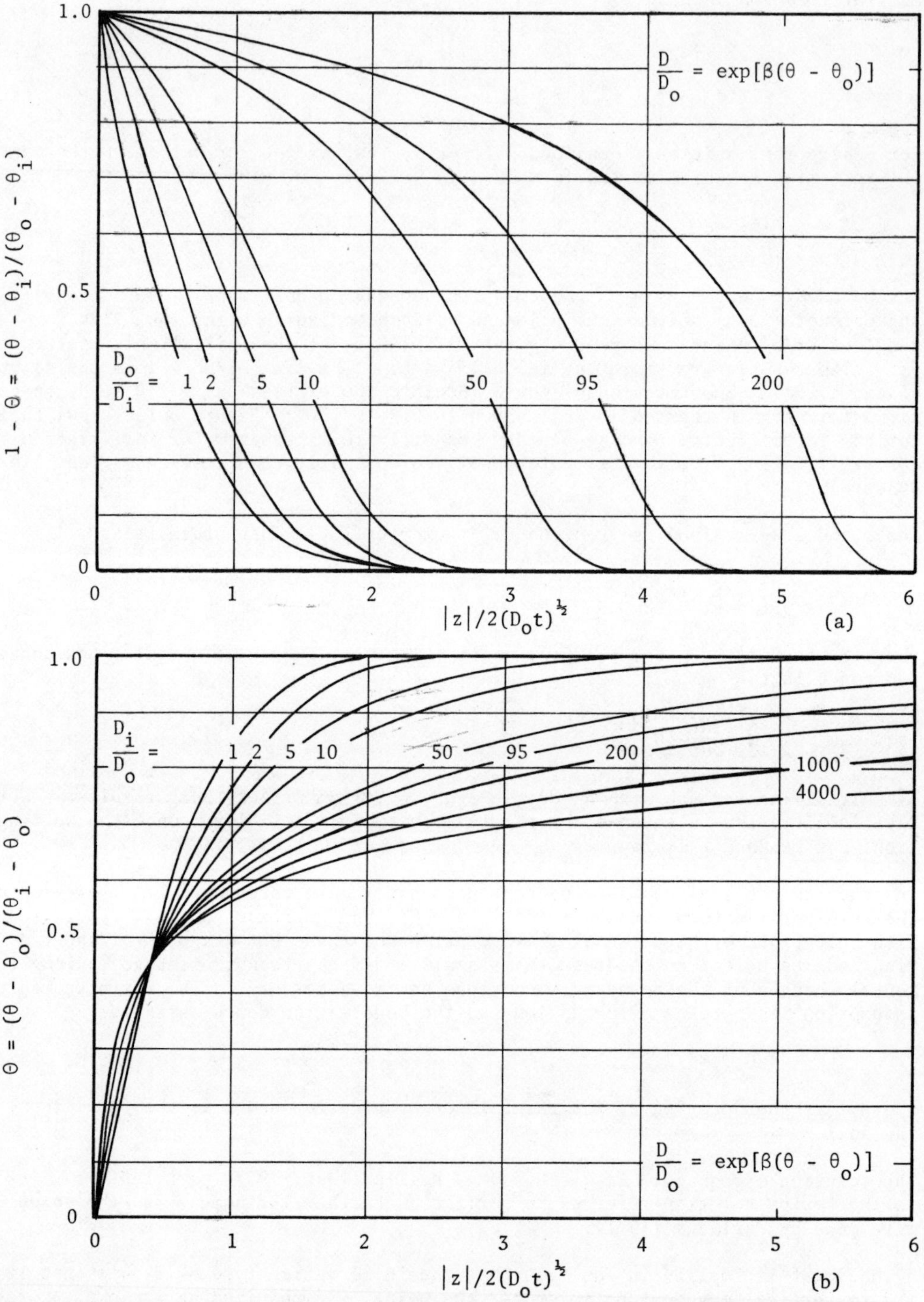

Fig. 6.13. Calculated moisture distribution in soils using $D(\theta)/D_0 = \exp[\beta(\theta-\theta_0)]$. (a) Infiltration, (b) Desorption (Crank, 1956:270-271; Gardner, 1959).

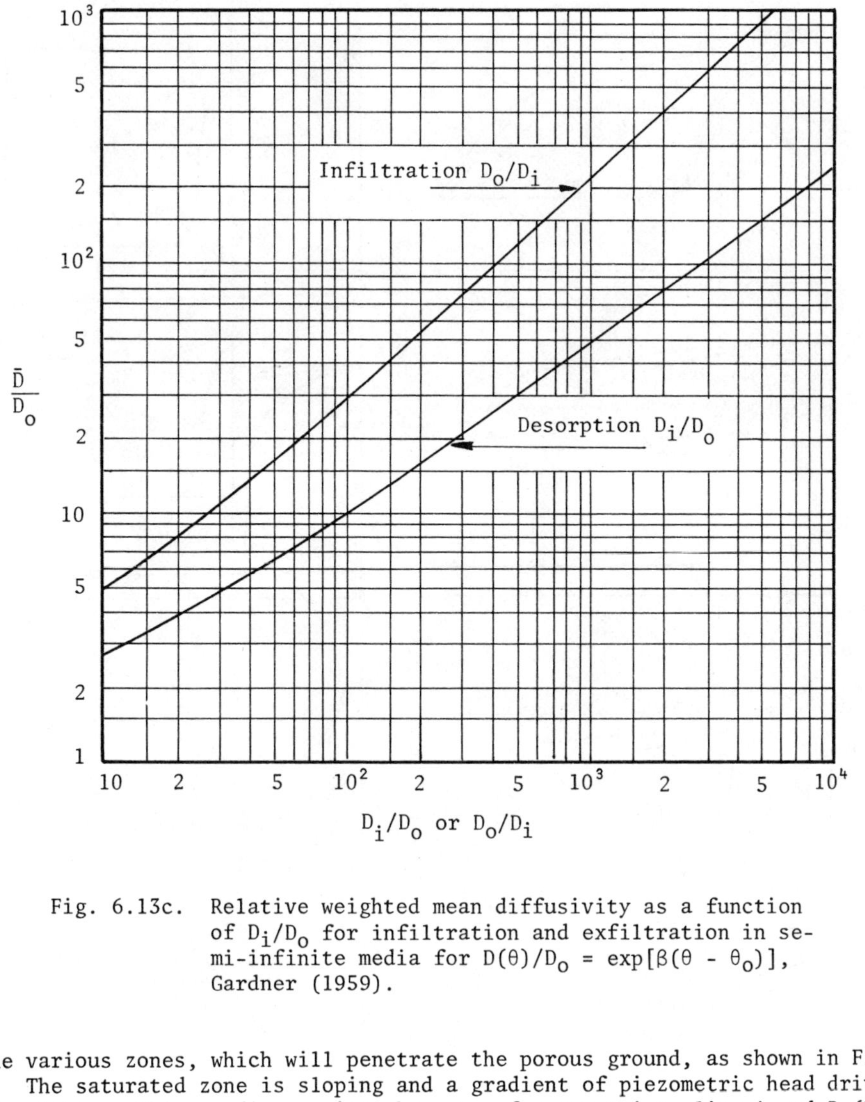

Fig. 6.13c. Relative weighted mean diffusivity as a function of D_i/D_o for infiltration and exfiltration in semi-infinite media for $D(\theta)/D_o = \exp[\beta(\theta - \theta_o)]$, Gardner (1959).

with the various zones, which will penetrate the porous ground, as shown in Fig. 6.12b. The saturated zone is sloping and a gradient of piezometric head drives the water downhill. This gradient exists because, for two points line A and B (Fig. 6.12), $p = 0$ and $\Delta h = \Delta z$. In the zones of unsaturated flow, the saturation θ will also usually have non-zero components $\partial\theta/\partial x$ and $\partial\theta/\partial z$ and they drive the unsaturated flow in a sloping direction. The flow is described by eqn 6.10 with

$$\nabla = \underset{\sim}{i} \frac{\partial}{\partial x} + \underset{\sim}{k} \frac{\partial}{\partial z}$$

The expanded form of eqn 6.10 then contains a lateral component, as well as the relationship for vertical infiltration.

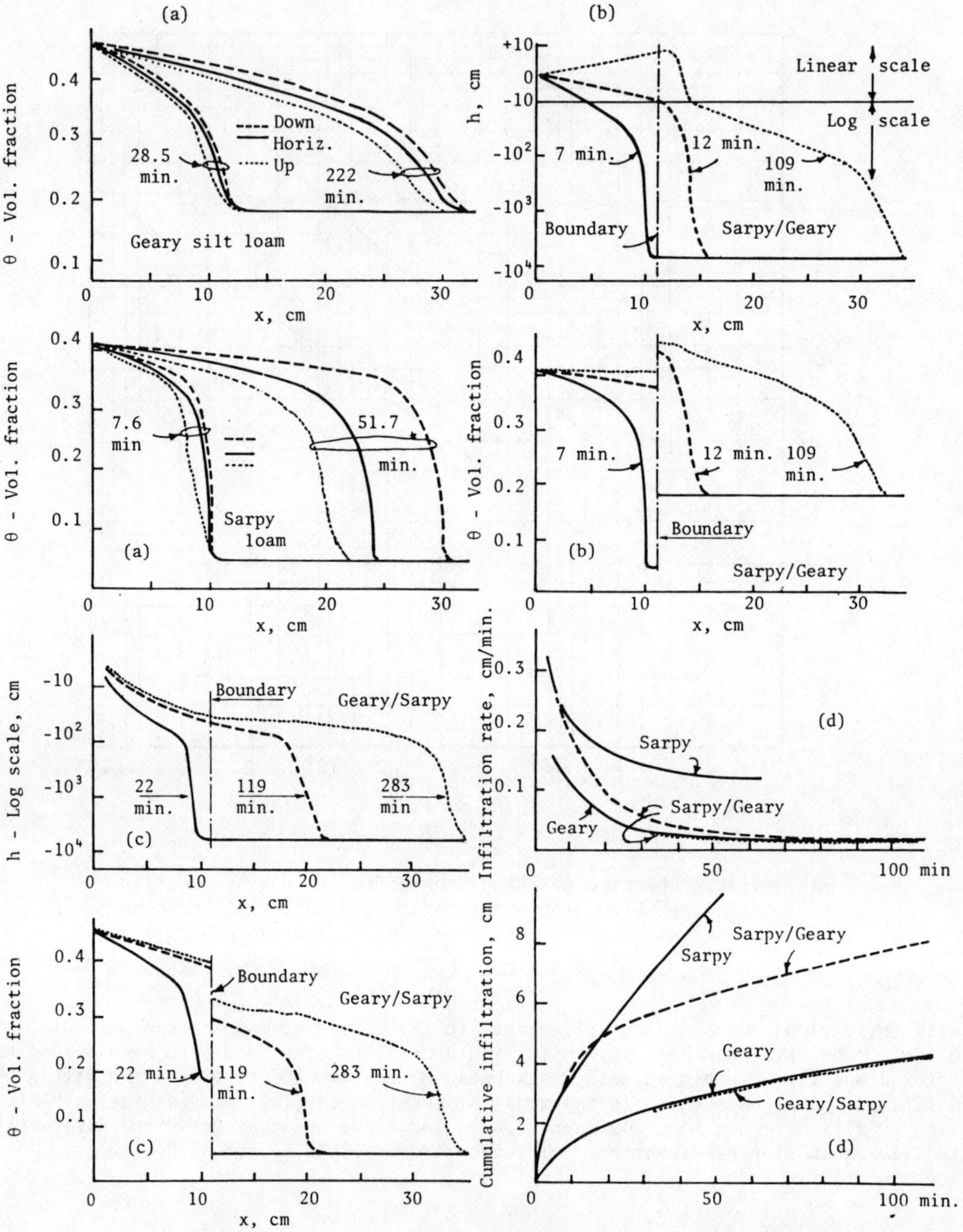

Fig. 6.14. (a) The effect of the gravitational term on θ versus x relationship in coarse textured soil (Sarpy loam, θ = 0.05 for x > 0, t = 0 and θ = 0.41 for x = 0, t > 0) and fine loam (Geary silt loam, θ = 0.184 for x > 0, t = 0 and θ = 0.460 for x = 0, t > 0). (b) The relationships of h versus x and θ versus x for layered soil of coarse over fine. Note the moisture discontinuity at the interface. (c) The same soils as in (b), but fine over coarse. (d) Influence of layer sequence on infiltration rate and cumulative infiltration. From Hanks and Bowers (1962).

Example 6.1. The surface of a loamy soil of porosity n = 0.35 and initial moisture content $\theta_i = 0.2$ is flooded. Assume constant diffusivity $D = 4.9 \times 10^{-2} cm^2/s$ and estimate the depth to which water will penetrate, the rate of infiltration, and the total volume of water that will infiltrate after 10 minutes.

From eqn 6.16c $f_c = (0.35 - 0.20)[4.9 \times 10^{-2}/(\pi \times 10 \times 60)]^{\frac{1}{2}} = 0.0765 \times 10^{-2}$ cm/s = 0.765×10^{-2} mm/s. The difference $\theta - \theta_i$ is negligible at $|z| = 4\sqrt{(Dt)} = 21.69$ cm. The volume of infiltration is given by eqn 6.16b as

$$V = 2(0.35 - 0.20)(4.9 \times 10^{-2} \times 10 \times 60/\pi)^{\frac{1}{2}} = 0.92 \text{ cm/cm}^2$$

i.e., a depth of 9.2 mm in 10 minutes.

Example 6.2. The infiltration parameters D_0 and β for a catchment are estimated to be $D_0 = 4 \times 10^{-7} m^2 s^{-1}$ and $\beta = 20$. Estimate the runoff distribution resulting from a three hour rainfall with 12 mm in the first, 6 mm in the second and 15 mm in the third hour. Porosity of the soil n = 0.4 and initial saturation is 25%.

Initial moisture content $\theta_i = 0.25 \times 0.4 = 0.1$. By eqn 6.18c $D(\theta)/D_0 = e^{\beta(\theta - \theta_0)}$. Thus, $D_i/D_o = \exp[\beta(0.1 - 0.4)]$ or $D_0/D_i = \exp(20 \times 0.3) = 403.4$ and Fig. 6.13c shows that $\bar{D}/D_i = 95$. Hence, $\bar{D}/D_0 = 95/403.4$ and $\bar{D} = 9.42 \times 10^{-8} m^2 s^{-1}$. The volume of infiltration is given by eqn 6.16b with $D = \bar{D}$ as

$$V = 2(0.4 - 0.1)(9.42 \times 10^{-8}/\pi)^{\frac{1}{2}} t^{\frac{1}{2}} = 1.04 \times 10^{-4} t^{\frac{1}{2}} \text{ m/m}^2 = 0.104 t^{\frac{1}{2}} \text{ mm}$$

t(s)	V_f(mm)	ΔV_f(mm)	Precip. (mm/Δt)	Runoff (mm/Δt)
0	0			
600	2.55	2.55	2.00	-
1200	3.60	1.05	2.00	0.95
2400	5.09	1.49	4.00	2.51
3600	6.24	1.15	4.00	2.85
4800	7.21	0.97	2.00	1.03
6000	8.06	0.85	2.00	1.15
7200	8.82	0.76	2.00	1.24
8400	9.53	0.71	5.00	4.29
9600	10.19	0.66	5.00	4.34
10800	10.81	0.62	5.00	4.38

Example 6.3. Calculate the infiltration parameters D_0 and β given a catchment with porosity n = 0.4, and

Storm 1: Initial moisture content $\theta_i = 0.15$. Precipitation intensity 25 mm/h for 2 hours. Total runoff 32 mm.

Storm 2: Initial moisture content $\theta_i = 0.25$. Precipitation intensity 14 mm/h for 3 hours. Total runoff 25.6 mm

Storm 1: $\theta_0 = n = 0.4$, $\theta_i = 0.15$, $\theta_0 - \theta_i = 0.25$, $V_f = 50 - 32 = 18$ mm
From eqn 6.16b $0.018 = 0.5\sqrt{(\bar{D} \times 7200/\pi)}$, $\bar{D} = 5.655 \times 10^{-7} m^2/s$.
Equation 6.18c gives $D_0/D_i = \exp(0.25\beta)$, $0.25\beta = \ln D_0/D_i = 2.3 \log D_0/D_i$, $\beta = 9.20 \log D_0/D_i$. Assume values of D_0/D_i. The corresponding values of $\bar{D}/D_i$ are obtained from Fig. 6.13c.

D_0/D_i	$\bar{D}/D_i$	$\bar{D}/D_0$	D_0	β	D_0/D_i	$\bar{D}/D_i$	$\bar{D}/D_0$	D_0	β
10	5	0.50	1.131×10^{-6}	9.20	100	29	0.29	1.950×10^{-6}	18.40
1000	210	0.21	2.693×10^{-6}	27.60	148	40	0.27	2.09×10^{-6}	20.00

Storm 2: $\theta_o = n = 0.4$, $\theta_i = 0.25$, $\theta_o - \theta_i = 0.15$, $V_f = 42 - 25.6 = 16.4$ mm, $0.0164 = 0.3\sqrt{(\bar{D} \times 10800/\pi)}$, $\bar{D} = 8.693 \times 10^{-7}$ m²/s, $D_o/D_i = \exp(0.15\beta)$. $0.15\beta = 2.3 \log D_o/D_i$, $\beta = 15.33 \log D_o/D_i$.

D_o/D_i	$\bar{D}/D_i$	$\bar{D}/D_o$	D_o	β	D_o/D_i	$\bar{D}/D_i$	$\bar{D}/D_o$	D_o	β
10	5	0.50	1.739×10^{-6}	15.33	1000	210	0.21	4.140×10^{-6}	46.00
100	29	0.29	2.998×10^{-6}	30.67	20	8.4	0.42	2.07×10^{-6}	20.00

Plot of D_o versus β shows that the two curves cross at $D_o \simeq 2.08\times10^{-6}$ m²/s and $\beta \simeq 2.0$.

6.4 Moisture Movement under Non-isothermal Conditions

Under *non-isothermal* conditions moisture movement can be both in the liquid and vapour form. In the vapour phase the water molecules diffuse towards the lower temperature region as a result of the thermally induced vapour concentration gradient. On reaching the cooler region the vapour may condense into liquid and create an extra difference in the soil moisture potential. This could lead to a return flow in the liquid phase. There are a number of other mechanisms for moisture transfer under temperature gradients (thermo-self-diffusion, thermo-osmotic flow, thermo-capillary flow, Knudsen flow, volume changes of entrapped air bubbles), but the total contribution of these is small. The non-isothermal conditions are generally more important in exfiltration (evaporation) and soil moisture studies than for infiltration.

The behaviour of water vapour is described with the aid of the gas law, $p = \rho RT$. In the treatments below, any changes in the soil matrix with moisture or temperature are usually assumed to be negligible. This assumption would be reasonable in most soils, but is obviously not acceptable if the soil is of the swelling type, for example, montmorillonite or bentonite. The vapour density, ρ, in the pore space can be related to the relative humidity, H_r, by $\rho = \rho_s H_r$, where ρ_s is the saturation vapour density at the same temperature. Edlefsen and Anderson (1943) showed that in the soil moisture region above the wilting point the equilibrium condition is described by the thermodynamic relationship

$$H_r = \exp(g\psi/RT) \tag{6.20}$$

or

$$g\psi = RT \ln H_r$$

where ψ is the matrix or capillary suction head and is proportional to the surface tension σ, R is the gas constant for water vapour (461.37 J kg^{-1} °K^{-1}), T is the absolute temperature in °K, and $g\psi$ is equivalent to the total specific energy or Gibbs' free energy Δf_s of the soil moisture. This energy can also be expressed in thermodynamic form as a chemical potential of water in soil, which according to Box and Taylor (1962) is also affected by the bulk density of the matrix. It was seen (Fig. 6.2) that in the capillary moisture range temperature effects on relative humidity are small.

Below the wilting point there is no capillary water and liquid water exists only in the strongly bonded adsorbed films. As much as 350 kJ kg^{-1} more of energy is required to evaporate water from an adsorbed film of a few molecular layers thick than from a free water surface. Equilibrium of water vapour with adsorbed water is described by

$$\frac{\partial(\ln e)}{\partial T} = \frac{L_e^*}{RT^2} \tag{6.21}$$

where L_e^* is the latent heat of evaporation of the adsorbed water and e is the partial vapour pressure.

In general, the movement of water and water vapour in soils under anisothermal conditions involves a diffusion process in a porous medium, coupled with evaporation and condensation in localised small areas, i.e., heat transfer. The net total moisture flux across a unit cross-sectional area normal to the direction of transport is the sum of the fluxes in vapour and liquid phase. Space here does not permit the development of the heat and moisture transport equations, but the reader will find an introduction to this topic in Raudkivi and Nguyen (1976), together with a list of references.

6.5 Measurement of Infiltration Rates

The complexities of the infiltration process and its dependence on a multitude of interacting processes make the measurement of infiltration rates and volumes under field conditions imperative. The basic methods of evaluation of the infiltration rates are
(i) the analysis of runoff hydrographs of known rainfall events on selected catchments, and
(ii) the use of infiltrometers which rely on artificial water supply to the sample area.

The analysis of runoff hydrographs is a good technique for small well-instrumented catchments. On large catchments the heterogeneity of the catchment characteristics and non-uniformity of rainfall distribution makes the method less useful. The surface runoff and precipitation are reasonably easily measured. The surface detention on well selected catchments should be small so that errors in its estimation would remain insignificant. Soil moisture distribution could be measured, for example, with neutron probes at the beginning and end of the rainfall event. The amount of interflow has to be estimated from the runoff hydrograph (See Hydrograph Analysis). This estimate is difficult to make and where interflow is not small, the errors in its estimate will make the estimate of infiltration rates unreliable. Where deep percolation is "lost", the difference between the precipitation and the sum of the terms of surface storage, runoff, change in soil moisture storage and interflow is assumed to be deep percolation. Where the groundwater table interacts with the stream flow and where groundwater may flow in or out of the catchment area, this method becomes difficult to apply. Changes in groundwater storage and the flows across the catchment boundaries are very difficult to measure.

The simplest and crudest of the methods based on the runoff hydrograph is the ϕ-index method. It is simply an estimate of the average value of infiltration over the period of the rainfall, which when added to the surface runoff makes the sum equal to the volume of precipitation. This infiltration estimate includes the surface retention and depression storage. For long periods of rainfall at rates greater than the infiltration capacity, particularly when falling on an already saturared ground, the ϕ-index becomes a good estimate of f_c. For short period storms the ϕ-index measures mainly the surface storage effect. The rainfall intensity and its distribution with time and area also affect the magnitude of the ϕ-index. The value of ϕ is higher for high intensity rains. This is due to the fact that the proportion of the area which produces the runoff throughout the rain becomes greater as the rainfall intensity increases, and not because of increased infiltration. The calculation of the ϕ-index is illustrated in Table 6.3. The rainfall excess, P_e, values obtained are plotted against the ϕ-values and a curve is drawn through the points. The actual surface runoff is determined from the measured runoff hydrograph (described later), for example, as 30 mm. Then, as seen from Fig. 6.15, the ϕ- value is 16 mm/h.

TABLE 6.3 Calculation of ϕ-index

Time from beginning of rainfall (hrs)	P mm/h	$P_e = P - \phi$ Assumed values of ϕ		
		10	15	20
1	4	-	-	-
2	14	4	-	-
3	18	8	3	-
4	30	20	15	10
5	30	20	15	10
		52	33	20

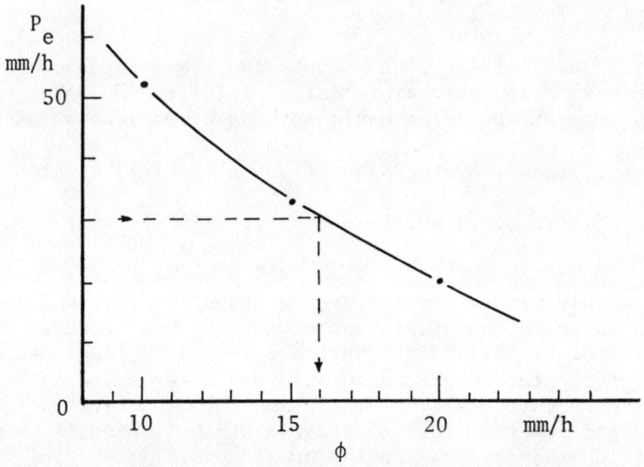

Fig. 6.15. Calculation of infiltration by the ϕ-index method.

Infiltrometers can be grouped into sprinkler and flooding types. Both use a small area of ground. The rate of supply of water and the runoff, in the case of sprinklers, are accurately measured. The difference is the rate of infiltration. This may be adjusted for evapotranspiration. The infiltrometers have a number of inherent disadvantages. It is difficult to eliminate boundary effects, natural action of raindrops cannot be reproduced, and the results are characteristic of the sample area.

The simplest of the infiltrometers is the flooding type shown in Fig. 6.16. It consists of two concentric rings which are driven about ½ m into the ground. The tubes can be of any diameter, but usually the inner ring is 230 mm (9 in.) and outer 356 mm (14 in.) in diameter. The water level is maintained constant at 5 mm above the ground. The feed rate to the inner ring is taken to give the rate of infiltration. The purpose of the infiltration from the outer ring is to confine the infiltration from the central area and prevent lateral spreading. The measured infiltration rates, because of lateral spreading and absence of the action of raindrops, can be twice that under natural field conditions. The results are of most value for flood irrigation calculations, where the field conditions are similar.

There are many different sprinkler type infiltrometers as well as plot sizes. The North American practice uses mostly the type F and type FA infiltrometers, where the latter has a lower nozzle pressure. The F-type is used with a 6 ft (1.83 m) wide and 12 ft (3.66 m) long plot and the FA-type with a 1 ft (30.5 cm) wide and 2.5 ft (76.2 cm) long plot. Both types have a row of nozzles along both sides of

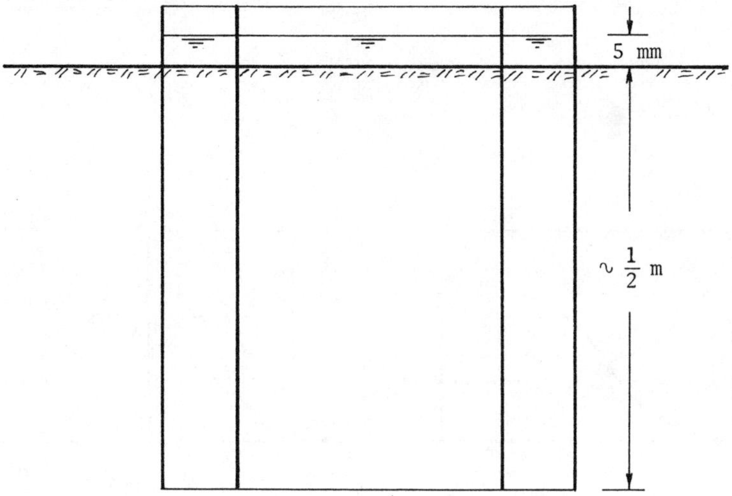

Fig. 6.16. The flooding type infiltometer.

the plot. The nozzles spray upwards and towards the plot and are designed to cover the area with a uniform simulated rainfall. The intensity of the simulated rainfall is determined by placing an impervious cover over the plot and measuring the runoff from it. The test commences when the cover is removed and continues until the runoff from the plot becomes constant. The nozzles are then shut off but runoff is measured until it stops. The moment the water disappears from the plot into the soil, the nozzles are turned on again and runoff is measured until a constant rate is reestablished. The nozzles are then turned off and the runoff measurements continued until the runoff ceases. The depth of water that has infiltrated is

$$F = P - R - S - S_d \tag{6.22}$$

where P is the depth of water supplied (precipitation), R is the depth of total runoff, S is the depth of water on the soil surface and S_d is the depth of water in depression storage. The duration of the tests is usually short enough for evaporation to be insignificant.

Figure 6.17 illustrates the surface runoff record and method of analysis. The value of S is determined from the recession part of the runoff record. During the recession period the surface storage is being reduced by continuing infiltration and by the recession runoff. The storage depth S times the area of the plot equals the recession runoff volume which is represented by the area under the recession curve. Hence, working in reverse on the time scale, the accumulated detention volume, as depth of water, can be plotted or tabulated against the rate of runoff. The infiltration rate during the recession period, f_r, has been found to decrease during the recession period and is usually estimated from

$$f_r = \frac{f_c}{Q_p} Q_r \tag{6.23}$$

The infiltration f_r is added to the observed Q_r for a revised estimate of S. During the repeat run (called the analytical run), the rate of infiltration is assumed to be constant at f_c. Therefore, the difference between $(P - t_c)$ and Q is due to surface detention S and depression storage S_d. The value of S_d can be computed from eqn 6.22.

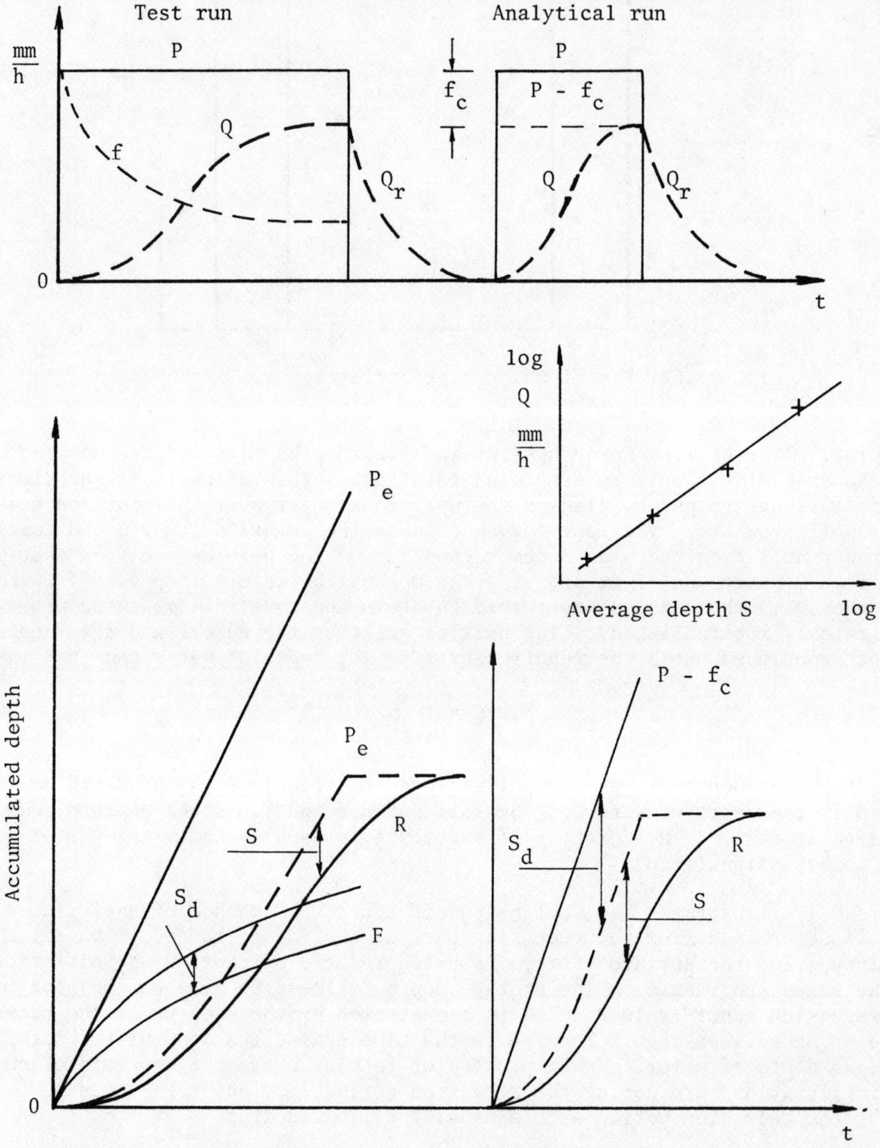

Fig. 6.17. Schematic illustration of the analysis of infiltrometer data, after Sharp and Holtan (1940, 1942).

6.6 Groundwater

Groundwater hydrology is a complex topic and includes geologic, hydrologic, hydraulic, mechanical and chemical aspects of water in the zone of saturation below the earth's surface, both in connected and unconnected pores and openings. Water in the unsaturated zone of aeration, or vadose zone, has already been discussed. The two zones together are referred to as interstitial water. Below the groundwater zone water exists only in chemical combination with rock. Although groundwater commands a literature of its own, it is still a vital part of the hydrologic cycle and cannot be divorced from hydrological studies in general.

Groundwater occurs in all formations and any of these may be an aquifer if sufficiently porous and permeable. An aquifer could be defined as a geological formation or stratum which contains water in its voids and pores and from which water could be extracted continuously in usable quantities. Alluvial gravel and sand deposits and sandstones are good examples. A geologic formation which is so impervious that for all practical purposes it inhibits movement of water is called an aquiclude. Porosity is a measure of the quantity of water that can be stored in an aquifer but not of the rate of yield of water. The yield rate is governed by pore size and the nature of interconnections. The water which can be drained from the aquifer by gravity is called the specific yield; the rest is known as specific retention. The sum of these two, expressed as percentages of volume, must equal the porosity. The specific retention of deposits with grain size, d_{90}, greater than 1 mm is of the order of 5% and increases rapidly with decreasing grain size. A 0.2 mm sand can retain about 30%. Specific yield is greatest for sand and shingle deposits and is of the order of 30%. The porosity of fine grained materials is of the order of 40% falling off to about 20% for coarse gravel. In a number of rock formations the ability to yield water is a function of the number of fractures and openings which have been enlarged by solution. Limestone, for example, has so low a permeability that its yield of water is of little practical value, yet karstic country can contain large underground rivers. There are many forms of glacial deposits, ground moraines, glacial fans, deltas and buried valleys. Buried valleys form some of the highest yielding aquifers. Table 6.4 shows some specific yield values as given by Morris and Johnson (1967).

TABLE 6.4 Specific Yield of Aquifer Materials

Aquifer material	No. of analyses	Range	Arithmetic mean
Sedimentary materials			
Sandstone (fine)	47	0.02 - 0.40	0.21
Sandstone (medium)	10	0.12 - 0.41	0.27
Siltstone	13	0.01 - 0.33	0.12
Sand (fine)	287	0.01 - 0.46	0.33
Sand (medium)	297	0.16 - 0.46	0.32
Sand (coarse)	143	0.18 - 0.43	0.30
Gravel (fine)	33	0.13 - 0.40	0.28
Gravel (medium)	13	0.17 - 0.44	0.24
Gravel (coarse)	9	0.13 - 0.25	0.21
Silt	299	0.01 - 0.39	0.20
Clay	27	0.01 - 0.18	0.06
Limestone	32	0 - 0.36	0.14
Wind-laid loess	5	0.14 - 0.22	0.18
Eolian sand	14	0.32 - 0.47	0.38
Tuff	90	0.02 - 0.47	0.21
Metamorphic rock, schist	11	0.22 - 0.33	0.26

In general, the aquifer is a porous medium formed by a network of interconnected voids, passages and fissures. The medium is usually assumed to consist of discrete particles which form the voids but frequently fractured rock is included in the above description. Voids which are not connected to others do not take part in groundwater flow. The flow through these three-dimensional passages is subject to repeated expansions, contractions, bifurcations and confluences. It is also subject to accelerations, decelerations and dissipation of energy. The velocity distribution across any passage is non-uniform but taken macroscopically, over an area large compared with that of individual pores, the discharge per unit area normal to the direction of flow is essentially uniform, i.e., the areal average velocity across a plane is uniform. It is in terms of these average velocities that groundwater flow is described. The detail of the flow in the pores is only considered in connection with the physics of dispersion in porous media. Classification of porous media is difficult and is based mainly on particle size. The bulk of usable groundwater is in porous media in which the movement of water is governed by pressure gradients, gravity forces, inertia and friction, that is, by mechanical forces. Beds of gravel, sand and fractured rock are examples of aquifers of this type. If the porous medium is clay, for example, then, in addition to the mechanical forces, molecular and electrochemical forces become important as well as the chemistry of the groundwater itself.

Almost all groundwater is derived from precipitation (meteoric water). Small amounts of connate water, which has been present in the rock since its formation and is often highly saline, is found in some areas. A third type is the juvenile water which is formed chemically within the earth and brought to the surface in intrusive rocks. Its quantities are small. Connate and juvenile water frequently contain minerals, particularly undesirable minerals, and may contaminate other groundwater.

Deep percolation to groundwater can lead to piezometric gradients and flow of groundwater in aquifers and fissures. The discharge may be in the form of springs or into rivers, lakes and the sea. The direction of flow in these aquifers and their discharge points may bear no relation to the topography of the land surface, a feature which makes water balance studies a very difficult task. Figure 6.18 illustrates some of these features and terminology.

It should be evident from this brief introduction that assessment of groundwater resources is a difficult problem. The geometry of the aquifers has been moulded by crustal movements, glacial activity, volcanic activity, and erosion and deposition. Many rock formations are extensively fractured, faulted and folded. The permeability of the aquifers varies widely and the recharge-discharge areas my impose complicated geometrical patterns. Therefore, extensive mapping of the region in question is necessary down to the bedrock. The surface detail is usually well known and documented by topographic maps. The subsurface information has to be obtained from field surveys. These surveys employ a great variety of techniques, such as seismic surveys, magnetic surveys, drilling, electrical resistivity measurements, γ-ray logging, use of isotope tracers, and study of the variation of tritium and oxygen-18 content. In addition, the isopach maps are used which show the equal thickness of the aquifer contour lines.

The flow of groundwater has to satisfy the continuity requirement and the equations of motion. The latter are usually in the form of the Navier-Stokes equations because turbulent flow of groundwater occurs only in large fissures and at very steep piezometric gradients, such as may arise near a well. For isothermal conditions the Navier-Stokes equations in the x-direction, when the z-axis of the Cartesian coordinate system is vertical, is

$$\frac{Du}{Dt} = \frac{\partial u}{\partial t} + u\frac{\partial u}{\partial x} + v\frac{\partial u}{\partial y} + w\frac{\partial u}{\partial z} = -\frac{1}{\rho}\frac{\partial}{\partial x}(p + \rho gz) + \nu\nabla^2 u \qquad 6.24$$

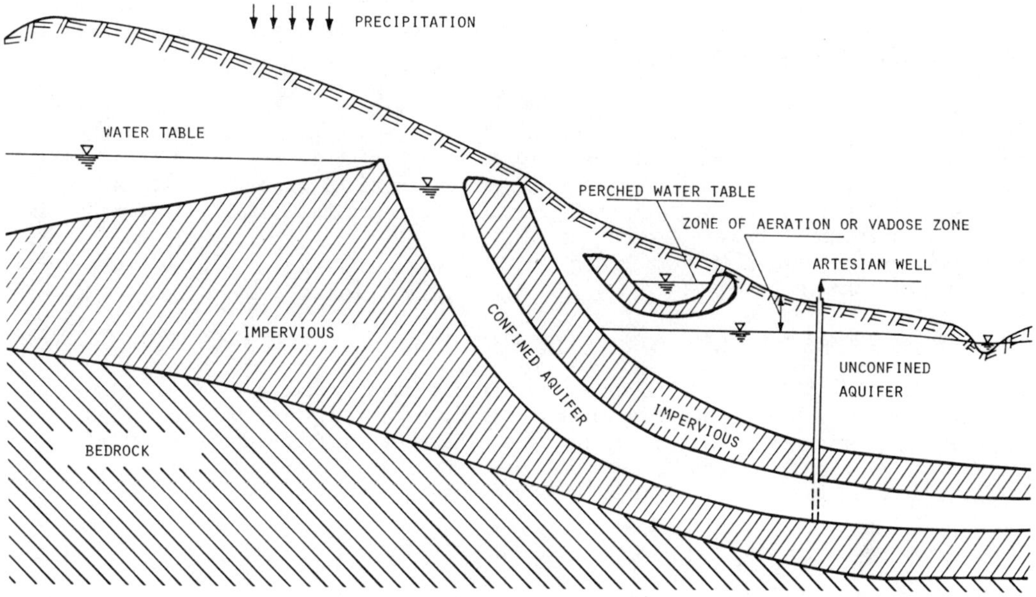

Fig. 6.18. Illustration of some of the groundwater terminology.

and similar equations hold in the y and z directions. The majority of groundwater flow problems are described in terms of the Darcy law $\underset{\sim}{q} = - K \text{ grad } h$ (eqn 6.6) or $u = - K\, \partial h/\partial x$, etc., where the piezometric head h is defined by $h = p/\gamma + z$. If now a quantity ϕ, the velocity potential, is defined by

$$\phi = - Kh + \text{const} \tag{6.25}$$

and K is a constant throughout the medium, then

$$\underset{\sim}{q} = - \text{grad } \phi \tag{6.26}$$

Thus, when the Darcy law applies, there is an analogy between the flow in a porous medium and the potential flow of classical hydrodynamics. This analogy is extensively exploited in solutions of groundwater flow problems (for introduction see Raudkivi and Callander, 1976). It must be borne in mind, however, that not all groundwater flows satisfy the Darcy law. Darcy's law may not be valid even in laminar flows. Non-Darcian behaviour has been reported in clay soils at very low Reynolds numbers and, of course, when the Reynolds number exceeds one (or ten, approximately). As the pore size or the piezometric gradient or both increase, the linear relationship between the velocity and piezometric gradient no longer holds. This non-linear relationship is frequently encountered in practice at the vicinity of well screens or filters.

The continuity equation for flow through a porous medium is readily written down as

$$\left[\frac{\partial}{\partial x}(\rho u) + \frac{\partial}{\partial y}(\rho v) + \frac{\partial}{\partial z}(\rho w)\right] = - \frac{\partial}{\partial t}(n\rho) \tag{6.27}$$

where n is the porosity of the medium. The right-hand side of this equation, for a unit control volume, is equal to $n\,\partial\rho/\partial t + \rho\,\partial n/\partial t$. The usual assumption at this stage is to assume an elastic aquifer in which the porosity varies with changes in total stress. The elastic aquifer concept was introduced to the groundwater literature by Jacob (1950) and is essentially the same as that used by Biot (1941) in connection with consolidation of soils.

The total pressure is composed of the pressure in water, p, (with an appropriate accounting for the atmospheric pressure) and that caused by the weight of the soil and overburden. If the contact area between the grains is expressed as a fraction, m, of the total cross-sectional area then, after Terzaghi, the total stress in soil is $\sigma = (1 - m)p + \sigma_s$, where σ_s is the pressure from grain to grain. Terzaghi introduced the effective stress $\sigma_e = m\sigma_s$ and put $(1 - m) \simeq p$. Thus, the total stress, σ, caused by the weight is balanced by the effective stress, σ_e, in the soil matrix and the pressure in pore space, i.e.

$$\sigma = \sigma_e + p \tag{6.28}$$

The relationship between the porosity, n, or the voids ratio $e = n/(1 - n)$, and the effective stress, σ_e, has so far eluded analytical formulation but in the light of experimental evidence it has been suggested that for small changes in effective stress $e = e_0 - \alpha_s\sigma_e$, where α_s is a coefficient of soil compressibility (not of solids) and e_0 is a constant. Then $\partial e/\partial\sigma_e = -\alpha_s$ or

$$\frac{\partial e}{\partial t} = -\alpha_s\frac{\partial\sigma_e}{\partial t} = -\alpha_s\left(\frac{\partial\sigma}{\partial t} - \frac{\partial p}{\partial t}\right) = \frac{\partial}{\partial t}\left(\frac{n}{n - 1}\right) = \frac{1}{(1 - n)^2}\frac{\partial n}{\partial t} \tag{6.29}$$

which implies that e is a function of σ_e only.

If the density of water is a function of pressure only, $\rho = f(p)$, then from the definition of the modulus of elasticity

$$E_w = \frac{1}{\beta} = -\frac{dp}{dV/V} = -\frac{dp}{d\rho/\rho}$$

and

$$\frac{\partial\rho}{\partial t} = -\beta\rho\frac{\partial p}{\partial t} \tag{6.30}$$

Introducing $\partial n/\partial t$ and $\partial\rho/\partial t$ from eqns 6.29 and 6.30 respectively, into eqn 6.27 and observing that the load usually remains constant ($\partial\sigma/\partial t = 0$), yields for the right hand side

$$-\rho[n\beta + (1 - n)^2\alpha_s]\partial p/\partial t$$

However, if the compressibility is related to an arbitrary bulk volume element V, then

$$E_s = \frac{1}{\alpha} = -\frac{d\sigma_e}{dV/V} \tag{6.31}$$

where V is composed of the volume of solids, V_s, and of water, V_w, and $V = V_s + V_w$. If it is assumed that the volume of solids $V_s = (1 - n)V = V/(1 + e)$ remains constant during consolidation (only porosity changes) and the volume element remains essentially stationary during the consolidation, i.e. fixed in space (not an integral part of an expanding or contracting aquifer) then $\alpha\partial\sigma_e = -\partial V/V$ or

$$\alpha\frac{\partial\sigma_e}{\partial t} = \frac{1}{V}\frac{\partial t}{\partial V}$$

Since $V_s = (1 - n)V$ is assumed to remain constant

$$(1 - n)\frac{\partial V}{\partial t} - V\frac{\partial n}{\partial t} = 0$$

Substituting $-V\,\alpha\,\partial\sigma_e/\partial t$ for $\partial V/\partial t$ and $(\partial\sigma/\partial t - \partial p/\partial t)$ for $\partial\sigma_e/\partial t$ and again assuming that $\partial\sigma/\partial t = 0$ yields

$$\frac{\partial n}{\partial t} = \alpha(1 - n)\frac{\partial p}{\partial t}$$

Substituting this and eqn 6.30 into eqn 6.27 leads to

$$\frac{\partial}{\partial x}(\rho u) + \frac{\partial}{\partial y}(\rho v) + \frac{\partial}{\partial z}(\rho w) = -\rho[\beta n + \alpha(1 - n)]\frac{\partial p}{\partial t} \qquad 6.32$$

Jacob assumed that under load the control volume compresses in height by $d(\Delta z)$. This yields for right-hand side of eqn 6.27, $-\rho(\beta n + \alpha)\partial p/\partial t$ or $-\rho g\beta(n + \alpha/\beta)\partial h/\partial t$. This is true for the two-dimensional case when $\rho \simeq$ constant. Then, the left-hand side of eqn 6.27 is

$$b\left(\frac{\partial u}{\partial x} + \frac{\partial v}{\partial y}\right) + \int_0^b \frac{\partial w}{\partial z}\,dz$$

where b is the thickness of the aquifer.

The integral is equal to $n\,\partial b/\partial t$ and $\partial b/\partial t = \alpha b\,\partial p/\partial t$ is the rate of expansion or contraction of the aquifer. Thus the equation becomes

$$\frac{\partial}{\partial x}(\rho u) + \frac{\partial}{\partial y}(\rho v) = -\rho(\alpha + n\beta)\frac{\partial p}{\partial t}$$

Jacob effectively assumed that $w = 0$. The differences in these models are of theoretical nature. The elastic soil model is only a reasonable approximation for well consolidated soils, when the variations in stresses are limited to small values, and for rocks but it has serious shortcomings when applied to unconsolidated soils and clays. Deformation of soils under load is generally the result of many small irreversible movements between the grains rather than an elastic compression. It is more a problem of dislocation. Consolidation of granular soils is also a function of the loading history. In clays there may be a considerable time lag between application of the load and consolidation. In general, an increase in load causes an immediate increase in pressure p, a slow increase in σ_e and a slow decrease in porosity as water drains from the pore space. Some models also consider the movement of the porous matrix (due to consolidation) but in most groundwater problems this effect on the groundwater velocity vector is very small.

Jacob introduced $p = \rho gh - \rho gz$ and Darcy's law $q = -K$ grad h, where K is a constant and not a function of ρ, and obtained

$$-\rho K\left(\frac{\partial^2 h}{\partial x^2} + \frac{\partial^2 h}{\partial y^2} + \frac{\partial^2 h}{\partial z^2}\right) - \beta\rho^2 gK\left[\left(\frac{\partial h}{\partial x}\right)^2 + \left(\frac{\partial h}{\partial y}\right)^2 + \left(\frac{\partial h}{\partial z}\right)^2 - \frac{\partial h}{\partial z}\right] = -\beta\rho^2 g\left(n + \frac{\alpha}{\beta}\right)\frac{\partial h}{\partial t}$$

Generally, the second group of terms on the left-hand side is small compared with the first group (for details of the development see e.g. Raudkivi and Callander, 1976). The right-hand side should actually be $-\rho^2 g[\beta n + (1 - n)\alpha]\partial h/\partial t$. Thus, with good approximation the continuity equation reduces to

$$\nabla^2 h = \{\frac{\rho g}{K}[\beta n + (1 - n)\alpha]\}\frac{\partial h}{\partial t} \qquad 6.33$$

The group of terms

$$\rho g(\beta n + \alpha) \qquad \text{or} \qquad \rho g[\beta n + (1 - n)\alpha] = S_s$$

is known as the specific storage with dimension 1/L. The term $\rho g\alpha(1 - n)$ gives the yield of water from storage as a result of compression of the porous medium and $\rho g\beta n$ is the yield resulting from a corresponding expansion of water.

It is also noted here that there are theoretical problems with the generalization of Darcy's law through a moving matrix of solids but these are of little practical significance. The major problems with the solution of practical problems are the definition of the aquifer geometry, the boundary conditions, and the determination of the values of the parameters involved.

Defining $S = S_s b$ as the storage coefficient and $T = bk$ as the transmissivity of the aquifer, where b is the thickness of the aquifer, then

$$\nabla^2 h = \frac{S}{T}\frac{\partial h}{\partial t} \qquad 6.34$$

When the flow is steady and two-dimensional

$$\nabla^2 h = 0 \qquad 6.35$$

or when using ϕ Kh + C then $\nabla^2\phi = 0$. If the thickness of the aquifer is constant (or approximately constant) then after introduction of the boundary conditions the Laplace equation is retained and analytical solutions to a number of flow problems are readily obtained. Equation 6.35 forms the basis of most of the treatments of well hydraulics. If the aquifer is unconfined on an impervious bed at a slope α, then the piezometric head, h, is $h = z \cos \alpha + x \sin \alpha$, where z is the depth of the aquifer normal to the base. This leads to $\nabla^2(\frac{1}{2}Kz^2) + (\partial/\partial x)(Kz \tan \alpha) = 0$, and if the slope is small

$$\nabla^2(\tfrac{1}{2}Kz^2) \simeq 0 \qquad 6.36$$

In unconfined unsteady flow the storage coefficient S is the sum of the elastic storage coefficient, S_e, and the storage coefficient due to partial drainage of voids, S_v, i.e., $S = S_e + S_v$. Generally, S_e is small compared with S_v. The partial drainage is from saturation down to field capacity of the soil.

Aquifers are often neither confined nor unconfined but leaky. This means that the confining layers are not entirely impervious. There may be several layers of aquifers separated from each other by layers whose permeabilities are much less than those of the aquifers but not zero. With symbols as defined in Fig. 6.19, and assuming that u and v do not vary with z, w is negligible and K/K_1 is large (i.e., the flow through the separating stratum is approximately vertical), it follows that

$$- b\frac{\partial u}{\partial x} - b\frac{\partial v}{\partial y} + \frac{K_1}{b_1}(H - h) = 0$$

or after introduction of the Darcy law

$$bK\nabla^2 h - (K_1/b_1)(h - H) = 0$$

For unsteady flow this becomes

$$\nabla^2 h - \frac{K_1}{Kbb_1}(h - H) = \frac{S}{T}\frac{\partial h}{\partial t}$$

After substitution of $s = H - h$ and $B = (Kbb_1/K_1)^{\frac{1}{2}}$ (known as the leakage factor) this becomes

$$\nabla^2 s = \frac{s}{B^2} = \frac{S}{T}\frac{\partial s}{\partial t} \qquad 6.37$$

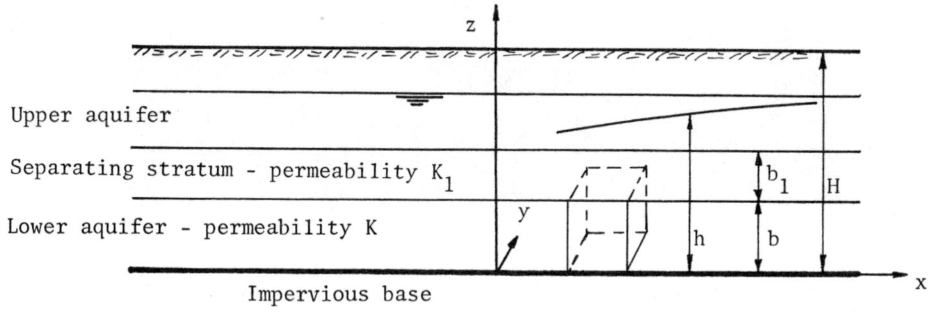

Fig. 6.19. Diagram of a leaky aquifer. H is the piezometric head in the upper and h in the lower aquifer.

These equations form the basis for analysis of groundwater flow. Equations 6.34 and 6.37 also form the starting point of well tests and determination of the formation constants S and T. The problems to be solved as well as the problems of analysis are many and varied. There are the problems associated with the extraction of groundwater by wells or by galleries from confined, unconfined and leaky aquifers, all of which may have complex geometrical shapes and boundary conditions, as well as those associated with recharging of groundwater. The planner needs to know the safe yield of the aquifer. The safe (meaning continuous) yield is governed by the amount of water available and by the nature and location of the aquifer. The rate of supply available depends on the natural or artificial recharge. If too much water is extracted the pressure in the aquifer may fall lower than that required to support the weight of the overburden and the matrix structure of the porous medium may collapse. Such a consolidation of the aquifer means a permanent loss of water bearing capacity. Excessive lowering of pressure in the aquifer may also lead to seawater intrusion and to contamination of the aquifer if it is near the sea. If groundwater reservoirs are to be used for storage of water then the usable storage volume has to be estimated. Numerical modelling of groundwater reservoirs is discussed by Kjaran (1976). The methods of analysis of groundwater problems form a large topic on their own. For introduction the reader is referred to Raudkivi and Callander (1976). Further reading: DeWiest (1965), Davis and DeWiest (1966), Walton (1970), Bear (1972) and Fried (1975). The last one deals with problems of groundwater pollution.

Chapter 7
RUNOFF

When the rate of rainfall or snowmelt exceeds the interception requirements and the rate of infiltration, water starts to accumulate on the surface. At first the excess water collects into the small depressions and hollows, until the surface detention requirements are satisfied. After that water begins to move down the slopes as a thin film and tiny streams. This early stage of overland flow is greatly influenced by surface tension and friction forces. With continuing rainfall the depth of surface detention and the rate of overland flow increase, Fig. 7.1, but the paths of the small streams on the surface of the catchment are still tortuous and full of obstructions. Every small obstruction causes a delay until the upstream level has

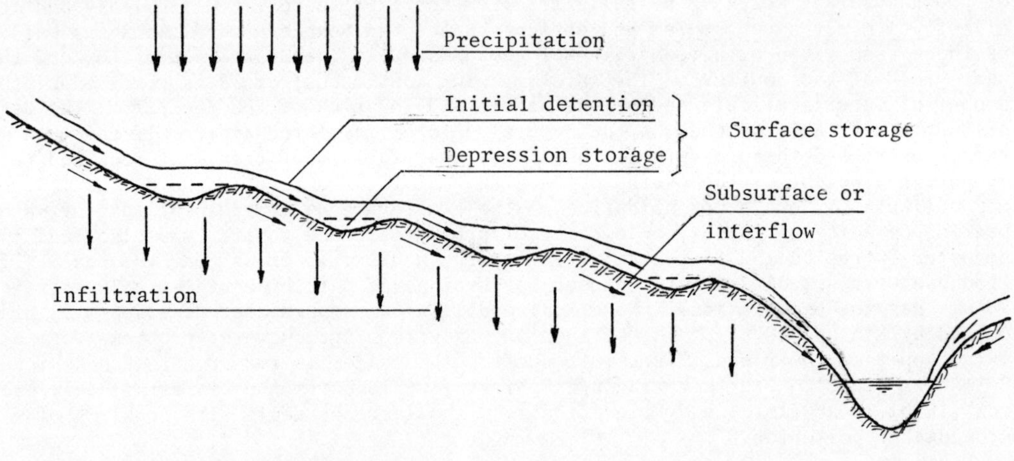

Fig. 7.1. Surface runoff.

risen to overflow the obstacle or to wash it away. On release a small wave speeds downstream and merges with another little rivulet. The merging of more and more of these little streams culminates in the river which drains the whole catchment in question.

This runoff process contains three elements. These are (a) overland flow as a thin sheet of water, (b) small stream flow, and (c) river flow. All of these have to obey the equations of continuity and momentum. The analysis of idealized flow problems yields useful information for understanding the runoff process. The complexities of the actual catchments, however, rule out this approach as a means of solving practical problems. For detailed discussion of the solutions based on hydrodynamic concepts the reader is referred to Iwagaki (1955), Henderson and Wooding (1964), Wooding (1965, 1966) and to the summarizing discussion by Eagleson (1970). A brief outline of this approach is given in Section 7.4.

In this discussion of runoff the word catchment is associated with surface runoff and the name drainage basin with the total discharge past a gauging station. It is important to realize that the boundaries of the surface and subsurface drainage systems need not be coincident in plan. The discharge of groundwater may be into or out of the catchment area, or both, depending on the geology of the region. The catchment is an area within a closed curve lying in the land surface, such that all the surface runoff is produced by precipitation falling on this area, and no other, and leaves the area as a concentrated stream at one point at the catchment boundary.

The manner in which rainfall is partitioned into the various forms of runoff is illustrated by the diagram in Fig. 7.2. The total discharge at the outlet from the

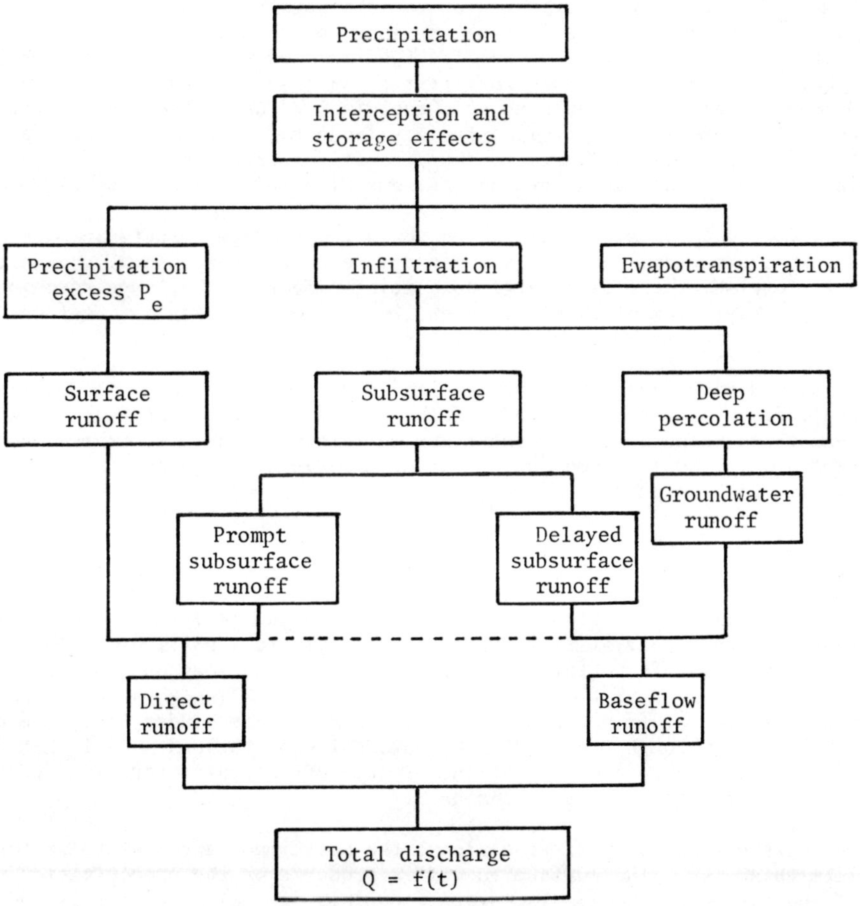

Fig. 7.2. Diagrammatic presentation of the runoff process.

drainage basin appears as Q = f(t) and is known as the *discharge hydrograph*. It is usually subdivided into three components. These are the surface and subsurface runoff, which together constitute the *runoff hydrograph*, and the groundwater runoff which maintains the dry weather runoff or *baseflow*. Not all streams have a baseflow component. If the groundwater table is lower than the river level the flow of seepage water is from the stream. Such streams usually run dry soon after the rainfall has ceased and are known as ephemeral streams. The route taken by the water, in terms of Fig. 7.2, depends on the nature of the drainage basin. On clay or rock surfaces with appreciable slopes most of the runoff will be in the form of surface runoff because the rate of infiltration is very small. At the other extreme no surface runoff occurs on a volcanic catchment covered by a deep layer of coarse pumice sand. Here the direct runoff is entirely a subsurface runoff.

The nature of the hydrograph on any stream is determined by two sets of factors:
(a) the physical characteristics of the catchment, and
(b) the climatic factors.

7.1 Physical Characteristics of the Catchment

The primary physical characteristics of the drainage basin are its area, shape, elevation, slope, orientation, soil type, drainage or channel system, water storage capability and vegetal cover. The effect of most of these factors is obvious. Soil type, for example, controls infiltration, surface storage, soil water and groundwater storage. Linked with soil type is land usage and plant cover. The combined effect of all the physical factors leads to the hydrological classification of catchments as small and large. The size of the area alone, however, is not the deciding factor. Two catchments of the same size may behave quite differently.

A *hydrologically large catchment* is one where the storage effects in water courses and in lakes, etc., dominate. This storage makes the response of the catchment to rainfall sluggish; there is a large equalizing effect. The large catchment is not sensitive to variation of rainfall intensity and to land usage. Most catchments large in size, with major rivers and lakes, fall into this category.

Small catchments are controlled by overland flow and land usage, slope, etc., have a strong influence on the magnitude of the peak discharge. Storage effects are small and the catchment is very sensitive to rainfall, that is, it responds quickly. Note that a quite small swampy catchment may behave hydrologically as a large one and have sluggish response.

The description of the physical characteristics of a drainage basin is not easily formulated. The study of the interaction of climate and geology forms a branch of science in its own right, *geomorphology*. The physical structure of the surface runoff pattern is open to view, and there are certain similarities in the land forms, which have evolved under similar geologic and climatic conditions. These similarities make it possible to introduce certain classifications and to introduce some formality into the description of catchment surfaces. The physical characteristics of the subsurface runoff pattern, on the other hand, are hidden from view and can be revealed only through very costly geophysical explorations. All that can be attempted here is to give a brief outline of some of the parameters and terminology in use for description of the catchment.

For the surface runoff, apart from slope, the most important feature is the system of stream channels. The surface runoff is produced by the precipitation over the area of the catchment and is discharged, together with all the solutes and sediments which it may pick up on its way, across the boundary of the catchment. In detail, the conditions vary with time because of changes in land surface, for example, tillage, changes in vegetation, changes in soil moisture, etc. Therefore, any description has to be confined to features which are more premanent, that is, describe a semi-equilibrium state. The erosion pattern of a catchment for a given climate and geology is characterized by a network of channels, some of which may be perennial streams. Geomorphologists have given names to these stream patterns, and Fig. 7.3 illustrates some of these classifications.

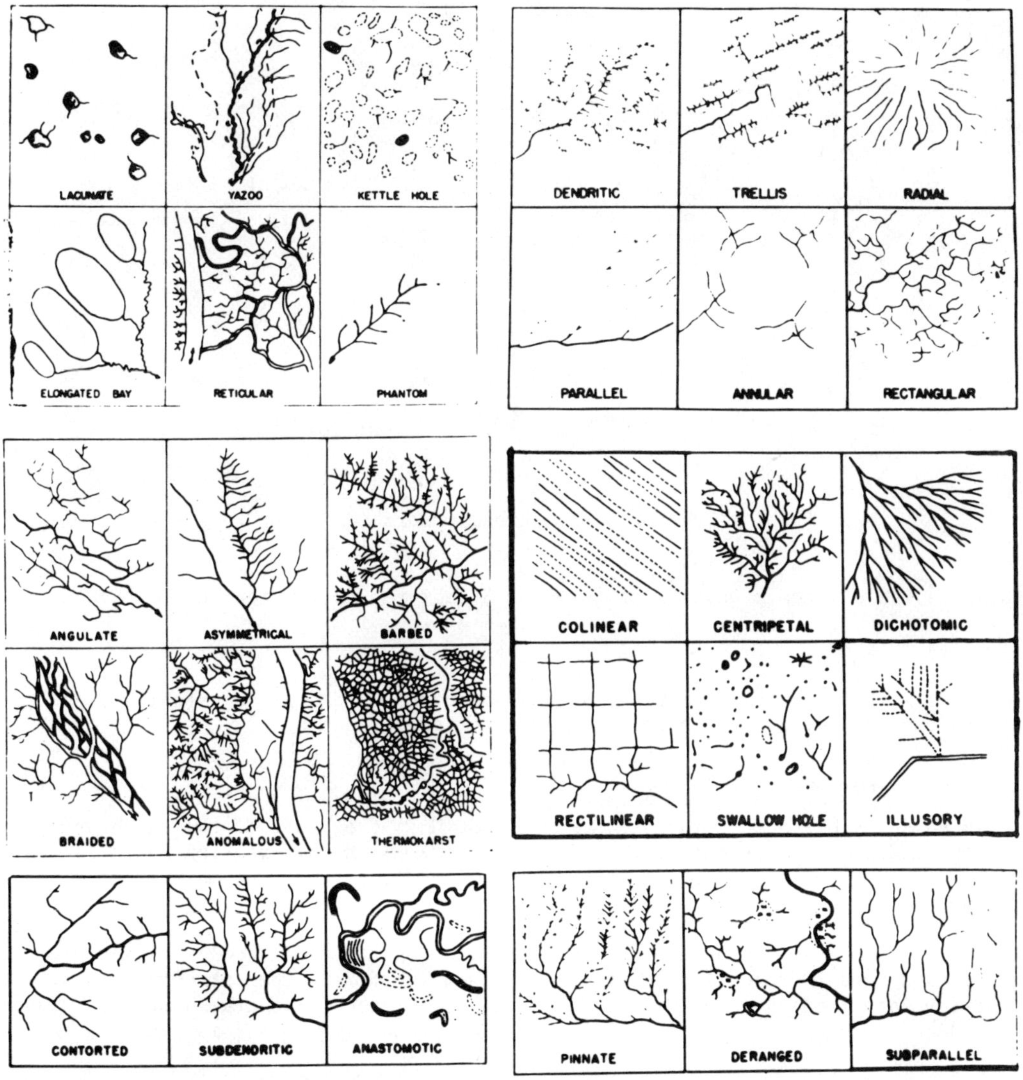

Fig. 7.3. Drainage patterns recognizable on aerial photographs, (Howe, 1960).

Horton (1945) published what is now known as Horton's law of stream numbers. It states that "the number of streams of different orders in a given drainage basin tends closely to approximate an inverse geometric series in which the first term is unity and the ratio is the bifurcation ratio". Strahler (1952) defined the stream orders as follows: "The smallest, or "finger-tip", channels constitute the first-order segments. For the most part these carry wet-weather streams and are normally dry. A second-order segment is formed by the junction of any two first-order streams,..." The channel system is a "tree" with the outflow station forming the root, Fig. 7.4.

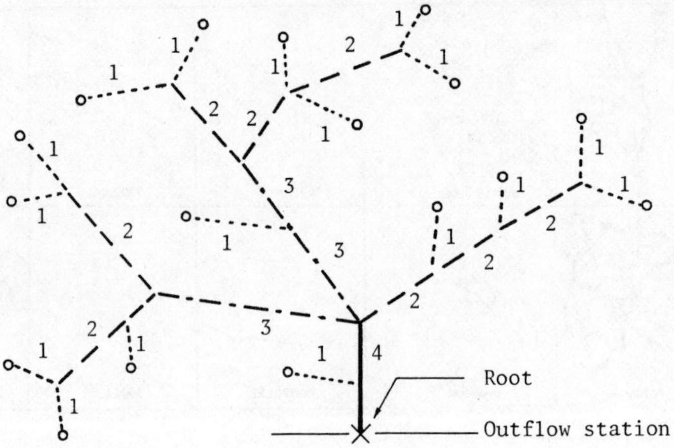

Fig. 7.4. Illustration of stream orders.

Horton's law of stream numbers in terms of Strahler's definition states that

$$S_\omega \simeq s_\omega \qquad 7.1$$

where S_ω is the number of Horton's streams of order ω in a given catchment, and s_ω is the corresponding term in the geometric series defined by

$$s_\omega = R_b^{\Omega-\omega} \qquad 7.2$$

in which R_b is Horton's bifurcation ratio and, Ω is the order of the catchment, that is, the order of the stream at the outlet or root. The bifurcation ratio is the ratio of the number of stream segments of a given order to the number of segments of the next higher order.

Shreve (1966) investigated the stream system as a problem of topology. Accordingly, every first-order branch starts at a source and two first-order streams join at a junction into a second-order stream, that is, after the first branch two new branches are added for each new source. The number of stream segments ℓ is given by

$$\ell = 2n - 1 \qquad 7.3$$

and

$$f = n - 1 \qquad 7.4$$

where n is the number of sources and f is the number of junctions or forks. The total number of combinations N(n) for linking of n sources is given by the recursion formula

$$N(n) = \frac{1}{\ell}\frac{\ell!}{(\ell - n)!\, n!} \qquad 7.5$$

(For details see Shreve, 1966). The number of different channel systems of order Ω having n sources is given by

$$N(n; \Omega) = \sum_{i=1}^{n-1} [N(i; \Omega - 1)N(n - i; \Omega - 1) + 2N(i; \Omega)\sum_{\omega=1}^{\Omega-1} N(n - i; \omega)] \quad 7.6$$

$N(1; 1) = 1$, $N(n; 1) = 0$, $N(1; \Omega) = 0$, $n = 2, 3, \ldots$ $\Omega = 2, 3, \ldots$ The first product of the sum i accounts for the establishment of systems of order Ω by combination of two systems each of order $\Omega - 1$, and the second accounts for the combination of one system of order Ω with those of lower order. Figure 7.5 illustrates the topologically distinct systems that can be obtained from six sources. Figure 7.6 illustrates topologically identical systems.

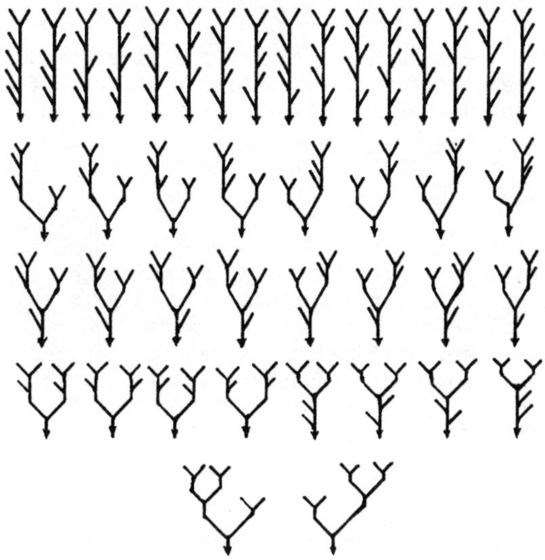

Fig. 7.5. Schematic diagrams of the N(6)=42 topologically distinct channel networks with ℓ=11 links and n=6 sources (first-order streams). In a topologically random population these networks would all be equally likely. The top row shows the possible second-order networks for which Ω=2, N(6; 2)=16, p(6; 2)=16/42=0.381, N(6, 1)=16 and p(6, 1)=16/16= 1.0. The other four rows show the possible third-order networks for which Ω=3, N(6; 3)=26, p(6; 3)= 26/42=0.619, N(6, 2, 1)=24, p(6, 2, 1)=24/26=0.929 (rows 2 to 4), N(6, 3, 1)=2 and p(6, 3, 1)=2/26= 0.077 (last row), Shreve (1966).

The probability distribution of the population of channel systems with specified n among the stream orders possible was calculated Shreve as

$$p(n; \Omega) = \frac{N(n; \Omega)}{N(n)} \quad 7.7$$

and was presented for a range of n and Ω as a table. He also calculated a table of sets of stream numbers for a range of n and Ω.

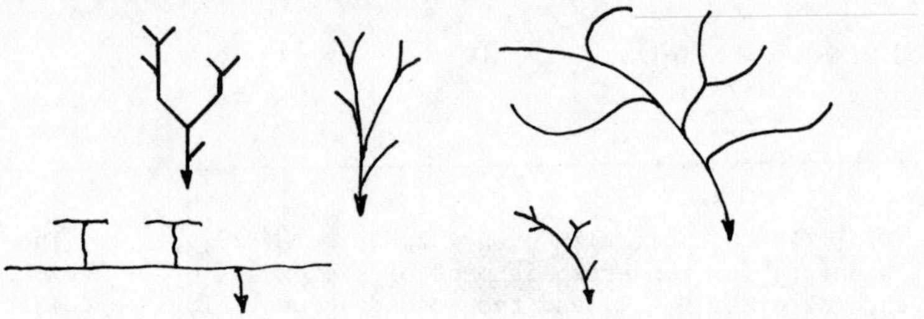

Fig. 7.6. Topologically identical channel networks having different stream patterns and drainage densities, Shreve (1966).

The mean length, L_ω, of stream channels of order ω gives the characteristic size of the components. The mean length increases with the order of the channel. Horton's law of stream lengths states that the mean lengths of stream segments of each of the successive orders of the catchment tend to approximate a geometric series in which the first term is the average length of the segments of the first order

$$L_\omega = L_1 R_1^{\omega-1} \qquad 7.8$$

where R_L is the ratio of mean length L_ω to the mean length of $L_{\omega-1}$, or if the logarithm of L_ω is plotted against ω, the ratio R_L is the anti-log of the regression coefficient of the line fitted to the plotted points. Horton also observed that the laws of stream numbers and stream lengths can be combined to a product to yield an equation for the total length of channels of a given order ω as

$$L_\omega = \sum_{i=1}^{n} \ell_\omega = L_1 R_b^{\Omega-\omega} R_L^{\omega-1} \qquad 7.9$$

by knowing the bifurcation and length ratios, the mean length L_1 and the order of the catchment Ω. The total length of channels for a catchment is estimated as

$$\sum_{i=1}^{\Omega} \sum_{i=1}^{n} \ell_\omega = L_1 R_b^{\Omega-1} \frac{r^\Omega - 1}{r - 1} \qquad 7.10$$

where $r = R_l/R_b$.

The flow paths from the water divide, (the boundary of the catchment), to the stream channel form a family of curves orthogonal to the topographic contours. These flow paths show the length of overland flow. The length of overland flow is defined as the length of the flow path (projected to the horizontal) of non-channel flow from a point on the catchment boundary to a point on the adjacent stream channel. The flow paths also delineate catchment areas which contribute to channel segments of the given order. Analogous to the earlier concept the area A_ω is defined as the projected area of the catchment contributing to the channel section of order ω and to all lower order channels. Areas which contribute directly to higher order channels are called interbasin areas. Horton suggested that the areas of progressively higher orders should increase in a geometric sequence. Schumm (1956) expressed this as

$$\bar{A}_\omega = \bar{A}_1 R_a^{\omega-1} \qquad 7.11$$

where $\bar{A}_\omega$ refers to mean areas and R_a is the area ratio analogous to the ratio of mean length R_L, eqn 7.8. Hack (1957) examined these above relationships and suggested that the area A_ω of a catchment of order ω can be related as

$$A_\omega = \bar{A}_1 R_b^{\omega-1} \frac{r^\omega - 1}{r - 1} \qquad 7.12$$

The area-length relationships and similarity of catchments have been the subject of intensive studies. The evolution of a catchment is subject to many random effects and any similarity must be based on values of average parameters. For similar catchments the ratio of the catchment area to the catchment length L squared should be a constant. For a number of catchments Gray (1961) showed that

$$L = 1.312\ A^{0.568} \qquad 7.13$$

where L is in km and A in km^2 (for L in mi and A in sq. mi, L = 1.40 $A^{0.568}$) with a 24.8 percent standard error. This error is indicative of the accuracy with which the relationship fits data in general. Rearranged to the dimensionless form

$$\frac{A}{L^2} = 0.58\ A^{-0.136} \qquad 7.14$$

shows that the ratio A/L^2 decreases as the area increases, that is, larger catchments tend to be more elongated. The catchment length, L, is taken to be the length of the main stream from the outlet to its limit at the upstream boundary. In a channel system the main stream is defined by moving upstream from the outlet along the stream of highest order. At the fork of two streams of the same order the branch which has the larger area is followed.

A further term introduced into the literature by Horton is the *drainage density*

$$D_\Omega = \left(\sum_1^\Omega \sum_1^n \ell_\omega \right) / A_\Omega \qquad 7.15$$

which is simply the ratio of the total channel lengths to the projected area of the catchment. For the entire catchment

$$D_\Omega = \frac{L_1 R_b^{\Omega-1}}{A_\Omega} \frac{r^\Omega - 1}{r - 1} \qquad 7.16$$

For large catchments the measurement of drainage density is a considerable task. Carlston and Langbein (referred to by Mark, 1974) proposed a quick method. This involves the superposition of a grid over the catchment and counting of the number of intersections, N, of streams with the grid lines. The drainage density is then empirically given by $D_\Omega = 1.571\ N/L$, where L is the total length of the grid lines. The average length of overland flow, L_s, is approximately equal to half the average distance between the stream channels and should therefore be also approximately equal to one half of the reciprocal of the drainage density, i.e.,

$$L_s \simeq \frac{1}{2D_\Omega} \qquad 7.17$$

Horton expressed this relationship as

$$L_s = \frac{1}{2D_\Omega \sqrt{1 - S_c/S_s}} \qquad 7.18$$

where S_c and S_s are the average channel and surface slopes, respectively. The inverse of D_Ω was called by Schumm (1956) the constant of channel maintenance. Its value increases with the size of the catchment. The constant gives an estimate of the area in m^2 of catchment required to maintain a metre of channel.

The number of stream segments per unit area was called by Horton the stream (or channel) frequency, F. Melton (1958), by analyzing 156 catchments, which covered a very large variation in size, climate, relief and cover, found that

$$F = 0.694\ D_\Omega^2 \tag{7.19}$$

or the dimensionless ratio $F/D_\Omega^2 = 0.694$. The scatter of points was remarkably small.

The area-altitude (hypsometric) analysis of catchment properties has been extensively developed by Langbein et al. (1947) and Strahler (1952, 1957). The method involves the measurement of the horizontal catchment area at each elevation above the outlet value. Figure 7.7 illustrates typical results.

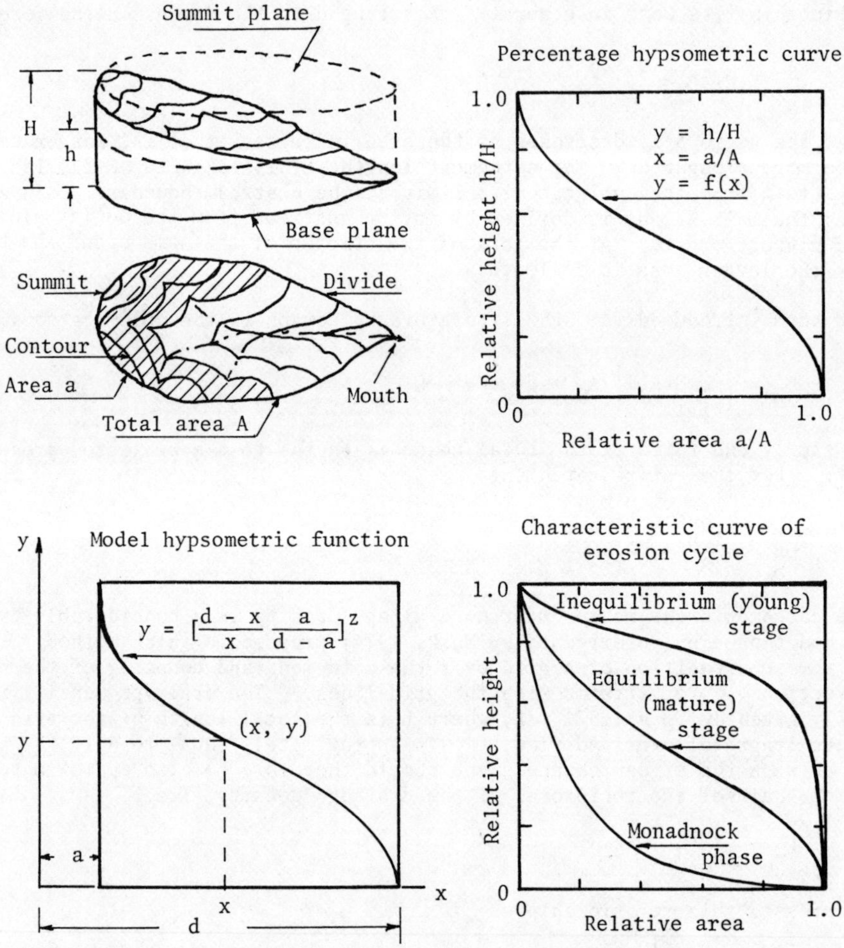

Fig. 7.7 Definition sketch for hypsometric analysis, Strahler (1957).

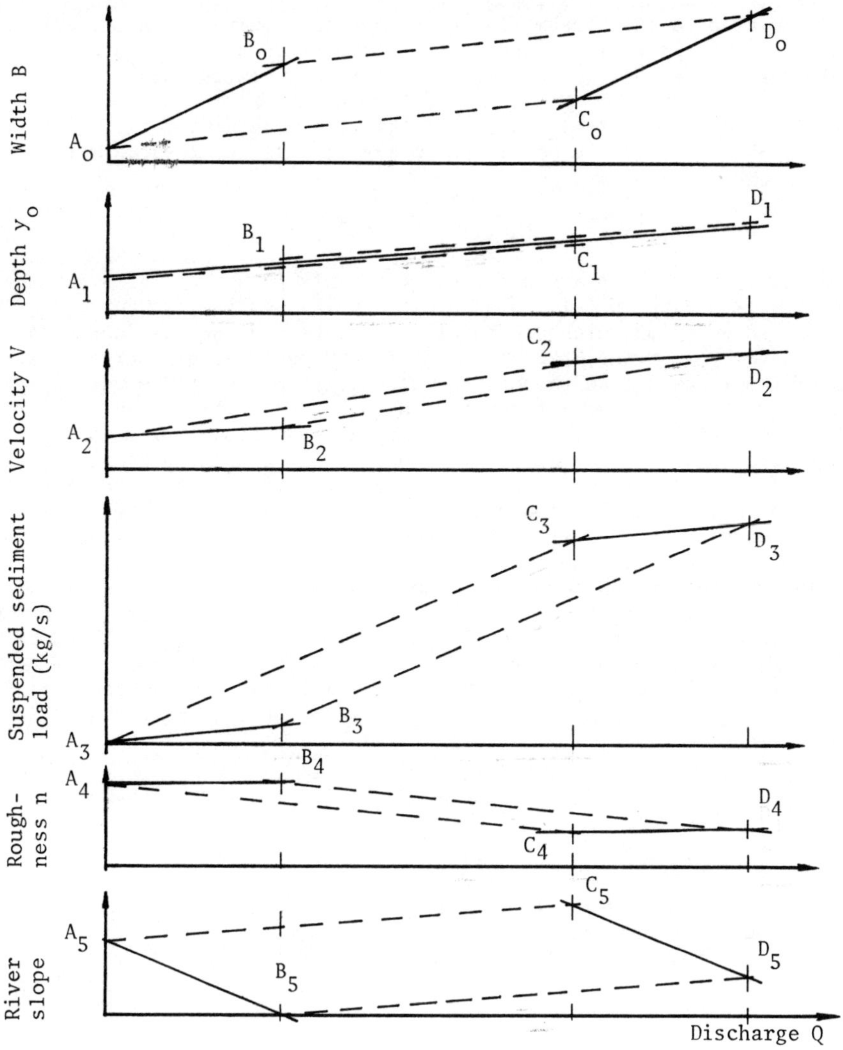

Fig. 7.8. Average width, depth, velocity, suspended sediment load, roughness and river slope expressed as a function of discharge at a station and downstream. All scales are logarithmic. Full lines show change in downstream direction for discharge of given frequency and dashed lines change at station for discharges of different frequencies. A_3-C_3 shows the suspended sediment relationship at a station for which width, depth and velocity are given by A_0-C_0, A_1-C_1 and A_2-C_2, respectively. For downstream dependence the respective relationships are given by B_3-D_3, B_0-D_0, B_1-D_1 and B_2-D_2. After Leopold and Maddock (1953).

The vertical distribution of area is most useful when the hydrological variables vary with elevation. It is important to realize that the catchment morphology is uniquely related to the stream flow only when the entire catchment contributes to the runoff. The partial response of a catchment to a single rainfall event is more the rule than the exception, particularly for large catchments.

Leopold and Maddock (1953) demonstrated that the mean depth, y_0, the surface width, B, and the mean velocity, V, of a stream are strongly correlated with discharge Q. They wrote the relationships as

$$B = aQ^n, \qquad y_0 = bQ^p, \qquad V = cQ^m$$

where a, b, c, m, n and p are numerical constants. Since $Q = VBy_0$ the exponents of Q must satisfy $m + n + p = 1$. The average values of the exponents, for discharges of streams in a region which are equalled or exceeded the same percentage of time, were obtained by plotting V, B and y_0 against Q on log-log paper and determining the slopes of the lines fitted through the respective plots. The values of the coefficients were taken as the intercept values at a value of discharge, very much smaller than those of the data, i.e. approximating zero flow. The values of these coefficients were found to vary widely from one cross section to another, but some consistency was found in the slopes. The slopes could be evaluated at a station or at different points in the downstream direction. Both at the given river station and in the downstream direction at mean annual discharge, width, depth and velocity increase with discharge. Only the exponent for depth, p, had approximately the same value at station and for downstream relationships. The velocity increases in downstream direction showing that the increase in depth compensates for decreasing in river slope. The measured depth, velocity and slope were related to roughness using the Manning formula. Figure 7.8 illustrates the tendencies of dependence on discharge, including a similar exponential relationship for suspended sediment load. For details of stream channel behaviour and sediment transport the reader is referred to specialized texts, e.g. Raudkivi (1976).

7.2 Climatic Factors

The climatic factors which influence runoff are:

(a) Nature of precipitation (rain, snow, sleet). The effect of a rainfall event is felt immediately but that of snow may be delayed for months. The eventual release of water from a snowmelt is usually much more gradual than from a fall of rain, because of the delays in the snowpack and diurnal temperature variations. The snowmelt hydrograph tends to have a lower peak and to extend over a longer period of time.

(b) Evapotranspiration and interception. These effects have been discussed in Chapters 4 and 5.

(c) Rainfall intensity. Only if the rainfall intensity exceeds the infiltration loss will any surface runoff occur.

(d) Duration of rainfall. The duration of the rainfall and its intensity are obviously the most important climatic factors. Their relationship to runoff will be discussed in more detail below.

(e) Areal distribution of rainfall. The areal distribution of a rainfall event affects the shape of the hydrograph. High intensity rain near the outlet leads to a rapidly rising and falling hydrograph with a sharp peak. Rainfall which is mainly concentrated in the upper reaches of the catchment produces a lower peak which occurs later, and a broader hydrograph. If all other conditions remain uniform throughout the catchment, a uniformly distributed rainfall will produce the minimum peak discharge. The more non-uniform the rainfall distribution the greater will the peak discharge be. The distribution coefficient, which is the ratio of the maximum point rainfall to the average rainfall over the catchment, is frequently used as an index.

(f) The distribution of rainfall with time. This parameter is significant on small catchments. On large catchments, the equalizing effect makes the hydrograph insensitive to rainfall distribution with time. In principle, a hyetograph which starts at a high intensity and decreases gradually to zero, produces a hydrograph with an upwards convex rising limb. A rainfall excess distribution, which gradually increases from zero to a maximum and stops, leads to a hydrograph with an upwards concave rising limb. On small catchments the rising limb is frequently followed by a flat peak (saturation segment). The recession limb of the hydrograph from a decreasing rainfall is concave upwards, and convex for an increasing hyetograph.

(h) Direction of storm movement. The direction of storm movement has the greatest effect on elongated catchments. It is obvious that the same amount of rain over the same period produces a much greater peak when the storm is moving down the valley. The rainfall from a storm moving up the valley becomes runoff long before the storm reaches the top of the catchment.

The effect of rainfall duration and the concept of the runoff hydrograph is best illustrated with the aid of Fig. 7.9. Assume that the rainfall starts all over the catchment at time $t = t_0$, and that the rainfall excess, i_e(mm/h), remains constant. The area ΔA_1 near the outlet after the start of the rainfall will soon contribute to the outflow at 0. As time goes on, more and more of the catchment will contribute to the outflow and the runoff at 0 will increase. The cumulative area of the catchment contributing to the runoff, plotted against time, leads to the area-time concentration curve. The dotted lines are the contours of equal travel time of surface runoff to the outlet, known as isochrones. When the duration of the rainfall excess equals or exceeds the maximum travel time the outflow becomes constant at the rate of (i_e x area). The time when this occurs marks the time of concentration for the given catchment, $t_c = t - t_0$. The S-curve is the response of the catchment to the constant rainfall excess.

Assume now that at $t = t_1$ a "negative" rainfall starts with the same rainfall excess and continues indefinitely. The effect of these two rainfalls is, by summation, a rainfall of finite period, $T = t_1 - t_0$ at i_e = constant. The "negative" rainfall excess has an identical reponse (an accumulative drainage, emptying out effect, of the catchment) to that of the positive rainfall excess. Clearly, the area between the S-curves is the runoff from the finite period rainfall T. Note that P_e, in general, refers to the rainfall excess in mm per given duration of rainfall T. If T is one hour, P_e becomes intensity but where intensity is implied the symbol i will be used.

7.3 Hydrograph Analysis and the Unit Hydrograph

Most practical techniques of forecasting runoff from rainfall are based on either correlation techniques between observed volumes of runoff and rainfall or on the unit hydrograph technique. The hydrograph method, as was indicated earlier, is a "black box" technique. It does not depend on physical laws but on an observed response function. It is therefore not a tool which will aid in the understanding and development of the physics involved. The method relies on the separation of the hydrograph into at least two components and this is at its best an empirical separation. It is indeed questionable whether there are any distinct components. The entire runoff may just be the sum of flows which reach the stream through a large number of different paths. Nevertheless the hydrograph technique has been one of the principal tools in hydrology and is likely to continue to give useful service, provided its limitations are understood.

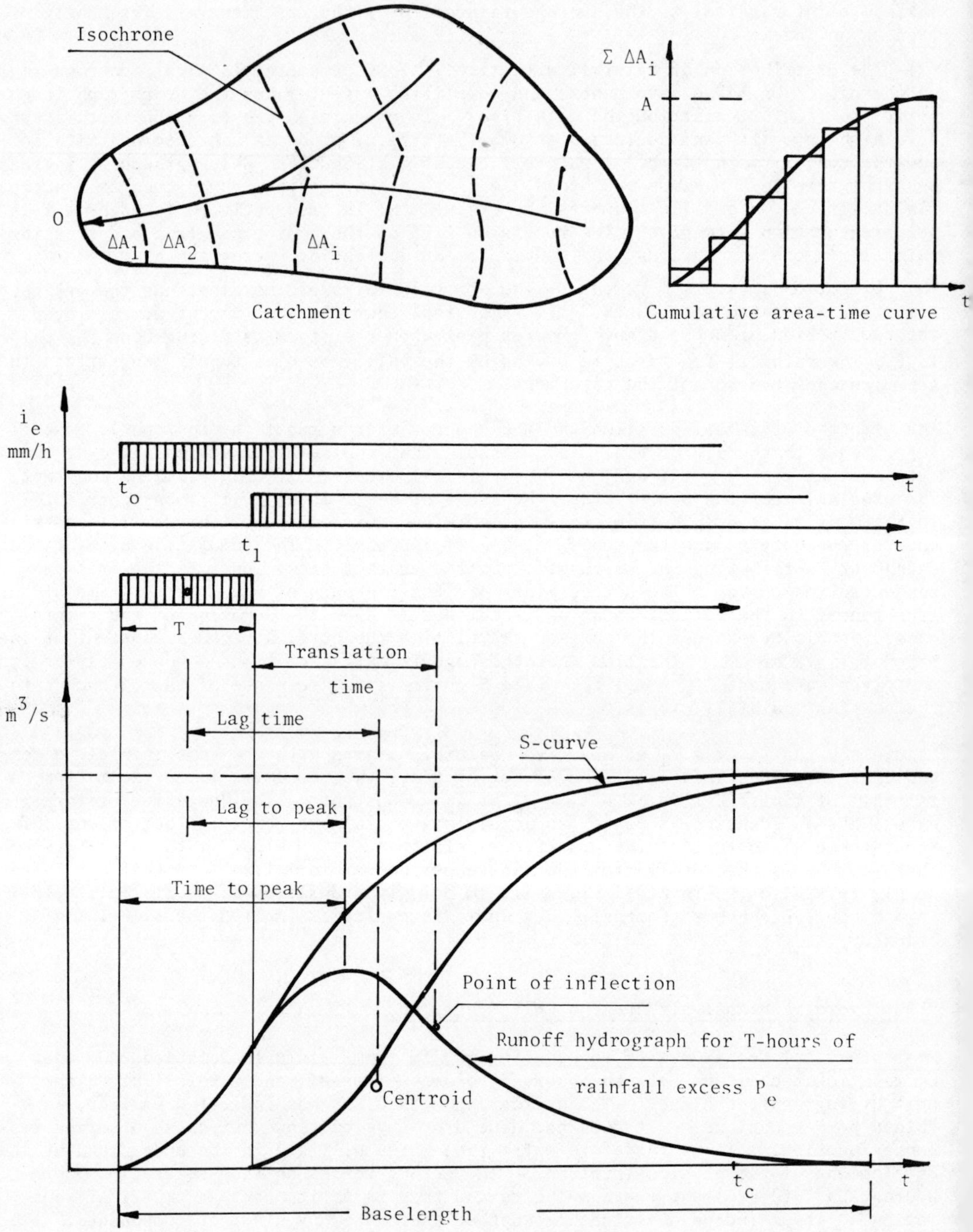

Fig. 7.9. Illustration of some of the terminology associated with the runoff hydrograph.

In perennial streams the baseflow is not assumed to be part of the runoff from a given rainfall and is separated first. The separation of the baseflow, however, is not an easy task. Figure 7.10 shows a river fed by groundwater. When the surface runoff begins, the river level rises rapidly. As a consequence the piezometric gradient reverses and flow occurs from the stream into bank storage. As the river level falls, the water from the banks starts to drain back into the river. Thus, the baseflow is as indicated in Fig. 7.10, but the actual form of this curve is difficult to evaluate.

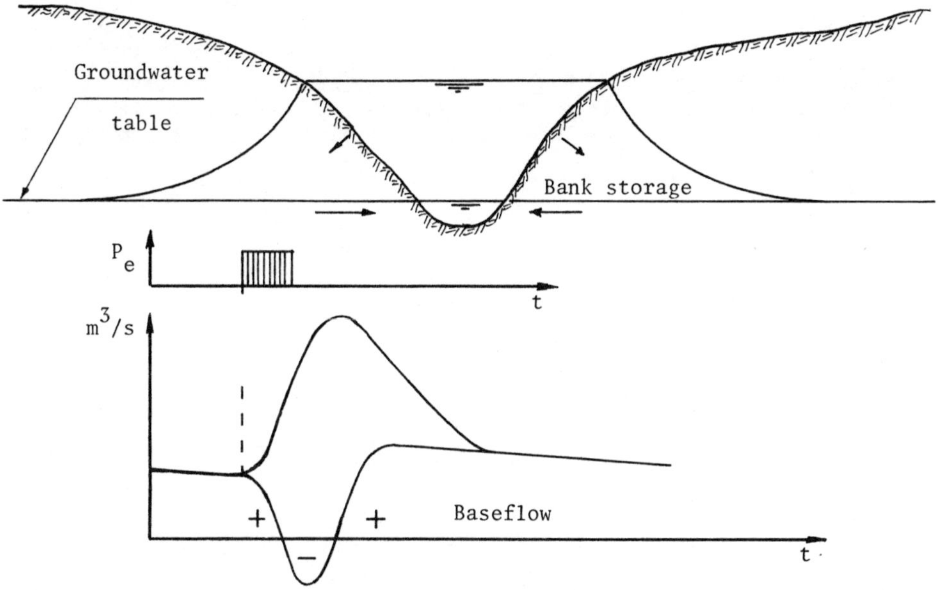

Fig. 7.10. Illustration of the variation of baseflow with time during the runoff process.

Once the hydrograph has passed its peak, it relates to drainage of surface waters, interflow and groundwater, all of which are exhaustion or decay processes (discharge from storage). Therefore, in principle, the falling part of the hydrograph can be subdivided into a number of recession curves: the *hydrograph recession* (channel flow derived from overland runoff) *curve*, the *interflow recession curves*, and the *baseflow recession curve*.

The interflow may consist of a number of components, each representing the discharge-time function for a particular layer of soil. If the storage-discharge relationship is assumed to be linear

$$S = KQ \qquad 7.20$$

where K is the storage coefficient (or delay time), then continuity requires that

$$I - Q = \frac{dS}{dt} = K\,\frac{dQ}{dt}$$

or

$$\frac{dQ}{dt} + \frac{1}{K}\,Q = \frac{1}{K}\,I \qquad 7.21$$

where I is the inflow. Equation 7.21 multiplied by $e^{t/K}$ yileds

$$\frac{d}{dt}\left(Qe^{t/K}\right) = \frac{1}{K}\, Ie^{t/K}$$

or

$$Qe^{t/K} = \frac{1}{K}\int Ie^{t/K}dt + \text{const}$$

If the inflow stops at time t_o, then the constant becomes $Q_o e^{t_o/K}$ and

$$Q = Q_o e^{-(t-t_o)/K}$$

or

$$\ln \frac{Q}{Q_o} = -\frac{\Delta t}{K} \qquad 7.22$$

Thus, if the logarithm of the discharge is plotted against time, each of these recession segments of the hydrograph should plot as a straight line with its own characteristic slope, Fig. 7.11. This type of recession curve is mathematically simple

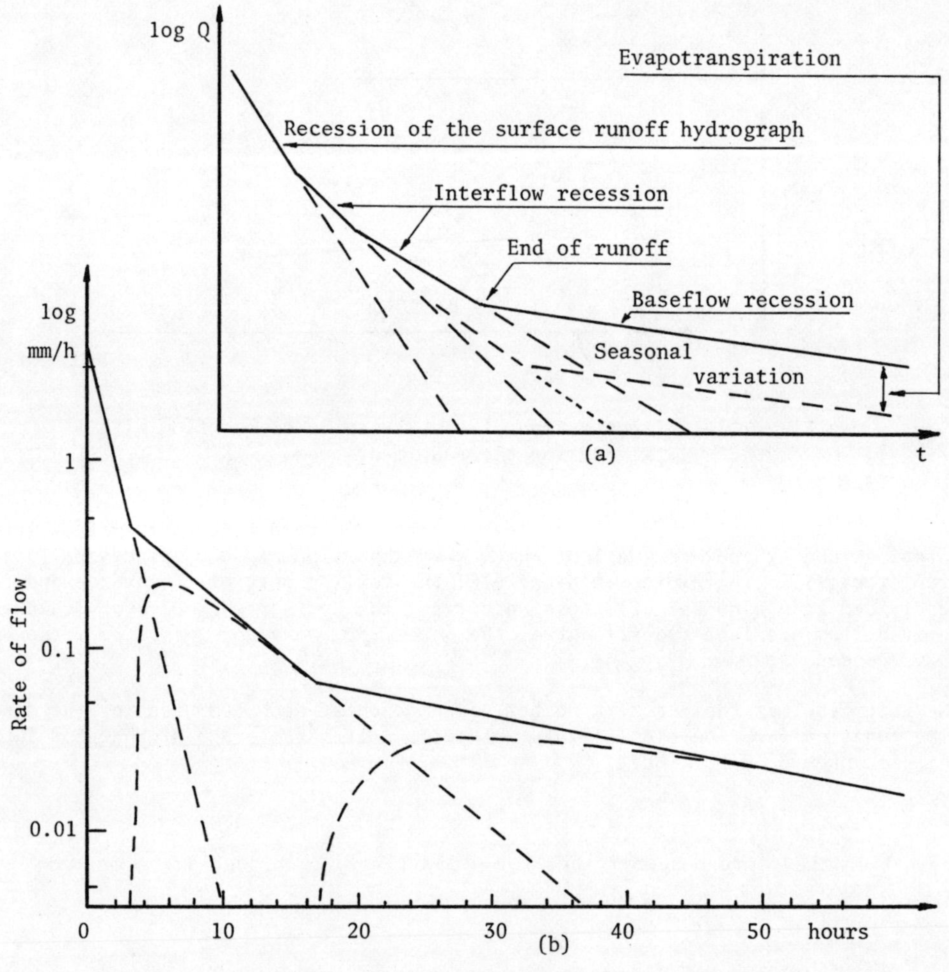

Fig. 7.11. (a) Illustration of recession curves. (b) Separation of flow regimes.

and in approximate agreement with observation. Depending on the soil type, the discontinuities in the semi-logarithmic plot may be distinct or merge into one. For long periods of dry weather the baseflow recession curves have with time decreasing slope, $[(d/dt)(\log Q)]$, mainly because of changing evapotranspiration. Indeed, baseflow recession curves may be decidedly different for different seasons. The plot of log Q versus time also yields an estimate of the time when the contributions from the various flow components cease. This is the intercept value at a negligibly small value of discharge, compared to data values, which approximates to zero value on the logarithmic plot.

In many studies there is no need to separate the runoff into the individual components. Where separation of flow regimes is required, it is best done by extending the recession of the first segment to $\log Q \simeq 0$ and subtracting it from the subsequent flows as shown in Fig. 7.11b. (The same procedure is generally used for separation of overlapping hydrographs of surface flows). The peak of the hydrograph thus obtained is at $t = (\ell n\, K_i - \ell n\, K_{i-1})/(K_{i-1}^{-1} - K_i^{-1})$.

The storage increment due to the flow under the recession curve is obtained by integration of eqn 7.22

$$\Delta S = Q_o e^{t_o/K} \int_{t_1}^{t_2} e^{-t/K} dt = - Q_o K e^{t_o/K}(e^{-t_2/K} - e^{-t_1/K})$$

or

$$\Delta S = - K(Q_2 - Q_1)$$

and the total storage S is obtained by integrating from t_o to ∞, simply $S = QK$. The total volume of the separated component hydrograph is therefore by superposition

$$S_i = Q_{oi}(K_i - K_{i-1})$$

where Q_{oi} is the rate of flow at the intersection with the previous recession segment, and K_{i-1} is the K value of the previous segment. The values of K are given by dividing the number of hours, required for the particular recession curve to cross one log-cycle, by 2.3, i.e., $\ell n\, 10$. The maximum outflow for a component can be determined from the maximum computed storage S_m as $Q_{max} = S_m/K$.

The baseflow recession relationship is frequently useful on its own, or it may be the piece of information required. The dry weather periods, however, are usually too short to define a baseflow recession curve for any length of time. If long term continuous records of Q versus time exist, a baseflow recession curve can be prepared by the sliding segment method. For this the hydrograph recession parts (segments beyond the end of the interflow and up to the beginning of the next surface runoff) are plotted as log Q versus t. Then, using a sheet of tracing paper, the hydrograph segment of the lowest discharge on record is traced at the right-hand end of the abscissa (time axis). The tracing paper is then moved along the time axis (keeping the axis coincident) until the segment of the next lowest discharge tangentially joins into the one drawn before, etc., as illustrated in Fig. 7.12.

If all the flow from groundwater storage were to flow into the river, the recession curve would be unique, but since evaporation and transpiration vary widely with seasons, it may be necessary to have a family of baseflow recession curves.

The baseflow recession curve can also be constructed by the correlation method. Values of Q at constant spacing Δt (e.g., daily values) are used from each of the individual dry weather flow hydrograph segments. Starting with an arbitrary value of Q on the segment, this value of Q is plotted against Q at Δt later (on log-log scales), the latter value of Q is again plotted against the Q value another Δt later, and so on until all the points of the segment have been plotted. This is repeated

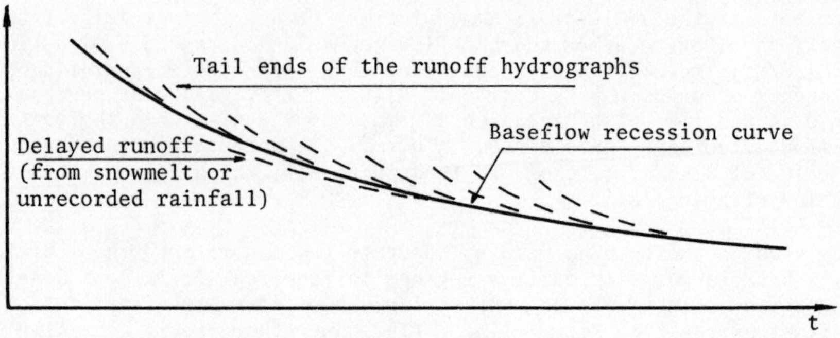

Fig. 7.12. Construction of the baseflow recession curve by the sliding segment method. (Ordinate is usually logarithmic in order to accommodate the range of discharges).

with all the other segments, Fig. 7.13. An envelope line is fitted to the points. Points above the line are again from parts of the segments which belong to surface runoff or interflow runoff.

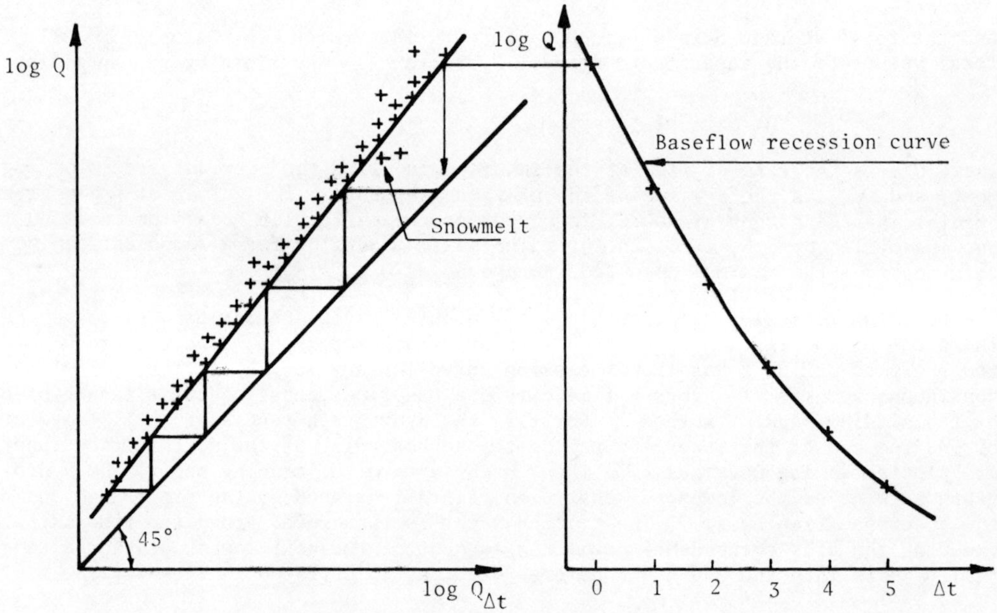

Fig. 7.13. Construction of the baseflow recession curve by the correlation method.

An obvious application of the baseflow recession curve is for prediction of low flows If the flow today is known, the flow, for example, 10 days later can be read off the curve, provided no rain falls in this period. The curve can also be used for filling in missing records using rainfall records for cross check. Integration of the baseflow rate yields the relationship between the flow rate and the volume of water in storage. This volume of water in storage also gives information about the geological features of the groundwater reservoir. Deep alluvial deposits would have large storage volumes whereas sound rock would have a very small storage.

The logarithmic plot of the tail end of the hydrograph is also used to define the end of the runoff (Fig. 7.11). The simplest approximation for separation of baseflow is to join the known starting point with the end point of the runoff, as determined by logarithmic plotting, by a straight line, Fig. 7.14a. The time between

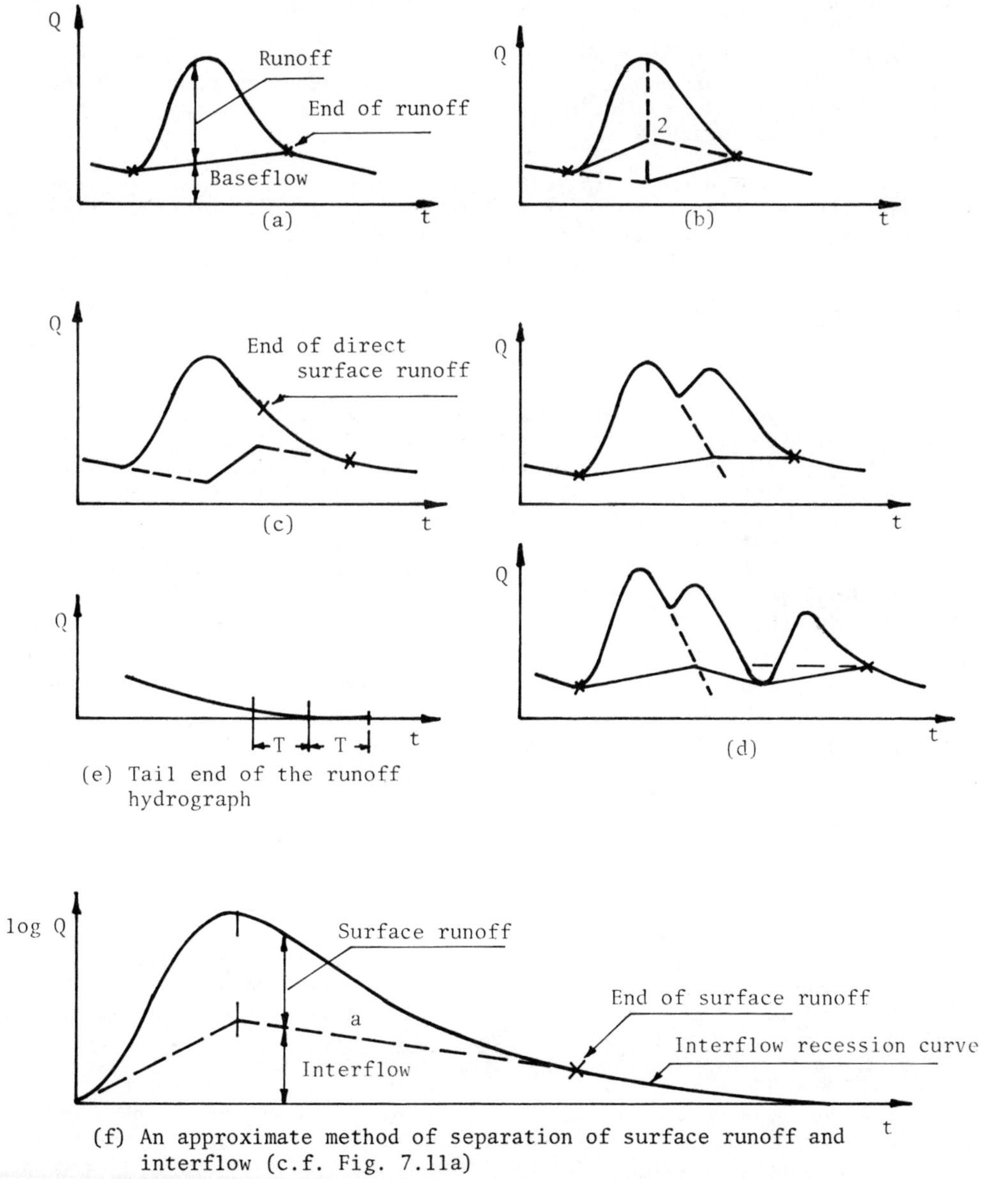

Fig. 7.14. Illustration of some of the approximate methods of separation of baseflow and interflow.

these two points is knwon as the hydrograph baselength. An alternative approximation is with the aid of the baseflow recession curve to extend the preceding baseflow until the crest of the hydrograph is reached as shown in Fig. 7.14b. When groundwater contributions are relatively large and reach the stream quickly, it may be advantageous to extrapolate the baseflow which follows the runoff hydrograph as shwon by line 2, in Fig. 7.14b. Where interflow is extensive the approximation shown in Fig. 7.14c may be used. Figure 7.14d shows methods in use with multiple peak hydrographs. Where the baseflow is a large fraction of the total flow a procedure as proposed by Linsley and Ackermann (1942) may be used.

The baselength of the hydrograph obtained by the separation of baseflow is not very accurate. Differences of one or two rainfall periods T, however, are not serious in practice because the area under this part of the runoff hydrograph corresponds to only a very small percentage of the total runoff volume, Fig. 7.14e.

The part of the hydrograph which is left after the baseflow has been separated refers to runoff from the given precipitation on the catchment. It is the response function of the catchment to the rainfall and refers to direct surface and subsurface runoff. Separation of the hydrograph into the three components is not commonly attempted. An approximate technique for the separation of interflow is to plot the direct runoff hydrograph as log Q versus time and then extrapolate (line a) as shown in Fig. 7.14f. 'An alternative approximation is to join the end of the interflow point by a straight line with the beginning of the runoff.

The direct runoff hydrograph resulting from a rainfall excess of one unit of rain uniformly distributed over the catchment over a period of T is called the *unit hydrograph*. The unit hydrograph is a widely used element of hydrological studies and applies to runoff from rainfall only, not to that from melting of snow or ice. All the essential elements of it were presented by Folse (1929), but his presentation was to complex. Sherman (1932) formulated the unit hydrograph (UH) in a simple form and since then it has been widely used. The UH refers to runoff from a rainfall excess uniformly distributed over the entire catchment. Sherman based his formulation on three postulates:

(a) *Constant baselength*. This means that for a given catchment the duration of runoff is essentially constant for all rainfalls of a given duration and independent of the total volume of runoff.

(b) *Proportional ordinates*. It is assumed that for a given duration and catchment the ordinates of the runoff hydrograph are proportional to the total volume of runoff, i.e., to the total *rainfall excess*.

(c) *Superposition*. This is the assumption of linearity. Accordingly the runoff hydrograph of a particular rainfall can be superimposed with concurrent runoff due to preceding rainfalls.

These postulates are empirical and can be shown to be not strictly correct. In many cases, however, the results obtained are in satisfactory agreement with observed data. The unit hydrographs are derived from measured hydrographs and therefore incorporate the integrated effect of all the catchment characteristics, such as infiltration, surface detention, physical features and vegetation of the catchment, as well as the effect of the actual distribution of rainfall.

The unit hydrograph is most easily derived from the hydrograph of a single isolated rainfall. After separation of baseflow the ordinates are adjusted to correspond with a convenient unit of rainfall excess, for example, 10 mm of rainfall excess. Thus, if the observed hydrograph originated from 15 mm of rainfall excess in T = 3 hours, all the ordinates are multiplied by 10/15. The result is a 3-hour unit hydrograph. The area under this unit hydrograph corresponds to the runoff volume from 10 mm of rainfall excess distributed uniformly over the given catchment. The duration of the rainfall associated with the UH must be clearly stated, because a 6-hour UH is not equal to a 3-hour UH, but is half of the sum of two 3-hour UH-s offset by 3 hours, Fig. 7.15, i.e., the superposition of two UH-s which yields two units of rainfall excess divided by two to bring it back to one unit of rainfall excess.

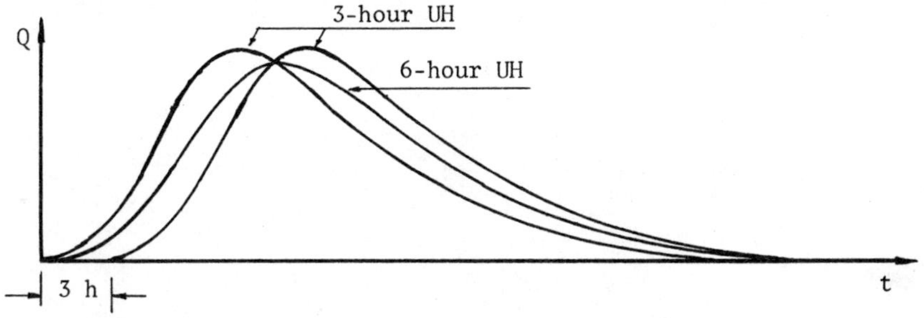

Fig. 7.15. Conversion of 3-hour unit hydrograph into a 6-hour unit hydrograph.

A given UH can readily be converted to a UH for a different period. If the desired period is a multiple of the period of the given UH, the process is as indicated above. The calculation is conveniently made in tabular form as shown in Table 7.1.

TABLE 7.1 Conversion of an 1-hour Unit Hydrograph into a 3-hour UH

Time hours	Ordinates of 1-hour UH ℓ/s	1-hour UH-s displaced by 1 h	2 h	Sum of ordinates	3-hour UH ordinates (Σx1/3)
0	0			0	0
1	50	0		50	17
2	140	50	0	190	63
3	320	140	50	510	170
4	750	320	140	1210	403
5	1100	750	320	2170	723
6	1050	1100	750	2900	967
7	950	1050	1100	3100	1033
8	830	950	1050	2830	943
9	700	830	950	2480	827
10	600	700	830	2130	710
11	440	600	700	1740	580
12	330	440	600	1370	457
13	250	330	440	1020	340
14	185	250	330	765	255
15	140	185	250	575	192
16	103	140	185	428	143
17	77	103	140	320	107
18	58	77	103	238	79
19	43	58	77	178	59
20	33	43	58	134	45
21	24	33	43	100	33
22	18	24	33	75	25
23	13	18	24	55	18
24	0	13	18	31	10
25		0	13	13	4

Conversion of the given UH to one of a shorter period, or to a period which is not a multiple of given period, can be achieved by the S-curve method. The S-curve is the hydrograph which would result from a continiuous rainfall excess of one unit per hour. Thus, if a large number of UH-s of the given rainfall duration T_g are added, with each succeeding one displaced by T_g hours from the preceding unit hydrograph, an S-curve is obtained (c.f. Fig. 7.9) as shown in Fig. 7.16.

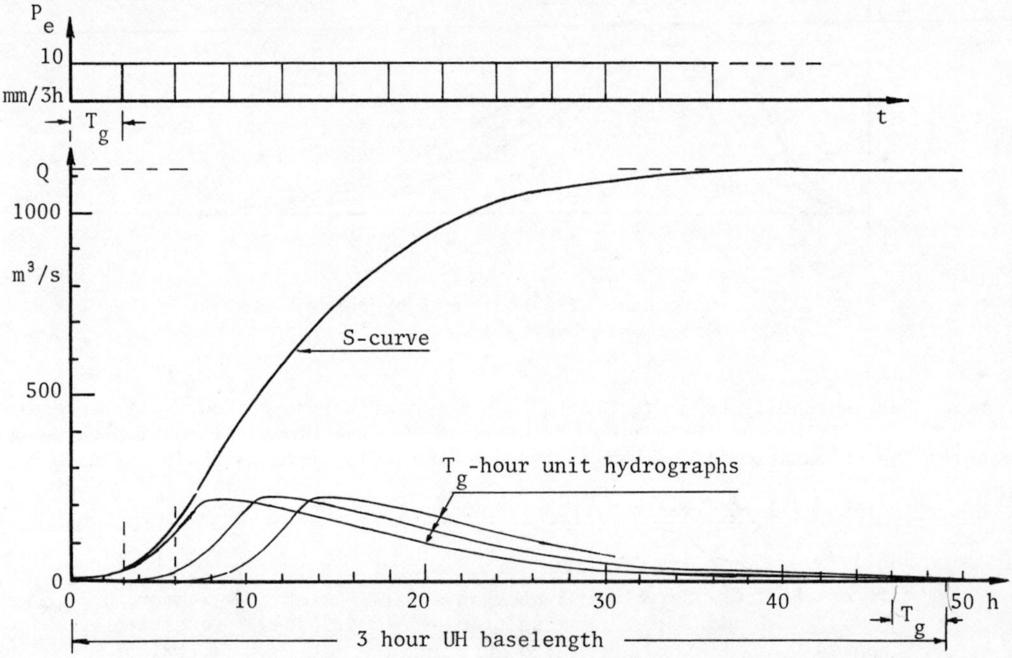

Fig. 7.16. Conversion of a given T_g-hour unit hydrograph to an S-curve. Shifting the S-curve by T_W-hours and multiplying the differences between the S-curves by T_g/T_W yields at a T_W-hour UH. The constant value of the S-curve is at Q = area x depth of excess rainfall/time.

The S-curve will rise to its constant runoff value in time one period T_g less than the T_g unit hydrograph baselength. If now the S-curve is offset by the wanted period T_W, then the difference (as in Fig. 7.9) between the two S-curves represents a hydrograph (runoff volume) of T_W/T_g units of rainfall excess, i.e., the S-curve represents runoff at the rate of one unit in T_g hours. Thus the ordinates of the T_W - UH are obtained by multiplying the S-curve differences by the ratio of T_g/T_W. The actual calculations are again conveniently done in tabular form. It is usually found, however, that when, for example, 3-hour UH ordinates defined at one hour intervals are added the S-curve tabulation does not yield a smooth curve. Oscillations with a period of T_g are generally present on both the rising and the horizontal part of the curve. The oscillations arise because the sum of the unit hydrograph ordinates at regular intervals, multiplied by this interval, is not exactly equal to the area of the unit hydrograph, nor is that sum necessarily equal to the sum of a parallel series starting at some other time. Furthermore, the ordinates of the observed hydrographs are not exact values. Therefore, in practice, the S-curve obtained by summation is smoothed out graphically before taking the S-curve differences. Table 7.2 shows the use of the S-curve method. Some further smooth-

TABLE 7.2 Conversion of a Given Unit Hydrograph by the S-Curve Method. Rainfall Excess 10 mm/3h, Catchment Area 1204 km^2

Time hrs.	Ordinates Of 3 h UH m^3/s	Ordinates of the displaced 3 h UH	Σ	Smoothed S-curve ordinates	Smoothed S-curve shifted	Δ S-curve difference	1 hour UH ordinates Δx3/1 m^3/s
0	0		0	0		0	0
1	2		2	2	0	2	6
2	8		8	8	2	6	18
3	21	0	21	21	8	13	39
4	45	2	47	47	21	26	72
5	81	8	89	89	47	42	126
6	135	21 0	156	156	89	67	201
7	190	45 2	237	233	156	77	231
8	220	81 8	309	309	233	76	228
9	220	135 21 0	376	382	309	73	219
10	215	190 45 2	452	452	382	70	210
11	207	220 81 8	516	518	452	66	198
12	198	220 135 21 0	574	580	518	62	186
13	187	215 190 45 2	639	637	580	57	171
14	175	207 220 81 8	691	691	637	54	162
15	164	198 220 135 21 0	738	742	601	51	153
16	153	187 215 109 etc.	792	788	742	46	138
17	141	175 207 220	832	830	788	42	126
18	130	164 198 220	868	868	830	38	114
19	118	153 187 215	910	903	868	35	105
20	107	141 175 207	939	934	903	31	93
21	96	130 164 198	964	962	934	28	84
22	85	118 153 187	995	987	962	25	75
23	74	107 141 175	1013	1009	987	22	66
24	62	96 130 164	1026	1027	1009	18	54
25	52	85 118 153	1047	1043	1027	16	48
26	44	74 107 141	1057	1055	1043	12	36
27	37	62 96 103	1063	1065	1055	10	30
28	31	52 85 118	1078	1073	1065	8	24
29	26	44 74 107	1083	1080	1073	7	21
30	21	37 62 96	1084	1086	1080	6	18
31	18	31 52 85	1096	1091	1086	5	15
32	15	26 44 74	1098	1095	1091	4	12
33	13	21 37 62	1097	1099	1095	4	12
34	11	18 31 52	1107	1102	1099	3	9
35	9	15 26 44	1107	1105	1102	3	9
36	7	13 21 37	1104	1107	1105	2	6
37	6	11 18 31	1113	1109	1107	2	6
38	5	9 15 26	1109	1110	1109	1	3
39	4	7 13 21	1108	1111	1110	1	3
40	3.3	6 11 18	1116.3	1112	1111	1	3
41	2.7	5 9 15	1114.7	1113	1112	1	3
42	2.3	4 7 13	1110.3	1114	1113	1	3
43	1.7	3.3 6 11	1118.0	1114.6	1114	0.6	1.8
44	1.3	2.7 5 9	1116.0	1114.9	1114.6	0.3	0.9
45	0.9	2.3 4 7	1111.2	1115.0	1114.9	0.1	0.3
46	0.6	1.7 3.3 6	1118.6	1115.0	1115.0	0	0
47	0.4	1.3 2.7 5	1116.4				
48	0.2	0.9 2.3 4	1111.4				
49	0	0.6 1.7 3.3	1111.6				
50		0.4 1.3 2.7	1116.4				

ing may be required on the final unit hydrograph, particularly near the tail end where the S-curve differences are small differences of large values. With very peaked hydrographs small time increments are necessary in order to avoid large fluctuations. Large amplitude fluctuations in the asymptotic values of the S-curve can be associated with an incorrect hydrograph baselength, i.e., (incorrect separation of baseflow).

In keeping with the instantaneous unit hydrograph concept, to be introduced later, the ordinates of the given T_g - UH could be multiplied before the displacement and taking of the S-curve differences to give the S-curve for one unit of rainfall excess per hour, i.e., the ordinates of column 2 in Table 7.2 are multiplied by 3. When this unit intensity S-curve is displaced by n-hours, the differences are divided by n to yield the n-hour unit hydrograph.

The unit hydrograph technique discussed above, although simple, has a serious limitation. Runoff hydrographs resulting from a single period of rainfall are in nature exceedingly rare. The hydrographs are usually produced by a sequence of rainfalls and therefore one has to develop procedures by which the unit hydrograph can be derived from data produced by a multi-period rainfall, for example, as shown in Fig. 7.17a. Separation of losses yields the histogram of rainfall excess, Fig. 7.17b. For simplicity let the three rainfall periods (r = 3) be equal and the excess yields be 1, 2 and 1 unit of rain, i.e., $P_1 = 1$, $P_2 = 2$ and $P_3 = 1$. This rainfall (excess) distribution produces a surface runoff (after separation of baseflow) with a baselength of m periods. The ordinates of this runoff hydrograph at the ends of the rainfall periods are X_1, X_2, X_3 ... X_m, where $X_m = 0$. The unknown ordinates of the unit hydrograph at these corresponding times are u(T, 1), u(T, 2), u(T, 3) ... or u_1, u_2, u_3 ... in short, where u(T, t) signifies the unit hydrograph produced by a rainfall of duration T, and t the ordinate at time t.

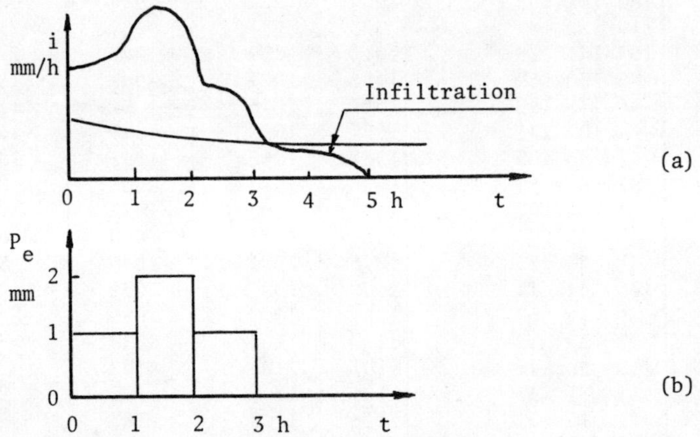

Fig. 7.17. Multi-period rainfall hyetograph.

During the first period only P_1 affects the hydrograph and at the end of first period

$$X_1 = P_1 u_1$$

from which u_1 can be calculated since both X_1 and P_1 are known, Fig. 7.18. At the end of the second period

$$X_2 = P_1 u_2 + P_2 u_1$$

from which u_2 can be calculated, etc.

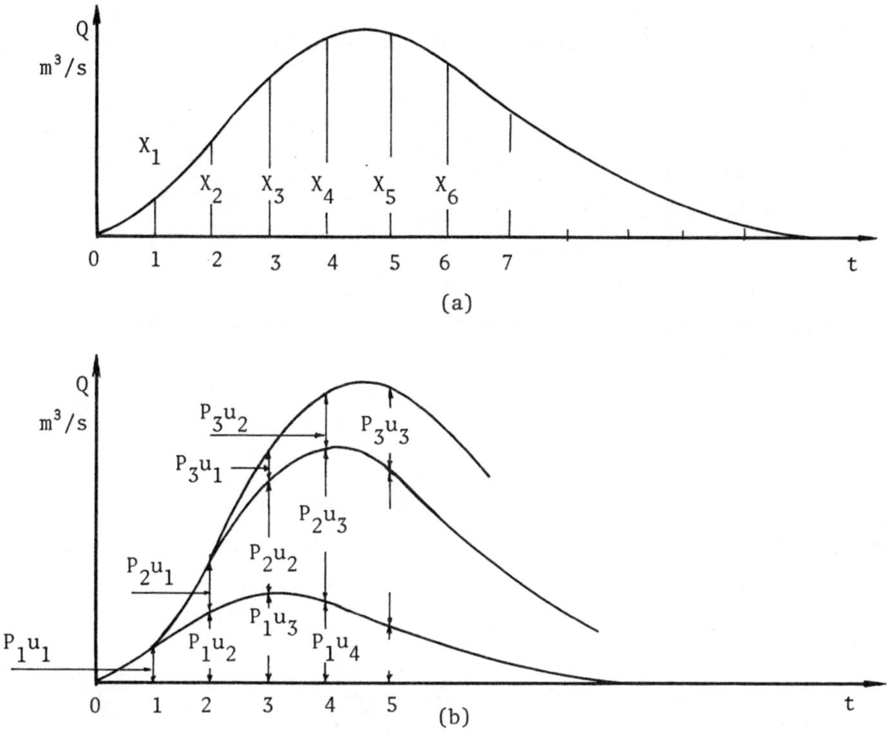

Fig. 7.18. Observed hydrograph (a) and the superposition of unit hydrographs (b).

It is readily seen that the number of UH periods n is $n = m - r + 1$ or $m = n + (r - 1)$. If, e.g., $m = 7$ then $n = 5$ and $u_5 = 0$ at the end of the UH while $X_7 = 0$ at the end of the runoff hydrograph. Thus,

$$\begin{aligned}
P_1u_1 &= X_1\\
P_2u_1 + P_1u_2 &= X_2\\
P_3u_1 + P_2u_2 + P_1u_3 &= X_3\\
+ P_3u_2 + P_2u_3 + P_1u_4 &= X_4\\
+ P_3u_3 + P_2u_4 + P_1u_5 &= X_5\\
+ P_3u_4 + P_2u_5 &= X_6\\
+ P_3u_5 &= X_7 = 0
\end{aligned}$$

Such a set of simultaneous equations is readily solved. In practice, however, the results obtained between rows 1-4, 2-5 or 3-6 will differ from each other because observational results include errors and do not exactly satisfy the algebraic equations. If P_1 is in error then the computed u_1 is in error and this error is magnified by subsequent calculations and may at times lead to nonsensical results. It is also likely that the rainfall distribution during the separate periods r is not exactly uniform or identical and this introduces a corresponding error since the theory is based on the assumption that the three unit hydrographs are identical. The effect of just one small error in the observed hydrograph ordinates is demonstrated by the example below.

Let $P_1 = 1$, $P_2 = 2$, $P_3 = 1$ and the set of simultaneous equations be

$$1u_1 = 10 \qquad 1$$
$$2u_1 + 1u_2 = 50 \qquad 2$$
$$1u_1 + 2u_2 + 1u_3 = 90 \qquad 3$$
$$+ 1u_2 + 2u_3 + 1u_4 = 82 \qquad 4$$
$$+ 1u_3 + 2u_4 = 40 \qquad 5$$
$$+ 1u_4 = 10 \qquad 6$$

where only 82, instead of 80, is different from the exact fit. The unit hydrograph ordinates obtained from rows 1 to 4 are

$$u_1 = 10, \quad u_2 = 30, \quad u_3 = 20, \quad u_4 = 12$$

rows 2 to 5

$$u_1 = 10.8, \quad u_2 = 28.4, \quad u_3 = 22.4, \quad u_4 = 8.8$$

and rows 3 to 6

$$u_1 = 6, \quad u_2 = 32, \quad u_3 = 20, \quad u_4 = 10$$

or in tabular form

rows	1-4	2-5	3-6	Ideal fit
u_1	10	10.8	6	10
u_2	30	28.4	32	30
u_3	20	22.4	20	20
u_4	12	8.8	10	10

A least-squares solution could be used to obtain the best fit to the data. The normal equations can be written down directly from the original ones, i.e., the first normal equation is obtained by multiplying each of the original equations by the coefficient of u_1 in that equation and adding the results, etc, Thus,

$$P_1(P_1u_1 \qquad = X_1)$$
$$P_2(P_2u_1 + P_1u_2 \qquad = X_2)$$
$$P_3(P_3u_1 + P_2u_2 + P_1u_3 \qquad = X_3)$$

$$u_1(P_1^2 + P_2^2 + P_3^2) + u_2(P_2P_1 + P_3P_2) + u_3(P_3P_1) = P_1X_1 + P_2X_2 + P_3X_3$$

and likewise for the 2nd, 3rd, etc. This yields for the estimate the set of equations:

$$(P_1^2+ P_2^2+ P_3^2)u_1+ (P_2P_1+ P_3P_2)u_2+ P_3P_1u_3 = P_1X_1+ P_2X_2+ P_3X_3$$

$$(P_1P_2+ P_2P_s)u_1+ (P_1^2+ P_2^2+ P_3^2)u_2+ (P_2P_1+ P_2P_3)u_3+ P_1P_3u_4 = P_1X_2+ P_2X_3+ P_3X_4$$

$$P_1P_3u_1 + (P_1P_2 + P_2P_3)u_2 + (P_1^2 + P_2^2 + P_3^2)u_3 + (P_1P_2 + P_2P_3)u_4 = P_1X_3 + P_2X_4 + P_3X_5$$

$$P_1P_3u_2 + (P_1P_3 + P_2P_3)u_4 + (P_1^2 + P_2^2 + P_3^2)u_4 = P_1X_4 + P_2X_5 + P_3X_6$$

Substitution of the assumed values used before yields

$$\begin{aligned} 6u_1 + 4u_2 + 1u_3 \qquad\qquad &= 200 \\ 4u_1 + 6u_2 + 4u_3 + 1u_4 &= 312 \\ 1u_1 + 4u_2 + 6u_3 + 4u_4 &= 294 \\ + 1u_2 + 4u_3 + 6u_4 &= 172 \end{aligned}$$

from which

$$u_1 = 8.4 \qquad u_2 = 29.8$$
$$u_3 = 21.4 \qquad u_4 = 9.5$$

The method of superposition and solution of a set of linear simultaneous equations (by a computer) is only one of the many possible methods of solution.

In general a function g(t) can, for all values of t, be represented exactly by the sum of an infinite series of other functions

$$g(t) = \sum_{m=0}^{\infty} c_m f_m(t) \qquad 7.24$$

in which c_m are constants, e.g., the polynomial series

$$g(t) = c_o + c_1 t + c_2 t^2 + \ldots$$

or the Fourier series

$$g(t) = a_o + a_1 \cos \omega t + a_2 \cos 2\omega t + \ldots$$
$$+ b_1 \sin \omega t + b_2 \sin 2\omega t + \ldots$$

Amongst many such series there are some in which the $f_m(t)$ functions have the property

$$\int_a^b f_m(t) f_n(t) dt = \begin{cases} 0 & \text{if } m \neq n \\ K & \text{if } m = n \end{cases} \qquad 7.25$$

and these are known as orthogonal functions.

The importance of the orthogonal functions stems from the ease with which the coefficient c_m in eqn 7.24 can be obtained. If we multiply both sides of eqn 7.24 by $f_m(t)$, integrate between the limits a and b and observe eqn 7.25, the result is

$$\int_a^b g(t) f_m(t) dt = 0 + 0 + \ldots + c_m K + 0 + \ldots$$

$$c_m = \frac{1}{K} \int_a^b g(t) f_m(t) dt \qquad 7.26$$

The Fourier series method of analysis was discussed by O'Donnell (1960) and the use of the Laguerre functions and Laguerre polynomials by Dooge (1965). The Laguerre method of derivation appears to lead to very large errors with late-peaked hyetographs. These and data error effects on unit hydrograph derivation were discussed by Laurenson and O'Donnell (1969). If computer programs or computers are not available the problem could be dolved by the trial coefficients or Collins' method. By this method the set of linear simultaneous equations is solved by trial and error by successive approximations. The Collins' (1939) method uses those four equations (in the example below) which include the maximum rainfall excess. The procedure is as follows:

(1) Assume the UH ordinates u at T-hour intervals, in the form of a histogram in percentages. Estimation of these initial UH ordinates is aided by the observation that the recession part of the runoff hydrograph is a decay process. Thus, if it is assumed that

$$Q = Q_o e^{-\Delta t/K}$$

then this part of the hydrograph would plot as a straight line on log-normal paper and the plot can be used to proportion the ordinates.

(2) Apply this assumed hydrograph to all periods of rainfall excess, except the greatest and determine the hydrograph that would result in the absence of P_e(max).

(3) Subtract the hydrograph so obtained from the observed one. The difference should be the hydrograph for P_e(max) alone, if the assumed distribution was correct.

(4) Calculate the next trial coefficients by using weighted averages for $\bar{u}_i$

$$\bar{u}_i = \frac{u_i + Fu_i'}{1 + F} \qquad 7.27$$

$$F = \frac{\Sigma(Q - \Sigma P_e u)}{\Sigma(P_e u)} = \frac{\Sigma \text{ Residuals}}{\Sigma(\text{sum } P_e u)} \qquad 7.28$$

where the summation is over the column corresponding to the omitted rainfall, i.e., 2 to 5 in the example.

Time unit period	P_e mm	Observed hydrograph Q m^3s^{-1}	Trial coeff. % u_1	u_2	u_3	u_4	$\Sigma P_e u$	Residual $\frac{Q}{A} - \Sigma P_e u$	$u_i' = \frac{\text{Resid.}}{P_e(\text{max})}$ *	u_i
0		Q_o					0			
1	P_1	Q_1	P_1u_1				P_1u_1			
2	P_2	Q_2		P_1u_2			P_1u_2			
3	P_3	Q_3	P_3u_1		P_1u_3		$P_3u_1 + P_1u_3$			
4		Q_4		P_3u_2		P_1u_4	$P_3u_2 + P_1u_4$			
5		Q_5			P_3u_3		P_3u_3			
6		Q_6				P_3u_4	P_3u_4			

* The calculations may be carried out in terms of depth of rainfall excess; then u_i' = Residual/Σ Residuals. Alternatively, each of the P_iu_i terms is multiplied by the area A. Then the residuals are Q - $\Sigma P_e uA$ and u_i = Residual/Σ Residuals ≃ $R/P_{e(max)}A$.

In general, when preparing unit hydrographs it is important to make sure that (1) the rainfall period is relatively short compared with the time to peak (preferably less than a quarter) in order to obtain an adequate definition of the rising

limb of the hydrograph,
(2) several records are analyzed and averaged into one unit hydrograph, and
(3) seasonal effects are accounted for.

Unit hydrographs are affected by the assumptions made in their derivation. The equal baselength assumption, for example, makes no allowance for the amount of water on the catchment. Usually, by the time the peak of the hydrograph has been reached, the rain has stopped and all the water is in storage on the catchment. The time taken to empty this reservoir would vary with the amount of water. The area under the tail end of the runoff hydrograph, however, represents a relatively very small volume of water so that a variation of the baselength by one or two rainfall periods does not cause any significant error in the derived unit hydrograph.

It has already been shown that the separation of the baseflow is an approximation. Usually, however, the baseflow is only a small fraction of the peak ordinate of the runoff hydrograph and the error introduced by an incorrect separation of the baseflow is small. The extent of the error likely to be caused can be checked by making a too low and a too high baseflow separation and comparing the derived unit hydrographs.

The assumption of proportional ordinates is another approximation. The hydrographs are strongly affected by storage. For example, inflow and outflow hydrographs to and from a reservoir are shown in Fig. 7.19. If "proportionality" were to hold,

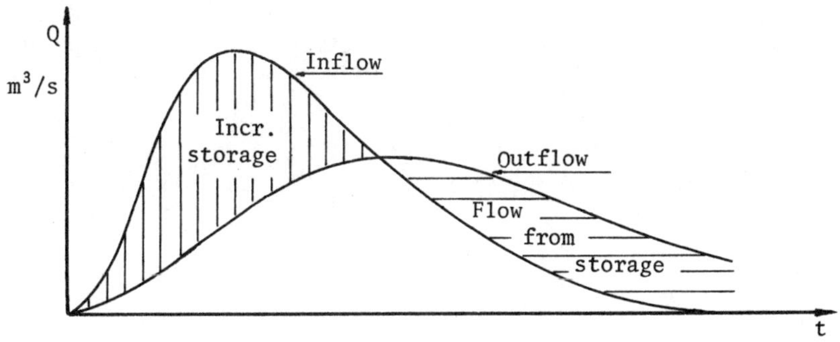

Fig. 7.19. Superposition of an inflow hydrograph to and an outflow hydrograph from a storage reservoir.

then doubling the rainfall excess would also double the storage volume. Storage, however, is not directly proportional to flow rate. Unless the banks of the reservoir are very flat the storage, S, is approximately proportional to the depth, H, over the outflow sill, i.e.,

$$S \propto H \qquad \text{or} \qquad S \propto H^{n>1} \qquad 7.29$$

If the outflow is over a spillway then

$$Q \propto H^{3/2} \qquad 7.30$$

and consequently

$$S \propto Q^{2/3} \qquad 7.31$$

Therefore, unit hydrographs derived from hydrographs of large floods will have greater peak ordinate values than those derived from hydrographs of smaller floods, i.e., the peak discharge will increase faster than the runoff volume.

If the outflow is through a channel then the velocity is proportional to the surface slope $S_o = (H/L)$ and

$$Q = \propto \sqrt{H} \qquad \text{or} \qquad S \propto Q^{2n} \tag{7.32}$$

where H is now the level difference. This shows that the peak discharge below the reservoir increases much more slowly than the total runoff volume. Hence, unit hydrographs derived from hydrographs of large floods will have lower peaks than those derived from hydrographs of small floods. Flooded overbank areas generally introduce this detention basin effect.

The unit hydrograph can be used to predict the runoff hydrograph for the same catchment at the same gauging station for any given sequence of rainfall excess. The major problem is the estimation of the rainfall excess because the amount of infiltration and interception varies widely with season and the antecedent weather. Analysis of past data with reference to the rainfall history of the preceding month will yield useful empirical information. In conjunction with large catchments use of index areas may be helpful. Index areas are small representative sub-catchments, which are well instrumented. The runoff from these small index areas is rapid and can be used to estimate the runoff coefficient, i.e., the rainfall excess. This value can then be applied to the total catchment.

Every catchment will also have its own *critical rainfall period*. The so-called *critical storm* will produce the maximum runoff rate from the given catchment. The rainfall yield for any given probability level (return period) increases at a decreasing rate with time. For example, the yield for a 25-year return period rainfall could be 1 hour rainfall 20 mm, 2 hour 30 mm, 3 hour 35 mm and 4 hour 39 mm. The 1 hour unit hydrograph for the given catchment has ordinates at 1 hour spacings as follows: 0, 16.6, 20.9, 14.0, 9.1, 6.4,... m^3/s with peak of 21.5 at 1 hour 45 min. Converted into 2, 3 and 4-hour unit hydrographs (10 mm as a unit) gives peak ordinate values of 19.8 m^3/s at 2 hours 24 min., 17.2 m^3/s at 3 hours 12 min., and 15.1 m^3/s at 4 hours, respectively. The peak runoff rates from the corresponding 25-year return period rainfall are then (assuming 100% runoff)

$$Q_{peak} = u_{max} P_e$$

or 43.0, 59.4, 60.2 and 58.89 m^3/s, respectively. A graphical interpolation shows that the critical rainfall period for this catchment is 2½ hours, approximately.

7.4 Modelling of the Hydrograph

When the duration, T, of the rainfall excess approaches zero, while the volume remains constant (unit depth of water uniformly distributed over the catchment), the input to the catchment becomes a pulse and the runoff produced by this pulse is known as the instantaneous unit hydrograph IUH. The IUH is the response function of the given catchment and is produced by an instantaneous rainfall yielding a unit rainfall excess; it is a mathematical abstraction. The ordinate of the IUH at time t is $u(0, t)$.

The UH concept presupposes a catchment which is characterized by a unique IUH, invariant in time t and independent of preceding rainfall events. The total discharge form such a catchment is the sum of all the elements of instantaneous inputs. This type of catchment forms a linear system with constant coefficients

$$A_n \frac{d^n y}{dt^n} + A_{n-1} \frac{d^{n-1} y}{dt^{n-1}} + \dots A_o y = x(t) \qquad 7.33$$

The output function is $y \equiv Q(t)$ and the input is $x(t) \equiv i(t)$. This system has a time-variant impulse response function $u(t)$. Thus, if the input is a succession of infinitesimal instantaneous inputs of volume $x(\tau)d\tau \equiv i(\tau)d\tau$, then each of these adds its contribution $i(\tau)u(0, t - \tau)d\tau$ to the rate of output $Q \equiv y$ at time t

$$Q(t) = \int_0^{\tau \leq T} i(\tau)u(0, t - \tau)d\tau \qquad 7.34$$

where $i(\tau)$ is the intensity of the effective rainfall (rainfall excess) at time τ, Fig. 7.20. The integral is a convolution integral and $u(0, t - \tau)$ is known as the kernel. The $u(t - \tau)$ can be looked upon as the memory of $Q(t)$ for the input $i(\tau)$ at time $(t - \tau)$ before.

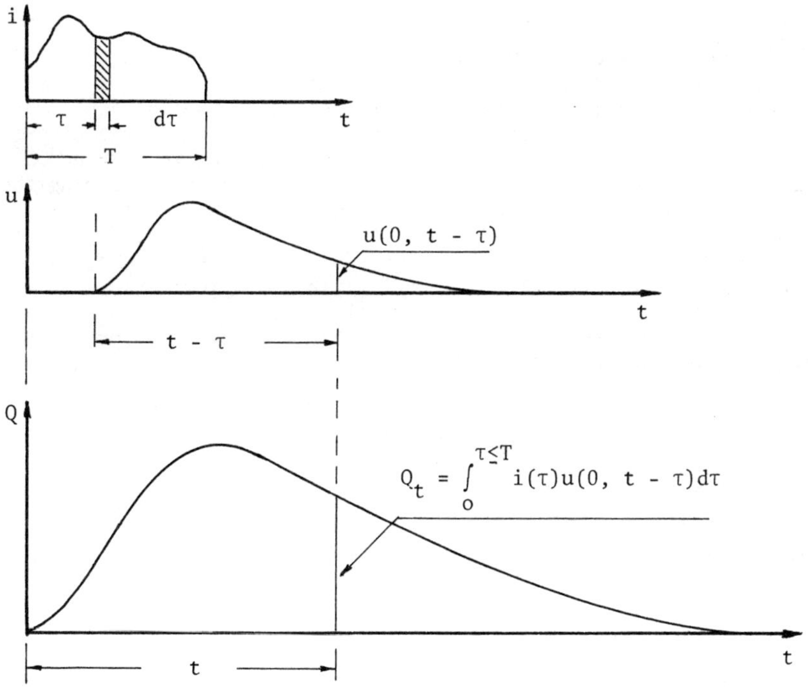

Fig. 7.20. Definition sketch of the runoff hydrograph. Note that i is the rate of rainfall excess.

The S-curve is the hydrograph due to a continuous effective rainfall of unit intensity. If this continuous rainfall is imagined to consist of a series of blocks of τ-hours each, then each of these blocks produces a τUH separated by τ-hours from one another. The sum of the simultaneously occurring ordinates is the runoff at that time. The S-curve ordinates are

$$S(t) = \tau[u(\tau, t) + u(\tau, t - \tau) + u(\tau, t - 2\tau) + \dots]$$

i.e., the sum of the τUH ordinates divided by the intensity $1/\tau$. As $\tau \to 0$

$$S(t) = \int_0^t u(0, t)dt \qquad 7.35$$

Hence, the S-curve is the integral of the IUH, or the IUH is the first derivative of the S-curve. The area under the IUH curve up to time t is proportional to the ordinate of the S-curve, and the slope of the S-curve is proportional to the ordinate of the IUH. The time at the point of inflection of the S-curve corresponds to the peak of the IUH.

Although the UH method is widely used, the strictly linear and time-variant relationship between the rainfall and the runoff does not exist. The hydraulic equations which express the component flow processes are all non-linear. A major source of the non-linearity is the subtraction of the time-varying losses and infiltration from the total rainfall in order to obtain the rainfall excess. Comparative studies, however, have shown that the combined effect of all the non-linearities is usually small. In view of the nature of hydrological data (large random errors) it is problematic whether the attempts to reduce the systematic error, arising from non-linearities, by the use of non-linear analysis is warranted. The non-linear analysis does, however, contribute to the understanding of the runoff process.

The various models of the hydrograph could be classified as
(1) the linear, time-invariant system, discussed above,
(2) the linear, time-variant system which is also described by eqns 7.33 and 7.34 except that now the coefficients of eqn 7.33 are variable and the kernel is a function of time but the very useful feature of superposition is retained. In this model the IUH and TUH change with time; short term changes occur during the rainfall and long term changes are associated with seasons.
(3) the non-linear time-variant system.
In non-linear systems the principle of superposition does not apply and the application becomes much more difficult. In principle, the time-variant non-linear models involve a systems analysis approach or functional forms which incorporate the memory of the system. One of the earliest systems analysis type approaches to the non-linear catchment was that by Kulandaiswamy (1964). He proposed a relationship for storage

$$S = \sum_0^n a_n(Q, I) \frac{d^nQ}{dt^n} + \sum_0^m b_m(Q, I) \frac{d^mI}{dt^m} \qquad 7.36$$

which, combined with the continuity equation, leads to differential equations used for dynamic systems for which there are known solutions.

Amorocho and Orlob (1961) expressed the output from a non-linear time-variant system as a series of functionals

$$\begin{aligned} Q(t) = & \int_0^t u_1(\tau)i(t - \tau)d\tau \\ & + \int_0^t\int_0^t u_2(\tau_1, \tau_2)i(t - \tau_1)i(t - \tau_2)d\tau_1 d\tau_2 + \dots \\ & + \int_0^t \dots \int_0^t u_n(\tau_1, \tau_2 \dots \tau_n)i(t - \tau_1)i(t - \tau_2) \\ & \quad \dots i(t - \tau_n)d\tau_1 d\tau_2 \dots d\tau_n + \dots \end{aligned} \qquad 7.37$$

Here the notation is reversed. The input is that at $(t - \tau)$ and the reponse function u has its time axis τ in the negative t-direction with $\tau = 0$ at t. The first term is the convolution integral of the linear sytem. The second term shows that the output at time t is affected by the contributions $i(t - \tau_1)$ and $i(t - \tau_2)$ whereas in the linear system the output from $i(t - \tau)$ is independent of any other $i(t)$. This kernel function $u_2(\tau_1, \tau_2)$ is a surface in a three-dimensional space whereas $u_1(\tau)$ is the IUH curve on a plane. The additional terms account for increasing interdependence of input events and their kernel functions are represented by surfaces in hyperspace. Because of the interdependence of inputs it is not necessary to satisfy the physical continuity. This is seen by considering the product $x(t)x(t - \tau)$. If either term is zero the product is zero and does not contribute to the output, even though the other term may have a non-zero value. Writing eqn 7.37 as

$$f(t) = f_1(t) + \dots + f_n(t)$$

the functional series representation can be illustrated as shown in Fig. 7.21. Analysis of such systems has been discussed by Brandstetter and Amarocho (1970) and Bidwell (1970, 1971).

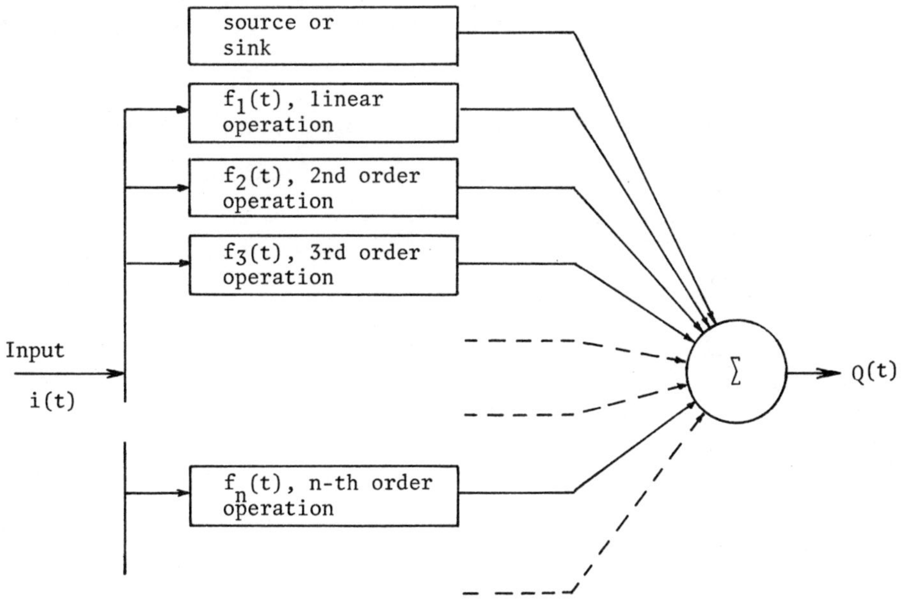

Fig. 7.21. Representation of functional series of a non-linear system.

The linear time-invariant models of the hydrograph have great practical advantages because of their ease of application. Their development can be divided into conceptual models and linear systems analysis models.

The conceptual models are made up of linear elements. These are linear storage (reservoirs) for which the storage S is proportional to the discharge Q and linear channels to account for the time of translation of flow through the catchment. A third element, which is sometimes used, is the time-area curve, which represents the distribution of the time of travel of the runoff from various parts of the catchment to the discharge point.

It should be realized, however, that the elaborate hydrograph models do not overcome the major deficiency of the hydrograph concept. This is the assumption of uniform distribution of rainfall excess over the entire catchment or a constant pattern from one rainfall to another. Such constancy or uniformity of rainfall excess does not occur and, therefore, the hydrograph method is not applicable to large catchments.

Linear reservoirs. The elements of this model are reservoirs in which the storage, S, is assumed to be proportional to the discharge, Q,

$$S = KQ \qquad 7.38$$

where K is a storage coefficient with the dimension of time, the average delay time imposed on the inflow by the reservoir. Storage has both a delaying effect and an attenuation effect on inflow. The delay, which is a translation in time or lag effect, tends to dominate when the storage is distributed. In the case of concentrated storage the attenuation effect is predominant. The distributed storage effect may be achieved by using, for example, $S = 0.5K(I + Q)$ rather than $S = KQ$, if the delay time is not long relative to the time of inflow, or by introducing a separate translation effect, or by routing the flow through a series of concentrated storages. The continuity requirement links the inflow with the outflow and change of storage

$$I - Q = \frac{dS}{dt} \qquad 7.39$$

It is important to realize that basically the storage-discharge relationhip is

$$S = KQ^x \qquad 7.40$$

Substitution from eqn 7.40 into eqn 7.39 yields

$$I - Q = KxQ^{x-1}\frac{dQ}{dt}$$

or

$$\frac{dQ}{dt} = \frac{I - Q}{KxQ^{x-1}} \qquad 7.41$$

For the recession limb of the outflow hydrograph I = 0 and

$$\frac{dQ}{dt} = -\frac{1}{KxQ^{x-2}} \qquad 7.42$$

The curvature of the recession limb is given by

$$\frac{d^2Q}{dt^2} = \frac{x - 2}{KxQ^{x-1}} = \frac{1}{K}\,\frac{x - 2}{x}\,\frac{1}{Q^{x-1}} \qquad 7.43$$

Shen (1963) used eqn 7.42 to check the reservoir (or catchment) for linearity, and it may be written as

$$\log(-\frac{\Delta Q}{\Delta t}) = -(x - 2)\log Q - \log(Kx) \qquad 7.44$$

which is a linear equation and a straight line on log-log paper. The slope of this line gives $-(x - 2)$ and the intercept at $Q \simeq 0$ is Kx. Hence, both x and K can be determined. At Q = 1, log Q = 0 and eqn 7.44 becomes

$$\log(-\frac{\Delta Q}{\Delta t}) = -\log(Kx)$$

or

$$\frac{\Delta Q}{\Delta t} = -\frac{1}{Kx} \qquad 7.45$$

When x = 2, $dQ/dt = -1/(2K)$ and the recession curve is a straight line on a linear scale; when $x = 1$, eqn 7.42 yields $dQ/dt = -Q/K$ and the reservoir is linear. Hence, from the measured outflow hydrograph the storage characteristics can be determined.

For a linear reservoir substitution for dS from eqn 7.38 into eqn 7.39 yields

$$Q + K\frac{dQ}{dt} = I \tag{7.46}$$

which, when multiplied by $e^{t/K}$, can be written as

$$\frac{d}{dt}(Qe^{t/K}) = \frac{1}{K} Ie^{t/K}$$

or

$$Qe^{t/K} = \frac{1}{K}\int Ie^{t/K}dt + \text{const} \tag{7.47}$$

This can be solved analytically or numerically if $I = f(t)$ is known (Yevjevich, 1959).

For instantaneous inflow of a unit volume of storage

$$Q = e^{-t/K}[\frac{1}{K}\int e^{t/K}\delta(0)dt + \text{const}]$$

$$Q = \frac{1}{K}e^{-t/K} \qquad t > 0$$

or

$$u(0, t) = \frac{1}{K}e^{-t/K} \tag{7.48}$$

where the integral is a Laplace transform of the δ-function which simply picks out the value of the function at $t = 0^+$, i.e.,

$$L[\delta(0)] = \int\delta(0)e^{-pt}dt = e^{p0^+} = 1$$

Thus, the response function of a linear reservoir to the pulse input is a sudden jump at the instant of inflow followed by an exponential decline. The linear reservoir model was introduced by Zoch (1934).

The routing equation for a single reservoir when I is constant and $Q = 0$, $t = 0$ is

$$Q_t = I\int_0^t \frac{1}{K}\exp[-\frac{t-\tau}{K}]d\tau = Ie^{-t/K}e^{\tau K}\Big|_0^t = I(1 - e^{-t/K}) \tag{7.49}$$

where $Q \rightarrow I$ as $t \rightarrow \infty$. If the inflow stops at time $t = t_0$, when $Q = Q_0$, then the outflow at time $t > t_0$ is $Q = Q_0\exp[-(t - t_0)/K]$. The inflow I_0 at $t = 0$ produces an outflow $Q_1 = I_0(1 - e^{-1/K})$ at the end of the first unit of time, $t = 1$. An inflow I_1 at $t = 1$ produces a similar outflow Q_2. By superposition, Fig. 7.22,

$$Q_2 = I_1e^{-1/K} + I_2(1 - e^{-1/K})$$

or in general

$$Q_t = I_{t-1}e^{-1/K} + I_t(1 - e^{-1/K}) \tag{7.50}$$

which expresses the flow rate at the end of an interval in terms of the discharge at the end of the preceding interval.

When one linear reservoir discharges into the next, the outflow from the first becomes the inflow for the second and

$$u(0,\ t) = \int_0^t I_2(\tau)\ \frac{1}{K_2} \exp[-\ \frac{t\ -\ \tau}{K_2}]d\tau = \int_0^t \frac{1}{K_1} e^{-\tau/K_1}\ \frac{1}{K_2} e^{-t/K_2}\ e^{\tau/K_2}\ d\tau$$

$$= \frac{1}{K_1 K_2}\ e^{-t/K_2} \int_0^t e^{\tau(K_1-K_2)/K_1K_2}\ d\tau$$

$$= \frac{1}{K_1\ -\ K_2}\ e^{-t/K_2}[e^{t(K_1-K_2)/K_1K_2}\ -\ 1]$$

$$= \frac{1}{K_1\ -\ K_2}\ (e^{-t/K_1}\ -\ e^{-t/K_2})$$

For two equal reservoirs

$$u(0,\ t) = \int_0^t \frac{1}{K}\ e^{-\tau/K}\ \frac{1}{K}\ e^{-t/K}\ e^{\tau/K}\ d\tau = \frac{t}{K^2}\ e^{-t/K}$$

This can be extended to linear reservoirs in series. From

$$Q_1 = I_2 \quad \text{and} \quad (1 + K_1D)Q_1 = I_1 \qquad 7.51$$

where $D = d/dt$

$$(1 + K_2D)Q_2 = I_2 + \frac{I_1}{(1 + K_1D)}$$

In general

$$(1 + K_1D)(1 + K_2D)\ldots.(1 + K_nD)Q_n(t) = I(t)$$

or

$$Q_n(t) = \frac{1}{(1 + K_1D)(1 + K_2D)\ldots(1 + K_nD)} I(t) = \frac{1}{\prod_{i=1}^{n} (1 + K_iD)}\ I(t) \qquad 7.52$$

where Π signifies a product series.

Dooge (1959) showed that for instantaneous flow the general solution can be expressed as a sum of n terms for a given volume of uniformly distributed rainfall excess V

$$\frac{Q_n(t)}{V} = \frac{K_1^{n-2}\ e^{-t/K_1}}{\Pi(K_1\ -\ K_i)} + \frac{K_2^{n-2}\ e^{-t/K_2}}{\Pi(K_2\ -\ K_i)} + \ldots + \frac{K_n^{n-2}\ e^{-t/K_n}}{\Pi(K_n\ -\ K_i)} \qquad 7.53$$

This expression is difficult to evaluate. For n equal reservoirs the solution simplifies to

$$Q_n(t) = \frac{V}{K}\ \frac{(t/K)^{n-1}}{(n\ -\ 1)!}\ e^{-t/K} \qquad 7.54$$

or allowing for non-integral values of n

$$u(o,\ t) = \frac{1}{K}\ (\frac{t}{K})^{n-1}\ \frac{1}{\Gamma_n}\ e^{-t/K} \qquad 7.55$$

which is the probability density function of the Gamma distribution and was derived by Nash (1957).

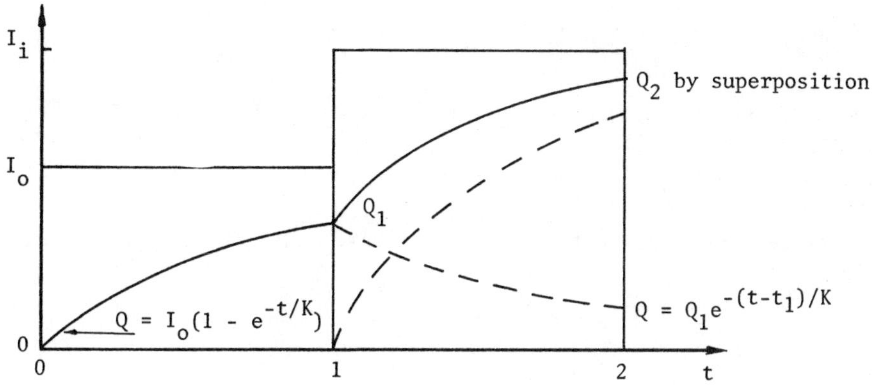

Fig. 7.22. Illustration of superposition for a linear reservoir.

Nash suggested that the series of n reservoirs is a sufficiently general model of the catchment and that eqn 7.55 represents the general equation of the IUH. The assumption of uniform reservoirs means that the effect of the shape of the catchment on the hydrograph shape cannot be found. Figure 7.23 illustrates schematically the effect of storages in series. A significant advantage is that the ouflow can be calculated with the aid of statistical tables. Note that for n = 1 the IUH is a decay function with its maximum at t = 0 and with an instantaneous rising limb. With increasing n the time to peak increases. McCuen (1973) showed by analysis of

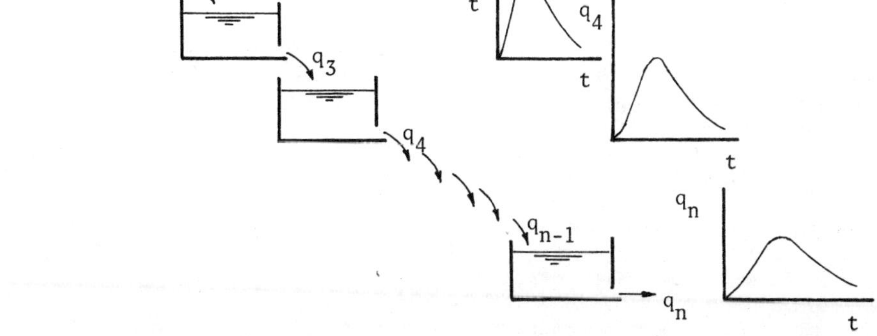

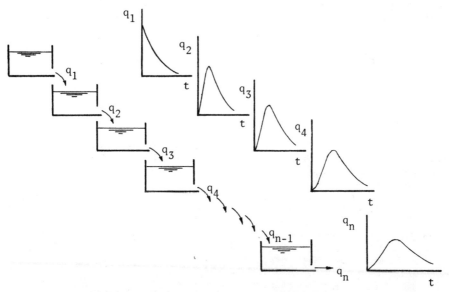

Fig. 7.23. Routing of instantaneous inflow through a series of linear storage reservoirs (Nash's model).

sensitivity that the rising limb is more sensitive to the values of n and K than the falling limb. The sensitivity of the peak was found to be one-third of the maximum change. For a total delay time of K the maximum outflow occurs at time (n - 1)K.

The S-curve from the Nash model is

$$S(t) = \int_0^t u(0, t)dt = \frac{1}{K}\int_0^t (\frac{t}{K})^{n-1} \frac{1}{(n-1)!} e^{-t/K} dt$$

$$= \Gamma(\frac{t}{K}, n - 1) \qquad 7.56$$

Thus, the ordinates of the finite duration unit hydrograph can be obtained from the tabulated values

$$u(T, t) = \frac{S(t) - S(t - T)}{T}$$

$$= \frac{V}{T} [\Gamma(\frac{t}{K}, n - 1) - \Gamma(\frac{t - T}{K}, n - 1)] \qquad 7.57$$

The IUH could be looked upon as the frequency distribution of the arrival times at the outlet of water particles, after an instantaneous uniform application of a unit volume of rainfall excess over the catchment. The average time, the expected value E(t), is the distance in time or lag between the centroids of the IUH and the input hyetograph (zero for instantaneous input). For the Gamma distribution

$$E(t) = \int_0^\infty u(0, t)tdt = \int_0^\infty t(\frac{t}{K})^{n-1} \frac{1}{(n-1)!} e^{-t/K} d(\frac{t}{K})$$

$$= nK \int_0^\infty (\frac{t}{K})^n \frac{1}{n!} e^{-t/K} d(\frac{t}{K}) = nK \qquad 7.58$$

where the integrand is another Gamma distribution of order (n + 1) and the area is equal to unity. The spread in the lag time, the variance, is

$$Var(t) = E(t^2) - [E(t)]^2 = \int_0^\infty u(0, t)t^2dt - (nK)^2$$

$$= K^2n(n + 1) - n^2K^2 = K^2n \qquad 7.59$$

The integrand in this case is a Gamma distribution of order (n + 2) and the area is equal to unity.

The values of K and n in the Nash model may be evaluated by the method of moments (Nash, 1959, 1960). Every element of the rainfall input $id\tau$, Fig. 7.24, produces an output a distance T_u later than τ and since the system is linear, the average of all values must satisfy

$$T_Q = T_i + T_u \qquad 7.60$$

Consequently, the first moments about the origin are also related as

$$M_{1Q} = M_{1i} + M_{1u} = nK \qquad 7.61$$

since the first moment is equal to the area times the distance to the centroid. Nash showed using convolution, and Diskin (1967) employing the Laplace tranform, that the second moments about the centroids of the respective areas are related by

$$M_{2u} = M_{2Q} - M_{2i} = nK^2 \qquad 7.62$$

Whence, from the simultaneous solution of eqns 7.61 and 7.62

$$n = \frac{(M_{1Q} - M_{1i})^2}{M_{2Q} - M_{2i}} \qquad 7.63$$

and

$$K = \frac{M_{2Q} - M_{2i}}{M_{1Q} - M_{1i}} \qquad 7.64$$

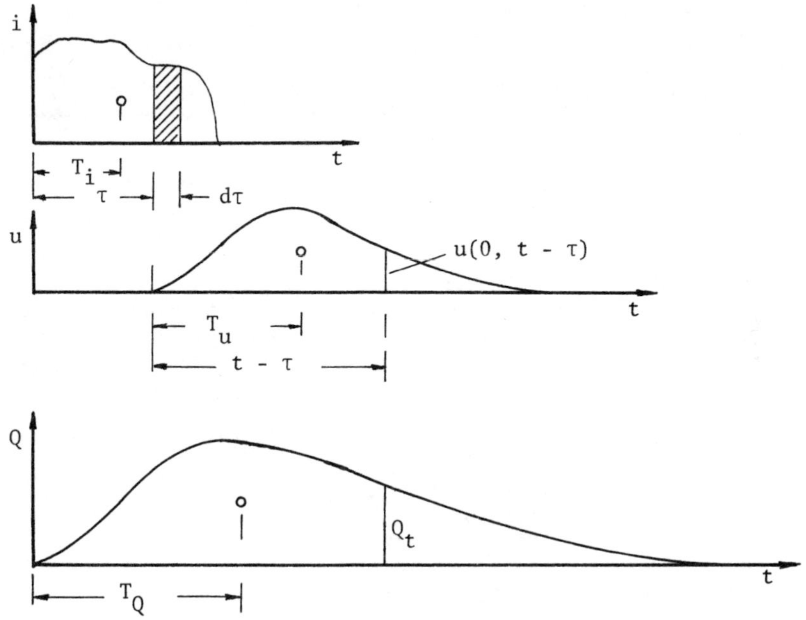

Fig. 7.24. Illustration of terms for the Nash method of moments evaluation of K and n. T_i, T_u and T_Q are the centroid distances of the rainfall excess hyetograph, the unit hydrograph and the hydrograph, respectively.

From the analysis of 90 catchments in the U.K. Nash found good correlation between n and the length of the main channel ($n = 2.4\ L^{0.1}$, L in miles), and between K and the area, slope and length ($K = 11\ A^{0.3}S^{-0.3}L^{-0.1}$, where A is in sq. mi, L in miles and slope in parts per thousand).

The *linear channel* is an analytical model postulating a channel in which the time of translation t_{tr} of discharge Q through a given length of the channel remains constant for all values of Q. The shape of the hydrograph routed through a linear channel remains unchanged. For inflow $I = f(t)$ the outflow is $Q = f(t - \tau)$. A linear reservoir and a linear channel yield

$$Q = \frac{1}{K} \exp[- \frac{(t - \tau)}{K}] \qquad t > \tau \qquad 7.65$$

In a linear system the effect of translation can be added before or after the reservoirs. In a real non-linear catchment, the translations have to be allowed for between the non-linear reservoirs, i.e., in the true sequence. Figure 7.25 illustrates the case of a linear channel and a linear reservoir in series. Although

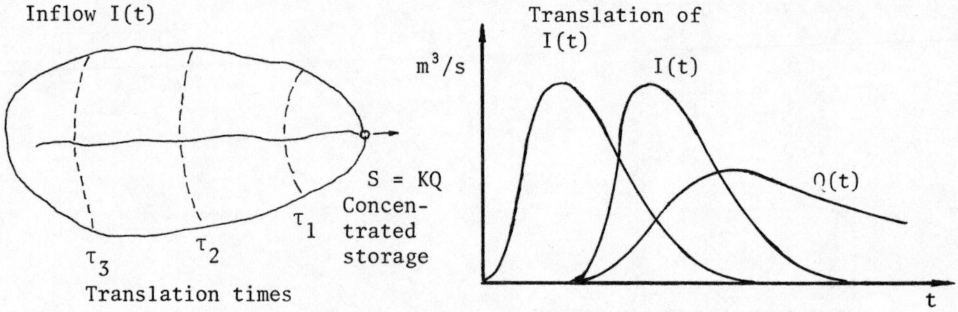

Fig. 7.25. Illustration of translation and concentrated storage in series.

convenient, the separation of translation from attenuation is an arbitrary analytical step, because any storage produces both. The true meaning of lag is the time it takes for the effect of an element of rainfall excess to reach the outlet and not the time of travel of a drop of water. The effect is produced by the wave movement and the translation of water.

The *area-time curve* is used to introduce the time of travel of the runoff to the discharge point. The name is misleading because it is not the travel time of a drop of water but the time for the effect of an input to be felt at the outlet, a storage delay. For a given element of input $i d\tau$, at time τ, every element of the catchment has a specific time period between the time of occurrence of the input and the centre of mass of the resulting outflow from the catchment. The contribution from the area dA, Fig. 7.26, to the discharge is

$$dQ(t) = i(\tau)dA(t - \tau) = \frac{dA(t - \tau)}{d\tau} i(\tau)d\tau$$

or

$$Q(t) = \int_0^t \frac{dA(t - \tau)}{d\tau} i(\tau)d\tau$$

Comparison with eqn 7.34 shows that

$$u(0, t - \tau) = \frac{dA(t - \tau)}{d\tau} \qquad 7.66$$

i.e., the IUH is of the same shape as the derivative of the area-time curve.

The delay time for any point varies with the discharge in the channel. Therefore, it is helpful initially to prepare a relative delay area-time diagram in which the delay time axis is in terms of $t/t_{max} = \kappa$, Fig. 7.26. The relative times are marked at their appropriate locations on the map and the isochrones are interpreted and drawn. This yields dA blocks which correspond with the gradation of the abscissa κ in Fig. 7.26. An estimate of the storage delay time K' for any element of the catchment can be obtained from $\Sigma(L/S^{\frac{1}{2}})$, where L is the length and S the slope of the elements of the flow path. The summation is along the flow path between the element and the outlet. The isochrone map of the relative delay time could be prepared as described by Laurenson (1964):

(1) A large number of points, uniformly distributed over the catchment, is marked on the contour map so that each point is on a topographic contour.
(2) For each point the distances between adjacent contours along the flow path to the outlet are tabulated and each is raised to the power of 3/2, i.e., $L/S^{\frac{1}{2}} = L^{3/2}/H^{\frac{1}{2}}$, where H is the contour interval (constant). If the outlet is not on a contour, the length of the lowest reach is multiplied by $(H/H_1)^{\frac{1}{4}}$, where H_1 is the fall of the lowest reach. The lengths to the power of 3/2 are summed for each point. These sums are proportional to the delay times and are reduced to give a delay time of unity for the upstream end of the catchment.
(3) The relative delay times are marked on the map at their respective locations. After this the isochrones can be drawn. These may have discontinuities at the ridge lines where the delay time depends on which of the paths the flow takes.

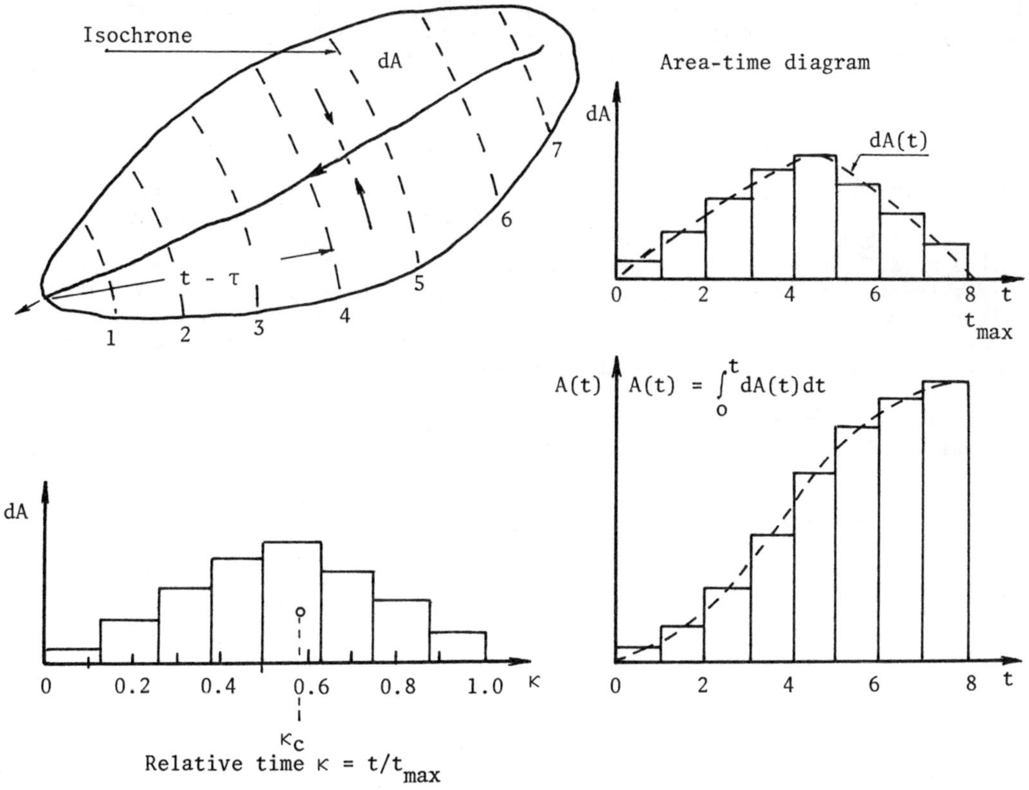

Fig. 7.26. Illustration of the area-time concept.

For the entire catchment the input is the total rainfall excess and the output, or response, is the total surface runoff. The time difference between their centres of mass is the average dimensionless delay time or lag

$$\bar{\kappa}_L = \frac{\int_0^t \int_A i_e \kappa \, dA \, dt}{P_e A}$$

where κ is the dimesnionless delay time for the input at dA, T is the duration of rainfall of excess intensity i_e, and P_e is the rainfall excess. For a uniform areal distribution of rainfall excess the average lag at a given instant is

$$\bar{\kappa}_{Lt} = \Sigma(A_i\kappa_i)/\Sigma A_i = \kappa_c \qquad 7.67$$

This is also the expression for the centroid of the dimensionless area-time diagram on the κ-axis, i.e., for any element of uniformly distributed rainfall excess occurring at any time t the lag is equal to the storage delay time of points on the catchment corresponding to the centroid of the area-time diagram. In dimensional terms the mean lag time

$$\bar{t}_{LT} = K_t \qquad 7.68$$

The lag time, $\bar{t}_{Lt}$, can be determined from rainfall and discharge records (Fig. 7.9) and the κ-axis can be made dimensional by multiplying it by the ratio of $\bar{t}_L$ to the centroid value of κ_c, if the effect of non-uniform rainfall excess on the lag is neglected. Thus, for any value of κ

$$t = \kappa \frac{\bar{t}_{Lt}}{\kappa_c} = K' \qquad 7.69$$

This method of distribution of the storage delay into K' values is based on two assumptions: uniform distribution of rainfall excess over the catchment, and linear catchment response. The first assumption can be satisfied by using only rainfalls with spatially approximately uniform rainfall excess. The second assumption is incorporated in eqn 7.67 which gives the average delay time in terms of the centroid of the dimensionless time diagram. The delay time is proportional to the area being drained and can only be independent of the volume of flow if the response is linear. In general, travel time varies with flow rate and $\bar{t}_{Lt}$ and K_t vary throughout the flood, but the major factor affecting the lag time of a linear catchment is L or $\sqrt{A}$. Askew (1970) fitted a variable lag time expression to the data from five catchments. The optimum lag time expressions were as follows:

	Imperial	S.I. Units	± St. error of estimate for t_d as % of calculated lag
1.	$t_d = 8.28A^{+0.57}q_{wm}^{-0.23}$	$t_d = 2.12A^{+0.57}q_{wm}^{-0.23}$	+22.9% -18.7%
2.	$t_d = 6.0.A^{+0.54}S_a^{-0.16}q_{wm}^{-0.23}$	$t_d = 1.60A^{+0.54}S_a^{-0.16}q_{wm}^{-0.23}$	+20.6% -17.1%
3.	$t_d = 2.91L^{+0.80}S_a^{-0.33}q_{wm}^{-0.23}$	$t_d = 0.877L^{+0.80}S_a^{-0.33}q_{wm}^{-0.23}$	+19.1% -16.0%

t_d	-lag time, C of M rainfall-excess to C of M direct runoff, in hours	... in hrs
q_{wm}	-weighted mean discharge, in cusecs	... in m^3/s
A	-catchment area, in sq. mi	... in km^2
L	-main stream length, in mi	... in km
S_a	-overland slope from grid/contour intersection method, dimensionless	... dimensionless

The weighted mean discharge was defined as "the mean rate of total discharge over the time of occurrence of direct runoff, weighted in proportion to the direct runoff discharged at that rate". Laurenson (1964) expressed lag in hours as $t = 64(\bar{q})^{-0.27}$, where $\bar{q}$ is the mean runoff in ft^3/s.

Laurenson (1962, 1965) used a series on non-linear concentrated storages. The storage relationship

$$S= KQ$$

was retained but K was made a function of Q

$$K = kQ^n \tag{7.70}$$

where k and n are constants and other symbols are as defined earlier. A constant value of n implies a constant non-linearity throughout the catchment. The continuity equation then in finite difference form is

$$K \frac{\Delta Q}{\Delta t} + Q = I$$

At the beginning of the time interval Δt, $Q = Q_1$, $K = K_1$ and at the end $Q = Q_2$, $K = K_2$. Thus,

$$\frac{K_2Q_2 - K_1Q_1}{\Delta t} + \frac{Q_2 + Q_1}{2} = \frac{I_2 + I_1}{2}$$

or

$$Q_2 = \frac{\Delta t}{2K_2 + \Delta t} I_1 + \frac{\Delta t}{2K_2 + \Delta t} I_2 + \frac{2K_1 - \Delta t}{2K_2 + \Delta t} Q_1$$

or

$$Q_2 = C_1I_1 + C_2I_2 + C_3Q_1 \tag{7.71}$$

Since $K_2 = f(Q_2)$ it cannot be calculated directly. Laurenson used the initial estimate of $K_1 = K_2$ and found that when $|n| < < 1.0$ two iterations gave satisfactory accuracy.

By analysis of observed hydrographs K can be related to Q. For one concentrated storage $K = t_L$ (the lag time) and Q is the mean flow during the period of runoff. It has been suggested that a better average to use is

$$\bar{Q} = \sum_{i=1}^{t} (Q_i^2)/\sum_{i=1}^{i} Q_i \tag{7.72}$$

which weights the rate of discharge in each time interval by the volume of flow in that interval. For distributed storage the catcment lag is divided between a number of concentrated storages in series. The total K must equal the sum of the K' values of the individual concentrated storages. The catchment lag is equal to the sum of the individual lags only when all the inflow is routed through all the storages. Equation 7.69 now applies for any value of κ

$$t = \kappa \frac{\bar{t}_{Lt}}{\kappa_c} = \frac{\kappa}{\kappa_c} kQ^n = K' \tag{7.73}$$

If the isochrones are at $\Delta\kappa = 0.1$ intervals the catchment will have ten concentrated storages with $K' = 0.1\ kQ^n/\kappa_c$ each. The error in κ_c (eqn 7.67) by not allowing for Q^n is small when $|n| < < 1.0$.

The feature of Laurenson's model is that each storage is related to an identifiable subarea of the catchment. This is particularly important with large catchments where the rainfall only rarely covers the entire catchment. The storages are assigned with the aid of the lag time, which is estimated from the contour map as $\Sigma(L/\sqrt{S})$.

Porter (1975) extended Laurenson's method by using sub-catchments, defined by internal watersheds instead of areas delineated by isochrones, and allowing for the translation time between the successive storages. The storages may occur both in series and parallel. By introduction of the translation time between the concentrated storages an input will not produce an instantaneous output and, hence, the model is in better agreement with nature. The catchment lag time is now

$$\bar{t}_L = K + T \qquad 7.74$$

where T is the translation time and is assumed to be independent of Q. Thus, all the non-linearity is assigned to the storages. The translation times T' between the storages are assumed to be equal.

Clark (1945) estimated the delay time from the continuity equation $K = -Q/(dQ/dt)$, evaluated at the point of inflection of the recession limb of the hydrograph. The inflection point marks the time of arrival at the outlet of the flood from the most distant point of the catchment.

A different approach to the description of the direct runoff hydrograph was introduced by Edson (1951). He described the hydrograph by the equation

$$Q = Ct^x e^{-yt} \qquad 7.75$$

where C, x, y are parameters which have to be determined, and t is time measured from the beginning of the rainfall excess. Both the area of the catchment contributing to the runoff and Q are assumed to increase proportionally with t^x for $x > 1$. This form of approach has been used by a number of subsequent researchers. The basic hydrograph profile is shown in Fig. 7.27 and is described by eqn 7.75 from 0 to C.

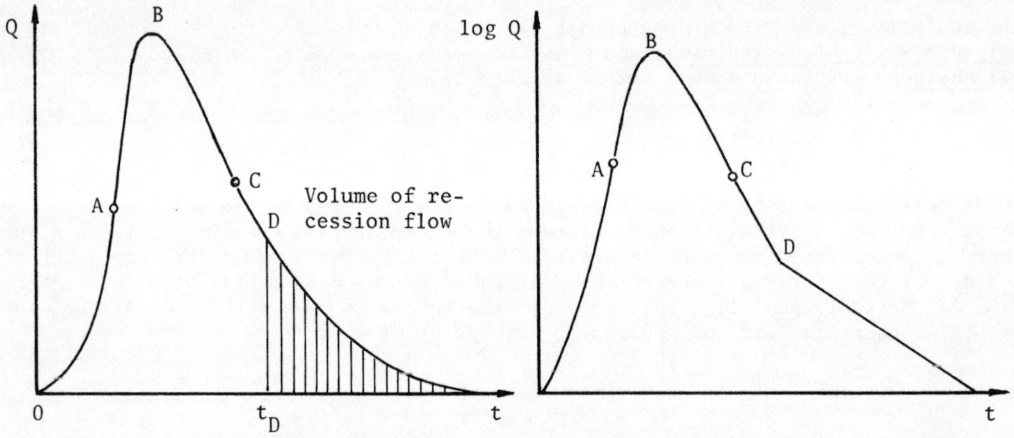

Fig. 7.27. Definition sketch of the surface runoff hydrograph in linear and semi-logarithmic representation. A and C are inflection points, B is the peak and D refers to the start of the recession curve.

The hydrograph recession is described by

$$Q = Q_D e^{-(t-t_D)/k}$$

where k is the recession constant given by the slope of the recession limb in the log Q versus time plot. From

$$\frac{dQ}{dt} = Cxt^{x-1}e^{-yt} - Ct^{x}ye^{-yt} = 0$$

the time to peak is

$$t_p = x/y \qquad 7.76$$

The minima correspond to t = 0 and t = ∞. With this value for the peak

$$C = Q_p e^{x}/t_p^{x} \qquad 7.77$$

The volume of recession flow is

$$V_r = \int_{t_D}^{\infty} Q\, dt = Q_D \int_{t_D}^{\infty} e^{-(t-t_D)/k} = kQ_D = nV \qquad 7.78$$

where V is the total runoff volume and $n = V_r/V$. The value of

$$Q_D = Q_p(t_D/t_p)^x\, e^{x-yt_D} = Q_p(t_D/t_p)^x\, e^{-x[t_D/(t_p-1)]} = nV/k$$

from which

$$x = \frac{\ln(nV/kQ_p)}{\ln(t_D/t_p) - t_D/t_p + 1} \qquad 7.79$$

Thus, all three parameters are known in terms of hydrograph properties. The inflection points are $t_A = t_p(1 - \sqrt{1/x})$ and $t_C = t_p(1 + \sqrt{1/x})$. When compared with observed hydrographs the calculated hydrograph tends to overestimate between 0 and A.

Henderson and Wooding (1964) and Wooding (1965, 1966) developed a hydrograph theory based on concepts of fluid mechanics. The treatment assumes that the Froude number Fr < 2 and will not apply to very steep catchments where the flow is in the form of a sequence of roll waves. The catchment, or a sub-catchment, is represented by a tilted V-shaped double-plane surface in which the valley forms the channel, Fig. 7.28. The runoff then consists of a thin sheet flow (the overland flow) and a small

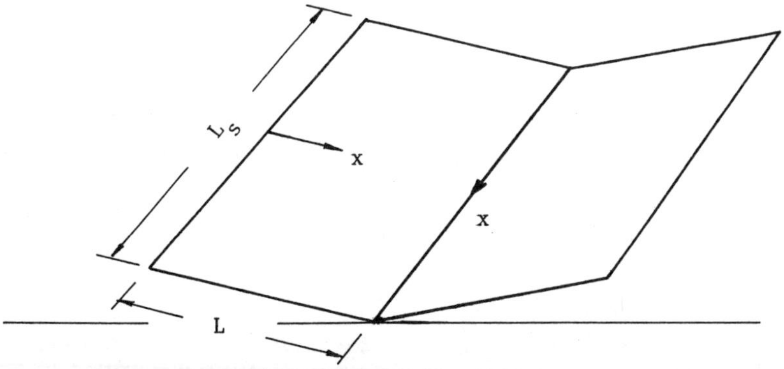

Fig. 7.28. Model of the catchment.

stream flow. The small streams combine and ultimately form the river which drains the catchment. All of these flows have to obey the equations of continuity and momentum. After neglecting minor contributions such as the momentum of the rain and the effect of the distribution of the velocity, and approximating the slope $S_o = \sin\theta \simeq \tan\theta$, the x-component of the equation of motion becomes

$$\frac{\partial y}{\partial t} + u\frac{\partial u}{\partial x} + g\frac{\partial y}{\partial x} = (i - f + \frac{2q_L}{B})\frac{u}{y} - (1 + \frac{2y}{B})\frac{\tau_o}{\rho y} + gS_o \qquad 7.80$$

and the continuity equation becomes

$$\frac{\partial y}{\partial t} + u\frac{\partial y}{\partial x} + y\frac{\partial u}{\partial x} = i - f + \frac{2q_L}{B} \qquad 7.81$$

where i is the rate of rainfall, f is the rate of infiltration, q_L is the lateral inflow, and y is the depth of flow. (A more detailed formulation of these equations is given by Grace and Eagleson, 1966). From Fig. 7.29

$$\frac{\partial H}{\partial x} = \frac{\partial}{\partial x}(h + \frac{u^2}{2g}) = - S_o + \frac{\partial y}{\partial x} + \frac{u}{g} + \frac{\partial u}{\partial x}$$

where S_o is of the order of 10^{-2} radians, i.e., $S_o = 0(10^{-2})$, $g = 0(10)$ m/s^2, $i - f = 0(10^{-6})$ m/s. For overland flow $B \sim \infty$, $q_L = 0$, $y = 0(10^{-2})$ m, $U = 0(10^{-1})$ m/s. For stream flow $y = 0(10^0)$ m, $U = 0(10^0)$ m/s, $q_L = 0[(i - f)L] = 0(10^{-6}L)$ m^2/s, where L is the half width of the area drained. Thus, for overland flow $[(i - f)U/y]/gS_o = 0(10^{-4})$, and for stream flow $[(2q_L/B)U/y]/gS_o = 0(10^{-5}L/B)$. It follows that the (i - f) terms are small in comparison to the gravity effect and may be neglected.

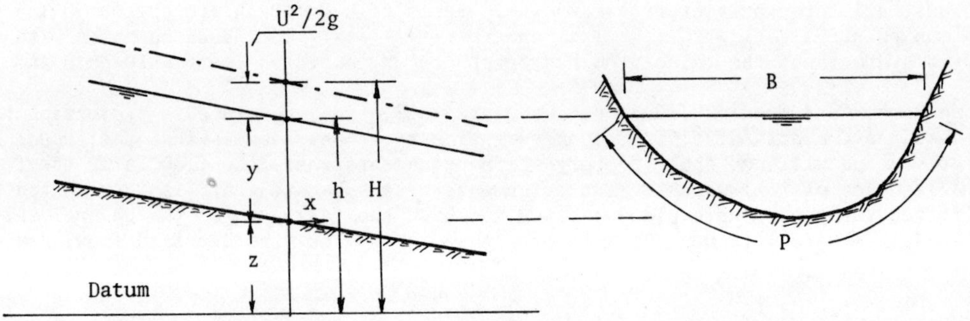

Fig. 7.29. Definition sketch.

Indeed, in overland flow all terms in eqn 7.80 are small in comparison to the slope and friction terms. The friction is described by

$$\tau_o = \gamma y S_o = c_f \rho U^2/2$$

or

$$U = \sqrt{2g/c_f}\,\sqrt{yS_o} \qquad 7.82$$

where c_f is a function of the Reynolds number and relative roughness. Introduction of C for $\sqrt{2g/c_f}$ leads to the Chezy formula and if C is assumed to be constant

$$q = Uy = \alpha\, y^{3/2}$$

where

$$\alpha = \sqrt{2gS_o/c_f} = C\sqrt{S_o} \qquad 7.83$$

or in general

$$q = \alpha\ y^a$$

For laminar flow $c_f = 4/Re = 4\nu/Uy$, and used in conjunction with eqns 7.82 and 7.83 yields

$$q = \frac{gS_o}{2\nu}\ y^3$$

or

$$\alpha = \frac{gS_o}{2\nu} \quad \text{and} \quad a = 3$$

For turbulent flow, using Manning's equation and eqn 7.82

$$c_f = 2g\ n^2\ y^{-1/3}$$

and

$$q = \frac{\sqrt{S_o}}{n}\ y^{5/3}$$

Experimental information indicates that $a \simeq 2$ satisfies a wide variety of surfaces (ranging from short grass to tar and gravel).

Thus, the continuity and momentum equations reduce to

$$\frac{\partial y}{\partial t} + \frac{\partial q}{\partial x} = i - f = i_e \qquad 7.84$$

and

$$q = \alpha\ y^a \qquad 7.85$$

This result gives the so-called kinematic wave equations, which were treated by Lighthill and Whitham (1955). Their solution is essentially that of the continuity equation (q is a function of y only and the friction slope $S_f = S_o$) and does not describe profile changes due to dynamic effects. The boundary conditions are

$$y = 0 \begin{cases} 0 \le x < L & t = 0 \\ x = 0 & t > 0 \end{cases}$$

and i_e is spatially constant but could vary with time.

Equations 7.84 and 7.85 are solved by the method of characteristics. The basic definitions

$$dq = \frac{\partial q}{\partial x}\ dx + \frac{\partial q}{\partial t}\ dt \qquad 7.86$$

$$dy = \frac{\partial y}{\partial x}\ dx + \frac{\partial y}{\partial t}\ dt \qquad 7.87$$

together with

$$\frac{\partial q}{\partial t} = \alpha\ ay^{a-1}\ \frac{\partial y}{\partial t} \qquad 7.88$$

and eqn 7.84 lead to

$$\left(\frac{dx}{dt}\right)^2 = \alpha\, ay^{a-1} \frac{dx}{dt} = 0 \qquad 7.89$$

This equation has the roots

$$\left.\begin{aligned} \frac{dx}{dt} &= c = \alpha\, ay^{a-1} = aU \\ \text{and} \qquad & \\ \frac{dx}{dt} &= c = 0 \end{aligned}\right\} \qquad 7.90$$

where c is the celerity of the wave. The zero wave speed corresponds to zero slope, infinite roughness, or zero depth. Equation 7.90 defines the characteristic, the path of an infinitesimal disturbance in the x - t plane (as seen by an observer moving with it), and the positive sign shows that the wave propagates downstream only at a celerity c. The celerity can also be expressed as $c = dq/dy = \alpha\, ay^{a-1} = U + ydU/dy$ and if U is defined by the Chezy equation, c = (3/2)U for a kinematic wave [(5/3)U when Manning's equation is used]. For a dynamic wave $c = (5/3)U + \sqrt{gy}$. Further, the same four eqns 7.84, 7.86, 7.87 and 7.88 yield the equations

$$\frac{dq}{dx} = i_e \qquad 7.91$$

$$\frac{dy}{dt} = i_e \qquad 7.92$$

$$\frac{dq}{dt} = i_e \alpha\, ay^{a-1} \qquad 7.93$$

$$\frac{dy}{dx} = \frac{i_e}{\alpha\, ay^{a-1}} \qquad 7.94$$

For overland flow along the characteristic defined by eqn 7.90

$$x - x_o = \alpha a \int_{t_o}^{t} y^{a-1}\, dt$$

where

$$y = \int_{t_o}^{t} i_e\, dt \quad \text{and} \quad i = f(t)$$

When $i_e = i_{ec}$ = constant, $y = i_{ec}t$ and

$$x - x_o = \alpha\, i_{ec}^{a-1} (t - t_o)^a \qquad t \leq T \qquad 7.95$$

where T is the duration of the rain. Equation 7.95 defines a curve on the x - t plane for each selected value of x_{oi}, Fig. 7.30. Characteristics above the limiting one imply inflow across the $x_o = 0$ boundary condition. At x = L the limiting characteristic defines the time of concentration t_c, which is seen to be dependent on i_e. The depth will increase everywhere at $dy/dt = i_{ec}$. For a given x, a different characteristic curve applies for each instant, and $q_{in} = q_{out}$ must be satisfied, except at x = 0 where $q_{in} = 0$. This disturbance at x = 0 travels downstream along the limiting characteristic and reaches a point x_j at t_j. At this point, and downstream of it, $y = i_{ec}t_j$ = constant, because x_j is the most downstream point reached by disturbances created by a lack of inflow at x = 0.

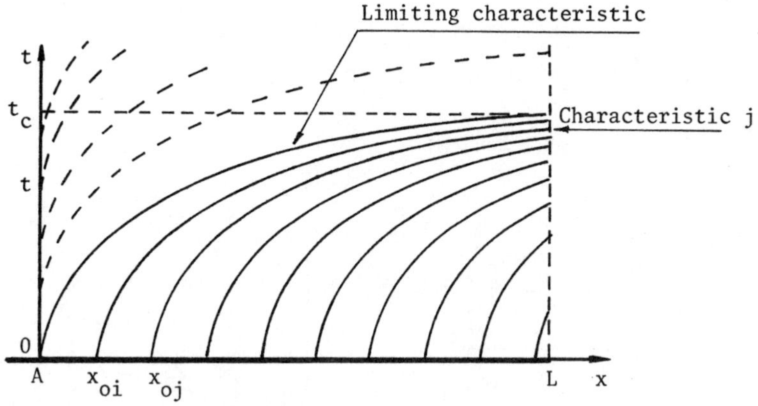

Fig. 7.30. Kinematic wave characteristics.

From eqn 7.95

$$t_j = \left(\frac{x_j i_{ec}^{1-a}}{\alpha}\right)^{1/a} \qquad 7.96$$

and $t_j = t_c$ when $x_j = L$. The water level, Fig. 7.31, at any point is given by

$$y_j = i_{ec} t_j = \left(\frac{x_j i_{ec}}{\alpha}\right)^{1/a} \qquad 0 \leq x \leq x_j$$

where

$$x_j = \int_0^{t_j} c \, dt = \alpha a \int_0^{t_j} y^{a-1} \, dt \qquad 7.97$$

and

$$y = i_{ec} t_j \qquad \text{for} \qquad x > x_j$$

After the rain stops all four equations (7.91-7.94) are equal to zero. This means that along a characteristic line the depth, discharge and wave celerity remain constant. Thus, a point at $t = T$ on the profile defines y, x and also t. Hence, this particular characteristic (x, t) can be located. All future locations of this particular depth are on the same characteristic and these locations are given by

$$\Delta x = c \, \Delta t = \alpha \, a y^{a-1} (t - T)$$

From eqn 7.97 at $t = T$

$$x = \frac{\alpha \, y^a}{i_{ec}} \qquad 7.98$$

and for $t > T$

$$x = \frac{\alpha \, y^a}{i_{ec}} + \Delta x = \alpha \, y^{a-1} [y i_{ec}^{-1} + a(t - T)] \qquad 7.99$$

When $T < t_c$, the depth $y = i_{ec} T$ is reached at $x_j = x_T$ and remains constant until x = L, which is at

$$t_p = T + \frac{L - x_T}{c} = T + \frac{L - \alpha\, y_{LT}^a / i_{ec}}{\alpha\, a y_{LT}^{a-1}} = T + \frac{t_c^* - T}{a} \qquad 7.100$$

where $t_c^* = L/(\alpha\, y_{LT}^{a-1})$ and $T = y_{LT}/i_{ec}$.

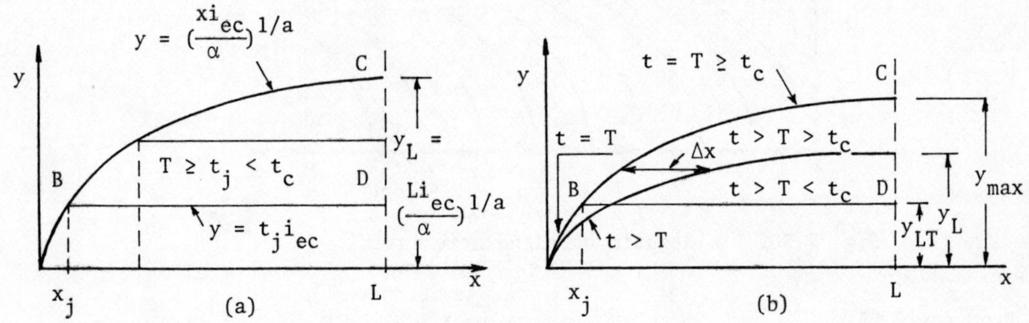

Fig. 7.31. Water surface profiles. (a) Increasing depth $t \le T$. For $x > x_j$ the depth is constant. (b) Decreasing depth $t > T$. Rain stopped at $t = T$ and disturbance travelled to B; y_{LT} is maximum constant depth for $x > x_B$. For $t > T$ depth remains constant and B moves towards D until they coincide.

Figure 7.32 illustrates typical outflow hydrographs for a constant depth of rainfall excess, i.e., $d = i_{ec}T$ = constant. When T = 0 the instantaneous input produces the instantaneous hydrograph 0 - M_o - N_o - P_o in Fig. 7.32, which has a constant outflow until t_c^*/a (eqn 7.95). After that y_L decreases as defined by eqn 7.99 with T = 0

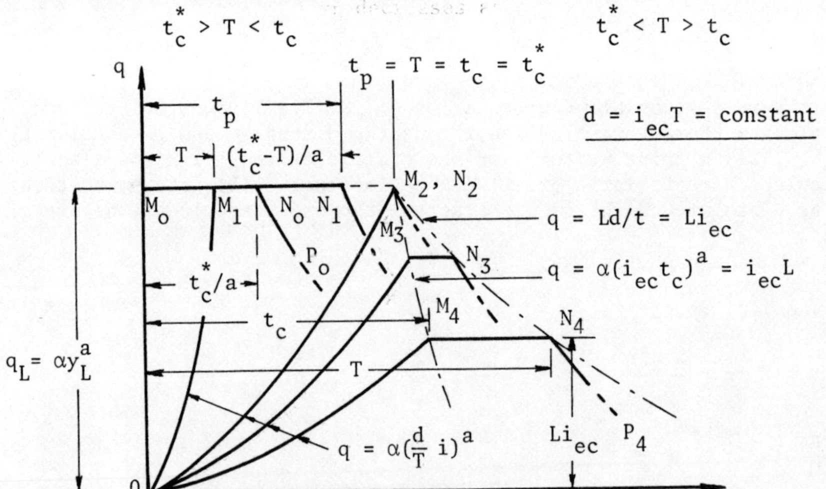

Fig. 7.32. Typical catchment outflow hydrographs after Henderson and Wooding (1964).

and $x = L$, i.e., $y_L = [L/(\alpha\ at)]^{1/(a-1)}$, and

$$q_L = \alpha\ y_L^a = \alpha(\frac{L}{\alpha\ at})^{a/(a-1)} \qquad 7.101$$

For $0 < T < t_c$ the peak discharge remains at this value because of the same total depth at $x = L$, but the peak is reached later at $t = T$. At the limit when $T = t_c$ the peak flow is reached only for an instant at $t = t_p = T = t_c$. For $T > t_c$ the peak discharge still occurs at $T = t_c$ and remains constant for $(T - t_c)$ at $q_L = Li_{ec} = Ld/T$. When the rain stops, $i = 0$, but the value of i_e is not necessarily zero. Infiltration continues as long as there is water on the surface. Figure 7.33 illustrates the effect of a steady infiltration rate.

The same approximate forms of the momentum and continuity equations are used for analysis of the behaviour of small streams, details of which are given by Wooding (1965, 1966) and Eagleson (1970). Figure 7.34 illustrates the results. The λ term was introduced by Wooding. It relates the concentration time of the stream to that of the constant uniform lateral inflow q_{Lmax} from the catchment as $t_s = \lambda t_c$. The value of $\lambda = 0$ when the isochrones are parallel to the stream; $\lambda \simeq 1$ when the isochrones are approximately concentric quadrants of a family of circles with the origin at the outlet; and $\lambda >> 1$ when the isochrones are perpendicular to the stream. Figure 7.34 corresponds to conditions where $a = 2$ for the catchment and $a_s = 3/2$ for the stream, and the rainfall is uniform and steady. The flow rate is $q = ix = \alpha\ y^a$ or $i = \alpha\ y^a/x = \alpha\ y_L^a/L = \alpha\ d^a/L$ for constant depth of flow at $y_L = d = it_c$. The time of concentration $t_c = L/c = L/\alpha\ y_L^{a-1}$. The dimesionless time $T' = T/t_c$ and the dimensionless ordinate $q' = q_L/\alpha\ y_L^a = q_L/\alpha\ d^a$ or $Q' = Q_{sL}/2\alpha d^a L_s$.

For $T' < 1$ the peak value is unity, or αd^2 in dimensional terms, and it persists over the time-interval $(T', \frac{1}{2}T' + \frac{1}{2})$. This "flat top" goes to zero width as $T' \to 1$, i.e., when $T \to t_c$, the time required for the catchment flow to reach a steady state. When $T' > 1$ the peak is again flat-topped over the interval $(\sqrt{T'}, T')$ and the peak flow rate is $1/T'$, or Ld/T in dimensional units.

Further details on unsteady, one-dimensional flow and the rising limb of the hydrograph are given by Woolhiser and Liggett (1967).

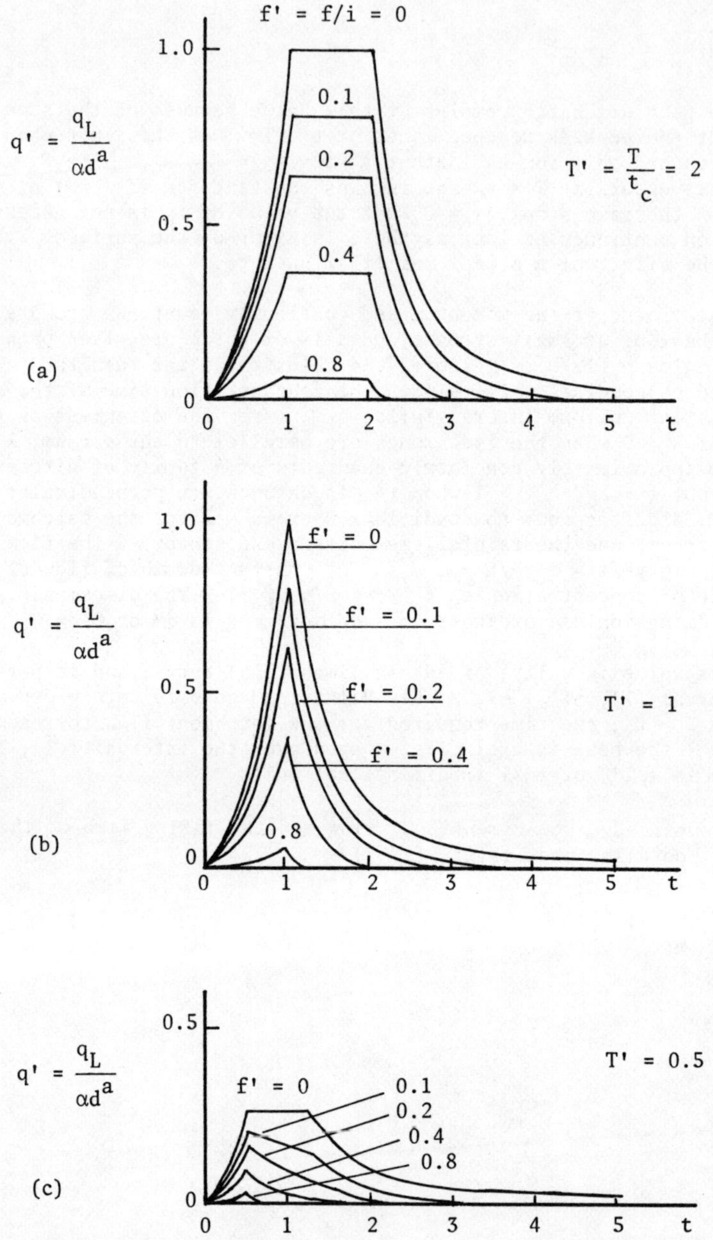

Fig. 7.33. Effect of steady infiltration $f' = f/i$ upon the catchment discharge $q' = q_L/(\alpha d^a)$ due to a steady rainfall of given intensity, according to Wooding (1965).

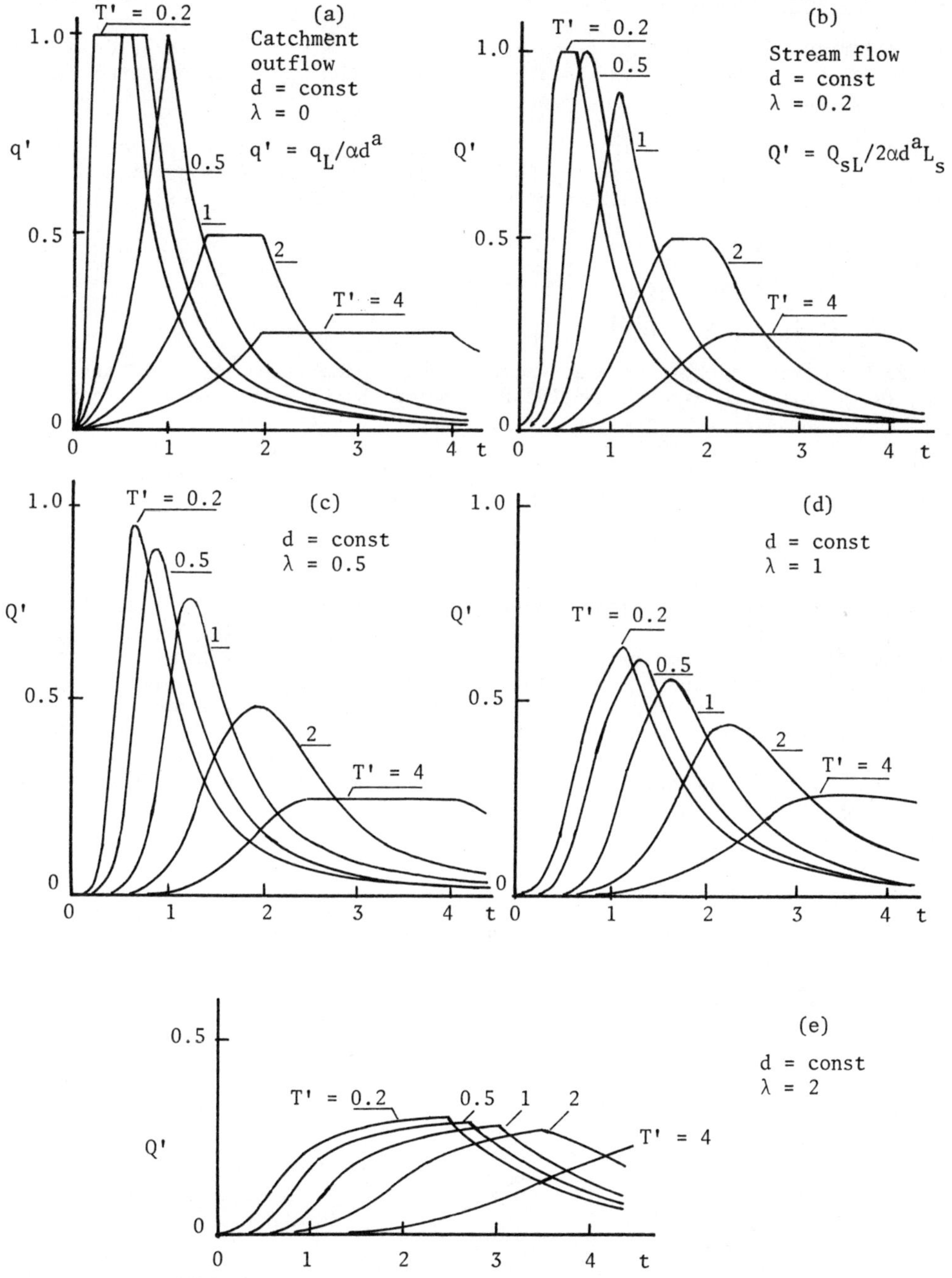

Fig. 7.34. Calculated forms of catchment (a) and stream flow hydrographs (b-e) for a steady rainfall of finite duration $T' = T/T_c$ as a function of λ. Total rainfall depth d = constant. According to Wooding (1965).

7.5 Synthetic Unit Hydrographs

The majority of catchments are ungauged, and hence it is of great practical importance to have a procedure by which a unit hydrograph may be constructed for an ungauged catchment.

Snyder (1938) used three parameters to describe the hydrograph. These were lag to peak t_L (Fig. 7.9), peak discharge Q_p, and baselength t_B. Using data from catchments in the Appalachian Highlands he expressed these as follows:

$$t_L = C_t(LL_{ca})^{0.3} \quad \text{hours}$$
$$Q_p = 640\ C_pA/t_L \quad \text{ft}^3/\text{s}$$
$$t_B = 3 + 3(t_L/24) \quad \text{days}$$

where L is the length of the main stream from the outlet to the catchment boundary, L_{ca} is the distance from the outlet to a point on the stream nearest to the centroid of the catchment (both in miles), C_t is a coefficient varying from 1.8 to 2.2, A is the area of the catchment in sq. mi, and C_p is a coefficient ranging from 0.56 to 0.69. These relationships were for a rainfall excess of duration $T_o = t_L/5.5$. For different durations the lag time was adjusted by

$$t_{LR} = t_L + (T - T_o)/4$$

With these values the shape of the hydrograph may be sketched. Note that the hydrograph baselength is always greater than 3 days and this implies a fairly large catchment. Similar empirical expressions were published by Langbein et al. (1947), Taylor and Schwarz (1952), Gray (1961) and others. Empirical expressions for the baselength are mainly of the form $t_B = T + aA^b$, where A is the area of the catchment, e.g., Linsley proposed a = 0.8 and b = 0.2 for A in km^2.

The Soil Conservation Service of the U.S. Dept. of Agriculture (1957) simplified the runoff hydrograph to a triangle, Fig. 7.35. The volume of runoff is $V = \frac{1}{2}Q_pt_B = \frac{1}{2}Q_p(t_p + t_e)$ or

$$Q_p = 0.75\ V/t_p \qquad 7.102$$

where $t_e \simeq 1.67\ t_p$ from observations and $t_p = \frac{1}{2}T + t_L$ is the time to peak.

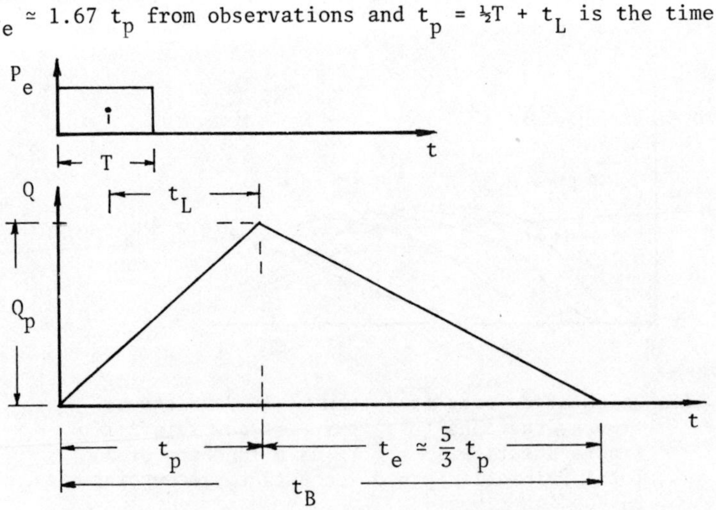

Fig. 7.35. The triangular hydrograph.

The lag to peak can be estimated from

$$t_L = \frac{\Sigma(AP_e t)_i}{\Sigma AP_e} \qquad 7.103$$

where $(AP_e t)_i$ refer to subareas with an appropriate area A each and runoff depth P_e, t is the time required for the water to travel from the centroid of the subarea to the outlet from the catchment, ΣA is the total area and ΣAP_e the runoff volume.

It is a major task to establish a rating curve for a stream, but the time to peak may be determined from simple measurements. A gauge, from which the river levels can be observed as a function of time, is easily installed. From the time history of the rise and fall of the stream the time to peak, as well as the baselength, can be estimated. Thus, with the known volume of runoff (A x P_e) the peak runoff rate is known.

The U.S. Soil Conservation Service also uses an average dimensionless hydrograph, Table 7.3. Once the values of Q_p and t_p have been estimated the unit hydrograph can be plotted.

TABLE 7.3 U.S. Soil Conservation Service Dimensionless Hydrograph*

t/t_p	0	0.1	0.2	0.3	0.4	0.5	0.6	0.7	0.8	0.9	1.0	1.1	1.2
q/q_p	0	0.030	0.100	0.190	0.301	0.470	0.660	0.820	0.930	0.990	1.000	0.990	0.930
Q_a/Q	0	0.001	0.006	0.012	0.035	0.065	0.170	0.163	0.228	0.300	0.375	0.450	0.522
t/t_p	1.3	1.4	1.5	1.6	1.7	1.8	1.9	2.0	2.2	2.4	2.6	2.8	3.0
q/q_p	0.860	0.780	0.680	0.560	0.460	0.390	0.330	0.280	0.207	0.147	0.107	0.077	0.055
Q_a/Q	0.589	0.650	0.700	0.751	0.790	0.822	0.849	0.871	0.908	0.934	0.953	0.967	0.977
t/t_p	3.2	3.4	3.6	3.8	4.0	4.5	5.0						
q/q_p	0.040	0.029	0.021	0.015	0.011	0.005	0.000						
Q_a/Q	0.984	0.989	0.993	0.995	0.997	0.999	1.000						

* Values of q/q_p refer to hydrograph ordinates and Q_a/Q to cumulative values, where q_p and t_p are peak discharge and time to peak, respectively, and Q_a is accumulated volume of runoff.

A number of attempts have been made to derive a dimensionless hydrograph as discussed by Chery (1967). Mitchell (1972) extended the triangular hydrograph approach into what he called the model hydrograph method, which has been applied with some success. The model hydrograph is intended to apply to a wide range of catchment sizes and types and may be linear or non-linear. The development of the model hydrograph starts with a dimensionless instantaneous translation hydrograph of baselength t_B. Fig. 7.36, which is routed through storage. If A is measured in km^2 and the rainfall excess P_e in mm, then the volume of runoff $V = 10^3 AP_e$ m^3 and the average flow rate in m^3/s is $\bar{Q} = (1/3.6)(AP_e/t_B)$, where the baselength t_B is measured in hours. Correspondingly, the peak ordinate of the triangular dimensionless hydrograph has the value of $Qt_B/(AP_e) = 1/1.8$ (or 1290 2/3 when Q is in ft^3/sec and P_e in in.).

The continuity equation is

$$Q_2 = 2S_1/\Delta t - Q_1 - 2S_2/\Delta t + (I_1 + I_2) \qquad 7.104$$

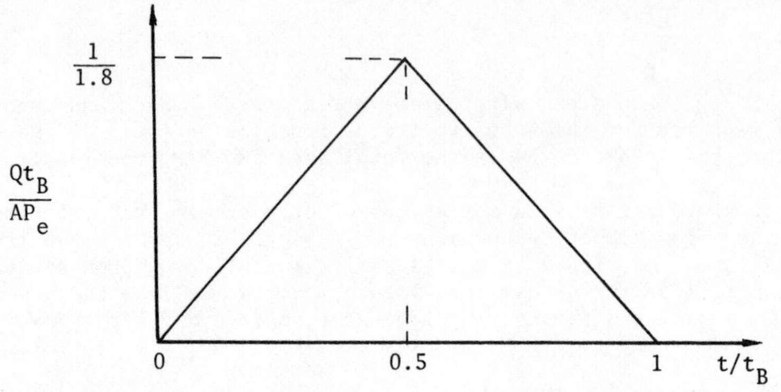

Fig. 7.36. Instantaneous translation hydrograph.

To solve for Q_2, the values of Q_1, I_1, I_2 and the storage as a function of discharge need to be known. The inflow information is given by the instantaneous translation hydrograph. By starting the computations when Q_1 is known, only S as a function of Q has to be defined, generally as

$$S = KQ^x \tag{7.105}$$

If x = 1 the storage is linear.

By introduction of the dimensionless ratios $r = K/t_B$ and $p = \Delta t/t_B$ the storage term becomes $S = rt_BQ$ and $2S/\Delta t = (2r/p)Q$. Substitution of these terms in eqn 7.104 yields

$$(2r/p + 1)Q_2 = (2r/p - 1)Q_1 + (I_1 + I_2) \tag{7.106}$$

For a given catchment with linear storage the value of r is a constant. The value of p depends on the value of Δt. Mitchell uses $\Delta t = 0.02\ t_B$, but 0.05 t_B or even 0.10 t_B is frequently adequate. The values of Δt should be small enough so that $\frac{1}{2}(Q_1 + Q_2)$ and $\frac{1}{2}(I_1 - I_2)$ are good estimates of the true mean values during the time interval.

The instantaneous hydrograph, routed through storage, is further modified by the duration T of the rainfall. First the instantaneous hydrograph is converted into a short period hydrograph, $z = T_S/t_B$. Then an S-curve is formed by summation of the short period hydrographs. The hydrograph for the desired period of rainfall excess T is obtained by the S-curve method. The S-curve is displaced by T hours, or in dimensionless terms by $y = T/t_B$, and the difference is divided by the ratio y/z. With increasing duration, as $T \to t_B$, the shape of the translation hydrograph approaches that of the normal probability curve, and for values of $T/t_B > 1$ that of a trapezoid. The baselength is $(T/t_B) + 1$ and the upper flat of the trapezoid has a length of $(T/t_B) - 1$. Mitchell gives tables of hydrograph ordinates for linear catchments for r values from 0.2 to 3.0. By using the nearest value of r in the table to the actual value of the catchment, the tabular values are within 5% of the true values

The parameters T, K and x for a given catchment can be determined from an observed hydrograph. The baselength, t_B, of the translation hydrograph is the time from the end of the rainfall excess to the inflection point of the recession limb of the hydrograph. For linear storage, $x = 1$ and r has the same value for all points equally spaced in time on the hydrograph recession curve. Both x and K can be determined with the aid of eqn 7.44. Thus, the model hydrograph approach is not strictly synthetic, unless the parameters are estimated without the use of field data.

Table 7.4 shows the procedure for computation. If, for example, $r = 0.2$ and $p = 0.02$, then eqn 7.106 reduces to

$$Q_2 = (19\,Q_1 + I_1 + I_2)/21$$

TABLE 7.4 Model Hydrograph Calculations

$\frac{t}{t_B}$	$\frac{Qt_B}{AP_e}$	$\frac{Q_1t_B}{AP_e} + \frac{Q_2t_B}{AP_e}$	Routed instant. hydrograph	$T_s = \Delta t$ hour hydrograph	Σ(col. 5)	Col.6 lagged by y	T-hour hydrograph
1	2	3	4	5	6	7	8

Columns 1 and 2 list the abscissae and ordinates of the triangular hydrograph in Fig. 7.36, respectively.
Column 3 gives the sum of the values in column 2 on the corresponding line and the preceding line, i.e., $I_1 + I_2$.
Column 4: The first value from column 4 is multiplied by 19 (in this example), added to the following line in column 3 and divided by 21 (in this example).
Column 5: Here the instantaneous hydrograph is converted into one of finite duration, T_s, for example, $T_s = \Delta t = 0.02\ t_B$. This is done by taking the average of two successive ordinates in column 4 and proceeding from row to row.
Column 6: Summation of column 5.
Column 7: Column 6 displaced by the desired period $y = T/t_B$.
Column 8: Difference between columns 6 and 7 divided by y/z. This gives the hydrograph for the desired period.

When the storage relationship is non-linear eqn 7.106 cannot be reduced to a simple form, except when $x = 2.0$ or 0.5. For any other values of x, the solution of eqn 7.106 requires a computer. The value of x for natural catchments is usually less than one and $x = 0.5$ is often a good approximation for small catchments. With non-linear storage the transformation for the duration effect has to be made before routing through the storage, i.e., in the natural sequence of the catchment. The S-curve method is again used to obtain the hydrograph for T hours of rainfall excess, which is then routed through the storage $S = K\sqrt{Q}$. This non-linear hydrograph is characterized by values of $K/\sqrt{AP_e/t_B}$. Again computed tables assist computation.

The model hydrograph method could also be applied to the hydrograph shown in Fig. 7.35, which may be based on data from simple observations.

Finally, it may be worth noting that the use of dimensionless hydrographs, storage elements, etc., is not restricted to conventional catchments. Swinnerton et al. (1972, 1973), for example, proposed a special purpose dimensionless hydrograph for drainage design of motorways running in cuttings, as well as a single linear reservoir model with two alternate values of the storage parameter.

Example 7.1. Measured at intervals of ½ hour the recession limb of an observed hydrograph yielded the following values:

Average Q m^3/s:	1.5	2.0	5.0	10	20	40
$\Delta Q/\Delta t$:	0.315	0.42	1.05	2.11	4.20	8.50

The estimated baselength of the instantaneous hydrograph on this catchment t_B = 10 hours. The area is 150 km^2. Calculate the discharge hydrograph for a rainfall excess of 100 mm in 4 hours.

Plotting of $-\Delta Q/\Delta t$ against Q on log-log paper yields a line at 45° slope, an intercept at Q = 1 of 0.222. Hence, the storage relationship is

$$S = KQ^x = 4.5\ Q$$

and the eqn 7.106 becomes

$$Q_2 = \frac{1}{19}(17Q_1 + I_1 + I_2)$$

using Δt = 0.5 hours, i.e., $p = \Delta t/t_B = 0.5/10 = 0.05$, $r = K/t_B = 1/0.222\text{x}10 = 0.45$, $z = T_s/t_B = 0.5/10 = 0.05$, $y = T/t_B = 0.4$ and $y/z = 8$ or $z/y = 0.125$. With these values the hydrograph calculations are carried out in tabular form as shown below:

$\frac{t}{t_B}$	$\frac{Qt_B}{AP_e}$	$\frac{Q_1t_B}{AP_e} + \frac{Q_2t_B}{AP_e}$	$Q_2 = \frac{1}{19}(17Q_i + I_1 + I_2)$	$T_s = \Delta t$ UHG	$\Sigma(5)$	$y = \frac{T}{t_B} = 0.4$	4 UHG Δ x 0.125
1	2	3	4	5	6	7	8
0	0	0	0	0	0	0	0
0.05	0.056	0.056	0.0029	0.0015	0.0015		0.0002
0.10	0.111	0.167	0.0114	0.0071	0.0086		0.0011
0.15	0.167	0.278	0.0249	0.0181	0.0267		0.0033
0.20	0.222	0.389	0.0427	0.0338	0.0605		0.0076
0.25	0.278	0.500	0.0645	0.0536	0.1141		0.0143
0.30	0.333	0.611	0.0899	0.0772	0.1913		0.0239
0.35	0.389	0.722	0.1184	0.1041	0.2954		0.0369
0.40	0.444	0.833	0.1498	0.1341	0.4295	0	0.0537
0.45	0.500	0.944	0.1837	0.1667	0.5962	0.0015	0.0743
0.50	0.556	1.056	0.2200	0.2018	0.7980	0.0086	0.0987
0.55	0.550	1.056	0.2524	0.2362	1.0342	0.0267	0.1259
0.60	0.440	0.944	0.2755	0.2639	1.2981	0.0605	0.1547
0.65	0.389	0.833	0.2903	0.2829	1.5810	0.1141	0.1834
0.70	0.333	0.722	0.2978	0.2940	1.8750	0.1913	0.2105
0.75	0.278	0.611	0.2986	0.2982	2.1732	0.2954	0.2347
0.80	0.222	0.500	0.2935	0.2960	2.4692	0.4295	0.2550
0.85	0.167	0.389	0.2831	0.2883	2.7575	0.5962	0.2702
0.90	0.011	0.278	0.2679	0.2754	3.0329	0.7980	0.2794
0.95	0.156	0.167	0.2485	0.2582	3.2911	1.0342	0.2821
1.00	0	0.056	0.2253	0.2369	3.5280	1.2981	0.2787
		0	0.2016	0.2134	3.7414	1.5810	0.2701
			0.1614	0.1708	4.1031	2.1732	0.2412
			0.1444	0.1529	4.2560	2.4692	0.2234
			0.1292	0.1368	4.3928	2.7575	0.2044
			0.1156	0.1224	4.5152	3.0329	0.1853
			0.1034	0.1095	4.6247	3.2911	0.1667
			0.0925	0.0979	4.7226	3.5280	0.1493
			0.0828	0.0876	4.8102	3.7411	0.1336
			0.0741	0.0784	4.8886	3.9323	0.1195
			0.0663	0.0702	4.9508	4.1031	0.1060
			0.0593	0.0628	5.0136	4.2560	0.0947
			0.0531	0.0562	5.0698	4.3928	0.0846
			0.0475	0.0503	5.1201	4.5152	0.0756
			0.0425	0.0450	5.1671	4.6247	0.0676
			0.0380	0.0402	5.2053	4.7226	0.0603

1	2	3	4	5	6	7	8
			0.0340	0.0360	5.2413	4.8102	0.0539
			0.0304	0.0322	5.2735	4.8886	0.0481
			0.0272	0.0288	5.3023	4.9508	0.0439
			0.0244	0.0258	5.3281	5.0136	0.0393
			0.0218	0.0231	5.3512	5.0698	0.0352
			0.0195	0.0206	5.3718	5.1201	0.0315
			0.0174	0.0184	5.3902	5.1651	0.0281
			0.0156	0.0165	5.4067	5.2053	0.0252
			0.0140	0.0148	5.4215	5.2413	0.0255
			0.0125	0.0132	5.4347	5.2753	0.0202
			0.0112	0.0118	5.4465	5.3023	0.0180
			0.0100	0.0106	5.4574	5.3281	0.0162
			0.0090	0.0095	5.4669	5.3512	0.0145
			0.0089	0.0085	5.4754	5.3718	0.0130
			0.0072	0.0076	5.4830	5.3902	0.0116
			0.0064	0.0068	5.4898	5.4067	0.0104
			0.0057	0.0060	5.4958	5.4215	0.0093
			0.0051	0.0054	5.5012	5.4347	0.0083
			0.0046	0.0048	5.5060	5.4465	0.0074
			0.0041	0.0043	5.5103	5.4574	0.0066
			0.0037	0.0039	5.5142	5.4669	0.0059
			0.0033	0.0035	5.5177	5.4754	0.0053
			0.0030	0.0031	5.5208	5.4830	0.0047
			0.0026	0.0028	5.5236	5.4898	0.0042
			0.0024	0.0025	5.5261	5.4958	0.0038
			0.0021	0.0023	5.5284	5.5012	0.0034
			0.0019	0.0020	5.5304	5.5060	0.0031
			0.0017	0.0018	5.5322	5.5103	0.0027

The discharge Q = Ordinate (Col. 8)$A(km^2)P_e(mm)/t_B(hours)$ = Ord x (150x100)/10 = 1500 x Ord and the peak discharge is

$$Q_p = 0.2821 \times 1500 = 423.15 \text{ m}^3/\text{s}$$

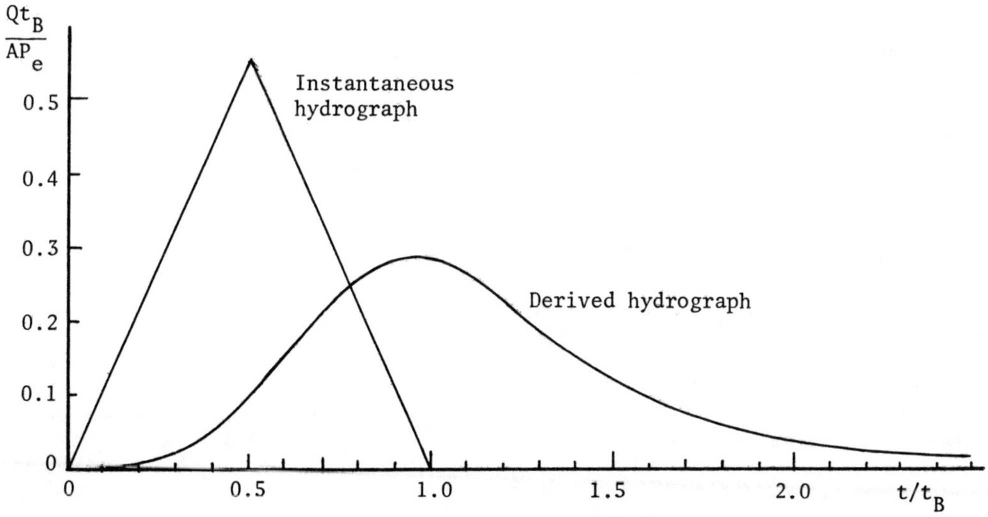

Instantaneous and derived hydrographs of Example 7.1.

7.6 Runoff from Snowmelt

For substantial areas of the Northern Hemisphere the runoff from melting snow is an important aspect of the control and utilization of water. Over the last few decades intensive research has led to a fairly good understanding of the physical process and to methods for estimation of seasonal yields. Methods for short range forecasting of peak flow rates, however, are less satisfactory. The estimates of seasonal and peak flows from snowmelt depend on information on the snow cover and its water equivalent. The melting of snow is basically a problem of thermodynamics. The amount of snowmelt produced depends on the net heat exchange between the snowpack and its environment. The problem is further complicated by the travel of melt water in the snowpack, which involves flow through a porous medium in which the porosity and permeability are dependent on the flow rate and time.

The heat transfers to and from the snowpack arise from
(a) net radiative energy exchange (longwave and shortwave),
(b) latent heat exchange (condensation, evaporation), and
(c) sensible heat exchange [with air, ground and water (mainly rain)].
The establishment of the daily energy budget for an active snowmelt is involved. It follows essentially the same procedure as that outlined in Chapter 4. Figure 7.37 is an illustration of such an energy budget.

The sum of the energy fluxes per unit area and time is

$$\Sigma H = 0 = R_n + H_c + H_p + H_g + H_r + H_s \qquad 7.107$$

where R_n is the net absorbed radiation from short- and longwave radiation, H_c is the convective transfer of heat by air (and flowing water), H_p is the heat flux due to phase changes (condensation, evaporation, sublimation), H_g is the heat conduction by soil, H_r is the heat carried by rain and H_s is the change in heat stored in the snowpack. At any time each of these terms must be inserted with its correct sign (positive for heat added to the snow). For the melting process to begin H_s has to be greater than the sum of all the other terms. The amount of melt water, d_m (mm), produced from the snowpack is then

$$d_m = \frac{H_s}{335\ \chi} \qquad 7.108$$

where χ is the thermal quality of the snow at the beginning of the time interval, and 335 kJ/kg is the heat of fusion. The thermal quality is defined as the ratio of heat required to melt a unit weight of snow to that required to melt a unit weight of ice at 0 °C. It averages 0.95-0.97 for snow with 3 to 5 percent of liquid water. It takes 335 J to produce 1 cm^3 of water from pure ice at 0 °C. Thus, if the heat input is measured in J/m^2, 335 kJ/m^2 of heat of fusion are required to produce 1 mm of melt water.

The properties of the snowpack change with time. The delicate crystals of snow initially form a pack of very low density. The density can be less than 50 kg/m^3 when the surface air temperatures are low. The density of new snow increases approximately from 50 kg/m^3, when falling at a surface air temperature of -8 °C, to 125 kg/m^3, when falling at 0 °C. The density of fresh snow also depends on wind during the snowfall. More than five-fold increases in density have been reported from falls during gale force winds as compared with calm conditions at the same temperature. With time snow compacts and changes into a coarser granular crystalline form. Its air, water and heat storage and transfer properties change. Occasional melting and subsequent freezing of the surface leads to the formation of ice planes in the snow. The compaction of snow is due to its own weight but the major changes in structure, called the ageing or ripening of snow, are caused by percolating water, movement of air in the snow and the various heat transfer processes. Of the heat

9-13 MAY 1949
SUMMARY OF ENERGY — W/m²

		Down	Up	Net
Radiation	Short-wave	387	-50	337
	Long-wave	299	-399	-100
	All-wave	686	-449	237
Transpiration		22†	-141	-119
Runoff				118

† Condensation

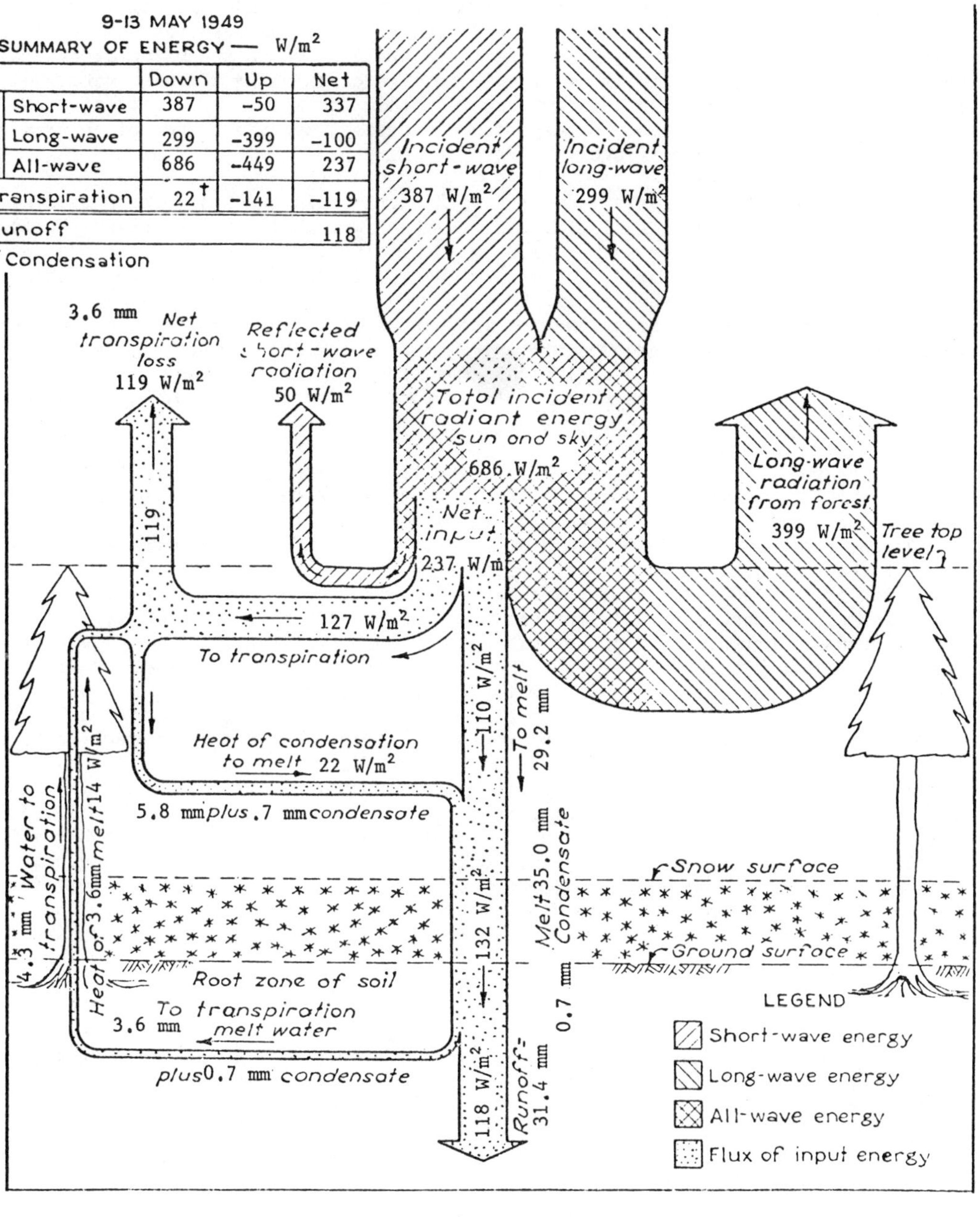

Fig. 7.37. Daily energy balance in heavy forest during active snowmelt. (Snow Hydrology, 1956, Plate 4-12).

transfer processes, those arising from the near surface air temperature and wind speed are the most important. Due to temperature gradients within the snow, air carrying heat and water vapour is in continuous motion. Condensation of water vapour releases the heat of fusion. Evaporation from and condensation onto crystal surfaces, together with percolating water, are believed to be the major agents in changing the structure of the snow crystals. The circulating air also tends to equalize the temperature and vapour pressure within the snow. The rate of transport of air is greatest when the temperature decreases upwards. The volume of air in the snow is given approximately by $V_a = 1 - \rho_s/0.92$.

The snowpack density generally increases with depth. In packs several metres deep the lower half may have densities of 350 to 400 kg/m^3, while in the younger surface layers the densities may be only 70-100 kg/m^3. With time the snowpack tends to become homogeneous, except for ice layers. The ice planes deflect or impede the movement of air and water but do not prevent it altogether. The areal extent of the impermeable layers is not great and the ice planes have many imperfections. Towards spring the snowpack becomes "ripe", a condition where changes are confined to the upper surface. In the hydrological sense the snow becomes ripe when it can hold no more liquid water against gravity, a condition analogous to the field capacity of soils.

For melting to occur the snow must be brought to 0 °C and given the heat of fusion. For appreciable snowmelt to occur, most of the snowpack must be brought to melting point temperature. The amount of heat required for this, as well as the rate of diffusion, depends on the specific heat, the thermal conductivity and the diffusivity of the snowpack. The transfer of heat may be by conduction, convection and diffusion of air and water vapour within the snowpack or by movement of liquid water from rain or snowmelt. In the snowpack the heat transfer is complicated by the simultaneous occurrence of a number of heat exchange processes. The temperature gradients create air and water vapour movement with the associated condensation, sublimation and heat of fusion. Due to the low conductivity of heat an increase in temperature applied to the surface diminishes very rapidly within the pack with distance from the surface and melt water (or rain) freezes in the subfreezing layers. The freezing process in turn releases the heat of fusion and warms up the pack. These two processes influence the conductivity and diffusivity of the snow. The structure, porosity, and liquid water content, etc., of the pack change. In addition, the snowpack, whose physical and thermal properties change continuously with variations in the weather, is an anisotropic crystalline substance. Therefore, the theoretical calculations of heat flow are much more difficult than those for isotropic solids with constant properties. Experimental work has shown that the density of snow, ρ_s, is a satisfactory index for its average thermal properties. The specific heat, c_p, of ice and snow of any density is half of that of water (water c_p = 4.1868 J/g °C, ice and snow c_p = 2.0934 J/g °C and air c_p = 1.0048 J/g °C. It is the heat required to raise the temperature of one gram of the substance by one degree Celsius). The heat conductivity, k_c, expresses the heat transmitted in one second per unit area when a 1 °C temperature difference exists between opposing faces a unit distance apart. The diffusivity, k_d (m^2/s), is related to k_c and density as follows

$$k_d = \frac{k_c}{\rho_s c_p} \qquad 7.109$$

Average values of k_c are given in Table 7.5.

The rate of transfer of heat through the snow is then $H = k_c \partial T/\partial z$. Thus, even at a steep gradient of 100 °C per m a snowpack of ρ_s = 130 kg/m^3 only transmits H = 0.0837(100/1) = 8.37 W/m^2. Temperature gradients are more pronounced in winter than in spring. Even after the snowpack has reached 0 °C throughout, cooling by radiation at night time still creates temperature gradients in the top layers of the snow (and ice crust). If it is assumed that the snowpack is homogeneous with constant values of k_c, c_{ps} and ρ_s, then it is easy to show that the total rate of heat storage in an element is

$$k_c\left[\frac{\partial^2 T}{\partial x^2} + \frac{\partial^2 T}{\partial y^2} + \frac{\partial^2 T}{\partial z^2}\right]\Delta x\ \Delta y\ \Delta z$$

which must equal the time rate of change of heat storage

$$\rho c_{ps}\frac{\partial \bar{T}}{\partial t}\Delta x\ \Delta y\ \Delta z$$

Hence,

$$\frac{\partial T}{\partial t} = \frac{k_c}{\rho c_{ps}}\left(\frac{\partial^2 T}{\partial x^2} + \frac{\partial^2 T}{\partial y^2} + \frac{\partial^2 T}{\partial z^2}\right) \qquad 7.110$$

where the average temperature $\bar{T} \to T$ as Δx, Δy and $\Delta z \to 0$ (c.f. equation of continuity for flow through a porous medium).

For a large snow field this equation reduces to a one-dimensional problem

$$\frac{\partial T}{\partial t} = k_d\frac{\partial^2 T}{\partial z^2} \qquad 7.111$$

The solution of this equation is treated in most texts on heat flow, for example, Carslaw and Jaeger (1959).

TABLE 7.5 Thermal Conductivity Values of Snow

Density ρ kg/m^3	Thermal conductivity k_c J °C^{-1} m^{-1} s^{-1}	Density ρ kg/m^3	Thermal conductivity k_c J °C^{-1} m^{-1} s^{-1}
1 000 (water)	0.5443	340	0.3140
900 (ice)	2.2399	330	0.2973
540	0.8541	250	0.2010
500	0.6573	130	0.0837
440	0.5401	50	0.0251
365	0.3852	1 (air)	0.0251
351	0.3391		

(After "Snow Hydrology" U.S. Army Corps of Engineers, North Pacific Div., Portland, Ore., 1956).

For the purpose at hand it is expedient to assume that the snowpack is initially at a constant temperature T_0(= 0 °C) and that from time t = 0 its surface temperature is maintained at T_s by the subfreezing air temperature T_a. This formulation avoids the introduction of the heat transfer coefficient, h, from air to snow, which is defined as the ratio of heat flux, Q, to the temperature difference, i.e., h = $Q/(T_s - T_a)$. With these boundary conditions the solution of eqn 7.111 is

$$\frac{T(z, t) - T_s}{T_0 - T_s} = \text{erf}\left[\frac{z}{2\sqrt{k_d t}}\right] \qquad 7.112$$

or if $T_0 = 0$

$$T(z, t) = T_s\left[1 - \text{erf}\frac{z}{2\sqrt{k_d t}}\right] = T_s\ \text{erfc}\frac{z}{2\sqrt{k_d t}} \qquad 7.113$$

where erf is the error function. (It is the area under a part of the curve $y = e^{-x^2}$ and is here defined as

$$\text{erf}\,(x) = \frac{2}{\sqrt{\pi}} \int_0^x e^{-u^2}\, du, \quad \text{i.e., } x \equiv z/2\sqrt{k_d t}\ ).$$

The heat flux, Q, is

$$- k_c \frac{\partial T}{\partial z} = k_c \frac{T_s}{\sqrt{\pi k_d t}} e^{-z^2/(4k_d t)} \qquad 7.114$$

or at the surface where z = o

$$Q = \frac{k_c T_s}{\sqrt{\pi k_d t}} = T_s \sqrt{\frac{k_c \rho_s c_{ps}}{\pi t}} \qquad 7.115$$

The quantity of heat lost per unit area is

$$H = \int_0^t Q dt = \frac{2k_c T_s}{\sqrt{\pi k_d}} \sqrt{t} = \frac{2T_s}{\sqrt{\pi}} \sqrt{k_c \rho_s c_{ps}}\, \sqrt{t} \qquad 7.116$$

For example, for ρ_s = 130 kg/m³, k_c = 0.0837 J °C^{-1} m^{-1}s^{-1}, c_{ps} = 2.0934 kJ kg^{-1} C^{-1} and a temperature drop T_s = 10 °C the heat loss is H = 1703$\sqrt{t}$, that is, 1.703 kJ m^{-2} per second or H ≃ 102.18 kJ m^{-2} per hour. Thus,

Time (hrs)	1	6	12	24	48	96
H(kJ m^{-2}), total	102.18	250.29	353.96	500.58	707.93	1001.16
H(kJ m^{-2}) during last hour*		21.81	15.07	10.54	7.41	5.23

* Obtained using $\sqrt{t_1 - t_2}$.

Compared with the 3350 kJ m^{-2} of heat of fusion required to produce 10 mm of melt-water, the rate of heat transfer into the snow is very small. Conversely, the rate at which snow can lose heat is also slow, whereas the rate at which snow can absorb heat at its surface is limited only by the rate of supply of heat for the fusion.

Equation 7.112 can be evaluated and plotted as shown in Fig. 7.38. The penetration of subfreezing temperatures is confined to the surface layers and the maximum depth of penetration is about 200-250 mm in a daily cycle. Heat conduction from the ground to fresh snow can be analysed in the same manner. The effect of ground heat diminishes very rapidly with time and becomes negligible after a few days. This type of analysis can be extended to the case of diurnal temperature variations within the surface layers of the snow and to the case where there is a time lag between the heat wave within the snowpack and the heat wave at the surface. The heat wave wihtin the pack is strongly attenuated. The heat required to raise the temperature of the snowpack to 0 °C is sometimes called the cold content, H_{cc} (i.e., a heat deficit), and is given by

$$H_{cc} = - \int_0^{d_s} \rho_s c_{ps} T_s\, dz \qquad 7.117$$

where ρ_s, c_{ps} and T_s are the density, specific heat and temperature in °C, respec-

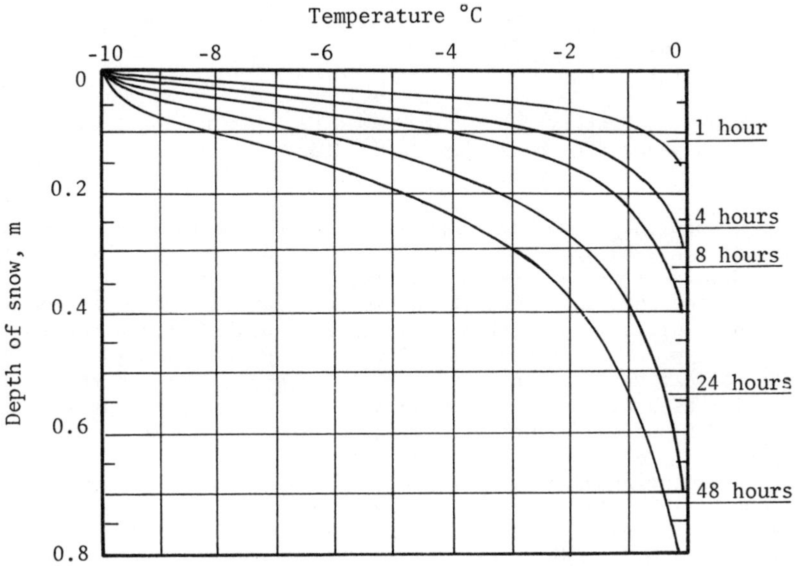

Fig. 7.38. Temperature profiles in snow. The initial temperature of the snowpack is 0 °C throughout, when the surface is suddenly cooled to -10 °C and maintained at this temperature.

tively, of the snowpack (all functions of z), and d_s is the depth of the snowpack. If average values are used

$$H_{cc} \simeq \rho_s c_{ps} T_s d_s \tag{7.118}$$

The heat deficit is then

$$H_d = \rho_s L_{fs} d_s + H_{cc} \tag{7.119}$$

where L_{fs} is the latent heat of fusion of snow. Because the specific heat of snow is small compared with the heat of fusion, very little heat is required to bring the snowpack to 0 °C compared with the heat of fusion needed to melt it. The cold content, H_{cc}, can be expressed in terms of an equivalent depth of water at 0 °C, which upon freezing wihtin the snow would, through release of the heat of fusion, raise the temperature of the snowpack to 0 °C. Thus

$$H_{cc} = L_f \rho_w d_w \tag{7.120}$$

Equating this to eqn 7.118 yields

$$d_w = \frac{\rho_s c_{ps} T_s d_s}{L_f \rho_w} \tag{7.121}$$

or simply $d_w \simeq 6.25 \times 10^{-3}(\rho_s/\rho_w)T_s d_s$. The cold content of a crust formed overnight under clear skies by radiation cooling is equivalent to about d_w = 2-3 mm. Alternatively, the time required to bring the snowpack to 0 °C by a rainfall of intensity i (mm/hr), which on reaching the surface has a temperature of 0 °C, is

$$t_c = 6.25 \times 10^{-3} \rho_s T_s d_s / i \tag{7.122}$$

The thermal quality, χ, can now be expressed as

$$\chi = \frac{H_d}{\rho_w d_m L_f} = \frac{L_{fs}}{L_f} + \frac{c_{ps} T_s}{L_f} \qquad 7.123$$

In ripe snowpacks some water is present, $L_{fs} < L_f$, $T_s \simeq 0$, and χ is about 0.97. In cold snowpacks, however, there is no liquid water and since $L_{fs} = L_f$, χ is greater than 1.0.

Liquid water is present where the temperature of the snowpack equals or exceeds 0 °C. The water may be *hygroscopic water* held as a thin film absorbed on the snow crystals, *capillary water* held by menisci in the pores, or *gravity water* which percolates through the snowpack under the action of gravity. The maximum amount of hygroscopic and capillary water that the pack can hold is termed the liquid water holding capacity, w. The holding capacity of a ripe snowpack is zero. For moist, non-ripe snow an estimate has to be made. Experimental data on the liquid water holding capacity of snow is scarce and mainly limited to spring snow of $\rho_s \sim 350$ kg/m^3. About 2-5 percent by weight of liquid water is the usual working assumption. The holding capacity determines the delay between the melt or rainfall and the runoff. The approximation used by snow hydrologists is that the snowpack is homogeneous and that it is filled from the top to capacity. If i is the intensity of the rainfall and $\bar{m}$ the average melt rate, then, for no ponding or lateral flow, the volume stored is $(i + \bar{m})t$. Assuming further that the pack is homogeneous and that the temperature of the rain water is 0 °C, then the volume, V, of water stored per unit area within the depth of penetration, d_p, in the initially cold snow (no water) can be written with the aid of eqn 7.121 as

$$V = d_w + \frac{\rho_s d_p w_{max}}{\rho_w 100} = S_s d_p \left(\frac{T_s}{160} + \frac{w_{max}}{100}\right) \qquad 7.124$$

Thus

$$d_p = \frac{(i + \bar{m})t}{S_s(T_s/160 + w_{max}/100)} \qquad 7.125$$

where $d_s = d_p$, i and $\bar{m}$ are in mm/h, t is in hours, w is in %, and S_s is the specific gravity of snow.

The water equivalent of the pack, after the cold content has been satisfied, is $S_s d_s + d_w$ or expressed as the deficiency of liquid water content, d_d, it is

$$d_d = \frac{w_d}{100}(S_s d_s + d_w) \qquad 7.126$$

and the time required to satisfy the liquid water deficiency is

$$t_d = \frac{d_d}{i + \bar{m}} \qquad 7.127$$

Any water in excess of the holding capacity, from rain or snowmelt, becomes gravitational water, which moves through the pack at a seepage velocity, v_s. If $(i + \bar{m}) = v_s$, the maximum transient storage depth is

$$d_t = \frac{i + \bar{m}}{v_s d_s} \qquad 7.128$$

and the transient storage delay is

$$t_t = d_s/v_s \qquad 7.129$$

The total depth of stored liquid water is

$$D = d_w + d_d + d_t \qquad 7.130$$

and the total delay time is

$$t = t_c + t_d + t_t \qquad 7.131$$

The seepage velocities in a granular ripe snowpack appear to be of the order of 100 to 200 mm/h.

The amount of melt water produced by any of the components of the heat balance equation can be determined from eqn 7.108. The components of the heat input have to be obtained by measurement or by calculations using the same methods as discussed for evaporation. The albedo for fresh snow is about 0.8, falling with time to about 0.7 after 10 days, and can be as low as 0.20 for dirty snow. For melting snow the albedo decreases rapidly from about 0.8 to 0.6 in the first two days and to 0.45 at 14 days. Thus, for the net shortwave or solar radiation, $R_{sn} = R_s(1 - r)$ (kJ/m^2 per day), the depth of meltwater is

$$d_{msr} = \frac{R_{sn}}{335 \times 0.97} \simeq 3.08 \times 10^{-3} R_{sn} \qquad 7.132$$

mm per day. For longwave radiation snow acts as a near perfect black body. The computations of net longwave radiation, however, involve estimation of back radiation from the atmosphere under clear or cloudy skies, and back radiation from snow under forest cover. For practical snow hydrology these complications are usually by-passed by use of empirical relationships, which use the air temperature over the snow surface. These are usually of the simple linear form

$$d_m = aT \pm b \qquad 7.133$$

For example, The U.S. Corps of Engineers "Runoff from Snowmelt" (1960) suggests:
(1) Under clear skies in the open

$$d_{m\ell r} = 0.0212(T_a - 32) - 0.84 \qquad 7.134a$$

or

$$d_{m\ell r} = 0.969\ T_a - 21.33 \qquad 7.134b$$

(2) Under forest canopy

$$d_{m\ell r} = 0.029(T_a - 32) \qquad 7.135a$$

or

$$d_{m\ell r} = 1.326\ T_a \qquad 7.135b$$

(3) Under complete cloud cover

$$d_{m\ell r} = 0.029(T_c - 32) \qquad 7.136a$$

or

$$d_{m\ell r} = 1.326\ T_c \qquad 7.136b$$

where in eqns (a) the daily snowmelt d_m is in inches per day, and the air temperature T_a and T_c are in °F. T_c is the temperature at cloud base. In eqns (b) d_m is in mm/day and T_a and T_c are in °C.

Similar semi-empirical formulae are in use for calculation of snowmelt from other sources of heat:

(a) *Condensation:* The heat released when water vapour condenses on snow surfaces is absorbed by the snow. The rigorous methods for calculation of the amount of condensate are involved (see Snow Hydrology, 1956) and usually fail because of lack of data. Since the most important parameters are the vapour pressure gradient and wind speed, the snowmelt produced may be expressed as

$$d_{m1} \propto (e_a - e_s)u_2 \tag{7.137}$$

where e_a and e_s are the vapour pressures of air at the elevation z_a and at the snow surface, respectively, and u_2 is the wind speed at a reference level. For every unit of condensate the amount of latent heat released is approximately equal to 7.5 times the heat of fusion. Hence, for a 100% efficient process, the right hand side of eqn 7.137 has to be multiplied by (1 + 7.5). From observed realtionships

$$e_a - e_s = (e_1 - e_s)z_a^{1/6} \tag{7.138}$$

and

$$u_2 = u_1 z_2^{1/6} \tag{7.139}$$

where subscript 1 refers to values at a reference level (1 ft above the snow surface in the U.S.A. practice). Thus,

$$d_{m1} \propto 8.5(z_a z_2)^{1/6}(e_1 - e_s)u_1 \tag{7.140}$$

or by substituting from eqns 7.138 and 7.139

$$d_{m1} = k_e(z_a z_2)^{-1/6}(e_a - e_s)u_2 \tag{7.141}$$

where the constant k_e has to be determined from field measurements. For d_{m1} in in./day, e in mb, u in mph and z in ft, the Central Sierra Snow Laboratory quotes 0.050 for k_e, whereas for the same units data by the Swiss Snow and Avalanche Research Institute (de Quervain, 1951) gives 0.0770. With z measured in m, u in m/s, and d_{m1} in mm/day, the values for k_e are 2.065 and 2.945, respectively. The formula used by de Quervain was essentially the same as that proposed by Trabert in 1896 for evaporation from snow

$$V = c\,\frac{T}{T_0}\sqrt{\frac{p_0}{p}}\,v^n(e_s - e_a) \qquad (\mathrm{g\ m^{-2}h^{-1}}) \tag{7.142}$$

(b) *Convection:* The heat transferred from warm air blowing over the surface of the snowfield, together with the associated heat from condensation, is the most important single factor causing snowmelt. The depth of snowmelt from convection is given in Snow Hydrology (1956) by the following empirical formula

$$d_{m2} = 0.00620\,\left(\frac{p}{p_0}\right)(z_a z_b)^{-1/6}(T_a - T_s)u_b \tag{7.143}$$

where d_{m2} is the depth of snowmelt in in./day, p and p_0 are the air pressures at the site and sea level, respectively, z_a and z_b are heights (in ft) above the snow surface for measurement of air temperature T_a (°F) and wind speed u_b (mph) respectively, and T_s is the temperature of the snow surface in °F. When z is measured in m, T in °C, u_b in m/s, and d_{m2} in mm/day, the proportionality constant becomes 0.433.

The depth of snowmelt (in./day) from condensation and convection is given by the U.S. Corps of Engineers in a combined equation as follows

$$d_{m1,2} = 0.00629(z_a z_b)^{-1/6}[(T_a - 32)\left(\frac{p}{p_0}\right) + 8.59(e_a - 6.11)]u_b \tag{7.144a}$$

Alternatively,

$$d_{ml,2} = 0.433(z_a z_b)^{-1/6}[T_a(\frac{p}{p_o}) + 4.769(e_a - 6.11)]u_b \qquad 7.144b$$

where the units for $d_{ml,2}$, T, u and z are mm/day, °C, m/s, and m, respectively. The ratio p/p_o varies from 1.0 at sea level to 0.7 at 3000 m (10 000 ft) and for areas where elevation changes are small this ratio may be taken to be constant. Further simplifications of the equation are possible by using standard heights for the measurement of wind speed and e_a.

(c) *Snowmelt from rain:* The contribution to snowmelt from rain is generally small and is approximately given by

$$d_{mr} = 0.0126\ PT_a \qquad 7.145$$

where the snowmelt d_{mr} and rainfall P are in mm/day, and the air temperature T_a is in °C. (Equation 7.145 comes from $d_{mr} = 0.007\ P(T_a - 32)$ where d_{mr} and P are in in./day and T_a is in °F).

During rain reflection from clouds and interception by clouds can reduce the radiation received at snow level to quite low values. In winter months it can be as low as 20 W m^{-2} in daily average. Assuming an albedo of 0.65 the net radiation $R_{sn} \simeq 7$ W m^{-2} and eqn 7.132 shows that about 1.8 mm of snowmelt is produced per day. This figure could be even less for densely forested areas. The melt from long-wave radiation could be estimated from eqns 7.134 to 7.136, where during the rainfall $T_c \simeq T_a$. During a rainfall from a given catchment, all the contributions to the snowmelt can be combined into one equation.

Total snowmelt relationships may also be assembled for a given catchment for rain-free periods. The U.S. Corps of Engineers recommends the following empirical relationships for ripe snow at 0 °C:

(a) Heavily forested areas with over 80% cover

$$d_m = 0.074(0.53\ T'_a + 0.47\ T'_d) \quad \text{(in./day)} \qquad 7.146$$

(b) Forested areas with 60-80% cover

$$d_m = k(0.0084\ u_b)(0.22\ T'_a + 0.78\ T'_d) + 0.029\ T'_a \quad \text{(in./day)} \qquad 7.147$$

(c) Partly forested areas with 10-60% cover

$$d_m = k'(1 - F)(0.004\ R_{sn})(1 - a)$$
$$+ k(0.0084\ u_b)(0.22\ T'_a + 0.78\ T'_d) + F(0.029\ T'_a) \quad \text{(in./day)} \qquad 7.148$$

(d) Open areas

$$d_m = k'(0.00508\ R_{sn})(1 - a) + (1 - N)(0.0212\ T'_a - 0.84)$$
$$+ N(0.029\ T'_c) + k(0.0084\ u_b)(0.22\ T'_a + 0.78\ T'_d) \quad \text{(in./day)} \qquad 7.149$$

where T'_a = the difference between the air temperature at 10 ft above the snow and the air temperature at the snow surface in °F.
T'_c = the difference between the cloud base temperature and the snow surface temperature in °F.
T'_d = the difference between the dewpoint temperature at 10 ft above the snow and dewpoint temperature at the snow surface in °F.
u_b = the wind speed at 50 ft above the snow in mph.
R_{sn} = the observed or estimated insolation on a horizontal surface in ℓy/day

(1 ℓy = 41.868 kJ/m^2).

k' = the catchment shortwave radiation melt factor (between 0.9 and 1.1) which depends on the exposure of the areas as compared with a horizontal surface, Fig. 7.39.

k = the catchment condensation-convection melt factor, which depends on the relative exposure of the area to wind. It varies from about 0.2 for densely forested areas to a little over 1.0 for exposed ridges or mountain passes.

F = the estimated basin forest cover.

N = the estimated cloud cover.

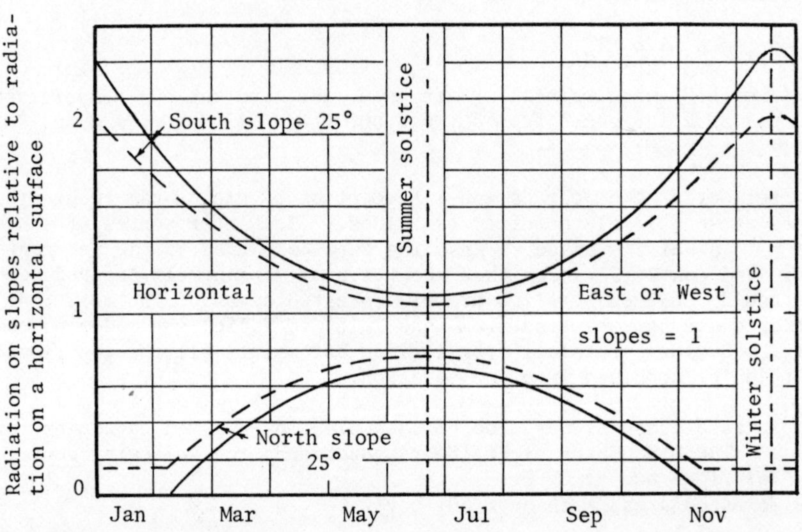

Fig. 7.39. Theoretical values of daily solar radiation on clear days on north and south slopes of 25 degree gradient relative to radiation on a horizontal surface, calculated for 46°30' N latitude and for an elevation of approx. 1585 m above MSL. Full lines refer to direct solar radiation only and dashed lines to direct solar plus diffuse sky radiation. (Snow Hydrology, 1956).

Methods for prediction of snowmelt vary with the purpose of the prediction, the accuracy required, and the time and data available. These methods could be listed in order of refinement and complexity as *temperature index or degree-day methods, degree-day and recession analysis methods, generalized snowmelt equations, index plots and regression analysis methods,* and *hydrograph synthesis and stream flow routing methods.*

The main advantage of the temperature index method is its simplicity. Air temperatures for the area of study are usually readily available from nearby stations. The prediction equations are of the form

$$d_m = C(T_a - T_b)$$

where d_m is the snowmelt per day, T_a is the mean daily (or maximum daily) air temperature, and C and the base temperature T_b are constants obtained by regression

analysis of the data (T_b is close to 0 °C). The U.S. Corps of Engineers recommends the estimation of daily snowmelt (in inches) from the relationships:
(a) open sites

$$d_m = 0.06(T_{mean} - 24)$$

$$d_m = 0.04(T_{max} - 27)$$

(b) forest sites

$$d_m = 0.05(T_{mean} - 32)$$

$$d_m = 0.04(T_{max} - 42)$$

In the above, $34\ °F < T_{mean} < 66\ °F$ and $44\ °F < T_{max} < 76\ °F$.

The degree-day is the name given to the amount of heat corresponding to one degree temperature increment averaged over 24 hours. The degree-days for a 24 hour period are given by the average positive temperature. Below zero temperatures are neglected and taken as 0 °C because temperature changes in the snow cover require very little heat compared with the heat of fusion of ice. The effect of refreezing of a wet snowpack should not be ignored. In some applications the daily maximum and minimum temperatures are averaged and used as an indicator of the temperature above freezing point. The degree-day-factor

$$DDF = \frac{\text{Volume of daily snowmelt (in mm or in.)}}{\text{Number of degree-days}}$$

and varies form season to season, Fig. 7.40.

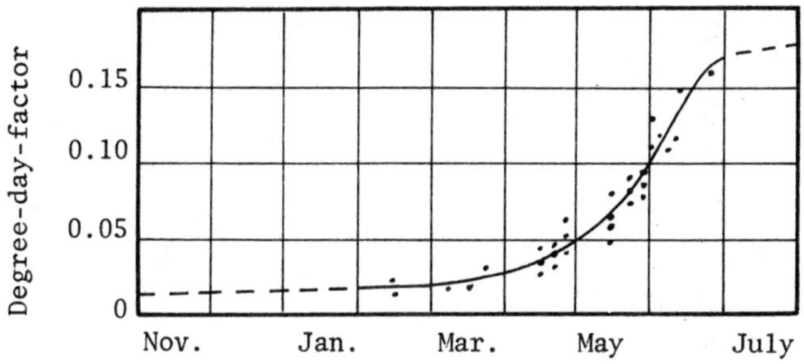

Fig. 7.40. Degree-day-factor in the Lower San Joaquin River Basin, California (U.S. Corps of Engineers).

Martinec (1975 a,b) developed the degree-day method further by taking into account the variability of the degree-day-factor and depth of snow cover. He found that $DDF = 1.1(\rho_s/\rho)$, where ρ_s/ρ is the specific gravity of snow. The recession analysis method requires the establishment of an equation for the recession limbs of the diurnal stream flow hydrographs, Fig. 7.41. This can be done by plotting the recession limbs on semi-logarithmic paper and determining the average slope. The following information was statistically derived:
(a) the volume of stream flow, V_1, from the day's snowmelt.
(b) the recession volume of stream flow, V_2, from the day's snowmelt,
(c) the height of the hydrograph peak, Q_1, above the previous day's trough, and
(d) the height of the hydrograph trough, Q_2, above the previous day's trough
From statistical analysis of daily observations it was established that

$$V_1 = b_1T_1 + b_3T_3 - C_1$$

$$Q_1 = bV_1 + C_2$$

$$Q_2 = b_4V_1 + C_3 \qquad 7.150$$

where T_1 is the maximum temperature at the selected site, T_3 is the accumulated daily maximum temperatures at the site, and b, b_1, b_3, b_4, C_1, C_2 and C_3 are empirical constants.

The U.S. Soil Conservation Service (National Engineering Handbook, 1972) uses a multiple regression technique to predict runoff from snowmelt for several months in advance. The regression formula is of the form

$$Y = b_o + b_1X_1 + b_2X_2 + b_3X_3 + b_4X_4 \qquad 7.151$$

where Y is runoff, X_1 is base flow, X_2 is autumn precipitation, X_3 is snow water equivalent, X_4 is spring precipitation and b_i are the regression coefficients which have to be determined from local data. The spring precipitation term, in particular, causes large fluctuations if forecasts (10-60% of variance in runoff).

Predictions of snowmelt based on a generalized theory of thermodynamics are presented in detail in Snow Hydrology (1956) and Runoff from Snowmelt (1960). The methods used in Russia for the prediction of snowmelt are outlined by Alekhin (1964). A collection of papers in "Floods and their Computation", Vol. II (1967) provides a good picture of the methods used in the various countries to predict yield of snowmelt.

As the melting season progresses the effective snow cover shrinks, and it becomes necessary to determine the actual area of the snow, which contributes to the snowmelt. The most reliable means of obtaining this information is from aerial photography using either satellites or aircraft or both. However, it may be necessary to obtain this information indirectly, for example, by using past correlations between accumulated stream flow and snow cover.

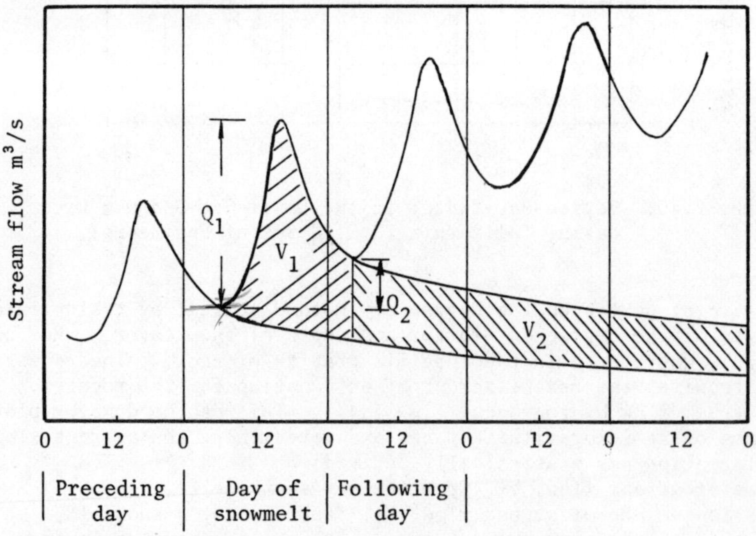

Fig. 7.41. Recession analysis with degree-days, (Garstka et al., 1959).

A reliable method for prediction is based on the use of index plots. These are small typical areas within the catcment area and are well instrumented. The information on runoff from these plots becomes available quickly and can then be extended to the whole catchment. Hydrograph synthesis and computer simulation techniques have been applied to specific areas, but are not easy to use in general prediction work.

For additional reading and references on snow and ice see Martinec (1976).

Chapter 8

FLOOD ROUTING

The hydrograph can be used to predict the discharge-time relationship at the gauging station from which the hydrograph was derived, but there are many problems where this information is required at some other location for which there is no hydrograph. The procedure by which the flood wave is followed along the stream is called flood routing. The two main types of problems that use routing methods are:

(a) The hydrograph is known at the gauging station and the discharge (level) - time history is required at a different downstream location. This problem includes forecasting of water levels in the stream after man-made changes have been made to the river channel, for example, after confinement of the flow between stopbanks.

(b) The assumption of uniform rainfall is seldom reasonable for catchments larger than 500 km^2. Therefore, the catchment is subdivided into a number of reasonably homogeneous subcatchments and the flows from these are combined by the methods of flood routing. The combination of the flood waves from tributaries is a similar problem.

Although surges created by the operation of power stations or control gates, and tidal waves running up estuaries are similar problems, these are seldom considered under the heading of flood routing.

Routing is a problem of unsteady flow. It is concerned with the translation of the flood wave and the associated changes in its shape. The various routing methods which have been developed could be divided in order of increasing complexity into three groups:

(1) hydrologic routing,

(2) routing based on a convection-diffusion equation, and

(3) methods based on the numerical solution of the equations of motion and continuity, also known as hydraulic routing.

The *hydrological routing* methods are the most numerous and can be further subdivided into *reservoir routing* and *stream* or *channel routing*. The hydrological routing methods concentrate on the storage of flood water and do not include the effects of the resistance to flow. These methods are, therefore, based on the continuity equation.

8.1 Reservoir Routing

Reservoir routing is the simplest of the routing problems. The velocities of the through flow in a large reservoir are very low and the water surface in the reservoir may be assumed to be horizontal. Consequently, both the storage and the discharge can be expressed as functions of the reservoir level. Simultaneous analytical solutions of the storage (continuity) and the storage-discharge equations are possible for simple geometric forms (Yevjevich, 1959), but usually the equations are solved numerically. Figure 8.1 shows an inflow hydrograph with an outflow hydrograph superimposed upon it.

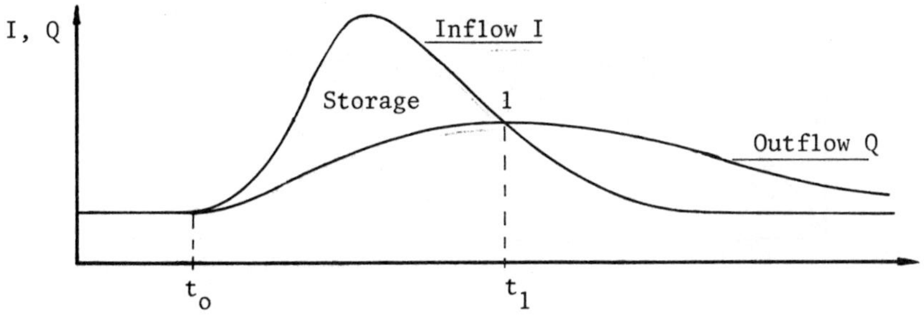

Fig. 8.1. Inflow and outflow hydrographs for a reservoir.

Prior to time t_o the conditions are steady and the inflow equals the outflow. After the arrival of the flood wave the inflow is greater than the outflow (i.e., for $t_o < t < t_1$). The reservoir level rises and maximum storage is reached at $t = t_1$. The area between the two hydrographs for $t_o < t < t_1$ represents the volume of water in storage. At $t = t_1$ the outflow Q reaches its maximum value and $dQ/dt = 0$. For $t > t_1$ the outflow is greater than the inflow and the reservoir level falls. The most dominant effects of the reservoir on the outflow hydrograph are attenuation and delay.

The usual procedures for reservoir routing are based on the equation of continuity

mean inflow = mean outflow + change in storage

$$\frac{1}{2}(I_1 + I_2)\Delta t = \frac{1}{2}(Q_1 + Q_2)\Delta t + (S_2 - S_1)$$

or

$$\frac{1}{2}(I_1 + I_2) + \left(\frac{S_1}{\Delta t} - \frac{Q_1}{2}\right) = \frac{S_2}{\Delta t} + \frac{Q_2}{2} \tag{8.1}$$

This equation has two unknowns and cannot be solved without a second independent relationship. The second independent equation is provided by the dynamic relationship between the outflow and the storage. The outflow is some function of head, for example,

$$Q \propto H^{3/2} \tag{8.2}$$

where H is the elevation of the reservoir level above the crest of the spillway. The storage is also a known function of this elevation, i.e.

$$S = \int_0^H A(H)\, dH \tag{8.3}$$

where A is the surface area of the reservoir. For large reservoirs with reasonably steep sides $S \simeq AH$. Hence, the discharge can be expressed as $Q = f(S)$ and eqn 8.1

could be solved by trial and error. It is, however, helpful to plot auxiliary graphs of Q versus $S/\Delta t$ and Q versus $(S/\Delta t + Q/2)$ as shown in Fig. 8.2. The curve of Q versus $(S/\Delta t - Q/2)$ may also be plotted but is not essential. With the aid of Fig. 8.2 the routing problem is readily solved in tabular (Example 8.1) or graphical form.

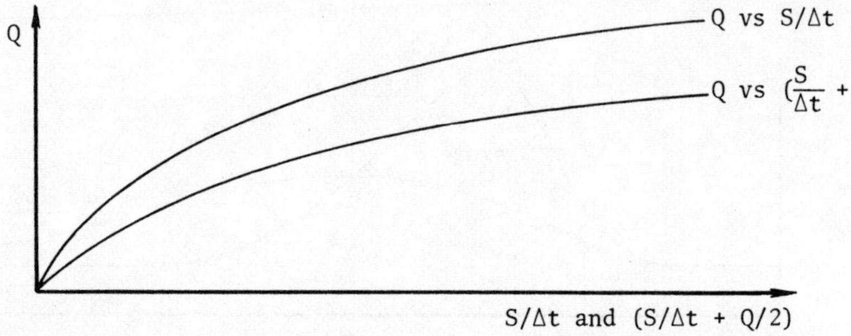

Fig. 8.2. Storage-discharge relationships for reservoir routing.

Example 8.1. Reservoir routing. A steep-sided reservoir with surface area A = 172.8 ha and spillway flow $Q = 5H^{3/2}$. The ordinate values I (m^3/s) of the inflow hydrograph at intervals of Δt = 12 hrs are as follows: 30, 50, 100, 210, 310, 350, 300, 220, 150, 95, 60, 50, 42, 35, 30. At the beginning of the routing procedure ($t = t_0$) the reservoir level is at the crest level of the spillway, i.e., H = 0. Storage S = AH,

$$\frac{S}{\Delta t} = \frac{172.8 \times 10^4 \times H}{12 \times 3600} = 40H \ m^3/s$$

H(m)	$Q = 5H^{3/2}$ (m^3/s)	$S/\Delta t$ (m^3/s)	$S/\Delta t + Q/2$ (m^3/s)
0	0	0	0
1	5.0	40	42.5
2	14.2	80	87.1
3	26.0	120	133.0
4	40.0	160	180.0
5	56.0	200	228.0
6	74.0	240	277.0
8	113.1	320	376.6
10	157.0	400	478.5
12	207.8	480	583.9
15	290.0	600	745.0
18	381.8	720	910.9
20	445.0	800	1022.5

Routing.

Period (½ days)	Inflow I (m^3/s)	$\frac{1}{2}(I_1 + I_2)$ +	$\frac{S_1}{\Delta t} - \frac{Q_1}{2}$ =	$\frac{S_2}{\Delta t} + \frac{Q_2}{2}$	Outflow Q_2 (m^3/s)
1	2	3	4	5	6
0	30				0*
1	50	40	0	40	5

1	2	3	4	5	6
2	100	75.0	37.5	112.5	20
3	210	155.0	90.0	245.0	62
4	310	260.0	183.0	443.0	141
5	350	330.0	301.5	631.5	231
6	300	325.0	400.5	725.5	280
7	220	260.0	446.0	706.0	271
8	150	185.0	438.5	623.5	226
9	95	122.5	395.0	517.5	174
10	60	77.5	341.0	418.5	130
11	50	55.0	285.0	340.0	98
12	42	46.0	241.0	287.0	78
13	35	38.5	209.0	247.5	63
14	30	32.5	184.5	217.0	52
15	30	30.0	164.0	194.0	44
16	30	30.0	148.0	178.0	40
...	...	...	...	...	... etc.

* The initial condition could be a steady state condition where $I = Q_0$. Alternatively, the routing could commence when the reservoir surface is at the spillway crest level. In the latter case at the beginning $Q_1 = Q_2 = 0$ in eqn 8.1 and $0.5(I_2 + I_2) \simeq (1/\Delta t)(S_2 - S_1)$ or $S_2/\Delta t = 40$. After that Q_2 becomes Q_1 for the next line for which $S_1/\Delta t$ is read from the graph, etc.

The designers of outlet structures also have to consider the effects of rainfall on the reservoir itself and the critical rainfall period of the catchment-reservoir system. Not all of the rainfall that occurs contributes to the observed inflow hydrograph. Some of it falls directly on the reservoir surface and some on the catchment which drains directly to the reservoir without becoming part of the inflow hydrograph. The rainfall on the reservoir surface is a direct contribution to the storage in the reservoir, and the storage-discharge relationship, for calculations of the outflow hydrograph, can be modified accordingly at each step of the computation. The direct inflow is more difficult to account for but an acceptable approximation is to *scale up* the contribution of the direct precipitation on the reservoir surface in proportion to the contribution by overland flow. In practice the approximations range from addition of the total direct inflow to the inflow hydrograph (which is then routed through the reservoir) to all the direct inflow being added to the routed outflow hydrograph. The effect of more than one inflow hydrograph is obtained by superposition. Local flows from unmeasured streams could be treated by using synthetic hydrographs.

The reservoir in a particular location has a critical rainfall period for a given probability level of occurrence. This period can be determined in a manner analogous to the method for determining the critical period for a catchment. Given the design criterion of an x-year return period rainfall, the hydrographs for this return period produced by T, 2T, 3T, ... etc. hour rainfalls are routed through the reservoir. The plot of the peak flows of the outflow hydrographs against these periods reveals the period which will produce the maximum outflow. The designer should bear in mind that the critical storm may arrive when the reservoir is already full, or the two critical storms may follow each other.

8.2 Stream Routing

Stream routing differs from reservoir routing because the storage in the reach of a stream channel is not a function of the stage (i.e., elevation of the surface at a given location) or the discharge only. The storage in the reach is not even defined in terms of the depths at both ends as shown in Fig. 8.3a. When the inflow varies and the discharge remains constant a wedge-shaped storage volume is introduced, as shown in Fig. 8.3b. Generally, both the inflow and the outflow depths vary with time. For sufficiently short reaches the storage may be subdivided into prismatic and wedge-shaped volume elements, as shown in Fig. 8.3c. The wedge-shaped volume elements increase the total storage volume during the rise of the river level and decrease the total storage volume during the falling stages. Thus, a plot of storage or stage versus discharge will show a loop, Fig. 8.3d, the loop rating curve (discussed in detail by Henderson, 1966). Related to this non-unique stage-discharge relationship is the attenuation of the peak discharge along the reach.

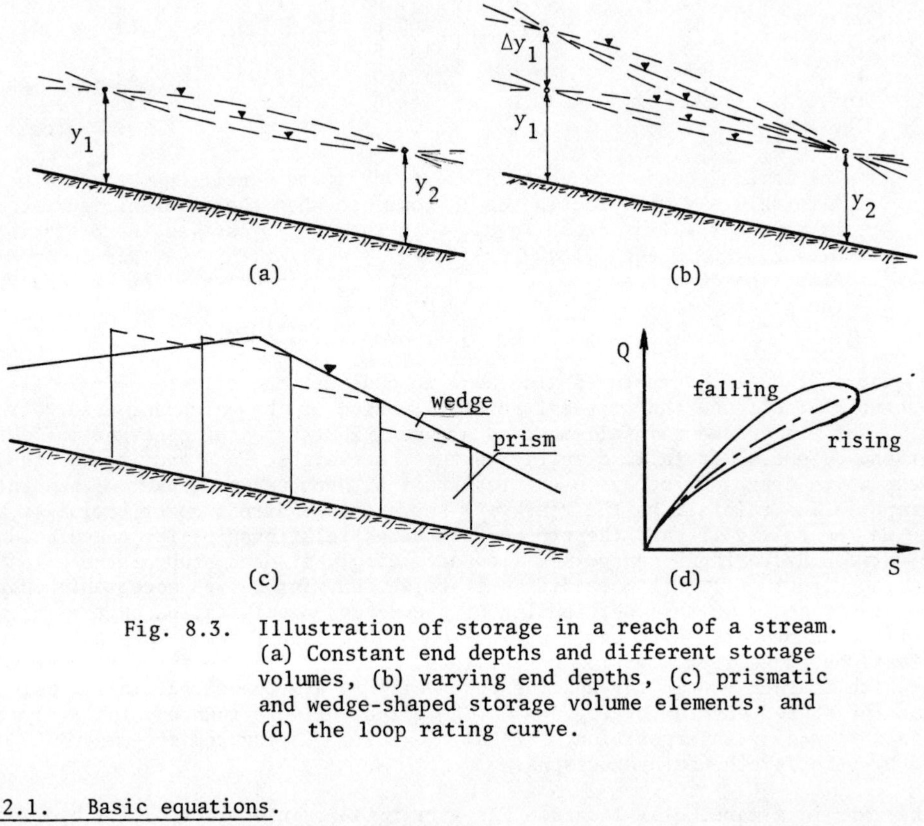

Fig. 8.3. Illustration of storage in a reach of a stream. (a) Constant end depths and different storage volumes, (b) varying end depths, (c) prismatic and wedge-shaped storage volume elements, and (d) the loop rating curve.

8.2.1. Basic equations.

The basic equations for flood routing are the equations of motion (momentum) and continuity. The continuity requirement is that the difference between the inflow and the outflow must equal the time rate of change in the volume of water in storage Hence, for the element shown in Fig. 8.4

$$\text{inflow} = (Q - \frac{\partial Q}{\partial x}\frac{\Delta x}{2})dt + q\,\Delta x\,dt$$

$$\text{outflow} = (Q + \frac{\partial Q}{\partial x}\frac{\Delta x}{2})dt$$

rate of change in volume of water in storage $= \frac{\partial A}{\partial t}\Delta x\,dt$ or

$$\frac{\partial A}{\partial t} + \frac{\partial Q}{\partial x} = q \qquad 8.4$$

where $q = f(x, t)$ is the rate of lateral inflow per unit length of the channel, $Q = AV$, $A = f(y_{x,\ t},\ x)$ is the cross-sectional area, and $V = f(x, t)$ is the mean velocity. If there is no lateral inflow $q = 0$ and eqn 8.4 reduces to the standard form.

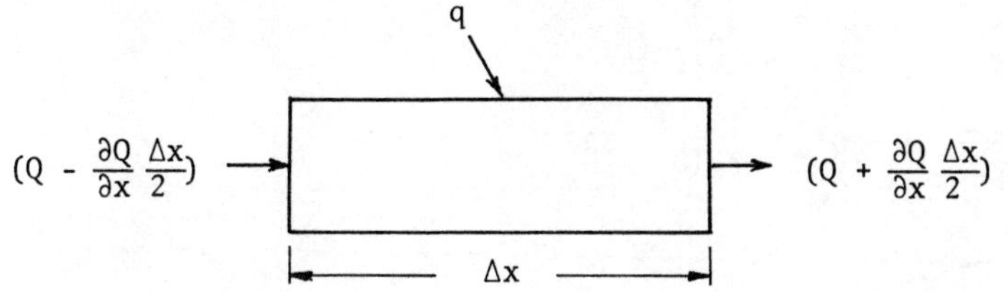

Fig. 8.4. Definition sketch for the continuity equation.

Equation 8.4 may be written as

$$V \frac{\partial A}{\partial y}\frac{\partial y}{\partial x} + V \frac{\partial A}{\partial x} + A \frac{\partial V}{\partial x} + \frac{\partial A}{\partial y}\frac{\partial y}{\partial t} = q \tag{8.5}$$

which for a rectangular channel becomes

$$y \frac{\partial V}{\partial x} + V \frac{\partial y}{\partial x} + \frac{\partial y}{\partial t} = \frac{q}{B} \tag{8.6}$$

where B is the width (constant), and y is the depth of the channel. The three terms on the left hand side are known as the prism storage, wedge storage and rate of rise, respectively.

The equation of motion in the x-direction is

$$\frac{\partial u}{\partial t} + u \frac{\partial u}{\partial x} + v \frac{\partial u}{\partial y} = -\frac{1}{\rho}\frac{\partial}{\partial x}\left(p + \gamma h + \frac{\tau_o P}{A}dx\right) = gS_o - g\frac{\partial y}{\partial x} - gS_f \tag{8.7}$$

where $\tau_o/\rho m = gU|U|/C^2 m = gS_f$ by the Chézy formula, S_f is the friction slope which may also be evaluated by the Strickler-Manning formula

$$S_f = \frac{Q^2 n^2}{A^2 m^{4/3}} \tag{8.8}$$

where n is the Manning roughness coefficient, $m = A/P$ is the hydraulic mean radius, P is the length of wetted perimeter and $S_o = -\partial z/\partial x$ is the bed slope, see Fig. 8.5. The friction slope S_f can be expressed (Henderson, 1966) as

$$S_f = S_o - \frac{\partial y}{\partial x} - \frac{u}{g}\frac{\partial u}{\partial x} - \frac{1}{g}\frac{\partial u}{\partial t} \tag{8.9}$$

Steady uniform flow

Steady non-uniform flow

Unsteady non-uniform flow

Assuming v is negiligible and putting $u = V$, eqn 8.7 becomes

$$\frac{\partial V}{\partial t} + V \frac{\partial V}{\partial x} = g\left(S_o - \frac{\partial y}{\partial x} - S_f\right)$$

and substituting from V = Q/A

$$\frac{\partial V}{\partial x} = (A \frac{\partial V}{\partial x} - Q \frac{\partial A}{\partial x})/A^2$$

and

$$\frac{\partial V}{\partial t} = (A \frac{\partial Q}{\partial t} - Q \frac{\partial A}{\partial t})/A^2$$

yields

$$\frac{\partial Q}{\partial t} + \frac{\partial}{\partial x} (\frac{Q^2}{A}) = Ag(S_o - \frac{\partial y}{\partial x} - S_f) \qquad 8.10$$

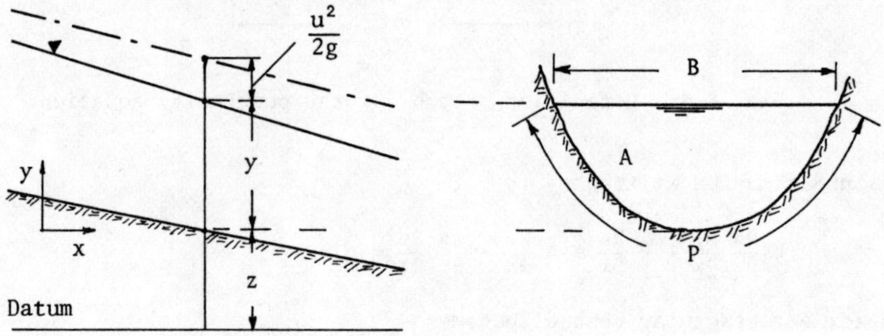

Fig. 8.5. Definition sketch of open channel parameters.

If there is lateral inflow the integral form of the momentum equation has to be evaluated

$$\Sigma F = \frac{\partial}{\partial t} [\int_{cv} \rho \underset{\sim}{q} \, d\forall] + \oint_{cv} \rho q(\underset{\sim}{n} \cdot \underset{\sim}{q}) dS$$

where q is the velocity vector, n is the normal to a surface element, $\forall$ is the volume, ρ is the density and cv stands for control volume. The first term on the right hand side is the time rate of change of momentum within the control volume and the second term is the net flux of momentum across the control surface. Expressed for the direction of flow in the main stream the right hand side becomes

$$\rho \ [\frac{\partial}{\partial t} (Q)\Delta x - QV_1 - qv_x \Delta x + (Q + q)V_2]$$
$$= \rho \ \{\frac{\partial}{\partial t} [(VA + vA)\Delta x] + Q(V_2 - V_1) + q(V_2 - v_x)\Delta x\}$$
$$= \rho \ [\frac{\partial}{\partial t} (Q) + \frac{\partial}{\partial x} (Q^2/A) - qv_x]$$

where q is the rate of lateral inflow at velocity v per unit length of the channel, v_x is the component of v in the streamwise direction, V_1 and V_2 are the mean velocities of flow into and out of the reach of length Δx, respectively, and A is the cross-sectional area of the stream flow. For the condition of lateral inflow eqn 8.10 becomes

$$\frac{\partial Q}{\partial t} + \frac{\partial}{\partial x} (\frac{Q^2}{A}) = Ag(S_o - \frac{\partial y}{\partial x} - S_f) + qv_x \qquad 8.11$$

In general, the lateral inflow is deficient in streamwise momentum and has a retarding effect on the stream flow. The lateral outflow loses its momentum outside the stream channel through friction and this component of momentum is not recovered by the stream.

Equations 8.4 and 8.10 form a non-linear system of equations which has no general analytical solution. The equations may be solved numerically or in the x-t plane by the method of characteristics. The available numerical methods can be divided into four groups:

1. Finite difference schemes that solve the characteristic equations for water level y and velocity V at x-t values defined by the curvilinear characteristics grid.
2. Explicit finite difference schemes for the characteristic equations using a rectangular x-t grid.
3. Explicit finite difference schemes for the original equations using a rectangular grid.
4. Implicit finite difference schemes for the original equations using a rectangular grid.

Methods of solution to the basic equations are available at various levels of approximation. These can be classified with the aid of eqn 8.9. Writing for the discharge $Q = Cm^x\sqrt{S_f}$ and for the steady uniform flow $Q_s = Cm^x\sqrt{S_o} = f(y)$ then

$$Q = Q_o \sqrt{1 - \frac{1}{S_o}\frac{\partial y}{\partial x} - \frac{V}{S_o g}\frac{\partial V}{\partial x} - \frac{1}{S_o g}\frac{\partial V}{\partial t}}$$

- Kinematic wave: $Q = Q_o\sqrt{1}$ (first term only)
- Diffusion analogy solution: up to and including the $\frac{1}{S_o}\frac{\partial y}{\partial x}$ term
- Complete dynamic wave solution: all terms

where u has been put equal to the mean velocity V. The above is also the expression for the looped rating curve which reduces to the Jones formula

$$Q = Q_o \sqrt{1 + \frac{1}{cS_o}\frac{\partial y}{\partial t}}$$

when the acceleration terms are insignificant, i.e., on substitution of dx = cdt, where c is the celerity of the flood wave.

The only known exact analytical solution of eqns 8.4 and 8.10 is for the monoclinal wave in a uniform channel of infinite width with zero lateral inflow. This solution is discussed in detail by Henderson (1966, in Chapter 9). The monoclinal wave (a step increase in discharge) travels at a constant speed of

$$c = V + A\frac{dV}{dA} \qquad 8.12$$

The wave-form is an S-curve asymptotic to a smaller depth downstream with the larger depth upstream. Equations 8.4 and 8.10 may be solved numerically for any particular problem using a finite difference scheme. This, however, is a formidable task for practical applications. In order to simplify the solution of these equation various approximations have been proposed.

Henderson showed that for rivers in steep alluvial country with fast-rising floods the orders of size of the terms on the right hand side of eqn 8.9 are approximately 5×10^{-3}, 1×10^{-4}, 3×10^{-5} and 1×10^{-5}, respectively. Neglecting lateral inflow, which is usually small compared with the main stream flow, the two terms on the left hand side of eqn 8.4 must be of the same magnitude. Using this result in eqn 8.10 gives

$$\left|\frac{\partial Q}{\partial t}\right| / \left|\frac{\partial}{\partial x}\left(\frac{Q^2}{A}\right)\right| = 1$$

i.e. the local and convective accelerations are of the same order of magnitude. An estimate of the length of the flood wave can be obtained by introduction of a length scale X and a time scale T which are related as $X = QT/A$. If it is now assumed, for example, that the duration of the flood is 3½ days ($\sim 3 \times 10^5$ s), $Q \simeq 1000$ m^3/s and $A \simeq 300$ m^2, then the length scale $X \simeq 1000$ km. This is greater than the length of most rivers. Assuming further that the Manning $n = 0.03$, the bed slope $S_o = 10^{-3}$, the hydraulic mean radius $m = 5$ and that $\partial y/\partial x \sim m/X$, then the ratios of terms in eqn 8.10 are

$$|gAS_f|/|gAS_o| = S_f/S_o \simeq 1.17$$

$$|gA \frac{\partial y}{\partial x}|/|gaS_o| = \frac{m}{X}/S_o \simeq 5 \times 10^{-3}$$

and

$$|\frac{\partial Q}{\partial t}|/|gAS_o| = \frac{Q/T}{gAS_o} \simeq \frac{1}{9} \times 10^{-2} \simeq 1.1 \times 10^{-3}$$

These values show that friction governs the momentum of the flow and that the effect of the water surface slope $\partial y/\partial x$, defined relative to the slope of the river bed, on the momentum of the flow is fairly small. Likewise, the effects of acceleration and convection on the momentum of the flow are small relative to friction, the ratio being approximately 1:1000. Assuming a lateral inflow of 10^{-3} m^3/s per metre at right angles to the stream, then $v_x = 3.3$ m/s,

$$\frac{|qv_x|}{|gAS_o|} \simeq 1.1 \times 10^{-3}$$

and

$$\frac{|q|}{|\partial A/\partial t|} = q/\frac{A}{T} \simeq 1$$

This shows that under usual conditions when the lateral inflow is small the contribution to the momentum of the flow from the lateral inflow may be ignored. When the effect of the lateral inflow on the continuity equation is significant the above order of magnitude analysis should be modified. However, usually the conclusions reached by ignoring the lateral inflow are acceptable for all but the extreme conditions of lateral inflow.

Thus, the approximate forms of the continuity and momentum equations given below do satisfy the majority of conditions

$$\frac{\partial A}{\partial t} + \frac{\partial Q}{\partial x} = q \qquad 8.13$$

$$0 = S_o - \frac{\partial y}{\partial x} - \frac{Q^2 n^2}{A^2 m^{4/3}} \qquad 8.14$$

Equation 8.14 may also be written as

$$Q = \frac{A}{n} m^{2/3} \sqrt{S_o - \frac{\partial y}{\partial x}} \qquad 8.15$$

where y as a function of x and A must be known. Although eqn 8.15 could be substituted into eqn 8.13, the resulting expression is not suitable for analytical flood routing because the value of A is strongly dependent on the local channel geometry.

Further complications arise from the distinctly different hydraulic characteristics of flow in the stream channel compared with flow over the flood plain, and from the fact that the channel itself is not straight. It is customary in hydraulics to subdivide the flow into channel flow, Q_c, and berm or flood plain flow, Q_f. This approach is also used in routing studies (e.g. Zheleznyakov, 1971). The method is further developed by Price (1973a, b) or in Vol. 3 of the Flood Studies Report (1975). Some aspects of Price's work are summarized here.

If it is assumed that the water level across the flood plain, normal to the mean direction of flow, is the same as that in the river channel, then eqns 8.13 and 8.14 can be written for the channel as

$$\frac{\partial A_c}{\partial t} + \frac{\partial Q_c}{\partial x} = q^* \qquad 8.16$$

$$0 = S_o - \frac{\partial y_c}{\partial x} - \frac{Q_c^2 n_c^2}{A_c^2 m_c^{4/3}} \qquad 8.17$$

and for the flood plain as

$$\sigma \frac{\partial A_f}{\partial t} + \frac{\partial Q_f}{\partial x} = - q^* + q \qquad 8.18$$

$$0 = \sigma^3 (S_o - \frac{\partial Q_f}{\partial x}) - \frac{Q_f^2 n_f^2}{A_f^2 m_f^{4/3}} \qquad 8.19$$

where q^* is the lateral inflow from the flood plain into the channel per unit length, σ is the sinuosity defined as the ratio of the length of the channel to the length of the flood plain in the general direction of the channel, and subscripts c and f refer to the channel and the flood plain, respectively. Price assumed the channel in the model to be straight and the plan area of the flood plain to be the true area. Therefore, the width of the flooded area in the model is $1/\sigma$ times the actual width, and A_f is the wetted cross-sectional area in the model. Equations 8.16 and 8.18 may be combined to give

$$\frac{\partial}{\partial t} (A_c + \sigma A_f) + \frac{\partial Q}{\partial x} = q$$

or

$$(1 + \sigma \frac{\partial A_f}{\partial A_c}) \frac{\partial A_c}{\partial t} + \frac{\partial Q}{\partial x} = q \qquad 8.20$$

and since the water levels are the same $\partial A_f/\partial A_c = B_f/B_c$, where B_f and B_c are the flood plain and the channel water surface widths, respectively. The schematic model of the river is shown in Fig. 8.6. Equations 8.17 and 8.19 yield

$$Q = Q_c + Q_f = (\frac{A_c m_c^{2/3}}{n_c} + \sigma^{3/2} \frac{A_f m_f^{2/3}}{n_f}) (S_o - \frac{\partial y_c}{\partial x})^{\frac{1}{2}} \qquad 8.21$$

from which

$$\frac{\partial Q}{\partial t} = [\frac{m_c^{2/3}}{n_c} (1 + \frac{2}{3} \frac{A_c}{m_c} \frac{\partial m_c}{\partial A_c}) + \frac{\sigma^{3/2}}{n_f} \frac{\partial}{\partial A_c} (A_f m_f^{2/3})] (S_o - \frac{\partial y_c}{\partial x})^{\frac{1}{2}} \frac{\partial A_c}{\partial t}$$

$$- \frac{\frac{1}{2}Q}{(S_o - \partial y_c/\partial x)^{\frac{1}{2}}} \frac{\partial}{\partial x} (\frac{1}{B_c} \frac{\partial A_c}{\partial t}) \qquad 8.22$$

Assuming that the channel width is large with respect to the channel depth ($A = By \propto S^{-3/10}$) and eliminating $\partial A_c/\partial t$ between eqns 8.20 and 8.22 yields

$$\frac{\partial Q}{\partial t} + c\left(1 - \frac{1}{S_o}\frac{\partial y_c}{\partial x}\right)^{3/10}\left(\frac{\partial Q}{\partial x} - q\right) =$$

$$- \frac{\frac{1}{2}Q}{(S_o - \partial y_c/\partial x)^{\frac{1}{2}}}\frac{\partial}{\partial x}\left[\frac{1}{\lambda B_c}\left(\frac{\partial Q}{\partial x} - q\right)\right] \qquad 8.23$$

Here the convection speed

$$c = \frac{Q}{\lambda(Q_c n_c)^{3/5} B_c^{2/5} S_o^{3/10}}\left(1 + \frac{2}{3}\frac{A_c}{m_c}\frac{\partial m_c}{\partial A_c} + \theta\right) \qquad 8.24$$

where

$$\lambda = 1 + \sigma\, B_f/B_c \qquad 8.25$$

and

$$\theta = \sigma^{3/2}\,\frac{A_f m_f^{2/3}}{n_f}\left[\frac{A_c}{A_f}\frac{B_f}{B_c}\left(1 + \frac{2}{3}\frac{A_f}{m_f}\frac{\partial m_f}{\partial A_f}\right)\right.$$

$$\left. - \left(1 + \frac{2}{3}\frac{A_c}{m_c}\frac{\partial m_c}{\partial A_c}\right)\right]\left(\frac{A_c m_c^{2/3}}{n_c} + \sigma^{2/3}\,\frac{A_f m_f^{2/3}}{n_f}\right)^{-1} \qquad 8.26$$

If the discharge is less than bank-full $\lambda = 1$ and $\theta = 0$. When q is approximately uniform along the river and $|\partial y_c/\partial x|$ is small compared to S_o, eqn 8.23 becomes

$$\frac{\partial Q}{\partial t} + c\,\frac{\partial Q}{\partial x} = Q\,\frac{\partial}{\partial x}\left(a\,\frac{\partial Q}{\partial x}\right) + \frac{aQ}{S_o}\frac{dS_o}{dx}\frac{\partial Q}{\partial x} + \frac{3}{5}\,a_c\left(\frac{\partial Q}{\partial x}\right)^2 + cq \qquad 8.27$$

where

$$a = 1/(2\lambda\, B_f S_o) \quad \text{and} \quad a_c = 1/(2B_c S_o) \qquad 8.28$$

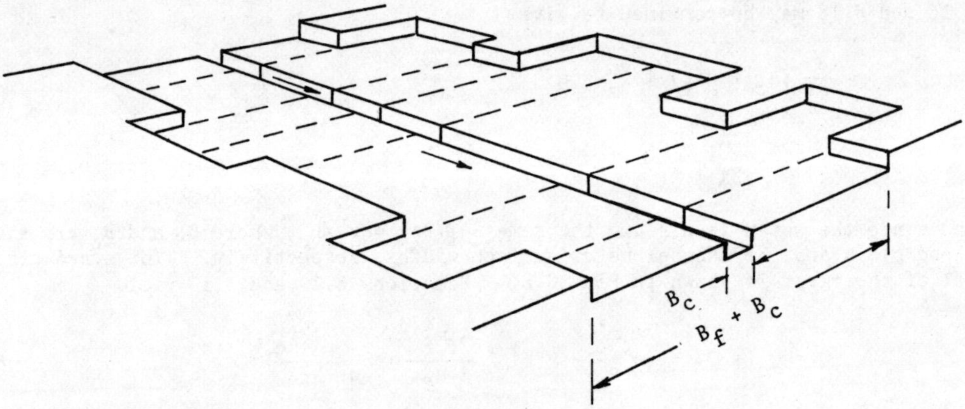

Fig. 8.6. Channel-flood plain model (Price, 1973).

Equations of the form of 8.27 form the basis of most of the flood routing methods. For simplicity the parameters c, a and aQ are usually assumed constant in space and time and the more of the terms on the right hand side are neglected the easier will be the solution. Omitting the third term makes the equation linear. In fact the parameters are functions of at least Q and x and vary considerably when the stream flow exceeds the bank-full discharge. Although eqn 8.24 gives a functional expression for c, the form of the equation makes the task of finding c as a function of Q and x very difficult. The Flood Studies Report recommends that where records exist c should be calculated from the travel times of the peaks of several flood waves. Given also a reliable rating curve, c can be determined for each flood peak and related to the peak discharges. Hence, the average value of c over the reach may be obtained as a function of Q only, i.e., $\bar{c} = f(Q)$. The value of a as a function of Q could be calculated form eqn 8.28 using the average value of the bed slope and the maximum width of the channel, but the value obtained would ignore variation of the parameters λ, B and S_o along the reach.

8.2.2. Diffusion methods

For a uniform rectangular channel the second term on the right hand side of eqn 8.27 becomes zero and the third term may be neglected. The remainder of this equation then forms the basis of the various diffusion methods of flood routing. This equation may be expressed either in terms of the river stage (Hayami, 1951, Hayashi, 1965) or in terms of the discharge (Price, 1973b). Following Hayami's work eqn 8.27 may be written as

$$\frac{\partial Q}{\partial t} + \omega \frac{\partial Q}{\partial x} = \mu \frac{\partial^2 Q}{\partial x^2} + \omega q \qquad 8.29$$

where ω and μ are assumed to be constant parameters which for $q = 0$ is also known as the Forchheimer formula. Customarily, ω is taken to be equal to the speed of the observed flood wave, but in fact ω depends on the attenuation of the flood wave along the river. In long rivers with extensive overbank flooding attenuation is substantial and ω can differ significantly from the celerity c of the flood wave. Hayami obtained the differential equation by using the Chézy equation $U = C[y_o(S_o - \partial y/\partial x)]^{\frac{1}{2}}$ (the equation of motion) and the diffusion (continuity) equation $\partial y/\partial t + \partial q/\partial x = \mu \partial^2 y/\partial x^2$.

The most sweeping approximation is to ignore all the terms on the right hand side of eqn 8.27. The justification for doing this is that these terms are significantly smaller than those on the left hand side. The resulting equation, which is called the first-order equation, is

$$\frac{\partial Q_1}{\partial t} + c_1 \frac{\partial Q_1}{\partial x} = 0 \qquad 8.30$$

where Q_1 signifies the first term in the expansion for Q and $c_1 = f(Q_1, x)$. Equation 8.30 describes the kinematic wave moving at celerity c_1 and was analysed by Lighthill and Whitham (1955). Equation 8.30 is essentially a continuity statement. It has one set of characteristics given by $dx = c_1\,dt$, and along each of these characteristic lines the discharge Q_1 is constant, i.e., for an observer moving at speed c_1 both Q and y appear constant. The peak discharge for the wave is unaffected by variations in the channel geometry. The celerity $c_1 = f(Q_1, x) = dQ/dA = V + A\,dV/dA$. Only one value of c_1 exists for a kinematic wave whereas for a dynamic wave two characteristic directions and two wave celerities are possible. For the kinematic wave Q is a function of y alone, and $S_o = S_f$ since the other slope terms in eqn 8.9 are negligible. Since Q is a function of y only c_1 increases with Q. Therefore, the wave crest travels faster than the rest of the wave thus affecting the shape of the wave along the river. Each cord line of the wave, parallel to the river gradient, will travel at a different speed and the wave becomes steeper as it moves downstream. Since the length of the cord cannot change, the kinematic wave cannot disperse and hence there is no subsidence of the wave as it moves downstream. The conclusion that the kinematic wave must steepen also applies to the river channel part of the flood wave which floods the flood plains. The celerity of the part of the wave which is above the bank-full level, however, is considerably reduced by the effect of the flood plains and this leads to flattening of the wave front above the bank-full discharge level, as shown in Fig. 8.7. The flattening is most pronounced when the flood plain is a flat valley. A similar but less pronounced modification occurs when the flow is receding and water flows from the flood plain into the river channel.

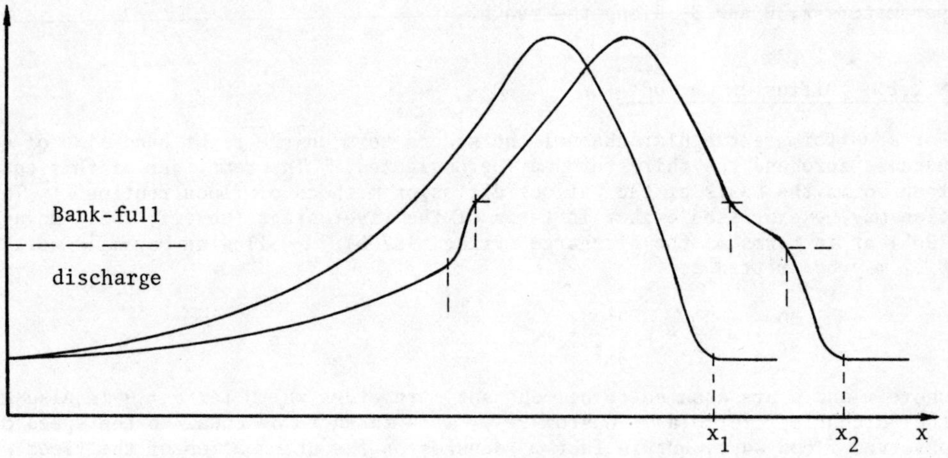

Fig. 8.7. Illustration of the deformation of the hydrograph for a kinematic wave.

The major failing of the kinematic wave solution is that it does not allow for attenuation of the flood wave. In order to account for attenuation at least some of the terms on the right hand side of eqn 8.27 have to be retained. Since attenuation may be substantial in long rivers and rivers with large flood plains, solutions which would account for attenuation of the flood wave were sought. This led to the study of the diffusion equation, eqn 8.29. For a constant ω and a constant diffusion coefficient μ, Hayami obtained the solution of eqn 8.29 for an elementary flood wave in the form

$$Q = Q_{om} \exp[(\frac{\omega}{2\mu} - p_m)x] \sin(\gamma_m t - q_m x) \qquad 8.31$$

where Q_{om} is the peak discharge at the upstream end of the reach where x = 0, $\gamma_m^2 Q_{om}$ is the curvature of the hydrograph peak at x = 0, and the parameters p_m and q_m are the roots given by

$$\begin{Bmatrix} p_m \\ q_m \end{Bmatrix} = \sqrt{\frac{\sqrt{(\frac{\omega^2}{4\mu})^2 + \gamma_m^2} \pm \frac{\omega^2}{4\mu}}{2\mu}} \qquad 8.32$$

The observed speed of the flood wave crest is $L/T_p = \gamma_m/q_m$, where L is the length of the reach and T_p is the travel time of the peak over the length L.

For the limiting case of very long wave $\omega^2/(4\mu) \gg \gamma_m$ and for this situation eqn 8.32 gives $p_m \simeq \omega/(2\mu)$, and since $\gamma_m/q_m \simeq \omega$ the value of $q_m \simeq \gamma_m/\omega$. With these values for p_m and q_m eqn 8.31 yields

$$Q = Q_{om} \sin(\gamma_m t - \frac{\gamma_m}{\omega} x) \qquad 8.33$$

which describes the discharge of a flood wave of constant amplitude (no attenuation) propagating at celerity ω. Therefore, eqn 8.33 is the kinematic wave solution.

For very short waves $\omega^2/(4\mu) << \gamma_m$ and eqn 8.32 yields

$$p_m \simeq q_m \simeq \sqrt{\mu_m/(2\mu)}$$

In this case the flood wave propagates with celerity greater than ω and attenuates very rapidly. Because of this rapid attenuation Hayami suggests that the second and higher order terms have only minor effect on the results obtained from the equations containing these higher order terms.

For long waves where $\omega^2/(4\mu)$ is appreciably larger than γ_m the binomial expansion of $\{[\omega^2/(4\mu)]^2 + \gamma_m^2\}^{1/2}$ in eqn 8.32 leads to

$$\frac{\gamma_m}{q_m} = \frac{L}{T_p} \simeq \omega + \frac{2\mu^2\gamma_m^2}{\omega^3} \simeq \bar{c} + \frac{2\mu^2\gamma_m^2}{\bar{c}^{\;3}} \qquad 8.34$$

and

$$p_m = \frac{\omega}{2\mu} + \frac{\mu\gamma_m^2}{\bar{c}^{\;3}} \qquad 8.35$$

If the diffusion coefficient is defined as

$$\mu = \frac{\alpha Q_p}{L} \qquad 8.36$$

where α is the attenuation parameter defined by eqn 8.41, then the attenuation along the reach is obtained from eqn 8.31 as

$$Q^* = Q_p\{1 - \exp[(\frac{\omega}{2\mu} - p_m)L]\} \simeq Q_p \frac{\mu\gamma_m^2 L}{\omega^3} = \frac{\alpha Q_p}{\omega^3}\gamma_m^2 Q_p \qquad 8.37$$

where $Q_p = Q_{om}$ is the peak discharge. From eqns 8,34 and 8.37 it follows that

$$\frac{L}{T_p} \simeq \omega + \frac{2\alpha Q^*}{L^2}$$

or

$$\omega \simeq \frac{L}{T_p} - \frac{2\alpha Q^*}{L^2} \qquad 8.38$$

where $\omega \simeq \bar{c}$.

8.2.3. Second order approximation

A second-order approximation of the flood routing equation was obtained by Hayashi (1965) and the second-order solution of eqn 8.27 is discussed by Price (1973). After some approximations Price obtained for the attenuation of the flood wave when $q = 0$

$$Q^* \simeq \frac{\alpha_p}{(L/T_p)^3} Q_p \left|\frac{d^2Q_p}{dt^2}\right| \qquad 8.39$$

where α_p is the attenuation parameter for the peak discharge Q_p and T_p is the time of travel of the wave crest over the length L of the reach. He suggested that the curvature of the hydrograph at the peak may be estimated from

$$\frac{d^2Q_p}{dt^2} = \frac{Q_1 + Q_{-1} - 2Q_p}{(\Delta t)^2} \tag{8.40}$$

where Q_1 and Q_{-1} are the discharges at Δt to either side of the peak and Δt is equal to one fifth of the time to peak of the hydrograph (to the nearest hour). This, as will be seen from Examples 8.2 and 8.3 gives satisfactory results when the flood hydrograph is reasonably symmetrical about the peak but not when the hydrograph is strongly skewed.

The theoretical solution by Hayami and the computations by Di Silvio (1969) showed that the rate of attenuation with respect to distance of the flood wave in prismatic channels decreased approximately exponentially with distance downstream. This is in accordance with eqn 8.39 since attenuation decreases the curvature of the hydrograph at the peak. Since eqn 8.39 is only a first-order approximation for the attenuation of the flood wave it can be expected to yield acceptable results only as long as the attenuation over the length of the reach is relatively small. An alternative formula, which is accurate to the second order, is given from the work by Hayami and Hayashi as

$$Q^* \simeq Q_p\{1 - \exp[\frac{\alpha_p}{(L/T_p)^3} \frac{d^2Q_p}{dt^2}]\} \tag{8.41}$$

Thus, eqn 8.39 should be limited to reaches along which the predicted attenuation is less than, for example, 10% of the original peak discharge Q_1. If $Q^*/Q_p > 0.1$, then a new estimate of Q^* is given by

$$Q^* = Q_p[1 - \exp(-\frac{Q^*}{Q_p})] \tag{8.42}$$

and $\omega = \bar{c}$ in eqn 8.38 is defined in terms of this Q^*.

Observing that $\lambda = 1 + \sigma B_f/B_c$ or $B_c\lambda = B_c + \sigma B_f$, and assuming that the reach is composed of subreaches as shown in Fig. 8.6 and that B_c is approximately uniform along the reach, then the attenuation parameter can be written as

$$\alpha(Q) = \frac{1}{2} \frac{\Sigma P_i^2/(L_i S_{oi}^2)}{[(1/L)\Sigma P_i/S_{oi}^{1/3}]^3} \tag{8.43}$$

where P_i is the plan area of the total flooded area of the i-th subreach, and L_i and S_{oi} are the length and bottom slope of the i-th subreach of the channel, respectively. This equation for α is readily evaluated when the areal limits of a given flood are known or for floods which fill a topographically defined flood plain. For inbank floods

$$\alpha = \frac{1}{2\bar{B}_c} \frac{\Sigma(L_i/S_{oi}^2)}{[(1/L)\Sigma(L_i/S_{oi}^{1/3})]^3} \tag{8.44}$$

The channel value and the maximum flood value of α are readily calculated but there is not usually enough information to calculate the intermediate values. Therefore, the function joining these two values has to be guessed. An example is shown in Fig. 8.8.

The question of what order approximation to use is not easily answered. It is a matter of compromise between the effort involved and the required level of accuracy. The more exact the solution the more stringent and detailed are the requirements for

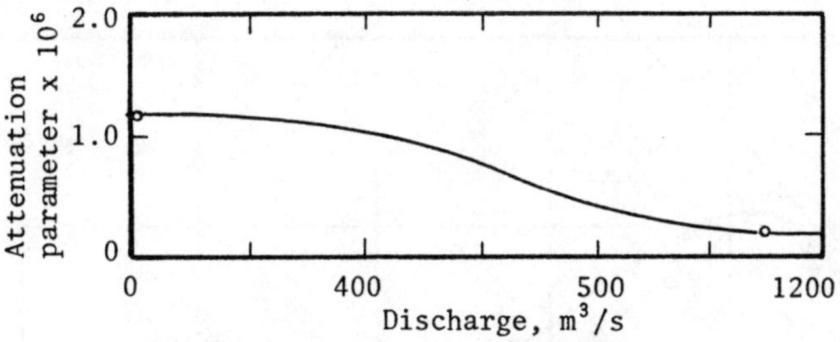

Fig. 8.8. Attenuation parameter for the Erwood to Belmont reach of the River Wye (Flood Studies Report, 1975, Vol. 3).

data, and data of the quality which will warrant the use of the higher order approximations are seldom available. There are also additional problems in that the peak discharge at the downstream station is expressed as an expansion in terms of the time τ when it occurred upstream as

$$Q_p = Q_o(0) + \frac{ax}{c^3} \left. \frac{d^2Q_o}{d\tau^2} \right|_{\tau=0} + \ldots \qquad 8.45$$

where a is a parameter and c is the celerity of the flood wave. This expression for $Q(\tau, x)$ may not be convergent or may converge slowly, in which case a large number of terms would be required.

8.2.4. Celerity of the flood wave

The celerity of the flood wave $\bar{c}(Q)$ is the average speed of a non-attenuating flood wave of peak discharge Q. For an attenuating flood wave the celerity of the peak depends on a number of variables. Hayami showed by his diffusion analysis, with constant ω and μ, that short period flood waves move at speeds greater than ω. The value of $\omega \simeq \bar{c}$ by Hayami's analysis is given by eqn 8.38. Since $\bar{c}$ is a function of Q and because $dQ_1/d\tau \simeq 0$ is not necessarily true at the peak of the hydrograph for a downstream section in longer reaches, the value of L/T_p is a function of $d\bar{c}/dQ$ and may also depend on $d\bar{\alpha}/dQ$. Price (1973b) found from numerical models with synthetic rivers that $\bar{c}$ depends strongly on $d(L/T_p)/dQ$ and the attenuation Q*, and he proposed the semi-empirical formula

$$\bar{c} = \omega + Q^* \frac{d}{dQ} \left(\frac{L}{T_p}\right) \qquad 8.46$$

Figure 8.9 shows the speed-discharge relationship for a reach of the River Wye, which has a shape typical of most natural rivers. The speed is a maximum for a discharge slightly less than bank-full.

The estimation of L/T_p is difficult when there are strong lateral inflows and large tributary inflows. If these inflows are reasonably steady it is advisable to plot L/T_p against the value of Q_p for the combined flow.

In the absence of any records eqns 8.21 and 8.24 could be used but this procedure is very indefinite because of the difficulties with the estimation of the roughness coefficients of the channel and the flood plain and the estimation of the shear between the channel flow and the overbank flow (c.f. Zheleznyakov, 1971). The Flood

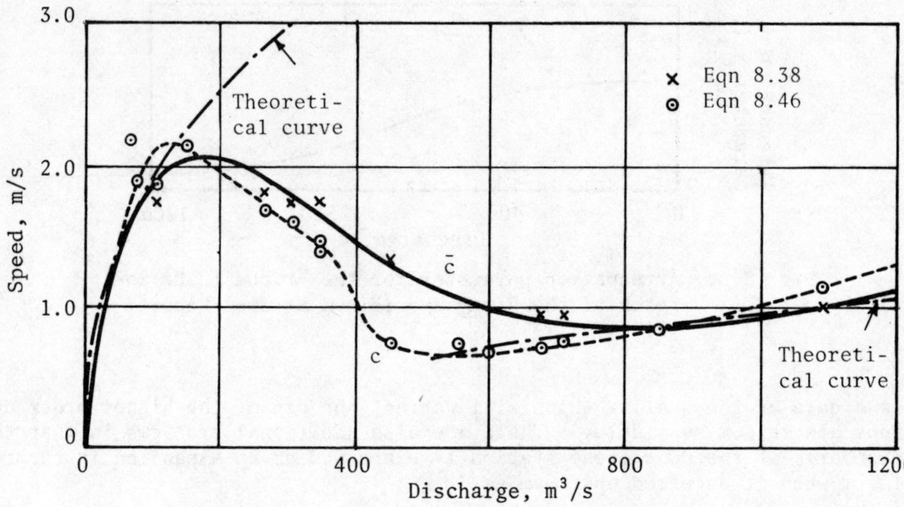

Fig. 8.9. Speed-discharge curves for the Erwood to Belmont reach of the River Wye. The theoretical curve corresponds to eqns 8.21 and 8.24 (Flood Studies Report, 1975, Vol. 3).

Studies Report (1975) recommends a procedure for estimation of $\bar{c}$ and gives auxiliary charts to simplify the calculations. The Report also discusses the numerical solution of what is called the variable parameter diffusion equation

$$\frac{\partial Q}{\partial t} + \bar{c}\,\frac{\partial Q}{\partial x} = \frac{\alpha}{L}\,Q\,\frac{\partial^2 Q}{\partial x^2} + \bar{c}q \qquad 8.47$$

and gives a computer program in FORTRAN IV.

8.2.5. The Muskingum-Cunge method

The most widely used method of hydrological stream routing is the Muskingum method originated by McCarthy (1938). The method uses a linear algebraic relationship between the storage and both the inflow I and the outflow Q, together with two parameters K and ε. The basic continuity or storage statement is

$$\frac{dS}{dt} = I - Q \qquad 8.48$$

The total storage, Fig. 8.10, is expressed as

$$S = KQ + K\varepsilon(I - Q) = K[\varepsilon I + (1 - \varepsilon)Q] \qquad 8.49$$

or

$$\Delta S = S_2 - S_1 = K[\varepsilon(I_2 - I_1) + (1 - \varepsilon)(Q_2 - Q_1)]$$

Combined with the continuity equation

$$\frac{1}{2}(I_1 + I_2)\Delta t - \frac{1}{2}(Q_1 - Q_2)\Delta t = S_2 - S_1$$

the latter expression yields

$$Q_2 = C'_1 I_2 + C'_2 I_1 + C'_3 Q_1 \tag{8.50}$$

where

$$C'_1 = \frac{\Delta t - 2K\varepsilon}{2K(1 - \varepsilon) + \Delta t}$$

$$C'_2 = \frac{\Delta t + 2K\varepsilon}{2K(1 - \varepsilon) + \Delta t}$$

$$C'_3 = \frac{2K(1 - \varepsilon) - \Delta t}{2K(1 - \varepsilon) + \Delta t}$$

By an algebraic modification

$$Q_2 = Q_1 + C_1(I_1 - Q_1) + C_2(I_2 - I_1) \tag{8.51}$$

where

$$C_1 = \frac{\Delta t}{K(1 - \varepsilon) + 0.5\ \Delta t}$$

$$C_2 = \frac{0.5\ \Delta t - K\varepsilon}{K(1 - \varepsilon) + 0.5\ \Delta t}$$

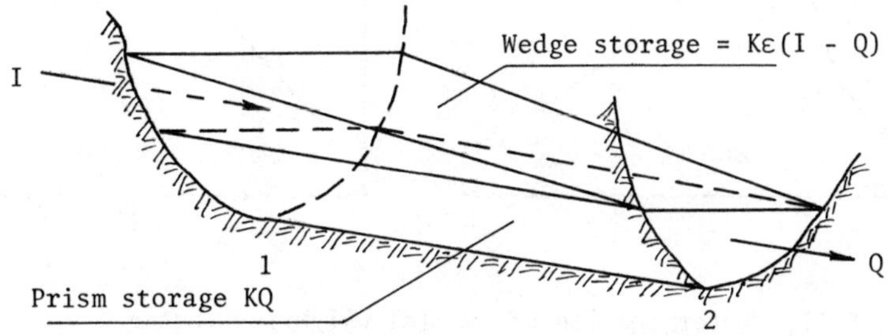

Fig. 8.10. Illustration of storage in a river reach.

The parameter ε with values between 0 and 0.5 is a weighting factor which expresses the relative influence of the inflow I and the outflow Q. K is a storage parameter with dimensions of time and expresses the storage to discharge ratio. Its value is approximately equal to the travel time through the reach. If the inflow and the outflow hydrographs for the reach are available the value of ε can be determined from the observation that the storage is maximum at the time when the inflow and the outflow hydrographs intersect, Fig. 8.11a. At this point $dS/dt = 0$. Differentiating eqn 8.49 and setting dS/dt equal to zero yields

$$\varepsilon \left(\frac{dI}{dt}\right)_c = -(1 - \varepsilon)\left(\frac{dQ}{dt}\right)_c \tag{8.52}$$

in which ε is the only unknown. With the known value of ε the value of K can be determined by plotting S versus $[Q + \varepsilon(I - Q)]$ or $[\varepsilon I + (1 - \varepsilon)Q]$, Fig. 8.11b. The slope of this line is the storage coefficient.

The volumes of storage in the river reach S_i at instants t_i, $i = 0, 1, 2, \ldots$ are represented by the area between the inflow and outflow hydrographs in Fig. 8.12a. These values plotted against $[\varepsilon I + (1 - \varepsilon)Q]$ for arbitrary values of ε give K as the slope

$$K = \frac{S_i}{\varepsilon I + (1 - \varepsilon)Q}$$

The value of ε which yields a loop closest to a single line, Fig. 12b, is taken to be the correct value. The loops are usually obtained by cumulative plots of the incremental values from

$$K = \frac{0.5\ \Delta t[(I_2 + I_1) - (Q_2 + Q_1)]}{\varepsilon(I_2 - I_1) + (1 - \varepsilon)(Q_2 - Q_1)} \qquad 8.53$$

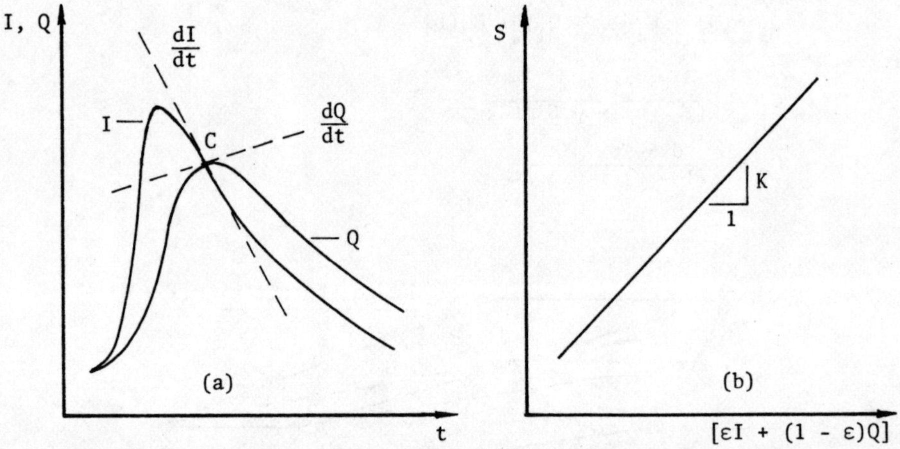

Fig. 8.11. Determination of the (a) weighting factor ε, and (b) storage parameter K for the Muskingum method of flood routing.

Over a given period the difference between the sum of inflows and the sum of outflows is the volume in storage and storage can be plotted as a function of discharge, Fig. 8.12c. The loops arise from the wedge storage effect. Figure 8.12c shows that the slope of the curve, i.e., the K value, may vary with the outflow. If this variation is substantial then the K = constant assumption is no longer satisfactory.

If ε and K are assumed to be constant, C_1 and C_2 are given constants and eqn 8.51 is readily solved, for example, in a tabular arrangement shown below where $C_1 = 0.3$ and $C_2 = 0.1$ are used:

Routing period Δt	I m^3s^{-1}	$I_2 - I_1$ (3)	$C_2(I_2 - I_1)$ (4)	$I_1 - Q_1$ (5)	$C_1(I_1 - Q_1)$ (6)	(4)+(6)	Q m^3s^{-1}
1	100	50	5	10	3	8	90
2	150	75	7.5	52	15.6	23.1	98
3	225						121.1
.	.						

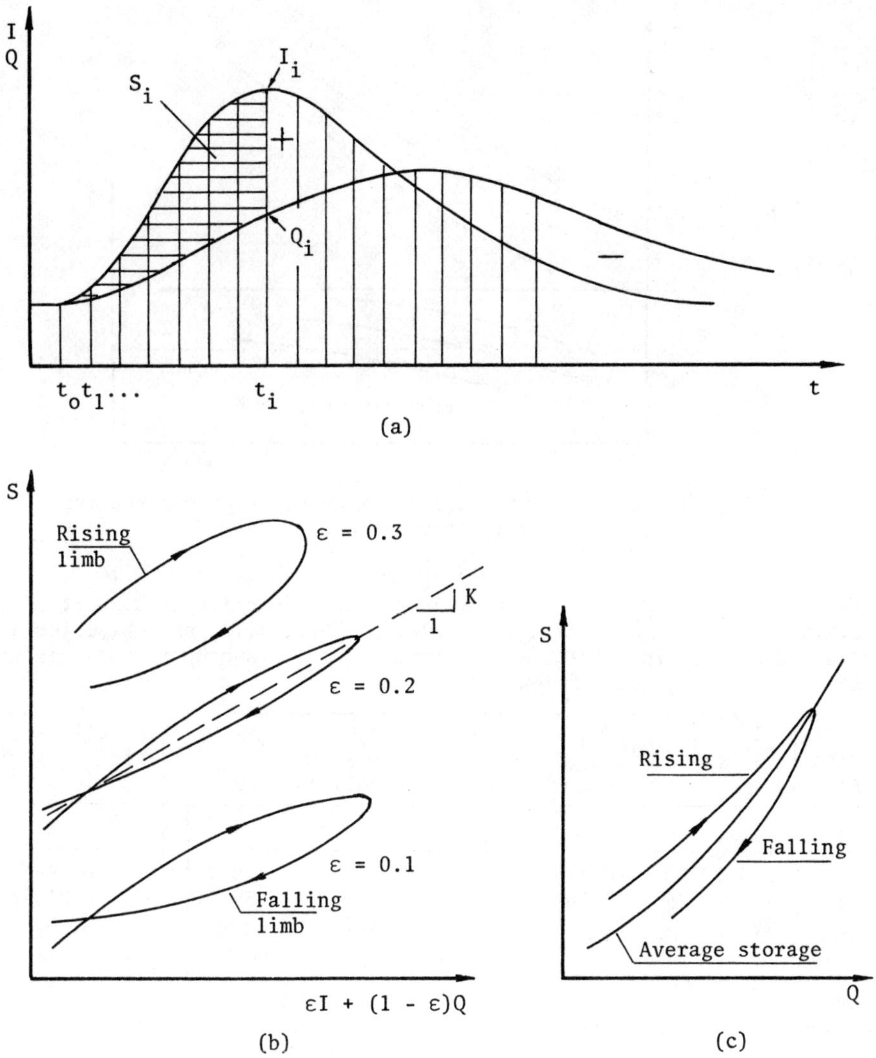

Fig. 8.12. Determination of ε and K for the Muskingum routing method.

If Δt is taken equal to $2K\varepsilon$ then eqn 8.51 reduces to

$$Q_2 = Q_1 + \frac{\Delta t}{K}(I_1 - Q_1)$$

which forms the basis of a very simple graphical method. If empirical information is available on the influence of tributary flows this can be readily incorporated in the routing process, Fig. 8.13.

When the assumption of K = constant is unsatisfactory (i.e., when the variation of the slope of the Q versus S relationship is appreciable) the values of K and of the

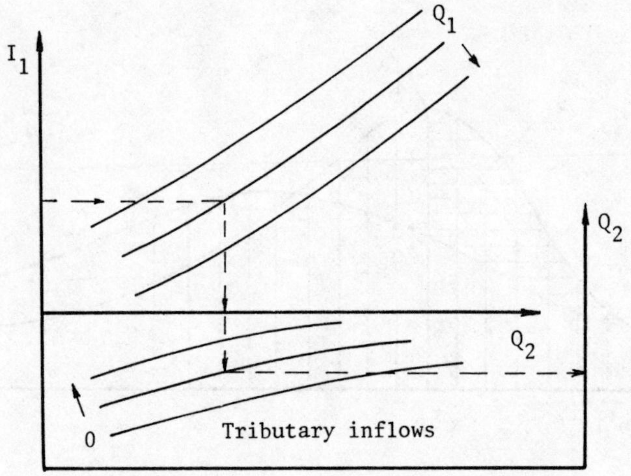

Fig. 8.13. Schematic illustration of graphical routing with tributary inflow.

coefficients C_1 and C_2 have to be evaluated as a function of Q. It is assumed that ε remains constant. The values are plotted, Fig. 8.14, and the calculations again may be carried out in tabular arrangement using the appropriate values of K, C_1 and C_2 at each step, as shown below:

Routing period Δt	I m^3s^{-1}	$I_2 - I_1$	C_2	$C_2(I_2 - I_1)$	Q m^3s^{-1} + col. 10	$I_1 - Q_1$	C_1	$C_1(I_1 - Q_1)$	$Q_2 - Q_1$ (5)+(9)
1	2	3	4	5	6	7	8	9	10
1	100	50	0.30	15	90	10	0.6	6	21
2	150	75	0.28	21	111	39	0.55	21.45	42.45
3	225				153				

The parameters K and ε determined from observed data can be expected to yield satisfactory results when routing is confined to the range of flows used to estimate K and ε, but extrapolation beyond the range of observations is risky.

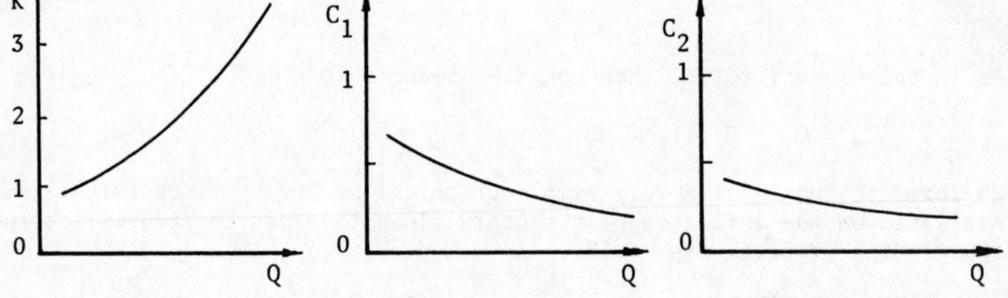

Fig. 8.14. Coefficients for Muskingum method of routing.

The Muskingum method and several other hydrological routing techniques are discussed in some detail by Chow (1964) and also by Linsley et al. (1949).

For floods with short times of concentration (steep rising limbs) the Muskingum method tends to give negative starting values for the outflow and it is then advisable to use some other method. Nash (1959a) expressed the Muskingum equation as

$$Q_1 = I_0[\frac{K}{\Delta t}(1 - c) - c] + I_1[1 - \frac{K}{\Delta t}(1 - c)] + Q_0 c$$

where

$$c = \exp[\frac{-\Delta t}{K(1 - \varepsilon)}]$$

The initial negative response occurs at large values of ε but can be avoided if $\Delta t > 2K\varepsilon$ but $\Delta t << K$. A very simple method for rapidly rising hydrographs is the unit reach method by Kalinin and Milyukov (1957). This method assumes the river to consist of n-equal linear reservoirs (ideal reservoirs for which the routing curve is a straight line). The travel time through the reach is assumed to be τ.

The outflow hydrograph for this model is given by

$$Q(t) = \bar{Q}\frac{\Delta t}{\tau^n(n - 1)!}t^{n-1}e^{-t/\tau} \tag{8.54}$$

where $\bar{Q}$ is now the mean inflow during the period Δt. The constants for the unit reach method can be found from those for the Muskingum method as

$$n = \frac{1}{1 - 2\varepsilon} \quad \text{and} \quad \tau = \frac{K}{n} \tag{8.55}$$

Cunge (1969) discussed the Muskingum method and extended it. He also showed that the attenuation of the flood wave obtained by the Muskingum method arises from the finite difference scheme which replaces the partial differential equations. From eqns 8.48 and 8.49

$$K\frac{d}{dt}[\varepsilon Q_j + (1 - \varepsilon)Q_{j+1}] = Q_j - Q_{j+1} \tag{8.56}$$

where Q_j is the inflow to the reach and Q_{j+1} is the outflow. Written in finite difference form

$$\frac{K}{\Delta t}[\varepsilon Q_j^{n+1} + (1 - \varepsilon)Q_{j+1}^{n+1} - \varepsilon Q_j^n - (1 - \varepsilon)Q_{j+1}^n]$$

$$= \frac{1}{2}(Q_j^{n+1} - Q_{j+1}^{n+1} + Q_j^n - Q_{j+1}^n) \tag{8.57}$$

If $K = \Delta x/\omega$, by definition, then eqn 8.57 is the finite difference form of the kinematic wave equation

$$\frac{\partial Q}{\partial t} + \omega\frac{\partial Q}{\partial x} = 0$$

Cunge expressed (Q_j^n) in terms of Taylor expansions and showed that eqn 8.57 is also a finite difference representation of

$$\frac{\partial Q}{\partial t} + \omega\frac{\partial Q}{\partial x} = \mu\frac{\partial^2 Q}{\partial x^2}$$

when

$$\mu = (\frac{1}{2} - \varepsilon)\omega\Delta x \tag{8.58}$$

With $\mu = (\alpha\bar{Q}_p)/L$, defined analogous to eqn 8.36

$$\varepsilon = \frac{1}{2} - \frac{\alpha\bar{Q}_p}{L\omega\ \Delta x} \qquad 8.59$$

where L is the length of the whole reach which is subdivided into subreaches of length Δx, and $\bar{Q}_p$ is the average peak discharge. The value of $\bar{Q}_p$ is the average of the values of Q_p at the upstream and downstream ends of the reach. It can be estimated using

$$\bar{Q}_p = Q_p - \frac{1}{2} Q^* \qquad 8.60$$

Equation 8.59 for ε is in the form given in the Flood Studies Report. In Cunge's version α/L was expressed in terms of the mean width of the flow and the resistance coefficient.

After K and ε have been estimated the discharge hydrograph at the downstream end of the reach is calculated using

$$Q_{j+1}^{n+1} = C_1 Q_j^n + C_2 Q_j^{n+1} + C_3 Q_{j+1}^n + C_4 \qquad 8.61$$

where

$$C_1 = \frac{K\varepsilon + \frac{1}{2}\Delta t}{K(1 - \varepsilon) + \frac{1}{2}\Delta t} \qquad C_2 = \frac{\frac{1}{2}\Delta t - K\varepsilon}{K(1 - \varepsilon) + \frac{1}{2}\Delta t}$$

$$C_3 = \frac{K(1 - \varepsilon) - \frac{1}{2}\Delta t}{K(1 - \varepsilon) + \frac{1}{2}\Delta t} \qquad C_4 = \frac{q\Delta t\ \Delta x}{K(1 - \varepsilon) + \frac{1}{2}\Delta t}$$

The length Δx of the subreaches is chosen so that with Δt equal to an integral number of hours the value of $\Delta x/(\omega\ \Delta t)$ lies below the curve shown in Fig. 8.15.

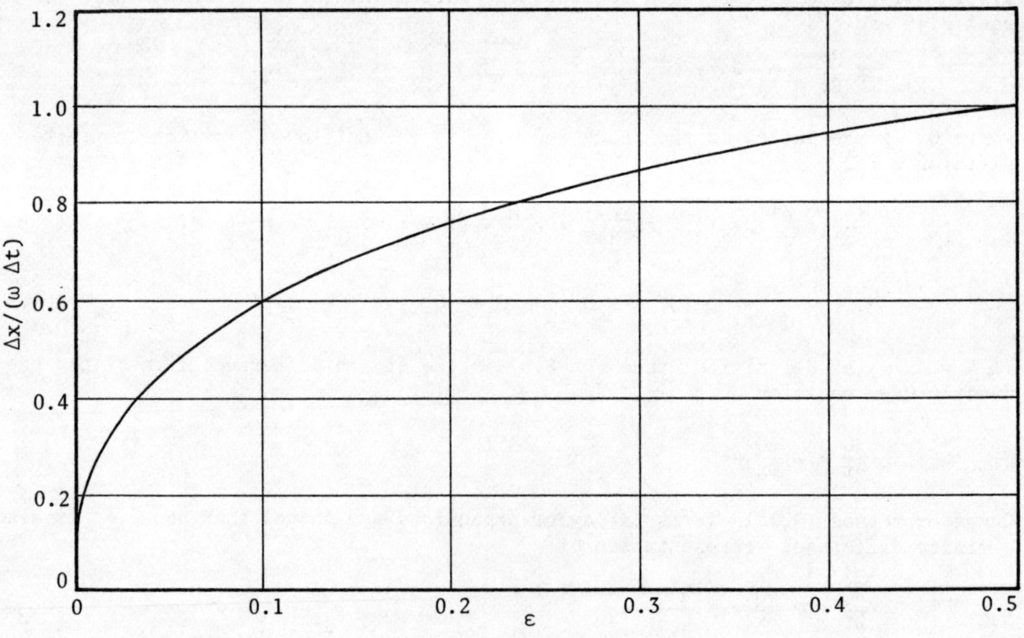

Fig. 8.15. Curve for $\Delta x/(\omega\ \Delta t)$, Cunge (1969).

The attenuation increases as ε decreases and $\varepsilon = \frac{1}{2}$ corresponds to zero attenuation. For a given value of ε attenuation increases as the wave period decreases. The celerity of the flood wave varies with ε, the wave length and the $\Delta t/\Delta x$ ratio.

The routing by the Muskingum-Cunge method for a wave in a uniform stream channel may be carried out as follows:

1. Determine or assume L/T_p
2. Select Δx
3. Calculate $\alpha_p = \frac{1}{2}(L/B)(1/S_0)$ from eqn 8.43
4. Calculate

$$\frac{d^2Q_p}{dt^2} = \frac{Q_{-1} + Q_1 - 2Q_p}{(\Delta t)^2}$$

5. Calculate

$$Q^* = \frac{\alpha_p}{(L/T_p)^3} Q_p \left| \frac{d^2Q_p}{dt^2} \right|$$

If $Q^* \leq 0.1\, Q_p$, $\omega = L/T_p$; if $Q^* > 0.1\, Q_p$, redefine Q^* as

$$Q^*_{new} = Q_p[1 - \exp(-\frac{Q^*}{Q_p})]$$

and ω as

$$\omega = \frac{L}{T_p} - \frac{2\alpha_p}{L^2} Q^*_{new}$$

6. $\bar{Q}_p = Q_p - \frac{1}{2}Q^*$; $(Q^* < 0.1\, Q_p)$

   $\bar{Q}_p = Q_p - \frac{1}{2}Q^*_{new}$; $(Q^* > 0.1\, Q_p)$

7. $K = \frac{\Delta x}{\omega}$; $\varepsilon = \frac{1}{2} - \frac{\alpha_p}{L\,\Delta x}\frac{\bar{Q}_p}{\omega}$

   where $\Delta x/\omega\Delta t$ must be less than the value given by the curve in Fig. 8.15.
8. Evaluate C_1, C_2 and C_3.
9. Calculate the outflow hydrograph at $x = j\Delta x$ and $t = n\Delta t$. The calculations may be carried out in a tabular arrangement as shown in Example 8.2. In the table first column is the inflow hydrograph for which the values are known. Likewise the values in the first row (i.e., the initial steady state condition) are known. Then from

$$Q^{n=1}_{j=1} = C_1Q^o_o + C_2Q^1_o + C_3Q^o_1, \text{ etc.}$$

   Plot the inflow hydrograph at $x = 0$ and the outflow hydrograph at $x = L$.
10. The attenuation is the difference between Q_p at $x = 0$ and Q_p at $x = L$.

8.2.6. Conclusions from the Flood Studies Report

The conclusions drawn in the Flood Studies Report from comparative studies are that (1) if there are enough data, particularly on roughness, the numerical solution of the equations of motion (eqn 8.27) can be expected to give the most accurate results, (2) if the parameters can be accurately defined the variable parameter diffusion equation gives a better prediction of the shape of the discharge hydrograph than the Muskingum-Cunge method,
(3) if the parameters cannot be defined accurately the Muskingum-Cunge method is preferred because of its simplicity and ease of application.

It was also found that "the curves α, L/T_p and $\bar{c}$ can usually be extrapolated with safety so that larger floods than have previously been recorded can be routed". Computer programs in FORTRAN IV for both routing models are given in the Flood Studies Report.

Example 8.2. The hydrograph at the upstream end of a river reach (x = 0) at time t = 0 is as follows:

t hours:	0	1	2	3	4	5	6	7	8	9	10
Q m^3/s :	10	12	18	28.5	50	78	107	134.5	147	150	146
t hours:	11	12	13	14	15	16	17	18	19	20	
Q m^3/s :	129	105	78	59	45	33	24	17	12	10	

Determine the hydrograph by the Muskingum-Cunge method at x = 18 km. Assume a constant river width of 50 m, no flood plains, a bed slope $S_o = 5 \times 10^{-4}$, and a wave speed L/T_p = 2 m/s. Use Δx = 6 km.

For a river of constant width eqn 8.45 reduces to

$$\alpha_p = \frac{1}{2}\frac{L}{B}\frac{1}{S_o} = \frac{1}{2}\frac{18 \times 10^3}{50}\frac{10^4}{5} = 3.6 \times 10^5$$

From eqn 8.42a

$$\frac{d^2Q_p}{dt^2} = \frac{Q_1 + Q_{-1} - 2Q_p}{(\Delta t)^2} = \frac{129 + 134.5 - 2 \times 150}{(7200)^2} = -\frac{36.5}{7.2^2} \times 10^{-6} = -0.704090 \times 10^{-6}$$

and from eqn 8.42b

$$Q^* = \frac{3.6 \times 10^5}{2^3} 150 \times 0.704 \times 10^{-6} = 4.75 < 0.1\ Q_p$$

Thus, from eqn 8.60

$$\bar{Q}_p = 150 - \frac{1}{2} \times 4.75 = 147.625\ m^3/s$$

$$\omega = 2\ m/s$$

$$K = \Delta x/\omega = 6 \times 10^3/2 = 3000\ s$$

Equation 8.59 yields

$$\varepsilon = \frac{1}{2} - \frac{3.6 \times 10^5}{18 \times 10^3 \times 6 \times 10^3}\frac{147.625}{2} = 0.2539583$$

for which Fig. 8.15 shows that

$$\frac{\Delta x}{\omega\,\Delta t} \ngtr 0.82 \qquad \text{or} \qquad \Delta t \nless \frac{6000}{2 \times 0.82} = 3658\ s$$

Take Δt = 7200 s, then

$$C_1 = \frac{3000 \times 0.254 + 3600}{3000(1 - 0.254) + 3600} = 0.747174$$

$$C_2 = \frac{3600 - 30000 \times 0.254}{3000(1 - 0.254) + 3600} = 0.486125$$

$$C_3 = \frac{3000(1 - 0.254) - 3600}{3000(1 - 0.254) + 3600} = -0.233299$$

$$\Sigma C_i = 1.0$$

From $$Q^{n+1}_{j+1} = C_1Q^n_j + C_2Q^{n+1}_j + C_3Q^n_{j+1}$$

n Δt hrs	j Δx 0	6	12	18 km
0	10	10	10	10
2	18	13.89	11.89	10.92
4	50	34.51	24.38	18.19
6	107	81.32	59.63	42.96
8	147	132.44	111.23	88.60
10	146	149.91	145.88	133.35
12	105	125.16	138.82	145.37
14	59	77.93	99.01	117.94
16	33	41.94	55.52	73.45
18	17	23.14	29.63	38.75
20	10	12.17	16.29	21.02
22	10	9.49	9.91	12.09
24	10	10.12	9.70	9.30
26	10	9.97	10.15	10.01
28	10	10.01	9.95	10.08

The plot of these shows that the attenuation at x = 18 km is $Q^* \simeq 4.6 \text{ m}^3/\text{s}$.

The widths of the hydrographs is:

	x = 0	x = 18 km
$Q = 130 \text{ m}^3/\text{s}$	4.20 h	3.60 h
$Q = 100 \text{ m}^3/\text{s}$	6.40 h	6.25 h
$Q = 50 \text{ m}^3/\text{s}$	10.60 h	10.85 h

Time for peak to travel 18 km is 2.8 h or speed of peak is 1.79 m/s.

Displacement times at various points on the wave for 18 km are:

	Rising limb		Falling limb	
$Q = 130 \text{ m}^3/\text{s}$	3.0 h	1.67 m/s	2.4 h	2.08 m/s
$Q = 100 \text{ m}^3/\text{s}$	2.8 h	1.79 m/s	2.5 h	1.92 m/s
$Q = 50 \text{ m}^3/\text{s}$	2.3 h	2.17 m/s	2.6 h	1.92 m/s

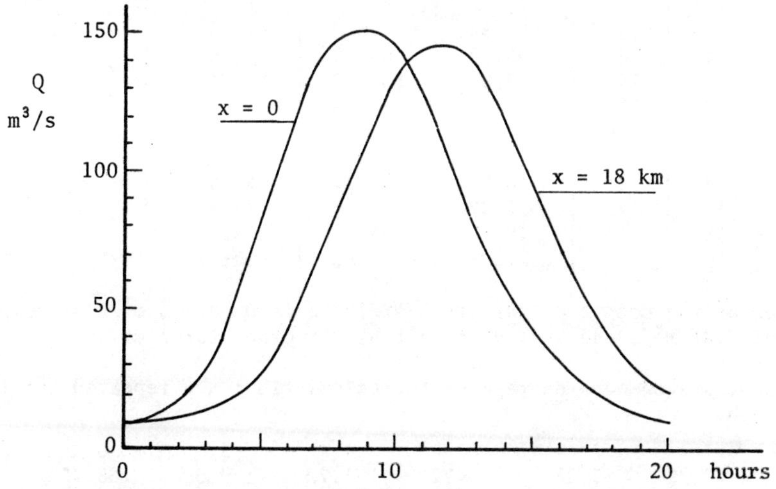

The initial and routed hydrographs of Example 8.2.

Example 8.3. Repeat the calculations of example 8.2 for the following hydrograph:

t hours:	0	1	2	3	4	5	6	7	8	9	10	11	12
Q m^3/s :	10	20	52	300	141	150	144	130	114	100	86	73	61

t hours:	13	14	15	16	17	18	19	20
Q m^3/s :	50	41	32	25	19	14	11	10

$$\alpha_p = 3.60 \times 10^5$$

$$\frac{d^2Q_p}{dt^2} = \frac{144 + 141 - 2 \times 150}{3600^2} = -1.157 \times 10^{-6}$$

$$Q^* = \frac{3.60 \times 10^5}{2^3} \times 150 \times 1.157 \times 10^{-6} = 7.81 \text{ m}^3/\text{s} < 0.1\, Q_p$$

$$\omega = 2 \text{ m/s}, \qquad \bar{Q}_p = 150 - \frac{1}{2} \times 7.81 = 146.1 \text{ m}^3/\text{s}$$

$$K = \frac{6 \times 10^3}{2} = 3000 \text{ s}$$

$$\varepsilon = \frac{1}{2} - \frac{3.60 \times 10^5}{18 \times 10^3 \times 6 \times 10^3} \times \frac{146.1}{2} = 0.256508$$

$$\frac{\Delta x}{\omega \Delta t} \not> 0.82, \qquad \Delta t \not< 3658 \qquad \text{take } \Delta t = 7200 \text{ s}$$

$$C_1 = \frac{3000 \times 0.257 + 3600}{3000(1 - 0.257) + 3600} = 0.749428$$

$$C_2 = \frac{3600 - 3000 \times 0.257}{3000(1 - 0.257) + 3600} = 0.485462$$

$$C_3 = \frac{3000(1 - 0.257) - 3600}{3000(1 - 0.257) + 3600} = -0.234891$$

$$\Sigma(C_1 + C_2 + C_3) = 1.0$$

n Δt hrs	j Δx 0	6	12	18 km
0	10	10	10	10
2	52	30.39	19.90	14.81
4	141	100.28	66.78	43.85
6	144	152.02	133.27	104.44
8	114	127.55	144.54	145.51
10	86	97.22	123.56	134.13
12	61	71.23	78.42	99.16
14	41	48.94	58.72	63.98
16	25	31.37	38.11	47.48
18	14	18.16	23.37	28.75
20	10	11.08	13.50	17.31

The attenuation of the computed hydrograph at 18 km is $Q^* \simeq 4.5$ m^3/s as compared to the estimate of 7.81 m^3/s and indicates that higher order terms may be required.

The problem becomes even more acute when the calculations are repeated for the following hydrograph:

t hours:	0	1	2	3	4	5	6
Q m^3/s :	10	25.28	65.28	114.72	154.72	170.00	168.25

t hours:	7	8	9	10	11	12	13
Q m^3/s :	163.08	154.72	143.53	130.00	114.72	98.36	81.64
t hours:	14	15	16	17	18	19	20
Q m^3/s :	65.28	50.00	36.47	25.28	16.92	11.75	10.00

Assume that the hydrograph at x = 50 km is required. Use other data as in above examples except Δx = 10 km.

For a river of constant width

$$\alpha_p = \frac{1}{2}\frac{L}{B}\frac{1}{S_o} = \frac{1}{2} \times \frac{5 \times 10^4}{50} \times \frac{10^4}{5} = 1 \times 10^6$$

$$\frac{d^2Q_p}{dt^2} = \frac{Q_{-1} + Q_1 - 2Q_p}{(\Delta t)^2} = \frac{154.72 + 168.25 - 2 \times 170}{3600^2} = -1.314 \times 10^{-6}$$

$$Q^* = \frac{\alpha_p}{(L/T_p)^3} Q_p \left|\frac{d^2Q_p}{dt^2}\right| = \frac{10^6}{2^3} 170 \times 1.314 \times 10^{-6} = 27.92 \text{ m}^3/\text{s} > 0.1\ Q_p$$

$$Q^*_{new} = Q_p[1 - \exp(-\frac{Q^*}{Q_p})] = 170[1 - \exp(-\frac{27.92}{170})] = 25.75 \text{ m}^3/\text{s}$$

$$\omega = L/T_p - \frac{2\alpha_p}{L^2} Q^*_{new} = 2 - \frac{2 \times 10^6}{25 \times 10^8} 25.75 = 1.98 \text{ m/s}$$

$$Q^* = \frac{10^6}{1.98^3} 170 \times 1.314 \times 10^6 = 28.77 \text{ m}^3/\text{s}$$

$$Q^*_{new} = 170[1 - \exp(-\frac{28.77}{170})] = 26.47 \text{ m}^3/\text{s}$$

$$\omega = 2 - \frac{2 \times 10^6}{25 \times 10^8} 26.47 \simeq 1.98 \text{ m/s}$$

$$\bar{Q}_p = Q_p - \frac{1}{2} Q^*_{new} = 156.77 \text{ m}^3/\text{s}$$

$$K = \frac{\Delta x}{\omega} = \frac{10^4}{1.98} = 5050 \text{ s}$$

$$\varepsilon = \frac{1}{2} - \frac{\alpha_p}{L\ \Delta x}\frac{\bar{Q}_p}{\omega} = \frac{1}{2} - \frac{10^6}{5 \times 10^4 \times 10^4}\frac{156.77}{1.98} = 0.342$$

From Fig. 8.15 it is seen that $\frac{\Delta}{\omega\ \Delta t} \not> 0.9$ or $\Delta t \not< 5612$ s.

Take Δt = 2 hrs = 7200 s. Then

$$C_1 = \frac{K\varepsilon + \frac{1}{2}\Delta t}{K(1 - \varepsilon) + \frac{1}{2}\Delta t} = \frac{5050 \times 0.342 + 3600}{5050(1 - 0.342) + 3600} = 0.769490$$

$$C_2 = \frac{\frac{1}{2}\Delta t - K\varepsilon}{K(1 - \varepsilon) + \frac{1}{2}\Delta t} = \frac{3600 - 5050 \times 0.342}{5050(1 - 0.342) + 3600} = 0.270537$$

$$C_3 = \frac{K(1 - \varepsilon) - \frac{1}{2}\Delta t}{K(1 - \varepsilon) + \frac{1}{2}\Delta} = \frac{5050(1 - 0.342) - 3600}{5050(1 - 0.342) + 3600} = -0.040029$$

$$\Sigma = (C_1 + C_2 + C_3) = 1.0$$

$$Q^{n+1}_{j+1} = C_1\ Q^n_j + C_2\ Q^{n+1}_j + C_3\ Q^n_{j+1}$$

n Δt hrs	j Δt 0	10	20	30	40	50 km
0	10.00	10.00	10.00	10.00	10.00	10.00
2	65.28	24.95	14.04	11.09	10.29	10.08
4	154.72	91.09	43.28	22.07	14.09	11.33
6	168.25	160.93	111.89	62.69	33.38	19.42
8	154.72	164.88	163.96	127.94	81.51	46.96
10	130.00	147.63	160.25	162.94	139.26	98.51
12	98.36	120.74	139.85	154.62	161.64	146.94
14	65.28	88.52	111.25	131.52	148.09	158.56
16	36.47	56.56	78.97	101.71	122.79	140.83
18	16.92	30.38	48.58	69.84	92.25	113.81
20	10.00	14.51	25.36	40.17	60.92	82.91
22	10.00	9.82	12.81	21.37	34.25	52.83
24	10.00	10.01	9.95	11.64	18.22	29.17

Attenuation of the peak discharge is $Q^* \simeq 10\ m^3/s$. The time of displacement of the peak is approximately 8 hours or speed of peak is 50 000/(8 x 3600) = 1.74 m/s.

In this example, the expansions converge slowly and the travel time of the peak over Δx is about 1 3/4 hours.

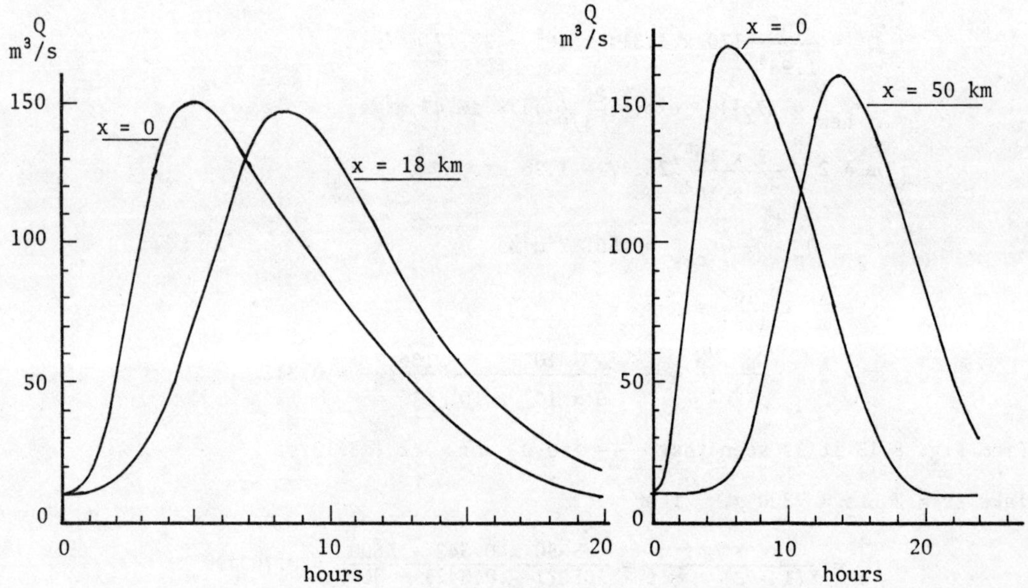

The initial and routed hydrographs of Example 8.3.

Chapter 9

EXTREME EVENTS, DESIGN FLOOD AND SMALL CATCHMENT RUNOFF

9.1 Extreme Event

The principal extreme events in hydrology are floods and droughts. A flood can be defined as a flow that overtops the banks of a river or a stream. This definition is not entirely hydrological since it also involves geomorphological, engineering and water management features. The bank-full capacity of a stream depends on the geology and topography of the area and it could be substantially modified by man-made structures, such as stopbanks or levees. In deep valleys and mountain gorges floods according to this definition would never occur.

Floods may be further characterized by the peak flow rate, flood elevation, flood volume and flood duration. The *flood discharge* is a convenient characteristic because it relates to the flow only and is not affected by the geometry of the river channel. *Flood elevation* is an important parameter in relation to the level of the banks and to human activities, but it is a difficult parameter to use because its value depends on the cross-section of the river channel and varies along the watercourse. For planning of flood protection *flood volume* is one of the most important parameters. It is defined as the volume of flood flow above some base discharge Q_o, Fig. 9.1.

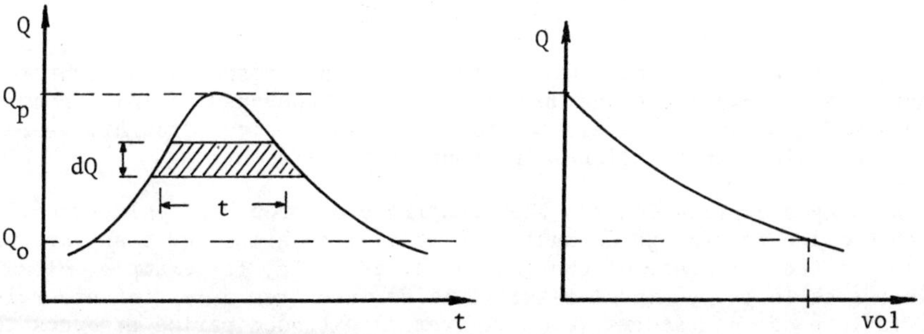

Fig. 9.1. Definition of flood volume.

The flood volume is given by

$$V = \int_{Q_o}^{Q_{peak}} t dQ \qquad 9.1$$

Flood volume is closely linked with *flood duration*. Flood duration at a given location may be based on observations, or it may be predicted by the methods of flood routing, provided the inflow hydrograph at some upstream location can be determined or estimated.

Duration is more important in agricultural than in industrial or commercial contexts. The damage to goods or homes will vary little with the duration of immersion. However, the damage to agricultural lands from a flood lasting 12 hours may be nil whereas a flood lasting a few days may ruin all crops. A flood of long duration may also raise the groundwater table to such an extent that harmful effects may extend over a much larger area than that inundated by the flood. Figure 9.2 illustrates a combined high peak and long duration of a flood.

The main feature of a flood, from the water management point of view, is the interference with human activities. The interference is measured in terms of the actual or potential economic loss and the danger to human life. The purpose of a flood analysis is to assess the magnitude and frequency of this interference. As noted before, the effects of a flood can be associated with various parameters. Of these the peak discharge and frequency are the most common, but it may be necessary to study the frequencies of flood volumes or of flood durations of various discharges.

Floods may arise from extreme rainfalls, from rapid melting of extensive snow deposits or from a combination of both. Where observed records of flood events are not available the flood flow estimates have to be derived from data on extreme rainfall events.

The term *drought* is much more difficult to define than flood. In one climate several months without rain may constitute a severe drought, in another it can be the norm as in winter seasons of the savanna lands. In a general sense a drought may be defined as an abnormal moisture deficiency in relation to some need, like the empty reservoir in Fig. 9.3. The engineer relates drought to a set of variables which describe rainfall, runoff and water storage. The economist relates drought to factors which affect human activities. Thus, there may be an agricultural drought related to shallow or deep rooted plants, a water supply drought, etc. The geophysicist talks of climatological droughts, and so on. A summary of definitions of droughts is given by Subrahmanyam (1967). A meteorological drought may be defined as a period of time with no rain or with rainfall less than some particular value, e.g., 1 mm per month. A climatological drought refers to long periods, such as sequences of years, with precipitation less than some base value, for example, less than 25% of the mean annual precipitation. An atmospheric drought refers to conditions of air temperature and humidity, etc. Sequences of low stream flows, lake levels and groundwater levels constitute a hydrological drought. Droughts may be further subdivided into regional and continental droughts.

There is no simple explanation for the occurrence of droughts. In general, the factors which combine to produce droughts are related to atmospheric and oceanic circulation, and to the influence of continental areas. If, for example, climatic conditions are such that the annual rainfall is derived from a few intense rainstorms, then the failure of such storms to occur over an extended period produces the drought. A temporary decrease in the number of rainfalls may arise from variations in the pattern of atmospheric circulation. Slower acting and longer lasting atmospheric effects can be produced by variation in the distribution of warm and cold water masses in the oceans. The occurrence of particularly cold or warm periods over continental areas may precede or produce droughts over adjacent continental areas. The

Fig. 9.2. Floods in Hungary, May 1970. Rivers Tisza and Szamos, tributaries of the Danube, flooded an area of about 520 km^2 for approximately 40 days. Photo by Dipl.-Ing. Gàbor Magyari.

Fig. 9.3. An empty reservoir, Clywedog, Wales, U.K., 1975.
Photo by Dr. M.D. Newson.

attempts at explanation of droughts are based on the physical relationships and interactions of the drought affecting factors, while the descriptions of droughts are based on statistical and analytical methods.

Prediction and analysis of drought conditions are of great importance in water resources management. The analysis may be concerned with one of the following aspects (Yevjevich, 1967): duration, probability of occurrence, severity (total deficiency of water relative to some reference level and duration), time of occurrence in the annual cycle, and areal extent. The studies are aimed at assessing the feasibility of using a given area for some particular purpose, for example, for growing certain crops, for water supply, or for recreation. The analysis also provides a basis for the assessment of the risks (also fire risk) associated with the proposed use, and provides the data for operational management.

Yevjevich divides the statistical techniques of drought analysis into four groups:
(1) Empirical methods. These are based on observed data from which deductions are made about the nature of the variables. For example, Joseph (1970) fitted distributions to river flow data and found that the Gamma distribution gave the best fit.
(2) Generation methods. Techniques, such as the Monte Carlo method and other more sophisticated methods, are used to produce long sequences of data with the same statistical properties (i.e., mean, variance, etc.) as the observed data.
(3) Analytical models. These are based mainly on the theory of probability.
(4) Analysis of runs. The periods when the flow is less than a base value Q_o are considered as a statistical variate, τ, and are associated with the duration of the drought, Fig. 9.4. The severity of the drought is measured by the total deficit in the volume of water with respect to Q_o during the period τ. The deficit in the volume of water is given by

$$S_\tau = \int_{t_i}^{t_i+\tau} (Q_o - Q)\,dt$$

where S_τ will have a certain probability distribution.

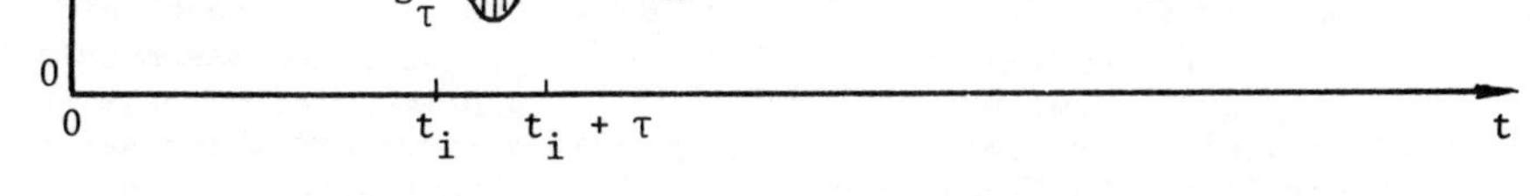

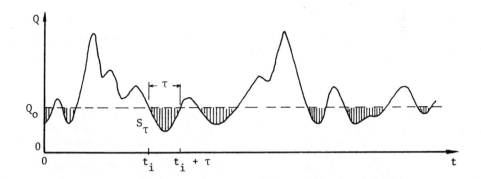

Fig. 9.4. Definition sketch for application of runs in drought analysis.

A relatively simple method of quantifying droughts was proposed by Herbst et al. (1966). They used the concept of monthly mean deficits MMD_t, t = 1,..., 12, which are defined as

$$MMD_t = \frac{1}{n} \sum_1^n (E_t - M_t)_i$$

and evaluated from the n-years of record. In it M_t is the monthly mean rainfall and E_t is the effective rainfall defined as

$$E_{t+1} = (A_t - M_t)W_{t+1} + A_{t+1}$$

where A_t is the actual monthly rainfall total and W_t is a weighting factor which allows for carry-over from one month to the next

$$W_t = 0.1(1 + 12M_t/MAR)$$

and MAR is the mean annual rainfall, ΣM_t. The carry-over is zero for the first month. The sum ΣMMD_t gives the mean annual deficit MAD. The beginning of the drought is determined with the aid of a sliding scale

$$x = (1/11)(MAD - MMMR)$$

where MMMR is the maximum of the twelve monthly mean rainfalls M_t. Sequential sums of deficits from t to t+11 are tested against this scale and

$$\sum_1^n (E_t - M_t) > MMMR + (n - 1)x$$

is deemed to mean that a drought has began from t. The first step on this sliding scale is the rare event when $(E_t - M_t) = MMMR$, the second is $(E_t - M_t) = MMMR + x$, etc. Where the algebraic sum of the differences becomes positive the potential drought is terminated. However, more severe tests for the end of a drought can be formulated.

A drought intensity index is defined as

$$Y = \frac{\sum_{t=1}^{n} [(E_t - M_t) - MMD_t]}{\sum_{t=1}^{D} MMD_t} \qquad 9.2$$

where D is the duration of the drought in months. A severity of the drought index is then given by YD.

9.2 The Design Flood

The failure of a dam, a stopbank, a drainage system, or whatever the project may be, during a flood may result in many kinds of damage to property, economic loss and danger to life. The design of a project, however, has to be based on a "design flood". The selection of the design flood is, in principle, an assessment of the risk involved against the cost of the failure of a structure which has been designed to prevent any loss with floods equal to or less than that of the selected frequency The "cost" includes the losses caused by the failure, the cost of repairs and the loss of revenue or service, etc. The most controversial factor in these cost estimates is the cost of human life. The basic statement is usually that human risk is intolerable. This implies that human life has an infinite value, an assumption contrary to all other expenditures for safety, c.f., roads, cars, etc. Projects which do not involve danger to life have generally been designed on the basis of economic cost-benefit considerations or justified on purely social or political grounds Larger projects, where the consequences of failure are of major concern, should be designed for floods estimated from probability considerations. This also applies to projects where "failure cannot be tolerated". The latter phrase, although widely used, is no more meaningful than saying that hurricanes or earthquakes cannot be tolerated (Benson, 1964). The consequences of failure depend also on the downstream conditions. The assessment of damage downstream is made by solving the dam-break problem and determining the level and time of travel of the flood wave as it moves down the valley. In order to carry out any risk or economic evaluation the probability of the selected flood must be assessed. The development of the methods for determining the design flood will depend on finding observed or theoretical relationships between the return period and the magnitude of rare floods and on the determination of the optimum return period to use. A return period of 10 000 years may appear to be more than safe until one recalls that there are more than 10 000 large dams of more than 15 m in height in the world. Consequently, one such flood on one of these dams could be expected every year on the average.

The procedures for the estimation of floods may be grouped as follows:
(1) Frequency analysis of flood flow records or frequency analysis of precipitation records and relating of precipitation to runoff from the catchment using the hydrograph or some other method.
(2) Transposition of storms.
(3) Transposition of depth-area-duration relationships.
(4) Probable maximum precipitation (PMP) method.
(5) Regional methods.
(6) Empirical methods.

9.2.1. Frequency analysis

Frequency analysis of flood flow records is, in principle, a good method. Once a distribution has been found which describes the record, then it can be used to predict the probable flood with a given return period. Unfortunately, records, if available at all, are almost invariably too short to obtain an adequate definition of the extreme floods. An extrapolation of a 25-year record to a 500-year return period flood yields a very unsatisfactory answer of low confidence value. If, for example, the Gumbel distribution is assumed then the theoretical 68% confidence limits of the 50, 100 and 500-year return period floods, in terms of years, are according to Bell (1969) as follows:
25 years of record: $12 \leq 50 \leq 200$; $15 \leq 100 \leq 400$; $16 \leq 500 \leq 2200$
100 years of record: $25 \leq 50 \leq 100$; $40 \leq 100 \leq 250$; $60 \leq 500 \leq 1500$
Further problems may arise with longer discharge records from the fact that catchment conditions have changed over the period of record.

Frequency analysis of precipitation data is usually more reliable because precipitation records have been kept for longer periods than flood flow records. However, much of this gain is lost after the rainfall is related to the catchment characteristics and expressed in terms of the runoff. The methods of estimation of the peak runoff on small catchments are further limited by the lack of suitable data. For example, for areas from 50 to 250 km^2 rainfall data must be available for unit times shorter than 2-4 hours for use with hydrograph methods. Such data seldom exist. The usefulness of frequency analysis of flood flows from small catchments is also restricted because observed flood flow records from small catchments are rare.

Most historic records of floods, where these exist, are usually too short and the results from frequency analysis may have to be extrapolated well beyond the observed range. The reader is referred to literature for techniques of extrapolation and methods for assessment of confidence limits of the extrapolated data. These techniques range from those derived from statistical mathematics to correlation of given data with data from longer sequences from climatically similar regions and checking against estimates from rainfall-runoff models.

9.2.2. Transposition of storms

The simplest of the transposition techniques is the transposition of isohyetal patterns without modification. It is not unusual to find that when a storm, which occurred somewhere in a hydrologically homogeneous area, is centred over a particular catchment in this area, that the resultant flood is greater than the maximum recorded one. The corollary to this is that the frequency analysis reflects only those storms which have occurred over the catchment and not those which may occur. The former are a subset in all transposable storms of a homogeneous region, i.e., for a fixed location. The probability of transposition is discussed by Alexander (1963). Briefly, if the area of the homogeneous region is A_h, then the probability of the centre of a storm falling within a specific catchment of area α is $P_t = \alpha/A_h$. Associated with this probability is a depth of rainfall. The probability of occurrence of a storm of rank r in a record of N storms is $P_r = r/N$, but since the depth of rainfall from a storm varies with the area covered by the storm the ranking has to be related to area. In addition, the observed maximum rainfall increases with the density of the rain gauge network. The probability of occurrence P of a transposable storm over the given catchment in any given year is the product P_tP_r. From this product it is easy to see why transposition tends to give floods much greater than those observed. If $P_r = 1/50$ and $P_t = 1/20$, then $P = 1/1000$ or a probability level of twenty times the length of the record.

The ranking of storms of a given duration in order of decreasing mean depths can be done in terms of the size of the area covered, for example, 100, 200, 500, etc., km^2. Alexander plotted these mean depths against the logarithm of the rank. He obtained a family of $\bar{P}$ versus r(A) curves. These curves were found to be linear and when they were extrapolated towards the lower values of $\bar{P}$ they were approximately concurrent. Designating this point by $\bar{P}_1$ and r_1, the depth for a given area can be expressed as

$$\bar{P} - \bar{P}_1 = b \log(r_1/r) \qquad 9.3$$

where b is a function of the area. Since $\bar{P}$ decreases as log A decreases these curves can be replotted as $\bar{P}$ versus log A to yield a family of $\bar{P}$ versus A(r) curves, which are concurrent at $\bar{P}_1$ and A_1, for all values of r. Hence,

$$\bar{P} - \bar{P}_1 = c \log(A_1/A) \qquad 9.4$$

for all ranks, where c is a function of r.

The duration effect has to be included with the aid of eqn 3.1, which describes a plot of log i versus log(t + c). As noted before, there is generally a break in the slope of this plot at about t = 2 hours because the mechanism producing short duration convective storms is different from that producing long period rainfalls. Hence, the appropriate slope has to be used. The three relationships may then be combined together to give

$$\bar{P} - \bar{P}_1 = c \log(A_1/A) \log(r_1/r) \log(t_1/t) \qquad 9.5$$

Hershfield (1962) found that the depth of a heavy rainfall increases as log T increases, where T is the return period of the rainfall of given depth and duration. This feature is characteristic of a number of distributions, such as extreme value, gamma and log-normal distributions, in which the slope is close to $\sqrt{2}$ (or variance equals ℓn 2). It is also a general observation that the slope of flood discharges on probability paper is always greater than that for the flood producing rainfalls.

Once a transposable storm has been identified the isohyetal pattern, usually that for the total storm precipitation, is transposed to the catchment under study. By trial positioning the isohyetal pattern is placed over the area of concern, so that it produces the maximum possible runoff from this storm. This position and orientation of the storm must be in keeping with meteorological requirements and experience in the area. The area enclosed by the isohyetal lines is usually greater than the study area. Details of transposition, orographic limitations and methods of adjustment were discussed in Chapter 3.

9.2.3. Transposition of depth-area-duration relationships

A convenient method of transposition of storm data from one area to another is by the use of precipitation data in the form of depth-area-duration relationships for individual storms, that is, in countries where these relationships have been developed. The principles of the method were discussed in Chapter 3 and additional details are available in the reports No 237, TP 129 (1969) and No 332 (1973) of the World Meteorological Organization.

9.2.4. Probable maximum precipitation method

The concepts of maximization of storms were introduced in Chapter 3. Under this heading are combined the techniques of moisture maximization, wind maximization and spatial maximization. The rainfall depths obtained from maximization of all contributing effects are usually several times the maximum observed value. These large values of the estimate are not necessarily unrealistic. These are estimates of extremes which have a very small and, unfortunately, unknown probability of occurrence. The term probable maximum precipitation should not be assumed to mean that the method yields the maximum value and removes the need to assess risk, that is, it does not provide a solution which removes the responsibility for the making a decision about the level of risk.

An illustration of the difference between the observed depths and those obtained by the maximization procedure was given by Ackermann (1964). He obtained for Urbana, Illinois, a 24-hour maximized rainfall depth of 775 mm where the maximum observed 24-hour rainfall depth was only 117 mm. The observed maximum 24-hour rainfall depth in the State of Illinois was 291 mm. As stated before the difference does not imply that the estimate of 775 mm is either a maximum or impossible. It is simply a value which has an unknown low probability of occurrence.

9.2.5. Regional methods

It is a common problem that flood-frequency information is required at a site for which there are no records. The design then has to be based on data available in the immediate hydrologically homogeneous region. The earliest regional methods were the empirical relationships between the runoff and area but, in a sense, most methods of forecasting are regional. Usually, however, reference to regional methods implies techniques such as *regional frequency distribution functions*, the *index flood method*, the *multiple regression methods* and the *square-grid method*.

Records are invariably too short and have large sampling errors. The fitting of distribution functions to such data leads to very unreliable results because several distribution functions may be fitted equally badly. However, if in the short period of observation data from ten gauging stations were available then the median of these values would give a better estimate of the event than a single observation.

By combining data from the region a better estimate can be made of the type of distribution likely to give the best fit.

The *regional distribution* curve is a frequency distribution of $Q/\bar{Q}$ for all stations in the region, that is, the relationship between $Q/\bar{Q}$ and the return period T is assumed to be valid for the entire region. The preparation of the regional distribution starts with the assembly of all available data. Data from streams with regulated flows are excluded and usually also those from very short records. Each set of data is then plotted, for example, on the EV1 (Gumbel) probability paper and a function is fitted to the points by eye. This yields the $\bar{Q}$ value at T = 2.33 years. The record of the annual maximum flows is then plotted as $Q/\bar{Q}$ versus the plotting position according to the Gumbel reduced variate y_1. In superimposing of records it is helpful to divide the y_1-axis into intervals, for example, -2.0 to -1.5, to -1.0, ... 4.5 to 5.0. All points within a given increment from all the records, which usually differ in length, are plotted as their mean value of $Q/\bar{Q}$ versus their mean value of y_1. (This does not mean that the data follows EV1 distribution. For EV1 distribution the plot is a straight line). A distribution could now be fitted to the plotted points and used as the regional curve for all locations, including ungauged sites. Either with the aid of the fitted function or extension by eye, a moderate amount of extrapolation is tenable.

In the Flood Studies Report (1975) the extension to higher values of y_1 was made by first grouping the data from all stations in the region into four or five groups, so that data from neighbouring stations did not appear in the same group. It was then assumed that the data within a group was statistically independent. (This assumption requires a substantial spatial separation of stations). The group containing N individual annual maxima from a number of stations is now regarded to be a sample of size N. The four highest values of $Q/\bar{Q}$ in each of the groups were then plotted at their appropriate plotting positions, i.e. the highest at N/(N + 1). A smooth curve was drawn through all the points, i.e. the earlier plot and those from the groups. It was found that all the points could be fitted by

$$Q/\bar{Q} = a + be^{cy}$$

or

$$Q/\bar{Q} = u + \beta\left(\frac{1 - e^{-ky}}{k}\right) \qquad 9.6$$

where u is the modal value of $Q/\bar{Q}$ when $y_1 = 0$, β is the gradient of the curve at $y_1 = 0$ (evaluated from the hand-drawn curve), and k is a curvature parameter obtained by using trial values of k to match the hand-drawn curve.

The grouping of data in effect assumes that concurrent records can be added on time scale, i.e. two 20-year records make one 40-year record. This is known as the station-year method of analysis. The length of the nominal record is greatly reduced if there is a cross-correlation within the stations, see eqns 9.14, 9.17 and 9.18.

Dalrymple (1960) proposed a procedure for testing of the data used in regional analysis for homogeneity. The test is based on the EV1 distribution which has the standard deviation

$$\sigma_{y_1} = (e^{y_1}/\sqrt{n})[1/(T - 1)]^{\frac{1}{2}} \qquad 9.7$$

In a large number of different but homogeneous records two thirds of the estimates should fall within the σ_{y_1} limits and 95% within $\pm\, 2\sigma_{y_1}$ of the most probable value of T. A return period T = 10 years was used with two standard deviation limits in the studies by Dalrymple. Since for T = 10 years $y_1 = 2.25$, the confidence limits are $2.25 \pm 6.33/\sqrt{n}$.

Carrigan (1971), following the work by Conover and Benson (1963), proposed a method by which advantage is taken of the spatial sampling of the existing records. This yields information for periods greater than the actual record. Assume that a set of n independent identically distributed concurrent records exist, each containing k observations of extremes, x_{ij}; i = 1, 2, ... n; j = 1, 2, ... k. These records can be arranged into a data matrix with n columns (stations) of data each having k observations. If the columns are arranged in descending order then the first row will contain the maximum of each record. The first element x_{11} is the largest value of all the nk observations and x_{21} (col. 2, row 1) is greater than x_{22} but not necessarily greater than x_{12}. Conover and Benson showed that, if the first row is considered to be a new series, the probability that another event x exceeds the i-th event in the series of maxima x_{i1} is

$$P(x > x_{i1}) = \sum_{m=0}^{i-1} n![(n-m)!k \prod_{j=0}^{m} (n + \frac{1}{k} - j)]^{-1} \qquad 9.8$$

For example, if n = 3 and k = 4 the probability that another event exceeds the second highest maxima a_{21} is

$$P(a > a_{21}) = \sum_{m=0}^{2-1} 3![(3-m)!k \prod_{j=0}^{m} (3 + \frac{1}{4} - j)]^{-1}$$

$$= \frac{3!}{3!4(3+\frac{1}{4})} + \frac{3!}{2!4(3+\frac{1}{4}-1)} = 0.180$$

By this method records from a homogeneous region can be combined to yield a record of the size of nk. However, this procedure is severely limited by the fact that flood records from a hydrologically homogeneous region are strongly cross-correlated, i.e., they are not independent. This reduces the maximum return period available to a function of the correlation coefficient. The probability for dependent data cannot be determined analytically but Carrigan (1971) developed a simulation technique by which this probability can be estimated.

The concept of the *index-flood method* is described by Dalrymple (1960). It consists of two parts, the dimensionless flood-frequency curves $Q/\bar{Q}$ versus T (the regional distribution curve) and the relationships between the characteristics of the catchment and the mean flood $\bar{Q}$. Dalrymple used only the catchment area and plotted $\bar{Q}$ versus A for stations within the region. Since then more independent variables have been added. These are basically of two kinds: physiographic characteristics, such as area, elevation, slope, main channel slope and length, stream frequency, percent of area covered by swamps, lakes, forest, urban areas, etc., and hydrometeorologic variables such as mean annual precipitation, temperature and humidity. Instead of plotting the drainage area against the mean annual flood *multiple regression* techniques can be used to establish the relationships between hydrologic and physiographic characteristics of the catchment. Multiple regression makes it possible to include more individual characteristics. The relationship for the T-year return period event has the general form

$$Q_T = f(A^a, B^b, C^c, \ldots)$$

where A, B, C,... are independent variables and a, b, c,... are constants derived by regression analysis. The exponents found for Q_T and for $\bar{Q}$ will, in general, be different and the ratio $Q_T/\bar{Q}$ will not be a dimensionless constant. Apart from inclusion of catchment characteristics the multiple regression does not require the assumption of a distribution for the flood peaks. Comparative studies have shown that the most important characteristics are the area and the mean annual precipitation.

An aspect of the regression analysis, when applied to a region, is that the river flows in a region are not independent of each other. Major floods and low flows are caused by large scale weather systems which usually cover the entire region. It has been shown that, if the records within a hydrologically homogeneous region are correlated to one another, the effective number of stations or the equivalent number of independent gauging stations n_e is strongly dependent on the correlation coefficient.

The mean and variance of data, $\bar{x}_j$ and S_j^2, for the j-th station are readily determined. If it is assumed that each station had the same number of observations, then the regional mean and its variance are

$$\bar{x}_r = \frac{1}{n} \sum_{j=1}^{n} \bar{x}_i \qquad 9.9$$

and

$$S_{\bar{x}_r}^2 = \frac{1}{n} \sum_{j=1}^{n} S_j^2 + \frac{2}{n^2} \sum_{j=1}^{n-1} \sum_{i=j+1}^{n} r_{ij} S_i S_j \qquad 9.10$$

where S_i and S_j are the standard deviations (columns and rows of data matrix, respectively) and r_{ij} is the correlation coefficient

$$r = \frac{S_{xy}}{S_x S_y} = \frac{1}{nS_x S_y} \sum_{i=1}^{n} (x_i - \bar{x})(y_i - \bar{y})$$

$$= \frac{\sum_{i=1}^{m} \sum_{j=1}^{n} f_{ij}(\Delta x_i \ \Delta y_j)}{[\sum_{i=1}^{m} f_i(\Delta x_i)^2]^{\frac{1}{2}} [\sum_{j=1}^{n} f_j(\Delta y_j)^2]^{\frac{1}{2}}} \qquad 9.11$$

In the latter form m and n are class intervals of x and y, respectively, Δx_i and Δy_j are the distances of class marks from $\bar{x}$ and $\bar{y}$, respectively and f_i and f_j are their marginal frequencies.

Defining the regional mean cross-correlation coefficient $\bar{r}$ as

$$\bar{r} = \frac{2}{n(n - 1)} \sum_{j=1}^{n-1} \sum_{i=j+1}^{n} r_{ij} \qquad 9.12$$

and standardizing the time series to a common mean of zero and variance S^2, leads to variance of the regional mean

$$S_m^2 = \frac{1}{n} S^2 [1 + \bar{r}(n - 1)] \qquad 9.13$$

The equivalent number of independent stations is then

$$n_e = \frac{n}{[1 + \bar{r}(n - 1)]} \qquad 9.14$$

A sequence of nk will correspondingly yield as much information about the mean as a sequence of $n_e k$ of uncorrelated observations.

The most common situation is that data at a station are serially correlated and between stations cross-correlated. For two serially correlated time series, with equal number of observations N, the effective number of points, N_e, is given by Yevjevich (1972) as

$$N_e = N/(1 + 2r_1 r_1' + 2r_2 r_2' + 2r_{n-1} r_{n-1}') \qquad 9.15$$

where r_k and r'_k are the k-th order auto-correlation coefficients of the two series, respectively, i.e.,

$$r_k = \frac{1}{(N - k)\sigma^2} \sum_{i=1}^{n-k} (x_i - \bar{x})(x_{i+k} - \bar{x}) \qquad 9.16$$

Equation 9.16 applies after the periodicities have been removed from both series. For equal length time series, described by first order Markov models, the effective number of observations is given by

$$N_e = N\left(\frac{1 - r_1 r'_1}{1 + r_1 r'_1}\right) \qquad 9.17$$

Thus, for a set of n gauging stations, with N observations each, which are both serially and cross-correlated, the effective number of stations becomes

$$N_e n_e = Nn\{[1 + \bar{r}(n - 1)](\ + 2\bar{r}_1 + 2\bar{r}_2 + \ldots\ 2\bar{r}_n)\}^{-1} \qquad 9.18$$

where $\bar{r}$ are the average serial correlation coefficients of the n time series.

The various correlation techniques do not apply to extreme events only. These are also used to transfer information from a station with long record to an adjacent station with short record or to study the effect of physical parameters, for example, on river flow. A very simple method of data transfer is the *flow-duration curves* or cumulative frequency curves of the flow rate. The cumulative frequency curve shows the percentage of time during which a specified discharge is equalled or exceeded. It combines into one curve a number of the flow characteristics. The curve can be prepared for any period but the most important flow-duration curve is the one for the complete year. The day is generally the most convenient time unit (the daily average flow rates), but for very large rivers longer periods may be adequate because of slow rates of change of flow, and vice versa for small steep streams.

The amount of work involved in preparing the flow-duration curve can be reduced by dividing the data into class intervals instead of handling each individual observation. The intervals should be chosen to give 20 to 30 evenly distributed points on the curve. The records of each year are tabulated first as shown in Table 9.1.

TABLE 9.1 Summary of Daily Average Flow Rates at...for...

m^3/s	Jan	Feb	Mar	Apr	May	June	July	Aug	Sept	Oct	Nov	Dec	Total
10	///	/			/								5
12	~~////~~ /	///	//	/							/	/	13
14	~~////~~ //	~~////~~ //	///	////	///	//	/	//	/	/	/	//	34
17													
.													
	31	28	31	30	31	30	31	31	30	31	30	31	365

Then the N-years of record are assembled as shown in Table 9.2. This cumulative curve of the percentage of time when the flow equals or exceeds a given flow rate Q is plotted as log Q versus percentage of time on log-probability paper for hydrological studies, because the probability paper gives a good definition of the extremities of the plot, Fig. 9.5. For water-supply or water-power studies a plot on natural coordinates is preferred because the area under this curve represents the total volume of water available and is easier to measure under a curve in natural coordinates. The flow-duration curve can also be prepared in terms of discharge per unit area of the catchment.

TABLE 9.2 Calculation of the Flow-Duration Curve at ... for the Period... in Terms of Average Daily Flow Rates.

Q m³/s \ Year	1951	1952	1953	...	1975	Total days	No. days flow equal or greater	% of time
10	5	7	11		3	105	9131	100
12	7	11	10		12	195	9026	98.85
14	12	15	19		20	474	8831	96.71
20	.	.	.		.	.	.	.
25	.	.	.		.	.	.	.
.								
600								
700	3	-	1		1	19	46	0.50
800	1	3	1		2	24	27	0.30
1000	-	-	1		-	3	3	0.03
	365	366	365		365	9131		

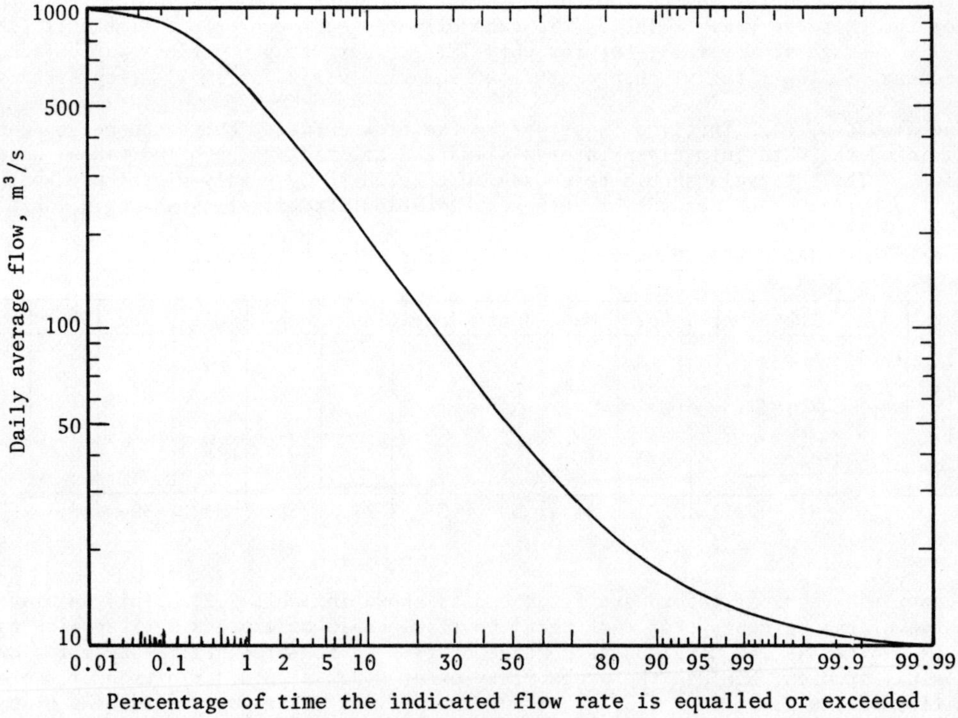

Fig. 9.5. Diagrammatic illustration of flow-duration curve.

Apart from the total volume of water available the flow-duration curve provides information on topographical and geological characteristics of the catchment. A flow-duration curve with a steep slope throughout indicates a stream with highly variable discharge. This is typical of conditions where the flow is mainly from surface runoff. A flat slope of the curve indicates equalising surface or groundwater storage or both. A flat low flow end of the curve indicates a large amount of groundwater storage and a flat high flow end is characteristic of streams with large flood plain storage, such as lakes and swamps, or where the high flows are mainly derived from snowmelt.

If a sediment rating curve is available for the given stream, then the flow-duration curve can be converted into a cumulative sediment transport curve by multiplying each flow rate by its rate of sediment transport. The area under this curve represents the total amount of sediment transported.

The flow-duration curve is in a sense a probability distribution curve. The median discharge is at 50% of time, etc. However, a flow-duration curve, for example, for a period of 25 years, does not represent the probabilities of the yearly flow. The flow lower than that equalled or exceeded 99% of the time might have occurred during one three month period of one year [(25 x 12)0.01 = 3 months]. Although this flow could be expected on the average only 1% of the time each year, it could occur on the average 25% of the time during one of the 25 years. It is, therefore, advisable to supplement the flow-duration curve with flow-duration curves for the year of the lowest and the year of the highest runoff.

The flow-duration curve technique is very useful for regional studies and for extrapolation of short records. The extrapolation of short flow records rests on the concept of regional hydrological homogeneity. If in a hydrologically homogeneous region there is a catchment A with a long record which also covers the period of the short record of catchment B in the same region, then it is assumed that the relationship which existed during the concurrent period is also valid for the long period of catchment A for which catchment B has no record. The extrapolation procedure is as follows. Flow-duration curves are prepared for both catchments for the concurrent period of record, Fig. 9.6a & b. Then a correlation curve is plotted from these two flow-duration curves by plotting the Q value from catchment A against the Q value from catchment B for a given value of percentage of time that both the flows are equalled or exceeded, Fig. 9.6c. Little weight should be given to the extreme points in drawing a function to the plotted points, because these points may be quite different in the next short sample. This correlation function will be at 45° slope if the yield per unit area is the same on both catchments, the equal yield line. If, for example, the high or the low flow characteristics are different, then the correlation function for the corresponding range of flows departs from the 45° slope. The correlation function established for the concurrent record is now used to extrapolate the record of catchment B. The correlation relationship is entered with values from the long period flow-duration curve and yields the corresponding values for catchment B, thus allowing plotting of the flow-duration curve.

The error in this type of correlation is not easy to define, but if, for example, the data approximately satisfies the log-normal distribution, then with n years of short record of flow rates q and N years of long record Q, $u = \ell n\, Q$ and $v = \ell n\, q$, the extrapolated value of q is described by

$$\bar{q} = \exp(\bar{v} + 0.5\, S_v^2) \qquad S_q = \bar{q}\sqrt{\exp(S_v^2) - 1}$$

$$\bar{v} = \bar{v}_n + r_{uv}\frac{S_{vn}}{S_{un}}(\bar{u}_N - \bar{u}_n) \qquad S_v^2 = S_{vn}^2 + r_{uv}^2\frac{S_{vn}^2}{S_{un}^2}(S_{uN}^2 - S_{un}^2) \qquad 9.19$$

where r is the correlation coefficient and S^2 is the variance of the sample. Hence,

$$S^2_{un} = \frac{1}{n-1}\Sigma(u_i - \bar{u})^2 = \frac{1}{n-1}[\Sigma u_i^2 - (\bar{u}_i)^2]$$

(S^2_{uN} is given by using N; S^2_{vn} by using v),

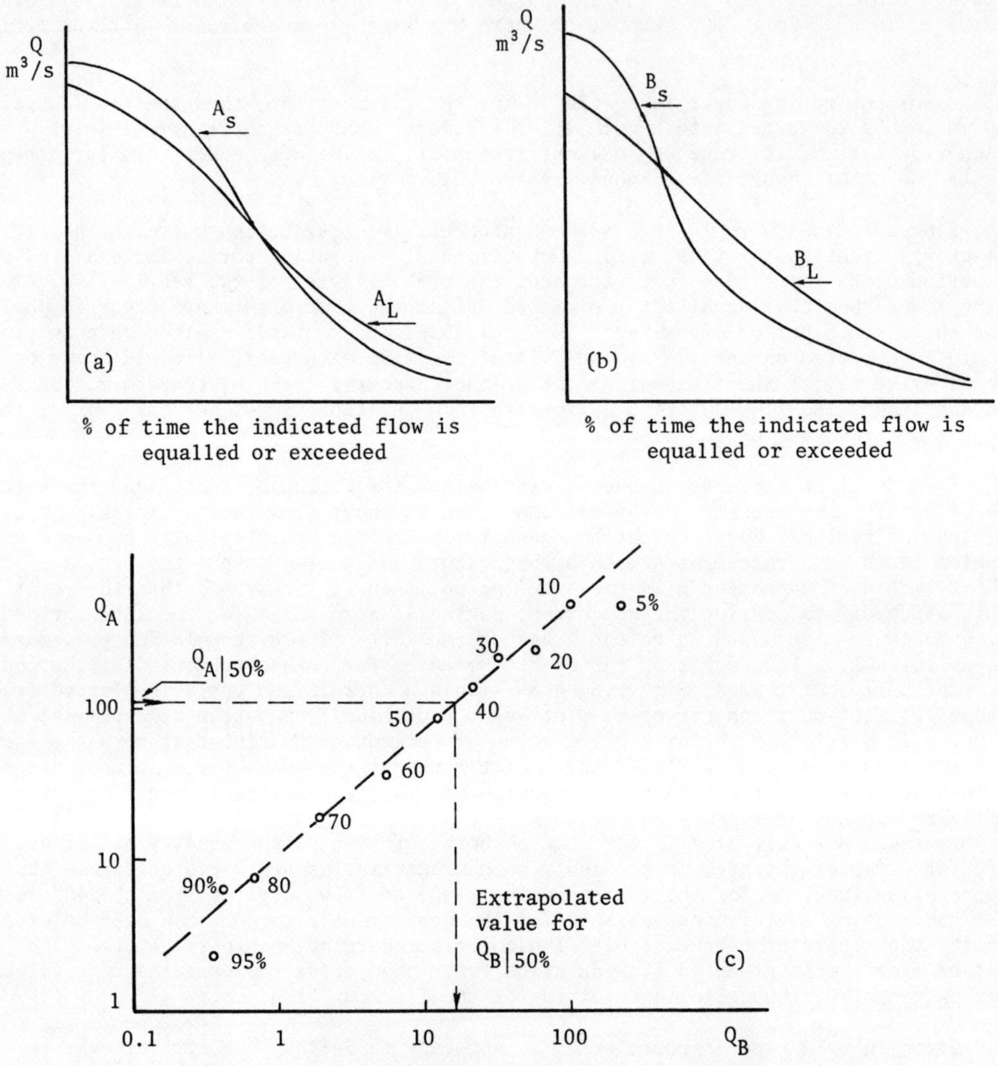

Fig. 9.6. Diagrammatic illustration of extrapolation of short term records with the aid of flow-duration curves. (a) Long period, A_L, and short period, A_S, flow-duration curves for catchment A. (b) B_S is the flow-duration curve for the with A_S concurrent short period and B_L is the extrapolated flow-duration curve for catchment B. (c) Correlation relationship from the concurrent flow records.

$$S_{uv}^2 = \frac{1}{n-1}\Sigma(u_i - \bar{x})(v_i - \bar{v}) = \frac{1}{n-1}(\Sigma u_i v_i - \frac{1}{n}\Sigma u_i v_i)$$

and

$$r_{uv} = \frac{S_{uv}}{S_u S_v}$$

It can be seen that the greater the correlation coefficient r_{uv} the smaller is the standard error of $\bar{v}$.

The *square-grid method*, proposed by Solomon et al. (1968), is an entirely computer orientated technique of regional analysis. For this method the study area is divided into a uniform grid. The squares are identified in cartesian coordinates and are associated with a set of parameters such as elevations, percentage of lakes, swamps, forests and urban areas, land use, soil type, etc. Squares which contain gauging stations and meteorological stations also have data on precipitation, temperature, etc. From these function surfaces covering the region, including ungauged areas, are developed. The square-grid method was extended for generation of synthetic streamflow records by Pentland and Cuthbert (1971).

9.2.6. Empirical methods

The numerous empirical relationships for flood estimation are also, in principle, regional methods in that these relationships can be applied with reasonable confidence only in the region from which the data has been collected.

Since the area of the catchment is the most deciding factor in determining the peak flow rate of the flood wave, it is only natural that the first empirical formulae to appear were of the form $Q \propto A^n$. When n = 1 the formula becomes

$$Q = CiA \tag{9.20}$$

where i is the rainfall intensity, C is a coefficient, and A is the area of the catchment. This is known as the rational or Lloyd-Davis formula and is still extensively used for urban stormwater calculations. The formula assumes that the runoff hydrograph is a rectangular block and linearly proportional to the rainfall intensity. The coefficient C is the same for all rainfalls. The rainfall intensity for a given catchment is based on the time of concentration and a specified probability level (return period).

When the peak discharge per unit area is plotted against area on log-log paper the resulting mean line through the plotted points can frequently be approximated by two straight lines. This leads to the two types of formulae in common use, i.e., $Q \propto A^{3/4}$ and $Q \propto \sqrt{A}$, where the former applies to $A < 1500$ km^2 and the latter to $A > 1500$ km^2, approximately. The literature contains a large number of equations for peak discharge in use around the world, where the area appears to some power between a half and one. Probably the most extensively used relationship is the so-called Creager curves (1950). These are envelope curves to observed extreme floods (discharge/unit area) plotted on log-log paper against the area of the catchment, and are described by

$$q = 46\ CA^{(0.894\ A^{-0.048} - 1)} \tag{9.21}$$

where q is in ft^3/s per square mile, A is in square miles, and C is a constant. The value of C = 200 is the envelope for all observed data points in the world and C = 100 is the envelope for most of the data points from the United States (and indeed for most of the points in the world). The curves in SI units are in Fig. 9.7.

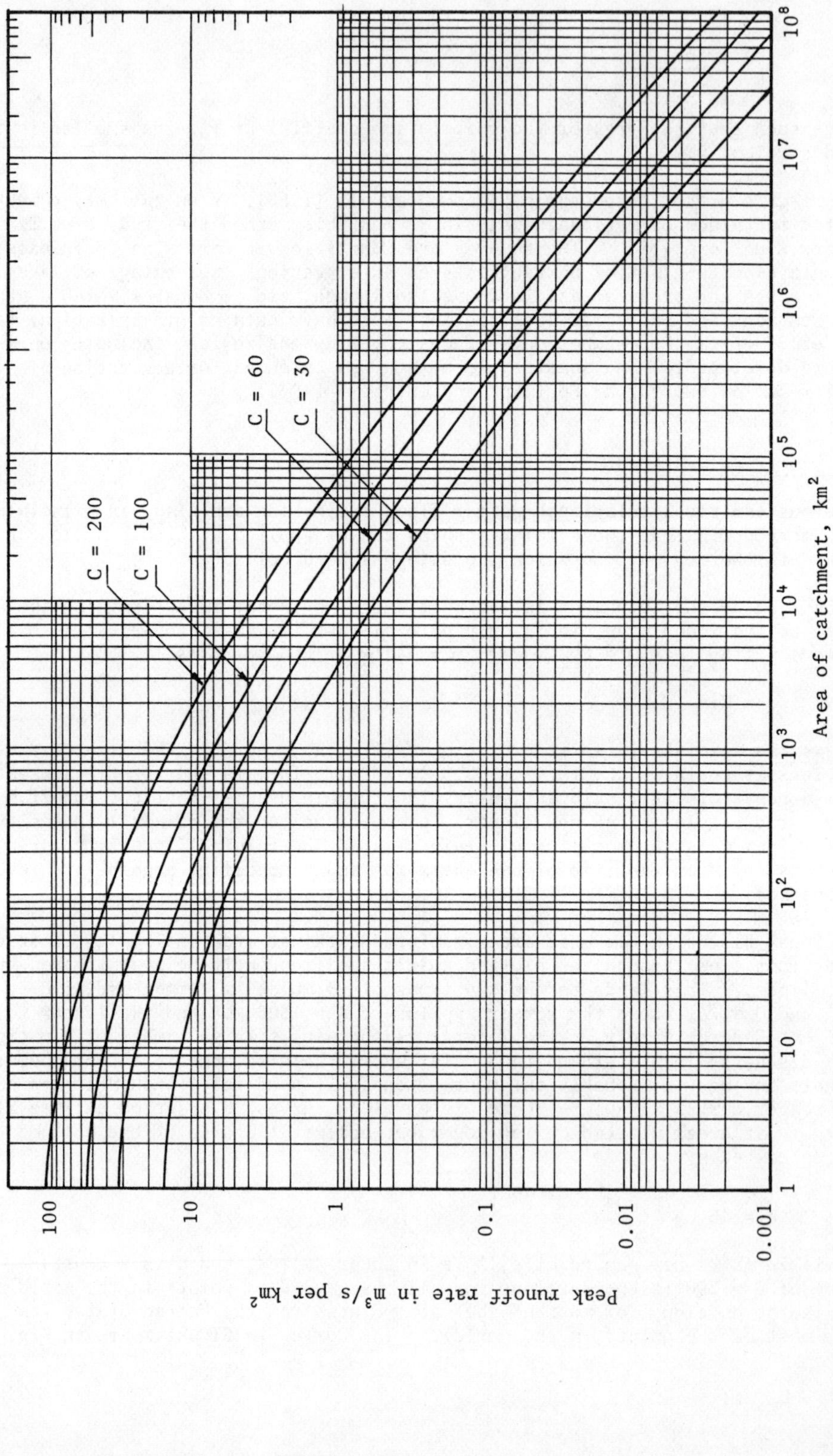

Fig. 9.7. Envelope curves to peak runoff rates after Creager and Justin (1950).

A somewhat similar technique of flood estimation, using extreme rainfall observations is based on the formula proposed by Richards (1955). The original derivation attempts to account for the rainfall intensity, duration and distribution, catchment area, shape, slope, permeability and antecedent moisture. Briefly, the depth, d, of runoff at time t from a catchment of unit width and length L, subject to uniform rainfall of intensity, i, is d = Kit, where K is a runoff coefficient. The velocity is expressed by Chézy's formula as $V = C\sqrt{KitS}$, where C, K, i and the slope, S, are constants for a particular storm. These two equations yield $t^3 = C'L^2/(KiS)$, where $C' = 9/(4C^2)$. Next, the rainfall intensity is expressed as $i = Rf(a)/(t + 1)$, with t in hours, and i and the coefficient of rainfall R in inches. The term f(a) is an areal rainfall reduction factor. After introduction of a correction factor N to account for the variations, such as the intensity of rainfall over the catchment and the path of storms over it, the equations were combined into

$$\frac{t^3}{t + 1} = \frac{NC'L^2}{KSRf(a)} \qquad 9.21$$

The value of N was found to range from 0.72 to 1.72, with 1.1 for a storm with central peak (Table 26, Richards, 1955). The coefficient of rainfall was expressed as

$$R = P\,\frac{T + 1}{T} \qquad 9.22$$

where P is the total precipitation and T is duration in hours. The value of P for a given duration can be taken from a regional envelope curve of maximum observed rainfalls or from the world maxima by Jennings, Fig. 3.4. Using the latter

$$P = 388.6\, t_{hrs}^{0.486} = Xt^{x} \qquad 9.23$$

in eqn 9.22 gives

$$R = \left(\frac{X}{25.4}\right)(t + 1)t^{x-1} \qquad 9.24$$

The KR versus C' relationship proposed by Richards (1955, Fig. 11) can be expressed by

$$C' = \frac{1}{54(KR - 0.4)^{0.42}} \qquad 9.24$$

Substituting from eqns 9.24 and 9.25 in eqn 9.21 yields

$$t^{2+x} = 0.0185\,\frac{Z}{Y}\,[Y(t + 1)t^{x-1} - 0.4]^{-0.42} \qquad 9.26$$

where $Z = NL^2/[2.59\ Sf(a)] = 0.386\ NL^2/Sf(a)$ and $Y = XK/25.4 = 0.0394\ XK$, where L is converted from miles to km and rainfall from inches to mm. The Richards expression for peak discharge, Q = KiA, can now be used to calculate the maximum flood for known duration $T \equiv t$ from

$$Q = 0.28\ AXKf(a)t^{x-1} \qquad 9.27$$

Equations 9.27 contains the runoff coefficient K and the area factor f(a). Richards gave a table of K values as follows:

Catchment	K
Rocky and impermeable	0.80-1.00
Slightly permeable, bare	0.60-0.80
Slightly permeable, partly cultivated or covered with vegetation	0.40-0.60
Cultivated absorbent soil	0.30-0.40
Sandy absorbent soil	0.20-0.30
Heavy forest	0.10-0.20

where the first K value refers to large catchments and the upper limit to small steep catchments. For a given catchment an average value of K has to be estimated.

In principle, one could fit a distribution function, f(a), to the observed areal distribution of rains in the region. A function $f(a) = 1 - aA^bT^c$ would satisfy the boundary conditions, but because of the variability of the areal distribution, as discussed in Chapter 3, this is not easy. If no data are available on areal distribution, Fig. 3.12 could be used.

9.3 Small Catchment Runoff

The primary hydrological characteristic of a small catchment is that the effects of overland flow, rather than channel flow and storage, dominate the peak runoff rate. The runoff from a small catchment is also very sensitive to the intensity of the rainfall and land usage. The size of a small catchment may vary from just a few hectares up to about 250 km^2. Small catchments may be grouped into two types: agricultural and urban catchments. As noted earlier, the prediction of the flow rates from small catchments is hampered by the relative lack of hydrological data, particularly on small agricultural catchments where the individual structures are usually small and are often designed by personnel inexperienced in hydrology.

9.3.1 Agricultural catchments

Researchers all over the world have been active in developing design methods which may be applied by non-specialist staff with reasonable consistency and reliability. Such methods are empirical in nature and, although the general approach can be used anywhere, the constants apply only for the region from which the original data were derived. The most widely known design methods are those developed in the United States. These are detailed by Chow (1964) and only a few examples are briefly discussed here.

Cook's method is named after its originator and was developed for the U.S. Agricultural Dept. Soil Conservation Service in the upper Mississippi region. The runoff is empirically related to the surface area, relief, soil infiltration properties, vegetal cover and surface storage. All of these, except the surface area, are combined into a parameter ΣW. The 50-year return period peak flow rate versus area relationship, with ΣW as a parameter, is given graphically in Fig. 9.8. The peak runoff Q_p is then modified by a geographical rainfall factor R and a frequency factor F, which adjusts the return period to a value different from 50 years if so desired. The peak runoff is then

$$Q = RFQ_{p50} \qquad 9.28$$

In some versions of the method F is also made to depend on infiltration and vegetation, but too many refinements make the method more difficult to apply and are of doubtful value considering the empirical nature of the method. Values of W for relief, infiltration, vegetal cover and surface storage, as well as F values, as given by Chow (1964), are shown in Tables 9.1 and 9.2.

The U.S. Bureau of Public Roads method (1961) is similar to Cook's method. It makes use of a topographic and a precipitation index for the given region. The parameters are related to the peak flow rate with the aid of a design nomogram.

The Chow method (1962) gives the peak discharge as Q = XYZA. Although empirical, the development of the method is strongly guided by hydrological principles. The factors affecting the runoff are divided into two groups. Group one includes factors which have a direct influence, such as the depth and duration of the rainfall, land usage, surface conditions and soil type. Group two contains the factors which affect the distribution of runoff and includes the size and shape of the drainage basin, the slope of the land, the lag time as a measure of the detention effect, and the distribution of the direct runoff in terms of the unit hydrograph.

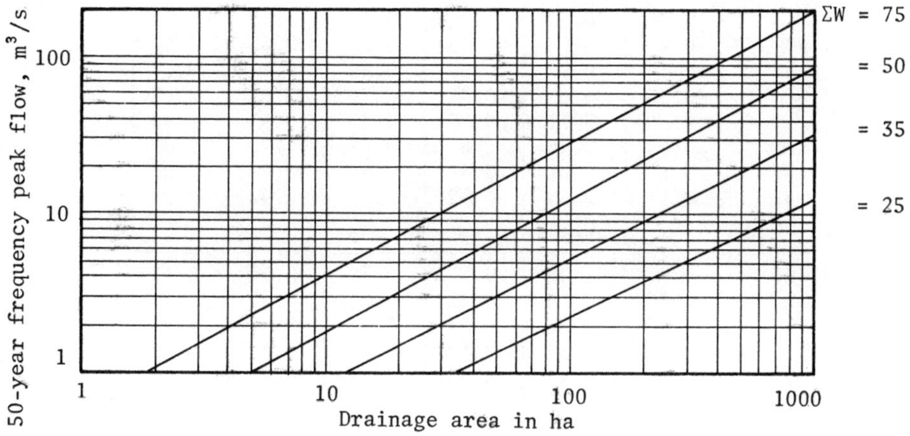

Fig. 9.8. W values for Cook's method.

TABLE 9.1 Incremental W Values in Cook's Method

Catchment characteristic	Extent or degree	W
Relief	Steep rugged terrain with average slopes generally > 30%	40
	Hilly, with average slopes 10 to 30%	30
	Rolling, with average slopes 5 to 10%	20
	Relatively flat land, slopes 0 to 5%	10
Infiltration (I)	No effective cover; either rock or thin soil mantle of negligible infiltration capacity	20
	Slow to take up water; clay or other soil of low infiltration capacity	15
	Deep loams with infiltration about that of typical prairie soils	10
	Deep sand or other soil that takes up water readily and rapidly	5
Vegetal cover (C)	No effective plant cover or equivalent	20
	Poor to fair cover; clean cultivated crops or poor natural cover; less than 10% of the watershed in good cover	15
	About 50% of watershed in good cover	10
	About 90% of watershed in good cover, such as grass, woodlands, or equivalent	5
Surface storage	Negligible; few surface depressions	20
	Well-defined system of small drainge	15
	Considerable depression storage with not more than 2% in lakes, swamps, or ponds	10
	Surface-depression storage high; drainage system poorly defined; large number of lakes, swamps, or ponds	5

TABLE 9.2 Frequency Factors for Use with Cook's Method

(I + C)	Average annual precipitation, mm 250	500	750	1000	1500	2000
	Ratio: 25-year/50-year					
5	0.31	0.38	0.41	0.44	0.48	0.51
10	0.41	0.50	0.55	0.58	0.63	0.66
15	0.50	0.59	0.64	0.69	0.73	0.77
20	0.55	0.65	0.71	0.76	0.82	0.87
25	0.60	0.71	0.78	0.83	0.90	0.92
30	0.64	0.76	0.83	0.89	0.92	0.92
35	0.67	0.81	0.89	0.92	0.92	0.92
40	0.71	0.85	0.92	0.92	0.92	0.92
	Ratio: 10-year/50-year					
5	0.05	0.08	0.10	0.12	0.15	0.17
10	0.10	0.16	0.21	0.24	0.30	0.34
15	0.16	0.25	0.31	0.37	0.45	0.51
20	0.21	0.33	0.42	0.49	0.60	0.68
25	0.26	0.41	0.52	0.61	0.75	0.80
30	0.31	0.49	0.62	0.74	0.80	0.80
35	0.36	0.58	0.73	0.80	0.80	0.80
40	0.42	0.66	0.80	0.80	0.80	0.80

The method used by the U.S. Soil Conservation Service (National Engineering Handbook, 1972) is similar to the Chow method and is based on many years of streamflow and rainfall records of agricultural catchments in the various parts of the United States. Their equation for the direct surface runoff, Q (in inches), is

$$Q = \frac{(P - 0.2\ S)^2}{P + 0.8\ S} \qquad 9.29$$

where P is the depth of rainfall (in inches) and S is the potential maximum retention (in inches), i.e. the maximum difference between rainfall and runoff as the duration of the rainfall becomes large. The initial abstraction by interception, infiltration and surface storage, before runoff starts, is approximated at 0.2 S, and the remaining 0.8 S is infiltration after the runoff begins. The potential maximum retention S is estimated with the aid of the empirical "curve numbers" CN, a number between 0 and 100, as

$$S = \frac{1000}{CN} - 10 \qquad 9.30$$

The curve number is determined from antecedent moisture conditions, soil type, land use and soil conservation practices. The soil type classification is shown in Table 9.3 and the curve numbers as a function of land use and soil type are shwon in Table 9.4. The curve numbers in Table 9.4 refer to a catchment with an initial antecedent moisture condition in State II as defined in Table 9.5. The hydrologic conditions poor, fair and good refer to surface runoff potential. The direct runoff from poor condition is greater than from good. For example, a ploughed land with furrows down the slope has a poor hydrological condition whereas the same land with deep contour ploughing would be in good condition. Below are a few extracts of the land use classification:

Crop rotations: The sequence of crops on a catchment must be evaluated on the basis of its hydrologic effects. Rotations range from poor to good largely in proportion to the amount of dense vegetation in the rotation. Poor rotations are those in which a row crop or small grain is planted year after year. Good rotations will include legumes or grasses to improve tilth and increase infiltration.

Native pasture or range: Poor pasture or range is heavily grazed, has no mulch or has plant cover on less than about 50% of the area. Fair pasture or range has between 50% and 75% of the area with plant cover and is not heavily grazed. Good pasture or range has more than about 75% of the area with plant cover, and is lightly grazed.
Farm woodlots: Poor woodlots are heavily grazed and regularly burned in a manner that destroys litter, small trees and brush. Fair woodlots are grazed but not burned, but usually these woods are not protected. Good woodlots are protected from grazing so that litter and shrubs cover the soil.

TABLE 9.3 Hydrologic Soil Group

Soil group	Description
A	Lowest runoff potential: Includes deep sands with very little silt and clay, also deep, rapidly permeable loess.
B	Moderately low runoff potential: Mostly sandy soils less deep than A, and loess less deep or less aggregated than A, but the group as a whole has above-average infiltration after thorough wetting
C	Moderately high runoff potential: Comprises shallow soils and soils containing considerable clay and colloids, though less than those of group D. The group has below-average infiltration after presaturation
D	Highest runoff potential: Includes mostly clays of high swelling percent, but the group also includes some shallow soils with nearly impermeable subhorizons near the surface.

TABLE 9.4 Runoff Curve Numbers

Land use or cover	Treatment or practice	Hydrologic condition	Hydrologic soil group			
			A	B	C	D
Fallow	Straight row	Poor	77	86	91	94
Row crops	Straight row	Poor	72	81	88	91
	Straight row	Good	67	78	85	89
	Contoured	Poor	70	79	84	88
	Contoured	Good	65	75	82	86
	Contoured and terraced	Poor	66	74	80	82
	Contoured and terraced	Good	62	71	78	81
Small grain	Straight row	Poor	65	76	84	88
	Straight row	Good	63	75	83	87
	Contoured	Poor	63	74	82	85
	Contoured	Good	61	73	81	84
	Contoured and terraced	Poor	61	72	79	82
	Contoured and terraced	Good	59	70	78	81
Close-seeded legumes or rotation meadow	Straight row	Poor	66	77	85	89
	Straight row	Good	58	72	81	85
	Contoured	Poor	64	75	83	85
	Contoured	Good	55	69	78	83
	Contoured and terraced	Poor	63	73	80	83
	Contoured and terraced	Good	51	67	76	80
Pasture or range		Poor	68	79	86	89
		Fair	49	69	79	84
		Good	39	61	74	80
	Contoured	Poor	47	67	81	88
	Contoured	Fair	25	59	75	83
	Contoured	Good	6	35	70	79

Meadow (permanent)		Good	30	58	71	78
Woodlands (farm woodlots)		Poor	45	66	77	83
		Fair	36	60	73	79
		Good	25	55	70	77
Farmsteads			59	74	82	86
Roads, dirt			72	82	87	89
Roads, hard-surface			74	84	90	92

TABLE 9.5 Classification of Antecedent Moisture Conditions

Condition	Description	5-day antecedent rainfall, mm Dormant season	Growing season
I	Catchment soils are dry, but not dry to the wilting point	up to 13	less than 35
II	The average case for annual floods	13-28	35 to 53
III	When heavy rainfall or light rainfall and low temperatures have occurred during the 5 days previous to the given storm	over 28	over 53

Conversion of the curve numbers to moisture categories I or III is given in Table 9.6. The runoff versus rainfall with the curve number as a parameter are shown in Fig. 9.9

TABLE 9.6 Runoff Curve Number CN, and Values of S and P for CN in Column 1

CN for condition II	CN I	CN II	S values mm	Curve starts where P = mm
1	2	3	4	5
100	100	100	0.00	0.00
98	94	99	5.18	1.02
96	89	99	10.59	2.03
94	85	98	16.21	3.30
92	81	97	22.10	4.32
90	78	96	28.19	5.59
88	75	95	34.54	6.86
86	72	94	41.40	8.38
84	68	93	48.26	9.65
82	66	92	55.88	11.18
80	63	91	63.50	12.70
78	60	90	71.63	14.22
76	58	89	20.26	16.00
74	55	88	89.15	17.78
72	53	86	98.81	19.81
70	51	85	108.71	21.84
68	48	84	119.38	23.88
66	46	82	130.81	26.16
64	44	81	142.75	28.45
62	42	79	155.70	31.24
60	40	78	169.42	33.78

CN for condition II	CN I	CN II	S values mm	Curve starts where P = mm
1	2	3	4	5
58	38	76	183.90	36.83
56	36	75	199.46	39.88
54	34	73	216.41	43.18
52	32	71	234.44	46.99
50	31	70	254.00	40.80
48	29	68	274.32	45.86
46	27	66	297.18	59.44
44	25	64	322.58	64.52
42	24	62	350.52	70.10
40	22	60	381.00	76.20
38	21	58	414.02	82.80
36	19	56	452.12	90.42
34	18	54	492.76	98.55
32	16	52	532.48	100.70
30	15	50	591.82	118.36
25	12	43	762.00	152.40
20	9	37	1016.00	203.20
15	6	30	1440.18	288.04
10	4	22	2286.00	457.20
5	2	13	4826.00	965.20
0	0	0	∞	∞

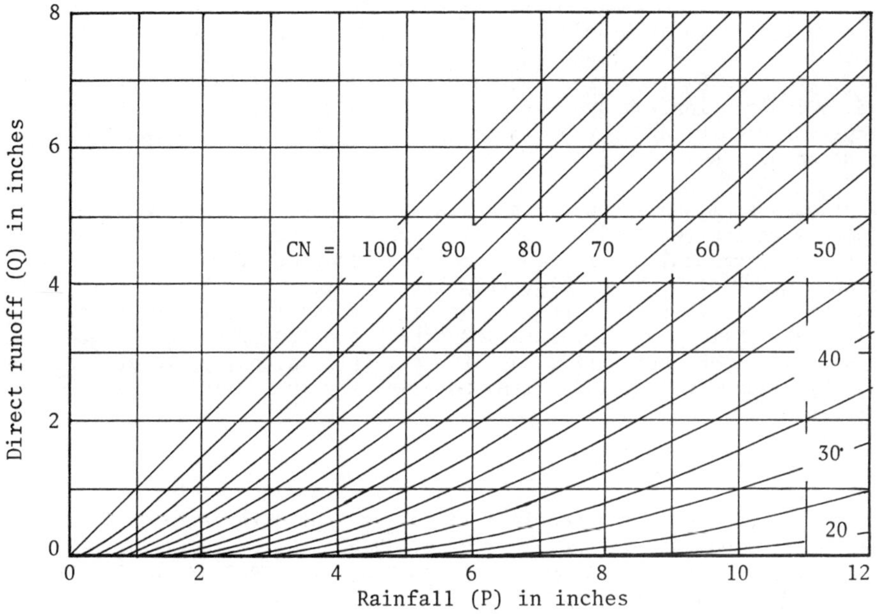

Fig. 9.9. Direct runoff as a function of rainfall and curve number (U.S. Soil Conservation Service).

Typical of the more regional methods is that by Gray (1961) who fitted the incomplete gamma distribution to runoff from areas under 200 km^2 (see also eqns 7.54, 9.41 and 9.42) as

$$Q_t = \frac{25\,\gamma'\,\alpha}{\Gamma(\alpha)}\,e^{-\gamma' t/t_p}\left(\frac{t}{t_p}\right)^{\alpha-1} \qquad 9.31$$

where Q_t is percent of flow at times equal to 0.25 t_p (t_p is time to peak of the hydrograph); γ' is a dimensionless parameter $\gamma' = \gamma t_p$, where γ is a scale parameter; $\alpha = 1 + \gamma'$ is a shape parameter; $\Gamma(\alpha)$ is the gamma function of α and $\Gamma(\alpha) = (\alpha - 1)!$ if α is an integer (if not $\Gamma(\alpha)$ is given for all α, not equal to zero or a negative integer, by $\Gamma(\alpha) = \Gamma(\alpha + k + 1)/[\alpha(\alpha + 1)\ldots(\alpha + k)]$, where k is the smallest integer such that $\alpha + k + 1 > 0$), and t is time in minutes. Gray found from data from Central-Iowa-Missouri-Illinois-Wisconsin region that

$$\frac{t_p}{\gamma'} = 9.27\left(\frac{L}{\sqrt{S}}\right)^{0.562} \qquad 9.32$$

where L is the length of the main channel in miles and S is its slope in percent, and that

$$\frac{t_p}{\gamma'} = \frac{1}{2.676/t_p + 0.0139} \qquad 9.33$$

However, it was also found that the factors and exponents were different for other districts. Consequently, such formulae can be used only in the regions for which the values have been obtained. The unit storm period in Gray's method is 0.25 t_p and the beginning of surface runoff is assumed to be at the centroid of the precipitation, i.e. the time points are 0.125 t_p, 0.375 t_p, 0.625 t_p,... The summation of the runoff ordinates is continued until the sum is equal to 100%.

As the amount of data available diminishes the methods used to calculate runoff have to become less and less refined. Where regional data is also lacking use has to be made of the synthetic hydrograph techniques discussed in Chapter 7. For example, the U.S. SCS dimensionless hydrograph, Table 7.3 could be used. In many instances the simple triangular hydrograph is adequate to estimate the peak discharge and time distribution of flow.

Runoff studies from small catchments have shown that the flood runoff is strongly correlated with the regional geological characteristics while, for other than forested catchments, vegetal cover, slope, moisture content of the soil, elevation, shape of the catchment, etc., have relatively little effect on flood flows. The influence of geological characteristics becomes negligible for areas larger than about 200 to 300 km^2. For small rural areas ($A < 200\ km^2$) the 100-year flood depends about 90% on the geological characteristics and the rainfall intensity; vegetal cover and the form of the catchment correlate with the remaining 10%. The simplest form of correlation is

$$q_{100} = CA^{-n} \qquad 9.34$$

where q_{100} is the 100-year flood in m^3/s per km^2, A is the area of the catchment in km^2, and n is a constant of the order of 0.3-0.5. C is a coefficient which depends on the geology of the subsurface, the peak rainfall intensity and the yearly average rainfall, and its value ranges from 1-20.

9.3.2 Urban catchments

There are several features which separate the urban catchment from the rest. Urbanization extensively alters the hydrological performance of the catchment. The object of urban drainage, in general, has been to get rid of the stormwater as quickly as possible. Impervious surfaces, gutters, drain pipes, etc., are all designed to provide fast unimpeded drainage. The effect of all these measures is to shorten the duration of the runoff, i.e., to shorten the hydrograph baselength. As a consequence the peak discharge must increase proportionally for the same volume of runoff. In general, the immediate runoff volume is also increased because the amount of infiltration is reduced. Many investigators have used the lag time t_ℓ as the principal hydrological measure of urbanization. Carter (1961) compared lag times on 22 streams in the Washington, D.C. area and concluded that $t_\ell \propto (L/S)^{0.61}$, where L is in km and the coefficient of proportionality is 1.203×10^{-2} (3.0), 4.812×10^{-3} (1.2) and 2.125×10^{-3} (0.53) for natural, partially sewered and completely sewered catchments, respectively. The numbers in brackets apply to Carter's formula in which L was measured in miles and the slope S in ft/mile. Carter's equation for the mean annual flood (2.33 year return period) is

$$Q = 2.8122\ KA^{0.85}\ t_\ell^{-0.45} \qquad 9.35$$

where Q is in m^3/s, A in km^2, t_ℓ in hours, and $K = (0.30 + 0.0045\ I)/0.30$, where I is the percentage of impervious cover (for ft^3/sec and mi^2 the factor is 223). Leopold (1968) correlated the increase in the average annual flood with urbanization as measured by the percentage of impervious area and by the percentage of the area drained by stormwater sewers, Fig. 9.10

An important consequence of urbanization is the sediment load imposed on streams and the reduction of water quality in streams and in the receiving waters. In general, the sediment production of a catchment increases with increasing peak flow rates, but this can be overshadowed by the effects of ground disturbances during develop-

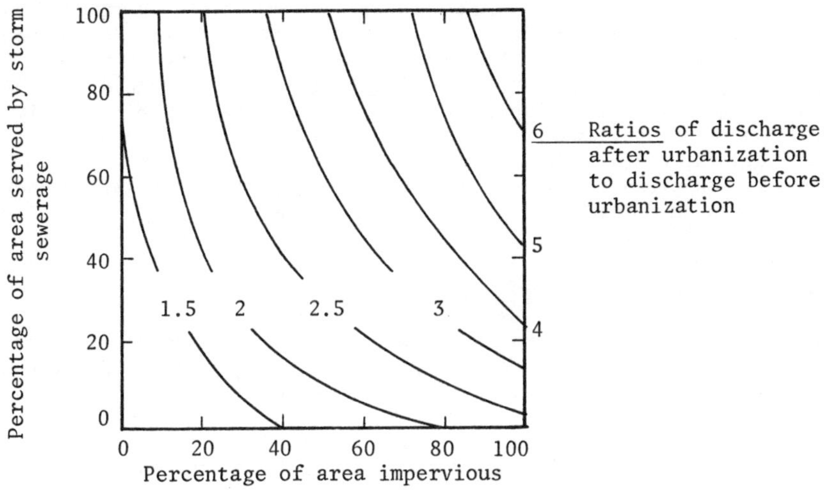

Fig. 9.10. Effect of urbanization on mean annual flood for a drainage area of 2.6 km^2.

ment and construction. The surface of forested catchments or catchments with a good cover of vegetation is usually well protected, and soil erosion is limited to scouring in the stream channel and to sediment derived from land slides in very steep country. When the catchment is converted into agricultural use the sediment yield increases steeply during the conversion process and settles down to a value which depends on the form of agricultural cropping. This value is normally higher than that of the original catchment and very sensitive to land management practices. The erosion rate is also related to the slope of the land. Musgrave (1947) found that the rate of erosion was proportional to the 1.35-power of the slope ($S^{1.35}$) and to the 0.35-power of the length of the slope ($L^{0.35}$). Meyer and Monke (1965) gave the erosion rate as $E \propto L^{1.9} S^{3.5} d^{-0.5}$, where d is the soil particle diameter. The sediment yield from small catchments during conversion into urban settlements can increase by more than ten thousand times. Typical values for small agricultural catchments are 100-400 tonnes/km^2 per year, depending on the slope and the type of agriculture. These values decrease as the catchment size increases. Suspended sediment runoff can be as high as 250 g/l from tilled land and zero from forested catchments. Generally, sediment yield from forested catchments and good lightly grazed pastures is very low. The sediment yield from mature urban catchments can also be quite low but the amount of dirt washed down by the rain increases with urbanization, particularly in areas used for industrial purpose. Urbanization also increases the magnitude and frequency of the peak flows and this also tends to increase the sediment yield. The sediment production during the construction from early groundworks to the completely developed stage can be reduced by appropriate construction methods, particularly when the entire catchment is developed as a single project. Where the construction and development stretches over years, control is much more difficult. The influx of dirt and soil into streams and receiving waters increases the dissolved solids content, decreases the dissolved oxygen content, and generally lowers the quality of the receiving waters. The problems associated with urbanization are thus threefold:

(1) handling of the increased flow rates,
(2) control of the sediment load, and
(3) pollution of the receiving waters.

The methods for the design of urban drainage systems are mainly based on empirical formulae. Ardis et al. (1969) reported on stormwater drainage practices of 32 cities in the United States and wrote: "Practically every city reports the use of the "rational" method in storm drainage design; however, only six cities use it correctly". It is important to realize that very few urban drainage systems are designed and built as a complete system. Drainage networks usually "grow" with the city over long periods of time; they are improved and renewed here and there as the need arises. At the same time it must be realized that vast sums of money are spent annually on urban drainage and that the efficiency (or inefficiency) of the drainage system can save (or waste) large sums of money.

9.3.2.1. The rational formula

The rational or Lloyd-Davis formula (eqn 9.20), in its original or modified forms, is used in the design of most urban stormwater systems. The formula requires the estimation of the value of the runoff coefficient C and the intensity of the uniform rainfall over the area involved. The intensity is based on the time of concentration t_c, which is defined as the time required for the surface runoff from the most remote part of the catchment to reach the point under consideration. The time of concentration is the sum of the time of overland flow t_e (also known as the time of entry) and the travel time in the drain t_t, so that

$$t_c = t_e + t_t \qquad 9.36$$

For relatively small urban catchment the S-curve plateau of the runoff hydrograph is rapidly reached. The runoff hydrograph by the rational formula is as shown in Fig. 9.11. The time of travel is estimated for the given slope and pipe size by dividing the length of the pipe by the full bore velocity. The time of overland flow is more difficult to estimate because it depends on the slope, surface roughness, infiltration characteristics, surface detention and rainfall intensity. Many empirical relationships are in use, such as

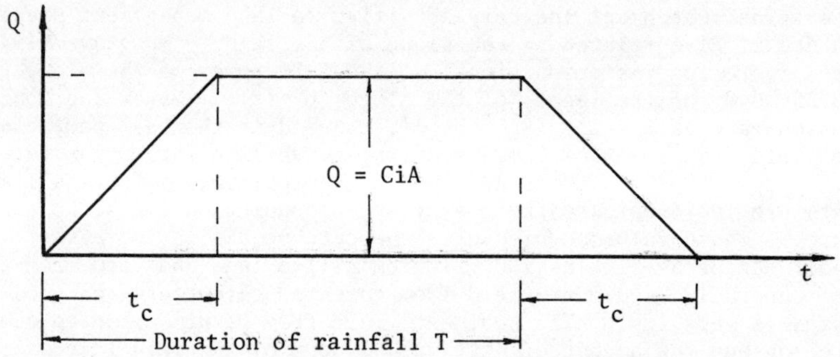

Fig. 9.11. Storm hydrograph by rational formula.

$$t_c = 0.0195\, L^{0.77}\, S^{-0.382} \qquad 9.37$$

by Kirpich (1940), where t_c is in minutes, S is the slope, and L in m is the maximum distance of travel of water along the watercourse (for L in ft the factor is 0.0078). Mockus (1957) put the overland flow concentration equal to the lag time divided by 0.60. Ragan and Duru (1972) expressed the overland flow time as

$$t_c = 6.917\ (nL)^{0.6}/(i^{0.4}\ S^{0.3}) \tag{9.38}$$

where n is the Manning roughness coefficient (n ≃ 0.02 for pavements and n ≃ 0.50 for grass), i is the intensity of the rainfall in mm/h, and the other terms are as above [t_c - 0.93 $(nL)^{0.6}/(i^{0.4}\ S^{0.3})$ when L is in ft and i in in./h]. Many design offices use empirical charts of the kind shown in Fig. 9.12 and discussed by Rantz (1971).

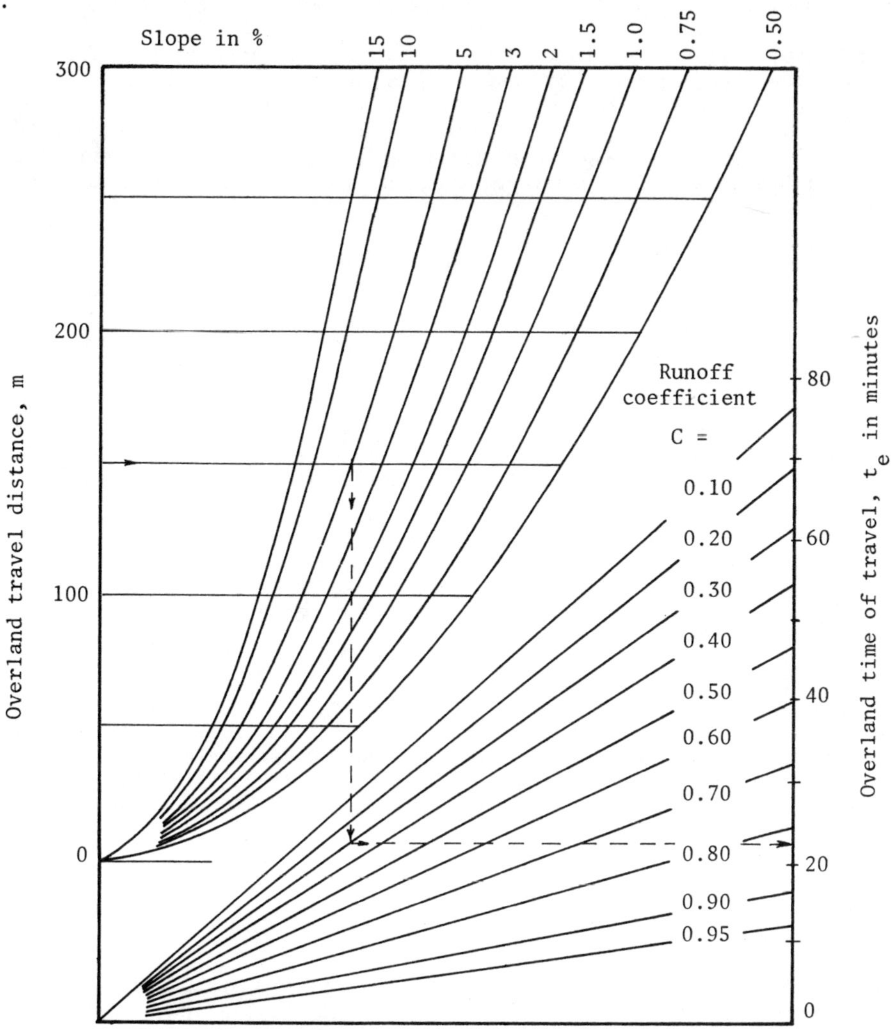

Fig. 9.12. Design chart for estimation of time of entry of overland flow.

It is important to note that the rainfall intensity for a given probability level (return period) is a function of the time of concentration and hence of the total area being drained. Therefore, as the design proceeds downstream and more and more of the subareas are added, the time of concentration increases and correspondingly the intensity of the rainfall decreases. The usual values for the time of concentration range from 5-30 minutes. For these durations the rainfall intensity versus duration curve is very steep, i.e., the intensity decreases rapidly with duration.

When using the rational method for urban drainage it is important to note that $Q_i = Ci\ A_i$ gives the flow at the outlet of the subcatchment A_i but as the design proceeds downstream the sum of the Q_i values does not give the total flow Q_{total} in the trunk stormwater sewer; Q_{total} is considerably smaller than the sum of the Q_i values.

The runoff coefficient C varies with the percentage of impervious surface, the type of soil and the slope of the catchment. Most cities have established their own empirical values for C. In regions with porous soils with very small surface slopes and low intensity rainfalls, runoff from pervious surfaces is ignored and the runoff coefficient is assumed to be given by the proportion of connected (continuous) impervious surface area. The usual practice is to relate the runoff coefficient to the soil type, for example:

Lawns:	sandy soil, slope	$S < 2\%$	C = 0.05 - 0.10
	" " "	$2 < S < 7\%$	0.10 - 0.15
	" " "	$S > 7\%$	0.15 - 0.20
	heavy soil, slope	$S < 2\%$	0.13 - 0.17
	" " "	$2 < S < 7\%$	0.18 - 0.22
	" " "	$S > 7\%$	0.25 - 0.35
Impervious areas			0.95

From the above a weighted average C value can be calculated on the basis of the proportions of the various types of surfaces. There are also tables of average values, such as:

Business areas	C = 0.70 - 0.95
Light industrial areas	0.50 - 0.80
Residential areas	0.25 - 0.50
Unimproved areas	0.10 - 0.30

The value of C can also be related to the parameters of the catchment and precipitation. For example, Miller (1968) expressed the coefficient as $C = C_b C_f C_c$, where the individual coefficients refer to surface cover, return period of storms and precipitation regime, respectively. The surface cover index is expressed as the sum of the percentage fraction of bare ground and half of the percentage fraction of the area covered with grass. This index is correlated with soil permeability and overland slope. The coefficient C_f is put equal to one for 50-year return period (0.6 for 2.33 and 1.11 for 1000-year return period). The coefficient C_c relates the mean annual rainfall and the 60-minute 50-year return period rainfall intensities for the given location. However, these refinements deprive the rational method of its simplicity and local values for the various coefficients, particularly C_b, are not readily available.

A different approach to the problem is to express C by a statistical description, e.g., Schaake et al. (1967), Aitken (1973). In the statistical approach S is defined as the ratio of the discharge per unit area for a certain return period to the peak rainfall intensity for the same return period. The respective discharge and rainfall intensity are obtained from separate frequency analyses of annual flood and rainfall peaks for durations equal to the time of concentration.

A modification of the rational formula includes an additional coefficient C_s, which accounts for channel storage. Then

$$Q = C_s CiA \tag{9.39}$$

The value of this coefficient is not easy to define. The storage could be simulated by routing the hydrograph of the rational method formed by the area-time diagram of the catchment, through a hypothetical reservoir at the design point, by the Muskingum method using S = KQ. Aitken (1968) found that $K \simeq 0.3\ t_c$ and proposed design values of C_s as shown in Fig. 9.13. The minimum value of C_s is a function of the storage delay time and is also shown in Fig. 9.13.

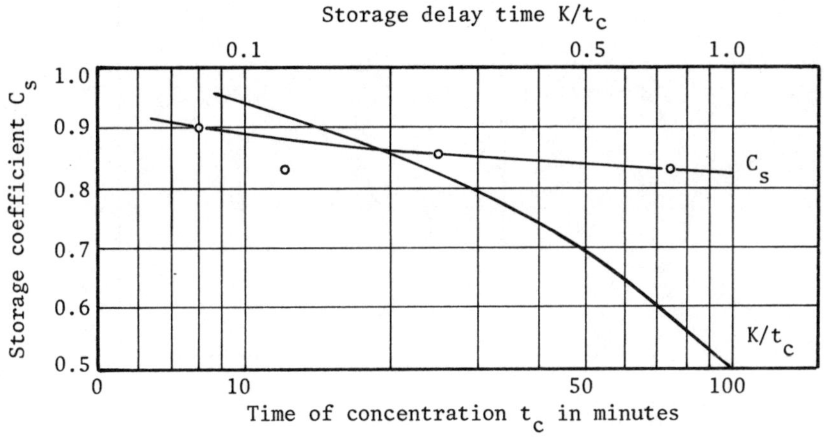

Fig. 9.13. Storage coefficient according to Aitken ($K = 0.3\ t_c$) and minimum values of storage coefficient C_s.

The minimum values relate to either large catchments or to inflow hydrographs with straight line area-time diagrams. An approximate value for the storage coefficient is

$$C_s = \frac{2t_c}{2t_c + t_t} \qquad 9.40$$

Thus, the design hydrograph is as shown in Fig. 9.14. Since the areas under the standard and modified hydrographs are equal the baselength b of the modified hydrograph is $b = 2t_c/C_s$. Substituting C_s from eqn 9.40 yields $b = 2t_c + t_t$.

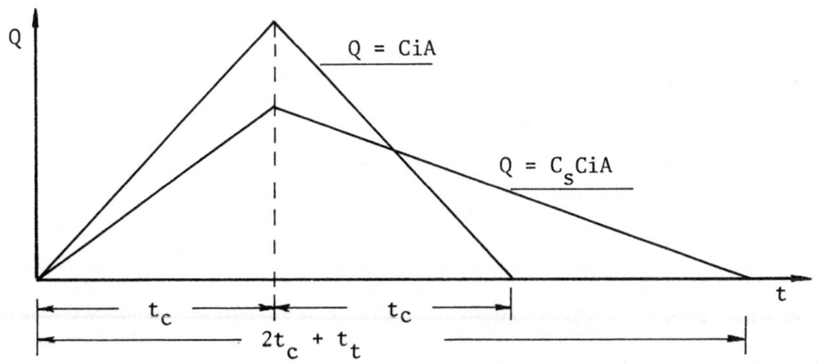

Fig. 9.14. The hydrographs by the rational method.

Watkins (1962) published the urban stormwater design method developed by the Road Research Laboratory in the United Kingdom. The RRL method uses the area-time diagrams and the excess rainfall hyetograph. After the drainage network has been decided on, the area-time curve can be developed. A straight line approximation for for the S-curve was used, Fig. 9.15.

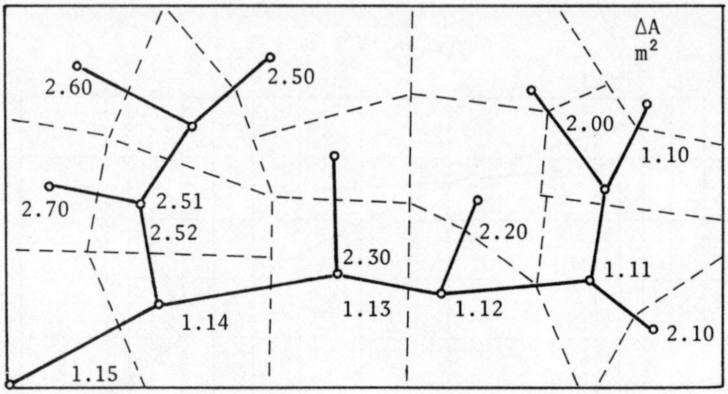

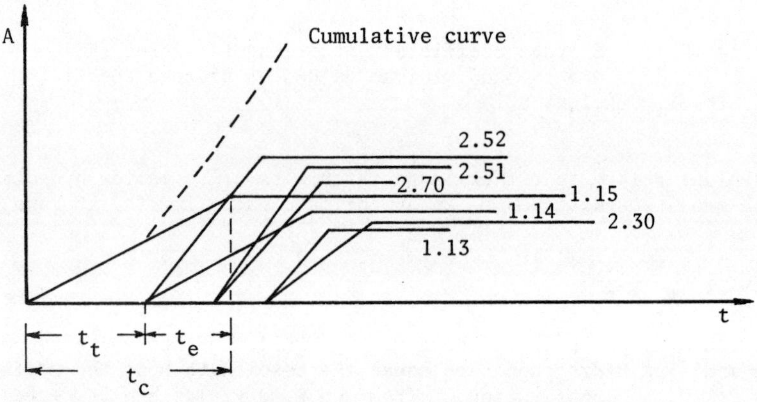

Fig. 9.15. Drainage network and development of the area-time curve.

The area-time curve and the rainfall hyetograph are divided into unit time slices, e.g., 1 minute intervals, Fig. 9.16. The hydrograph ordinates for the corresponding unit slices are then given by

$$Q_1 = i_1A_1 \qquad Q_2 = i_1A_2 + i_2A_1 \qquad Q_3 = i_1A_3 + i_2A_2 + i_3A_1 \qquad \text{etc.}$$

which can be solved in a tabular manner. A plot of these ordinates gives the unrouted hydrograph because the storage effect of the sewer system has not been taken into account. The maximum storage is the volume of water in the sewers at the peak rate of runoff and the storage volume at any subsequent time is given by the area under the hydrograph recession part. It has been found, however, that this integration can be avoided without significant loss of accuracy by assuming that the storage is proportional to the depth of flow at the outlet. Thus, a storage-discharge relationship can be prepared for the given drainage system, Fig. 9.17.

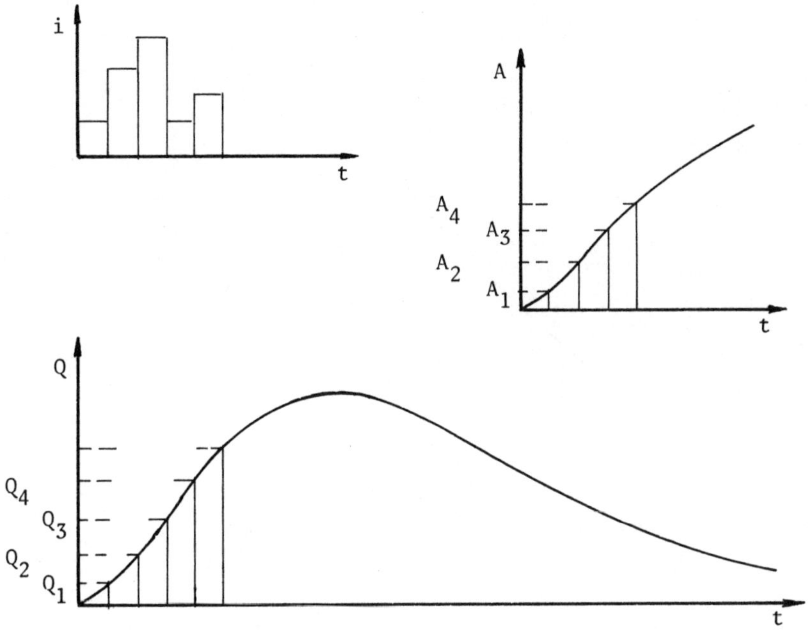

Fig. 9.16. The hyetograph, area-time curve and the discharge hydrograph.

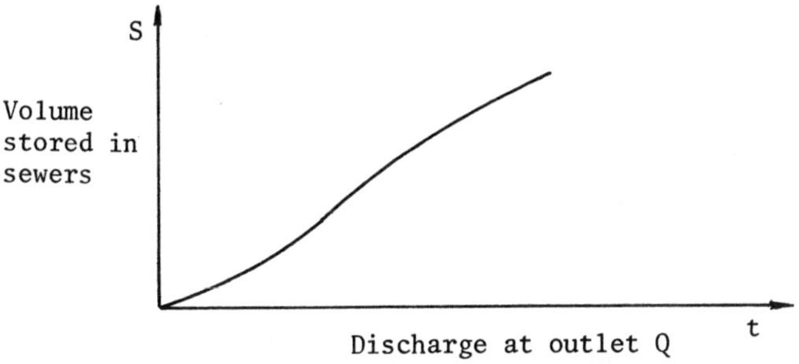

Fig. 9.17. Storage-discharge relationship.

In Fig. 9.18 Q and R are the unrouted and routed hydrograph ordinates, respectively, at each time increment. Designating the corresponding storages by S and assuming linear segments from one increment to the next, then for the interval 0 - Δt

$$\frac{1}{2}\Delta t\ Q_1 = \frac{1}{2}\Delta t\ R_1 + S_1$$

from which R_1, the only unknown, can be found. Similarly for the interval

$\Delta t - 2\,\Delta t$

$$\frac{1}{2}\,\Delta t(Q_1 + Q_2) = \frac{1}{2}\,\Delta t(R_1 + R_2) + (S_2 - S_1)$$

or

$$\frac{1}{2}\,\Delta t(Q_1 + Q_2 - R_1) + S_1 = \frac{1}{2}\,\Delta t\, R_2 + S_2$$

which yields R_2, etc.

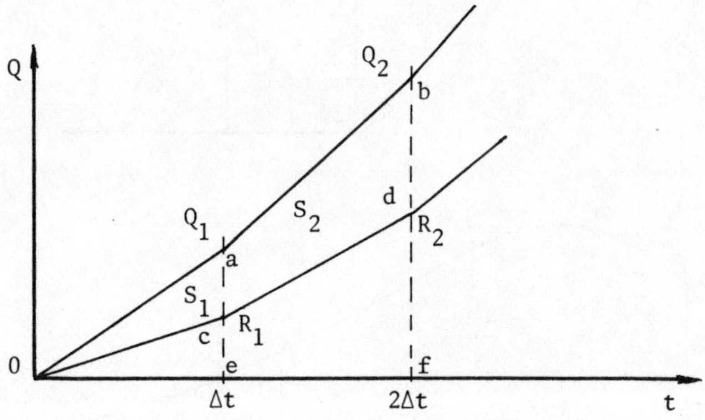

Fig. 9.18. Illustration of the calculation of the routed runoff hydrograph.

For the design of a sewer system or examination of an existing one this method requires the calculation of the hydrograph at each design point, i.e., at each junction (manhole). Where only the runoff at the outfall of the system is required, manual calculation, similar to reservoir routing, can be arranged in a convenient tabular form. For detailed analysis of the system a computer program is necessary. A programming aid was published by the Road Research Laboratory (1963).

9.3.2.2. Summary of additional methods

One of the most elaborate procedures for the design of urban drainage is that developed for the city of Chicago, Illinois, reported by Tholin and Keifer (1960). It has become known as the "Chicago hydrograph method". The stormwater sewers for a city of the size of Chicago are in effect underground rivers and this is reflected by the method, which is based on the concepts of hydrology and river hydraulics. The determination of the sewer hydrographs takes into account the shape of the hyetograph, infiltration, surface depression storage, overland flow and land usage, and the flow is routed through gutters and sewers. Although local design charts have been prepared, the method is basically for computer application.

Eagleson (1962) discussed the use of unit hydrographs for sewered areas. The rainfall excess was expressed as $P_e = (1 - c)P$, where the constant c was related to the percentage of impervious area. During the passage of the flood crest the sewer was assumed to flow full. The length of the full sewer was chosen as the mean travel distance, $\bar{L}$, and was defined as the area under the area versus distance curve divided by the total area. The area-distance curve is a plot of the areas enclosed by equal travel distance contours from the hydrograph station versus the travel distances. The mean velocity $\bar{V}$ was estimated by the Manning formula. It was found that the hydrograph was related to $S = \bar{H}/\bar{L}$, where $\bar{H}$ denotes the mean increase in elevation of the catchment, as given by the area under the area-elevation (hypsometric) curve, divided by the total catchment area. The peak discharge, $q_{max} = Q_{max}/\text{Area}$, was empirically related to S by $Q_{max}(\text{cusecs})/(\text{mi})^2 = (2.13 \times 10^5)S$ or $Q_{max}(\text{m}^3/\text{s})/A(\text{km}^2) = 1.948 \times 10^7$ S. The lag time t_p (distance between the centre of mass of the excess rainfall hyetograph and the hydrograph peak) was expressed as $t_p(\text{min}) = \bar{L}/(60\ \bar{V})$. The shape of the unit hydrograph was defined in terms of its width at 0, 50 and 75% of Q_{max}. Eagleson plotted the peak discharges per unit area against the widths in minutes. The correlations were well defined. The scatter was most pronounced for the base width (0% of Q_{max}), mainly because of its sensitivity to baseflow. The widths in minutes and q_{max} in m^3/s per km^2 correlated as follows: $w_0 = 1093/q_{max}$, $w_{50} = 388/q_{max}$ and $w_{75} = 230/q_{max}$. The data came from 27 storms and 5 urbanized areas in Louisville, Ky.

Espey and Winslow (1974) developed a number of flood frequency formulae for urban areas. Espey et al. (1969) also presented empirical equations for discharge and time to hydrograph peak in terms of area, length, slope, impervious cover, time to peak and a channel roughness factor.

Rao et al. (1972) discussed conceptual hydrological models for urban catchments and investigated linear storage models for urban use. For catchments with areas less than 12 km^2 the single linear reservoir model

$$u(t) = \frac{1}{K} e^{-t/K} \qquad 9.41$$

gave good results. K was related by regression analysis to catchment parameters as

$$K = 1.21\ A^{0.49}_{(\text{km}^2)} (1 + I)^{-1.683}\ P^{-0.24}_{e(\text{mm})}\ T^{0.294}_{(\text{hrs})} \qquad 9.41a$$

$$[K = 0.887\ A^{0.49}_{(\text{sq.mi})} (1 + I)^{-1.683}\ P^{-0.24}_{e(\text{in.})}\ T^{0.294}_{(\text{hrs})}]$$

where I is the ratio of impervious area to the total area and T is the rainfall duration in hours. For larger catchment the Nash model of n identical reservoirs, the cascade model

$$u(t) = \frac{1}{K_n} \frac{e^{-t/K_n}}{\Gamma(n)} \left(\frac{t}{K_n}\right)^{n-1} \qquad 9.42$$

was found to be the best model. Here $\Gamma(n)$ is the gamma function. The empirical relationships for K and the lag time t_ℓ in hours between the centroids were given as follows:

$$K_n = 0.56\ A^{0.389}_{(\text{km}^2)}\ (1 + I)^{-0.622}\ P^{-0.106}_{e(\text{mm})}\ T^{0.222}_{(\text{hrs})} \qquad 9.42a$$

$$[K_n = 0.575\ A^{0.389}_{(\text{sq.mi})} (1 + I)^{-0.622}\ P^{-0.106}_{e(\text{in.})}\ T^{0.222}_{(\text{hrs})}]$$

$$t_\ell = 1.275\ A^{0.458}_{(\text{km}^2)}\ (1 + I)^{-1.662}\ P^{-0.267}_{e(\text{mm})}\ T^{0.371}_{(\text{hrs})} \qquad 9.42b$$

$$[t_\ell = 0.831\ A^{0.458}_{(\text{sq.mi})} (1 + I)^{-1.662}\ P^{-0.267}_{e(\text{in.})}\ T^{0.371}_{(\text{hrs})}]$$

and

$$nK_n = t_\ell \qquad 9.42c$$

Wittenberg (1975) introduced a parallel cascades model. The two cascades represent the impervious and pervious area contributions to runoff, i.e., a fast and a slow response. The two cascades will also have differing rainfall excess hyetographs.

The parallel combination was already considered by Watkins (1962) but he concluded that the contribution to peak flows from the pervious areas was of secondary importance. The importance of the runoff from pervious areas will increase with the size of the catchment. The parallel cascade model could be useful as a prediction model of the effects of growing urbanization of a catchment.

An extensive discussion of the various rainfall runoff models, together with numerical comparisons and computer programs, was prepared by Dracup et al. (1973). Further comparisons, particularly on the urban catchment response time, were published by Schulz and Lopez (1974).

The estimate of runoff, on which the design of stormwater drainage is based, is a problem of economics. There is a level of protection beyond which it will be ecomically wasteful to design. Nevertheless, there are two practical aspects which the designer should keep in mind. The first is the assessment of the probable future expansion of the urbanization. It is all too common that a settlement has been developed at the bottom of a valley and then some years later, when the rest of the catchment is developed, the stormwater system is called upon to transmit the stormwater from all the additional areas, in addition to the runoff for which it was designed. The second is that the designer should consider the consequences of floods which exceed the design value. A good design should provide and safeguard a secondary drainage system for flows greater than those which are to be catered for by the primary stormwater system or for drainage in the event of a blockage of the primary system. Such a secondary system should minimize possible flood damage. This in effect means that the path defined by the valley floor should not be obstructed by buildings, causeways, etc.

9.3.2.3. Detention of stormwater

In connection with urban drainage one should also consider alternatives to the quickest runoff and the consequent high peak flows. These alternatives include surface ponding, storage tanks and storage ponds for delaying the runoff. The aim of these alternatives is to delay the runoff and spread it over a longer time at a lower rate of discharge. The suitability of the various methods of detention depends on the type and slope of the catchment, the type of urban use (e.g., residential villas, apartment blocks, industrial, commercial) as well as the characteristics of the precipitation. On flat residential land a considerable quantity of water can be ponded without any serious consequences. Where land is available storage ponds can be introduced. These will act as settling ponds and will remove some of the dirt carried by the stormwater. This will reduce the pollution of the receiving waters but introduce the need for periodic cleaning of the storage ponds. Settling ponds are an essential feature during land development works for removal of sediment carried by runoff from the disturbed catchment. Pond design should provide for maximum velocity reduction at the inflow to the pond and for adequate storage volume for the larger particles which settle out. The depth of flow should decrease towards the outlet while a constant forward velocity is maintained. This corresponds to a fan-shaped pond deeper at the inlet and decreasing in depth while increasing in width towards the outlet.

A more recent idea has been to use the basecourses of parking lots and lightly loaded streets for temporary storage of stormwater. A porous pavement will allow rain to soak into the basecourse from where it is collected and drained at a reduced rate into the stormwater drainage system. In some instances some of the water may be allowed to seep into the soil (subgrade). Where such seepage is likely to weaken the subgrade more than can be tolerated, the subgrade may be sealed off by an impervious membrane, leaving the basecourse as a porous storage tank. The Franklin Institute (1972) studied porous pavement systems and concluded that the technology exists to construct porous pavements that will satisfy strength and drainage requirements. Jackson and Ragan (1974) analyzed such a system.

Chapter 10

FLOW REGULATION, CATCHMENT YIELD, SEDIMENT YIELD

Storage may be natural or man-made. Natural storages in the form of lakes, swamps and rivers have a strong influence on the flow from the catchment but do not regulate the flow. Regulation in the water resources management sense requires man-made control structures. Man-made storage of water may be either for flood protection or for supply during periods of water deficiency. In the following discussion storage is confined to surface reservoirs, but it should be borne in mind that an important form of storage is in groundwater reservoirs (aquifers). Aquifers as reservoirs will increase in importance in the future because of the relative lack of suitable sites for surface reservoirs and their disadvantages, such as very high evaporation losses in semi-arid regions and loss of usually fertile valleys, etc. The analysis of aquifers as reservoirs incorporates the concepts of flow in elastic porous media. An example of this type of analysis is the work by Eliasson et al. (1973).

Reservoir storage refers to the part of the reservoir volume which can be used for flow regulation. Some of the terminology used is shown in Fig. 10.1. In general, a reservoir may be required to supply water for power generation, irrigation, water supply, flood protection and recreational use, etc. Each of these requirements of the reservoir imposes constraints on the operating policy. Many of these requirements contradict each other. For example, power generation requires a full reservoir or maximum head whereas flood protection requires an empty reservoir. In general, the outflow from the reservoir will be very different from the natural inflow.

For flood protection in a multi-purpose reservoir a certain volume of the storage reservoir is kept empty, ready to receive water during the peak of the flood. The stored water is discharged after the flood peak has passed. The reservoir receives all the water in excess of a design flow rate Q_d, i.e., the peak of the hydrograph is skimmed off and is delayed by storage, Fig. 10.2. Thus the flow rate downstream is limited to a predetermined maximum value. The stored volume is discharged when the flow rate downstream has fallen below the value of Q_d. This reduces the peak discharge rate of the downstream hydrograph but increases the baselength of the hydrograph. The task of the designer is to estimate the magnitude of the flood peak, its time of occurrence, and the volume of water to be stored. This volume depends on economic and environmental considerations. It is usually assumed, when designing flood control reservoirs, that the period between floods is long enough for the reservoir to be emptied before the arrival of the next flood. However, there are climatic conditions where two major floods may occur in succession (mainly in areas affected by tropical cyclones). Under these conditions the period between the floods cannot be ignored and has to be considered as a design variable.

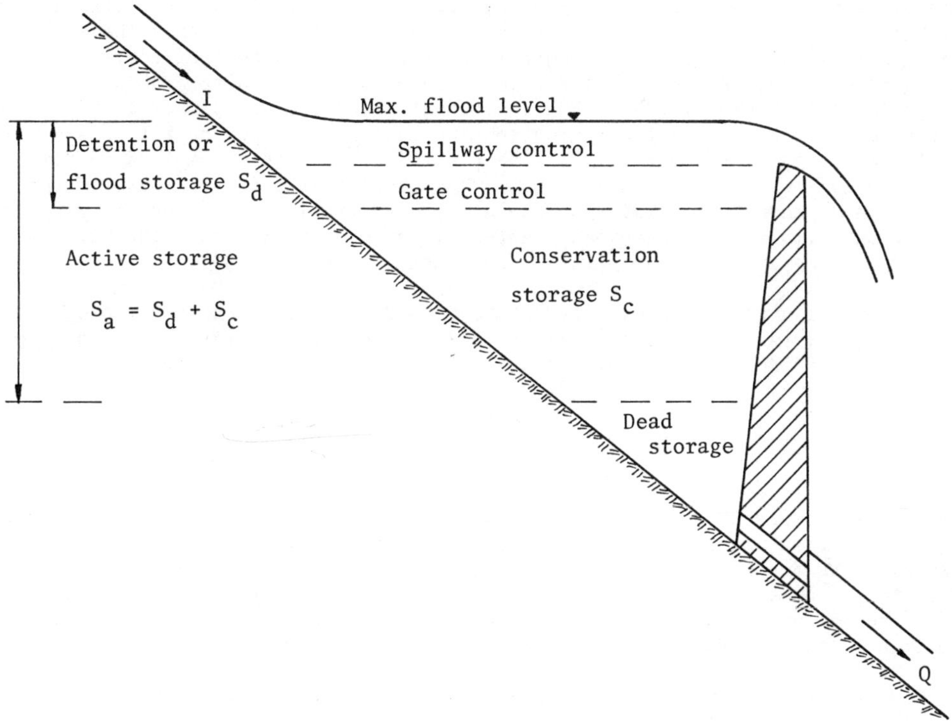

Fig. 10.1. Storages in a multi-purpose reservoir.

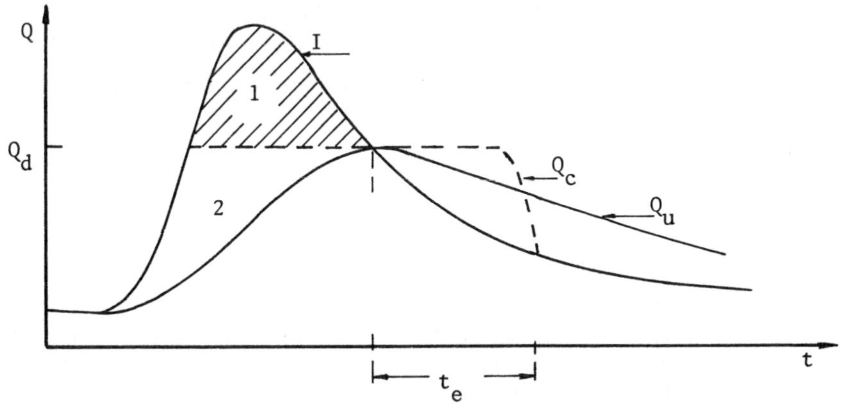

Fig. 10.2. Storage for flood control.

The reservoir storage may be controlled or uncontrolled. The storage above a spillway crest (without gates) is an uncontrolled storage and its effect on the outflow is determined according to the principles or reservoir routing. Note that even this type of reservoir has a control facility in the form of gated bottom outlets (sluice gates). For a fully controlled storage the outlet must be gated and the discharge capacity at zero storage must be equal to or greater than the maximum possible rate of inflow. A fully controlled storage is the most effective way of reducing flood peaks, since only the volume above the design discharge needs to be stored (area 1 in Fig. 10.2). For an uncontrolled storage volumes 1 and 2 have to be stored. In Fig. 10.2 the hydrographs Q_c and Q_u refer to controlled and uncontrolled discharges, respectively. The controlled storage is emptied in time t_e after the inflow has dropped below Q_d. The shape of the falling part of the Q_c hydrograph depends on the gate and operating characteristics. The emptying time for the uncontrolled storage is the interval between the intersection point of the I and Q_u hydrographs and the point where these curves coincide.

The higher efficiency of controlled storage is associated with some reduction in the overall safety, since gates may fail to open. Hence, an ungated emergency spillway is usually included in controlled storage schemes.

10.1 Storage for Flood Control

The capacity of the controlled storage to be provided depends on the volume under the inflow hydrograph above the design flow rate Q_d (area 1 in Fig. 10.2) and not on the peak discharge. The volume required is greatest for broad-crested hydrographs. The return period of the design flood by frequency analysis, however, relates to the peak discharge only and gives no information about the volume of water under the inflow hydrograph above the design discharge. A design procedure for estimation of the controlled storage volume for a N-year protection level at discharge Q_d is as follows (Klemes, 1973):

(1) Plot separate volume curves of all floods in each year and draw an envelope curve Fig. 10.3a. This gives n envelope curves for the n years of record.

(2) Plot the n envelope curves on one graph (n curves), Fig. 10.3b.

(3) Plot frequency curves of storage volumes S_1, S_2, ..., S_i... for discharge values of Q_1, Q_2, Q_3, ..., Q_i, ... The number of points on each of the volume-frequency curves is n, the number of years of record. These volume-frequency curves are fitted with a suitable theoretical distribution function which smoothes the data and enables extrapolation of the tail ends. With increasing discharge rate Q_i more and more of the n values of volume reduce to zero. These zero values must be retained for each frequency curve, Fig. 10.3c.

(4) Convert the frequency curves of the flood volumes to return period curves N_1, N_2,...., N_i,..., Fig. 10.3d.

(5) The vertical cross section of the family of return period curves defines the relationship between the discharge Q and the storage volume S for a T-year return period, Fig. 10.3e. These curves show the required control volume for a given design discharge Q_d.

Plots like Fig. 10.3e show that the volume required increases relatively little with the return period of protection.

A storage reservoir is not often used for flood protection alone. In general, the reservoir is designed for a multi-purpose use. Such reservoirs can be made more efficient by increasing the storage volumes through release of water just prior to the arrival of the flood. Flood control by a multi-purpose reservoir is discussed by Beard (1963).

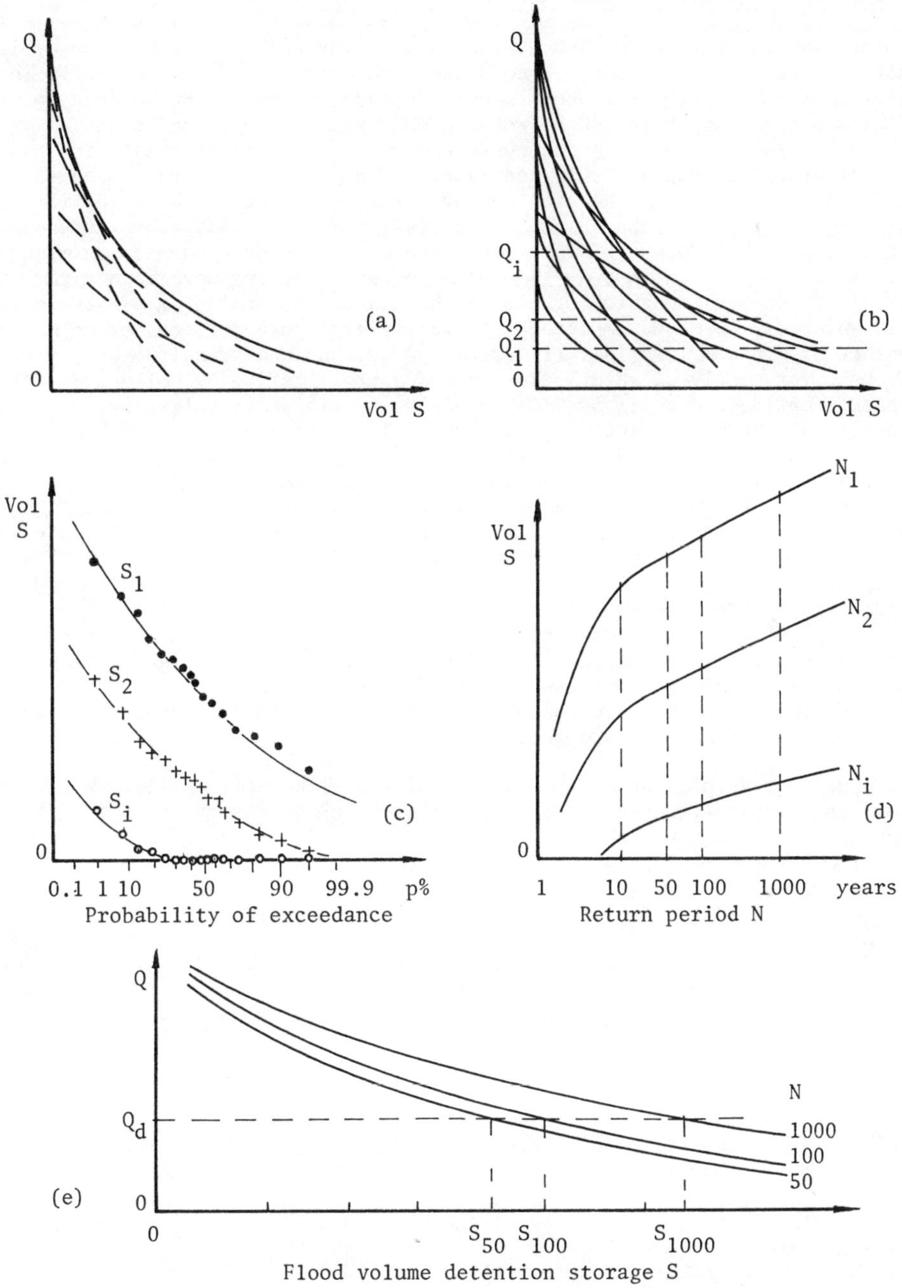

Fig. 10.3. Definition sketch for the method of estimation of reservoir storage volume. (a) Flood volume envelope curve for one year. (b) Flood volume envelope curves for n years. (c) Probability of exceedance. (d) Return period N. (e) Flood volume detention storage.

10.2 Regulation for Abatement of Water Shortage

A flood wave is a transient phenomenon and it may occur even where water is generally scarce. It is not to be confused with *excess of water*, which is related to water management with reference to human needs. From the human settlement point of view a swamp may have excess water even during drought conditions. *Water deficiency* is also related to human needs and may occur in some localities even during the wet years or seasons. In contrast, a *drought* is a natural phenomenon, which may of course aggravate the problems of water deficiency. Water deficiency arises only if there is not enough water to satisfy some need related to human interests or activities. The amount of water deficiency depends on economic and management considerations. The demand, or the amount of water deficiency, is not the sum of the individual demands because not all of these are consumptive or single-purpose uses. Some water may be used simultaneously for several purposes, for example, for maintenance of minimum low flow, fisheries and navigational requirements, for recreational use, for supply of water power, and for minimization of pollution effects, etc. The consumptive use of irrigation is 30-80%, municipal supplies 5-10%, industrial supplies 5% and hydro-electric power 0%. The difference between the water available and the concurrent demand, including losses, is the deficit. This deficit has to be satisfied from storage which has been accumulated during periods of water surplus or by the import of water from other regions. The problems with import of water are those of conveyance, i.e., hydraulic engineering, and will not be discussed here.

Regulation by storage involves two kinds of problems:
(1) The finding of relationships between the hydrological parameters of the catchment which define the inflow I as a function of time, and relationships between the form and nature of regulation which define the outflow.
(2) The finding of the solution which will satisfy all the imposed requirements (economic, environmental and social).

The hydrological problem of control associated with water deficiency (low flow control) consists of finding a relationship between the reservoir storage S_c, the demand rate Q_d, and the efficiency or reliability of the control, E, in one of the functional forms

$$S_c = f(E, Q_d), \quad E = f(Q_d, S_c), \quad Q_d = f(E, S_c) \qquad 10.1$$

It is assumed that the inflow I and the outflow Q are known functions of time, which may be either deterministic or stochastic in nature, The outflow is the actual flow rate of water released from the reservoir for all purposes, including seepage, evaporation, the demand flow rate Q_d and the spill of flood peaks. Hence the mean outflow $\bar{Q}$ is always greater than the mean demand $\bar{Q}_d$. The outflow is defined by a release or operating rule and may be either dependent on or independent of the inflow. For example, the water demand by industrial users is not related to the weather and catchment runoff whereas the irrigation demand in the region is strongly correlated with the lack of natural runoff. The demand may also vary with seasons.

The reliability or dependability of the flow control can be defined (after Klemes, 1973) in three forms as
(1) the occurrence-based dependability

$$E_f = \frac{n - m}{n} \times 100\% \qquad 10.2$$

where m is the number of years in which the demand cannot be met (failure years) and n is the total number of years considered,
(2) the time-based dependability

$$E_t = (1 - \frac{1}{T}\sum_T \Delta T) \times 100\% \qquad 10.3$$

where ΔT is the duration of a single failure (analogous to the length of run τ in drought analysis) and T is the total time considered, and
(3) the quantity-based dependability

$$E_q = (1 - \frac{1}{TQ_d}\sum_T \Delta V) \times 100\% \qquad 10.4$$

where ΔV is the quantity of water not available during a single failute period ΔT (Fig. 10.7). Usually $E_f < E_t < E_q$. The economically optimum values of the dependability cover a relatively narrow range. Typically $E_f \simeq 99\%$ for municipal water supply, 95-98% for industrial water supplies, 70-85% for irrigation in subhumid climates and 80-95% in arid climates.

10.3 Methods of Storage Calculation

The methods of solution of the reservoir storage problem may be grouped as *empirical*, *experimental* and *analytical* methods.

The *empirical approach* is based on the application of the mass curve concept, which was introduced into the design of reservoirs by Rippl (1883). The mass curve is the time integral of the inflow and represents the volume of inflow during the period, Fig. 10.4. The vertical distance between the tangents in Fig. 10.4 represents the volume of storage required to permit a continuous uniform release of water during the period t_1 to t_2. For maximum utilization the mean demand should equal the mean inflow, where the demand includes all losses caused by seepage and evaporation. From the mass (volume) curve and the superimposed demand curve a sto-

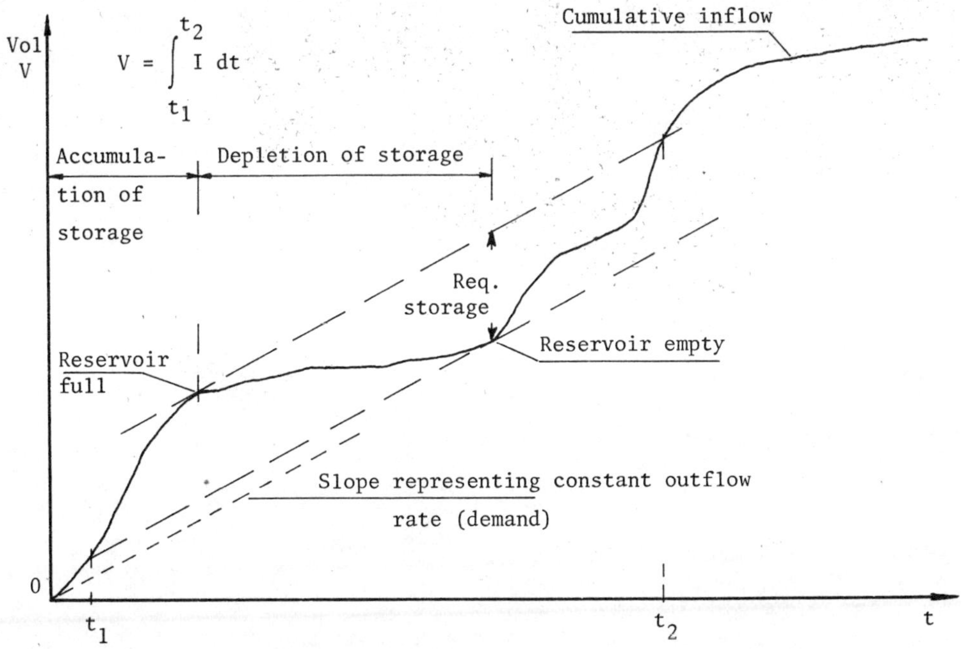

Fig. 10.4. Mass curve or Rippl diagram.

rage versus time curve can be plotted. From this curve a draw-off versus storage curve, which is useful for quick estimation of the possible dependable rate of draw-off at a given storage can be obtained.

It is convenient to define a new variable

$$\Delta S_j = (I_j - Q_j)\Delta t \qquad 10.5$$

where I_j and Q_j represent the mean inflow and outflow during the j-th period of Δt, respectively. If the objective is to draw continuously so that the mean outflow equals the mean inflow, i.e., the expected value $E(I) = E(Q) \simeq \bar{Q}$, then

$$\Delta S_j = (I_j - \bar{Q})\Delta t$$

is a random variable with zero expectation (or mean), and the average net storage is zero. The cumulative sum of these ΔS_j values represents the random variable

$$S_i = \sum_{t=0}^{t=i} \Delta S_j \qquad 10.6$$

as shown in Fig. 10.5a. The plot of S_i versus time is called the residual mass curve. The cumulative sum of the departures of the discrete series ΔS defines the state of storage at any instant as well as the maximum, S_n^+, and the minimum, S_n^-, values of the departures over a given period of $n\Delta t$. The vertical distance between

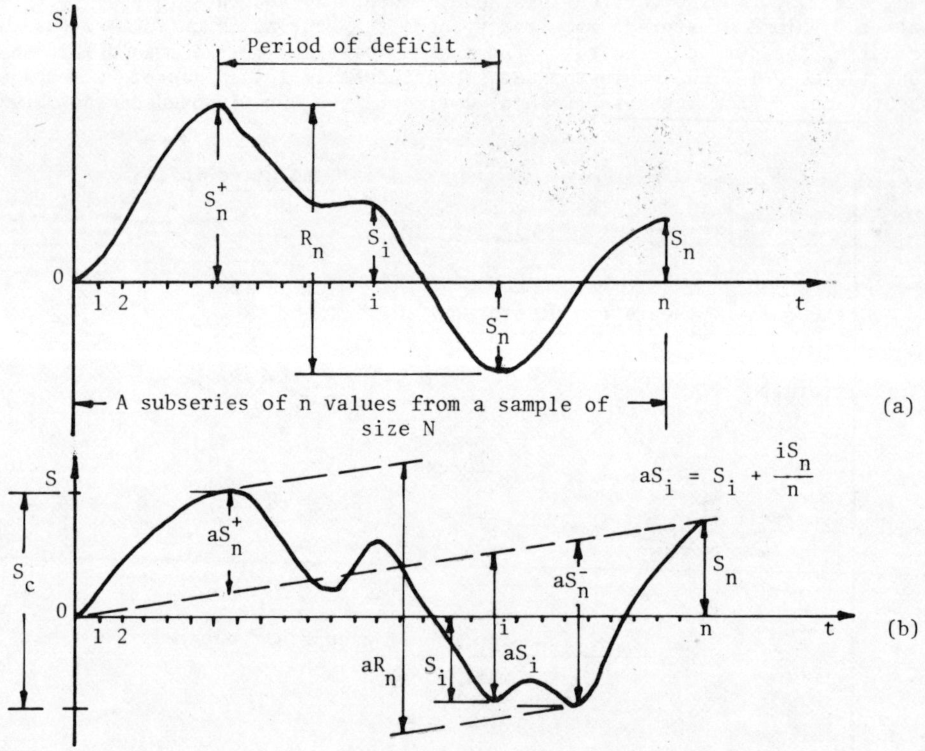

Fig. 10.5. Illustration of the residual mass curve. Definition sketch of storage and range (a), and of adjusted storage and adjusted range (b).

the extremes is the storage S_c required, also known as the *range R*. The slope of the curve at any point represents the flow rate (I - Q).

If a long record is subdivided into subseries, then the mean of any subseries is not, in general, equal to that of the record and is not known in advance. The values of of the storage S_n at n (Fig. 10.5) and $\bar{Q}_n$ are random variables even if $\bar{Q}$ is a constant. The mean discharge of the subseries is

$$\bar{Q}_n = \bar{Q} + S_n/n \qquad 10.7$$

Values of S_n^+, S_n^- and R_n defined with respect to $\bar{Q}_n$, Fig. 10.5b, were called by Feller (1951) the adjusted surplus aS_n^+, the adjusted deficit aS_n^-, and the adjusted range aR_n, respectively. In the treatment which follows the name deficit is reserved for a different definition, and the storage with respect to the mean of the record is referred to as the adjusted storage or range.

On the residual mass curve, Fig. 10.5a, the horizontal direction (zero slope) corresponds to the mean discharge rate, and the positive and negative slopes of the curve correspond to flow rates greater and smaller than the average flow rate, respectively. For a graphical solution of the storage relationship the scale is selected so that for a given volume of inflow I_j

$$\Delta S_j = (I_j - \bar{Q})k\,\Delta t \qquad 10.8$$

where k is the desired graphical scale factor. The graphical solution of $S_c = f(Q_d, E = 100\%)$ is shown in Fig. 10.6. For simplicity the demand $Q_d < \bar{Q}$ is assumed to be at a constant rate. The solution starts at zero surplus or deficit and proceeds as shown. The value of the storage given by the last tangent at the end of the design period (point A) is transferred to the beginning (point B) and constitutes the starting value for the second cycle of calculations. The conservation storage volume S_c needed for 100% dependability of demand Q_d is given by the largest

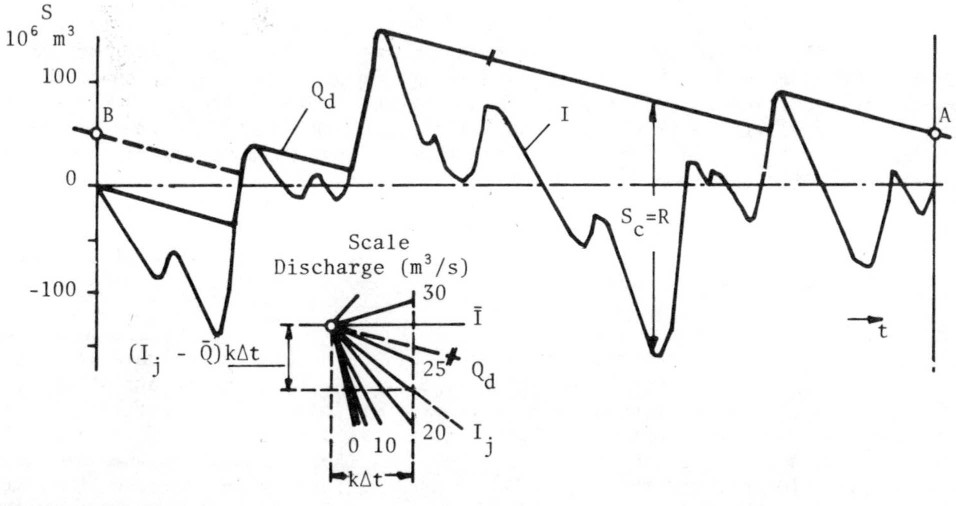

Fig. 10.6. Illustration of the use of the residual mass curve for finding the storage capacity for a given draft and 100% dependability (after Klemes, 1973).

vertical distance between the Q_d and I lines.

The solution of $E = f(S_c, Q_d)$ is shown in Fig. 10.7. The solution starts by drawing a vertically displaced residual mass curve I*. The vertical displacement equals

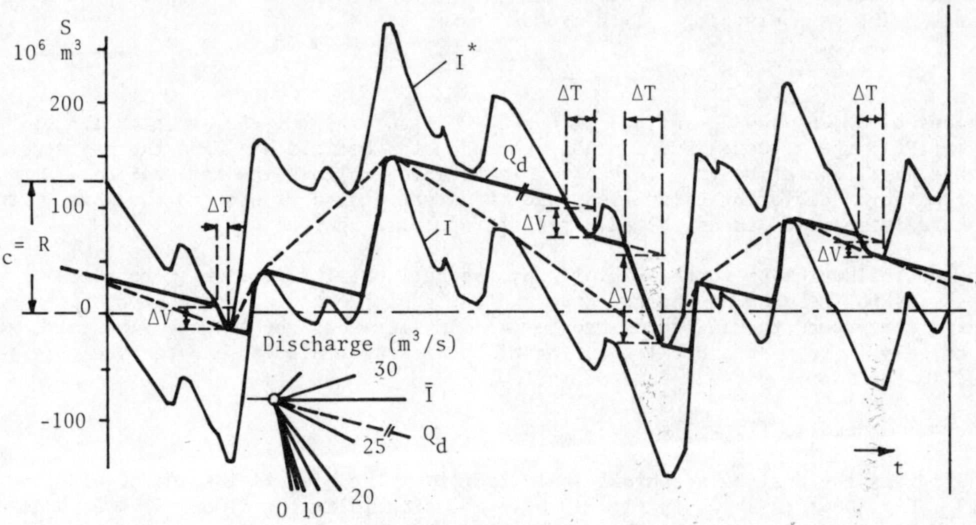

Fig. 10.7. Illustration of the use of the residual mass curve for finding the dependability, failure periods and resulting water deficits corresponding to given values of storage and draft, and for finding the draft for a given storage and a 100% dependability (after Klemes, 1973).

the given value of storage S_c. The demand is again, for simplicity, assumed to be at a constant rate $Q_d < \bar{Q}$. Lines parallel to Q_d are drawn in the space enclosed by the residual and shifted residual mass curves. The horizontal distance between the end of a Q_d line on the upper and the start of the next Q_d segment represents the length of the failure period ΔT, and the vertical distance gives the corresponding water deficit volume ΔV. Using the values of ΔT or ΔV the value of E_t or E_q can be found. The adjustment for the second cycle of calculations is analogous to that in Fig. 10.6. The solution of $Q_d = f(S_c, E = 100\%)$ is shown by the dashed line in Fig. 10.7. The desired value of Q_d is given by the slope of the most steeply falling segment of the dashed line.

The continuity statement written as $I - Q = \Delta Q$, where $\Delta Q = \Delta S_j/\Delta t$, forms the basis of the numerical solutions of storage problems. The storage calculations for a given draft $Q_d < \bar{Q}$ and 100% dependability, $S_c = f(Q_d, E = 100\%)$, are shown in Table 10.1. The solution commences with the reservoir full. The reservoir capacity is assumed to be so large that the solution is not limited by lack of water in storage. The draft used varies periodically with seasons, the time unit is the average month ($\Delta t = 2.63 \times 10^6$ s), and the required storage is given by the maximum absolute value of the sum $\Sigma\Delta Q$ multiplied by Δt. The outflow column in the table includes the draf the losses and the spillway discharge. The solution of $E = f(S_c, Q_d)$ for E_f, E_t or E_q for given values of S_c and Q_d is shown in Table 10.2. For these calculations the absolute value of $\Sigma\Delta Q$ cannot exceed the total storage value, i.e., $S_c/\Delta t$. Whenever this condition is reached a failure period occurs, for example, in February 193 the draft cannot exceed 9.6 when $|\Sigma\Delta Q| < 38.0$ m³/s.

The problems of $S_c = f(Q_d, E < 100\%)$ and $Q_d = f(S_c, E)$ can be solved by successive approximations of $E = f(S_c, Q_d)$, where S_c or Q_d are varied until the desired value of E is reached.

TABLE 10.1 Example of Computation of Storage S_c Given Draft Q_d (Column 4) and Dependability E = 100%

Year	Month	Inflow I m^3s^{-1}	Draft Q_d m^3s^{-1}	Outflow Q m^3s^{-1}	Instant change in storage $\Delta Q = I - Q$ m^3s^{-1}	Instant storage depletion $\Sigma\Delta Q$ m^3s^{-1}	Remark
1	2	3	4	5	6	7	8
						0.0	Full reservoir
	I	12.2	12.0	12.2	0.0	0.0	First cycle
	II	7.9	12.0	12.0	- 4.1	- 4.1	
	III	41.8	12.0	37.7	+ 4.1	0.0	
	IV	63.5	17.0	63.5	0.0	0.0	
	V	35.1	23.0	35.1	0.0	0.0	
1931	VI	20.3	28.0	28.0	- 7.7	- 7.7	
	VII	15.0	30.0	30.0	-15.0	-22.7	
	VIII	16.7	25.0	25.0	- 8.3	-31.0	
	IX	8.4	15.0	15.0	- 6.6	-37.6	
	X	10.4	13.0	13.0	- 2.6	-40.2	
	XI	28.6	12.0	12.0	+16.6	-23.6	
	XII	20.5	12.0	12.0	+ 8.5	-15.1	
	I	13.7	12.0	12.0	+ 1.7	-13.4	
1932	.	.	.	.	.	.	
.	.	.	.	.	.	.	
.	.	.	.	.	.	.	
1970	.	.	.	.	.	.	Final
	XII	12.8	12.0	12.0	+ 0.8	-37.8	depletion
	I	12.2	12.0	12.0	+ 0.2	-37.6	Second
	II	7.9	12.0	12.0	- 4.1	-41.7	cycle
	III	41.8	12.0	12.0	+29.8	-12.7	
1931	IV	63.5	17.0	50.8	+12.7	0.0	End of
.	.	.	.	.	.	.	computa-
.	.	.	.	.	.	.	tion*

Provided there is no number in column 7 with an absolute value greater than 41.7 then $S_c = 41.7\ m^3s^{-1} \times 2.63 \times 10^6$ s (average month) = $109.7 \times 10^6\ m^3$.

*All further computations are the same as in the first cycle (after Klemes, 1973).

The methods of storage analysis discussed above are all based on *range analysis*. When the demand Q_d equals the mean inflow the regulation is called *full regulation*. Such regulation, combined with the condition of 100% dependability, would be capable of storing all the surplus water within a given period of operation and could supply any deficit. The reservoir would neither spill nor experience deficits during the operation period and its storage capacity would be the same as the maximum value of the range R as shown in Fig. 10.5a or the required storage S_c in Fig. 10.5b, where Q_d is given by the horizontal direction. In practice reservoirs are seldom

TABLE 10.2 Example of Computation of Dependability and Storage*

Year	Month	Inflow I m^3s^{-1}	Draft Q_d m^3s^{-1}	Outflow Q m^3s^{-1}	Instant change in storage ΔQ= T-Q m^3s^{-1}	Instant storage depletion $\Sigma\Delta Q$ m^3s^{-1}	Failures in water supply: Length of failure periods ΔT month	Failures in water supply: Deficit $\Delta V = Q_d - Q$ m^3s^{-1}	Failures in water supply: Failure years denoted by F	Remark
1	2	3	4	5	6	7	8	9	10	11
.	.	.	.	.	.	.				First cycle
.	.	.	.	.	.	.				
1970										
.	.	.	.	.	.	.				
.	.	.	.	.	.	.				
	XII	12.8	12.0	12.0	+ 0.8	-36.5				
	I	12.2	12.0	12.0	+ 0.2	-36.3				
	II	7.9	12.0	9.6	- 1.7	-38.0	1	2.4		Second** cycle
	III	41.8	12.0	12.0	+29.8	- 8.2				
	IV	63.5	17.0	55.3	+ 8.2	0.0				
	V	35.1	23.0	35.1	0.0	0.0				
1931	VI	20.3	28.0	28.0	- 7.7	- 7.7			F	
	VII	15.0	30.0	30.0	-15.0	-22.7				
	VIII	16.7	25.0	25.0	- 8.3	-31.0				
	IX	8.4	15.0	15.0	- 6.6	-36.6				
	X	10.4	13.0	11.8	- 1.4	-38.0	1	1.2		
	XI	28.6	12.0	12.0	+16.6	-21.4				
	XII	20.5	12.0	12.0	+ 8.5	-12.9				
	I	13.7	12.0	12.0	+ 1.7	-11.2				
	.	.	.	.	.	.				
1932	.	.	.	.	.	.				
.							.	.	.	
.							.	.	.	
							$\Sigma\Delta T$	$\Sigma\Delta W$	$\Sigma m = \Sigma F$	

*Storage S_c = 100.0 x 10^6 m^3. If the average month is employed then max$|\Sigma\Delta Q|$ = 100/2.63 = 38.0 m^3s^{-1}, and Δt = 2.63 x 10^6 s.

**Outflow cannot exceed 9.6, otherwise $\Sigma\Delta Q > 38$, (after Klemes, 1973).

designed for full regulation because, even if the required dependability were less than 100%, the necessary storage capacity would be extremely large and the reservoir would not be economically tenable.

Figure 10.8a shows a residual mass curve with respect to the mean flow and the adjusted mean. Superimposed are sloping lines corresponding to mean discharges which are smaller than Q, for example, 0.1 σ and 0.2 σ below $\bar{Q}$, where σ is the standard deviation of the distribution of Q. It is seen that R_n is greater than aR_n and reckoned from the adjusted mean the maximum deficit is from point 1 to 2. Reckoned from the 0.1 σ 0.2 σ below $\bar{Q}$ lines the maximum deficits are from 3-4 and 5-4, respectively. When the mean regulated discharge is smaller than the average flow rate, the net average input is no longer zero but positive and the regulation is partial only.

The usual design case is one where the demand Q_d is lower than the mean flow, i.e., the Q_d line is sloping downwards as in Fig. 10.6. In these cases spillages are inevitable since the reservoir cannot store all the surplus water. In Fig. 10.6 the

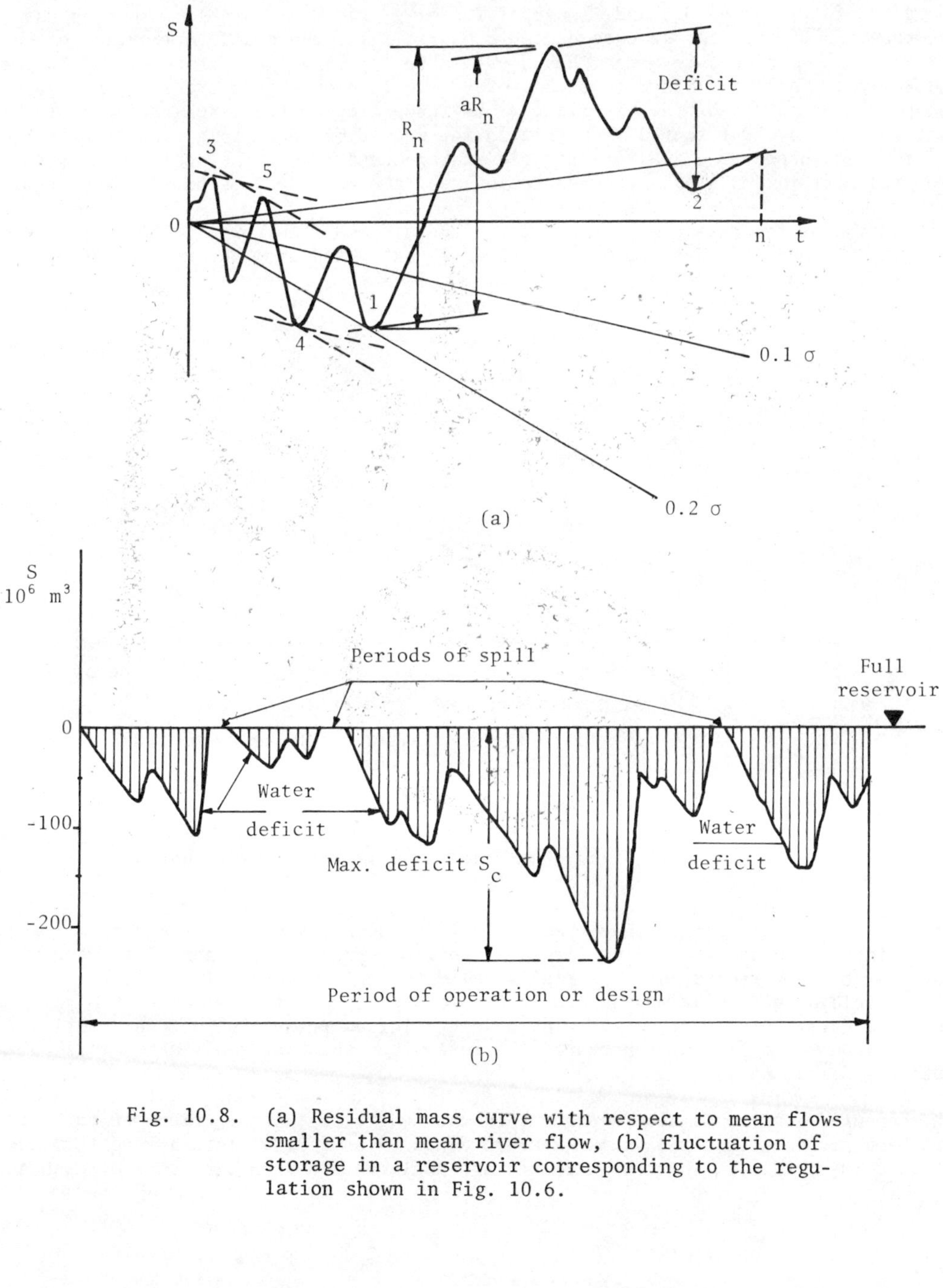

Fig. 10.8. (a) Residual mass curve with respect to mean flows smaller than mean river flow, (b) fluctuation of storage in a reservoir corresponding to the regulation shown in Fig. 10.6.

periods of spills are those between the points where the demand line Q_d meets the residual mass curve and the points from which a new demand line is drawn. This type of analysis is sometimes called *deficit analysis* since the storage capacity required (S_c in Fig. 10.6 or Table 10.1) is found as the largest deficit that would be encountered in a reservoir of unlimited depth which satisfies the demand Q_d. The deficits or reservoir drawdowns are indicated by the differences between the demand line Q_d and the residual mass curve in Fig. 10.6, and are usually plotted from a horizontal line representing a full reservoir as shown in Fig. 10.8b. Such a diagram shows the fluctuations, or behaviour, of storage in the reservoir throughout the period of operation for a demand Q_d, and is useful for many practical purposes, such as the construction of the storage exceedance (or duration) curve and determination of water levels, water surface areas, heads available for power generation or depths for navigation. Analyses based on such storage fluctuation charts are also known as *behavioural analyses* of the reservoir. By inspection of the discharge scale in Fig. 10.6 it is clear that a reduction in the demand Q_d leads to a reduction in the deficits and hence to a smaller required storage capacity. The smaller the regulation the more incomplete is the regulation, and the more periods of spillage occur.

A variation of the mass curve technique is the minimum flow curve technique. For this the minimum flow over selected periods is extracted from the records and plotted as volume against the period, Fig. 10.9.

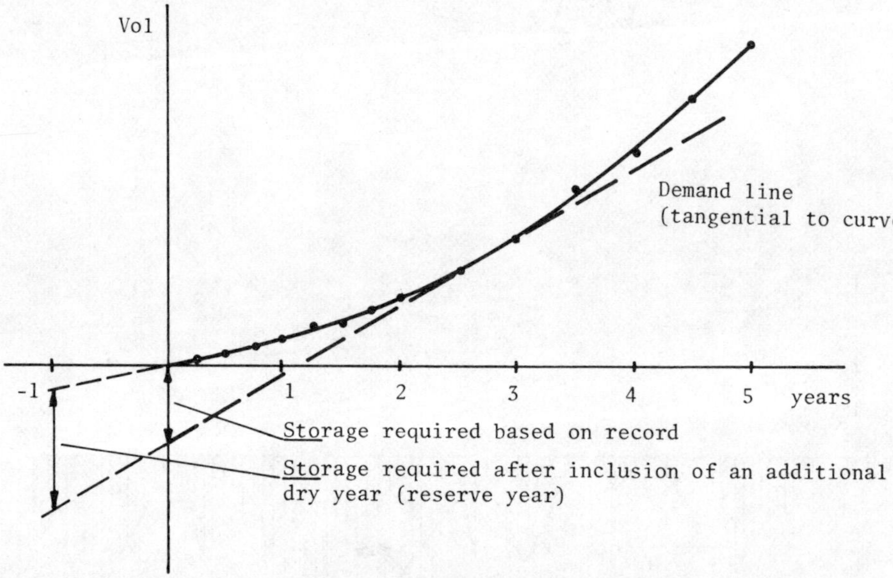

Fig. 10.9. Illustration of the minimum flow curve technique.

It is important to realize that these solutions (both graphical and numerical) are deterministic in nature. A particular solution represents the period of operation for the observed record and is strictly valid for this record only. It may have little relevance to what may happen in the future because the inflow to the reservoir is not a deterministic process. The observed inflow record is just one realization of the random process which produces the inflow and the next realization may look entirely different.

The computed storage is also dependent on the length of the operation period. It has been proposed that the active storage could be subdivided into a long term and short term storage. The long term storage could be found on the basis of mean annual flows and the short term storage from the seasonal or within year flow fluctuations. The basic idea is illustrated in Fig. 10.10.

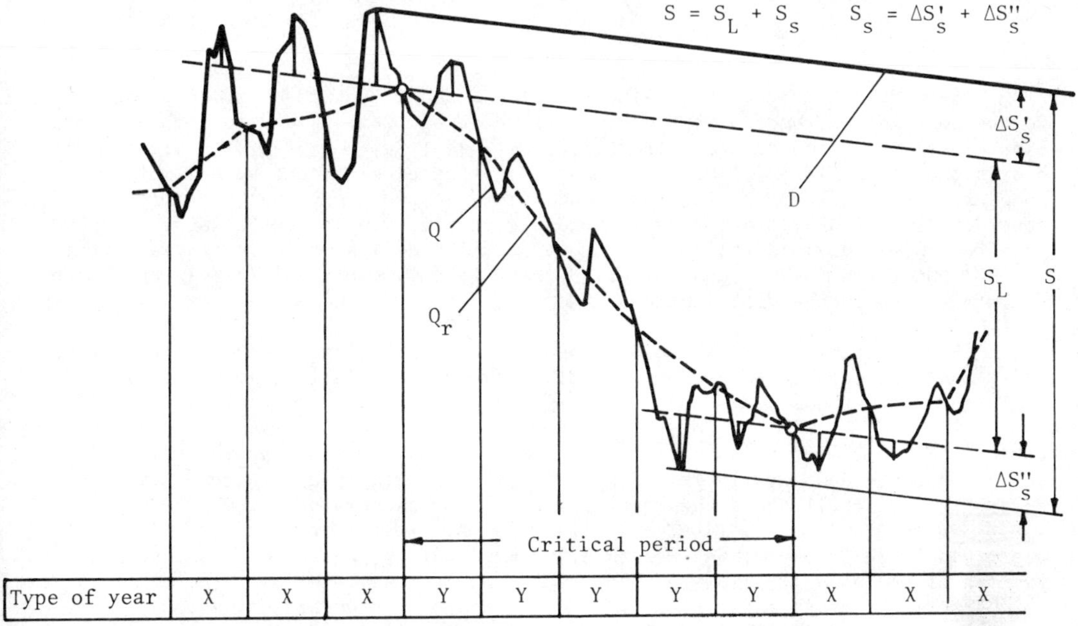

Fig. 10.10. Definition sketch showing long term (S_L) and seasonal (S_s) storage. (X, Y - years with above- and below-average mean flow, respectively; Q_r - residual mass curve for mean annual flows; Q - residual mass curve for mean monthly flows), (after Klemes, 1973).

Experimental methods: The main problem in calculating the required storage volume of a reservoir is the prediction of the time sequence of future inflows. This prediction could be made by a deterministic or a probabilistic model. The deterministic model is based on the physical laws and relationships which govern the rainfall runoff process. In general, the interactions of climatic and physical factors are too complex for an exact analytical description of the inflow. Therefore statistical or probabilistic descriptions have to be used. In the probabilistic model the actual physical process plays no part. The future flow sequence is described in terms of statistical parameters only. The values of these parameters have to be obtained from observed records. A further problem is that historical records are usually too short and seldom contain the extremes of floods and droughts. It was also pointed out earlier that a historic sequence is just one realization of the random process and that a future sequence may be very different. The sequence of inflows, however, has a controlling influence on the amount of storage required. In order to account for the effect of the flow sequence use has been made of the methods of data generation.

Data generation methods are discussed in Chapter 11. In brief, use is made of the statistical properties of the observed record, i.e., its mean, variance, skewness, persistence characteristics, etc. Then with the aid of a data generation model, which preserves these statistical properties, a long sequence of synthetic record is produced which can be subdivided into m non-overlapping sequences n years each. All these sequences have the same statistical properties but may differ appreciably in detail. Thus the generation methods enable the designer to use all the information in the record. It does not give additional information and if the record is short and of doubtful quality, then the generated records will be of doubtful quality also.

From the m generated records m values of S_n^+, S_n^- and R_n can be determined, according to the empirical methods discussed in the preceding section, and to each a probability density function can be fitted, i.e., $f(S_n^+)$, $f(S_n^-)$ and $f(R_n)$. For each distribution the statistical parameters (mean and variance, etc.) can be estimated. Consequently, the estimate of the required storage volume can be based on probability considerations. For statistically reliable results m needs to be large. A 50-year record would yield reliable information for a design period of 1 year (where m = 50) but for a 50-year design period there is only one value (m = 1) and hence no information on the distributions.

The data generation model can be combined with an analytical model of the reservoir and a function for the water demand. The computer simulations are basically experimental design methods.

Gould (1960) carried out numerical experiments. He assumed that the streamflow is an independent random variable and used the Monte Carlo type data generation procedure. Figure 10.11 shows the reservoir behaviour as obtained from 10 000 years of synthetic record. The inflows were normally and independently distributed and the draft was taken to be 0.1 S_Q less than the mean inflow, where S_Q is the standard deviation of the annual inflows. The results are for semi-infinite (bottomless) reservoirs with storage capacities of 2, 4, 6, 8 and 10 standard deviations of the annual inflow volumes. The probability of failure curve joins the end points of the

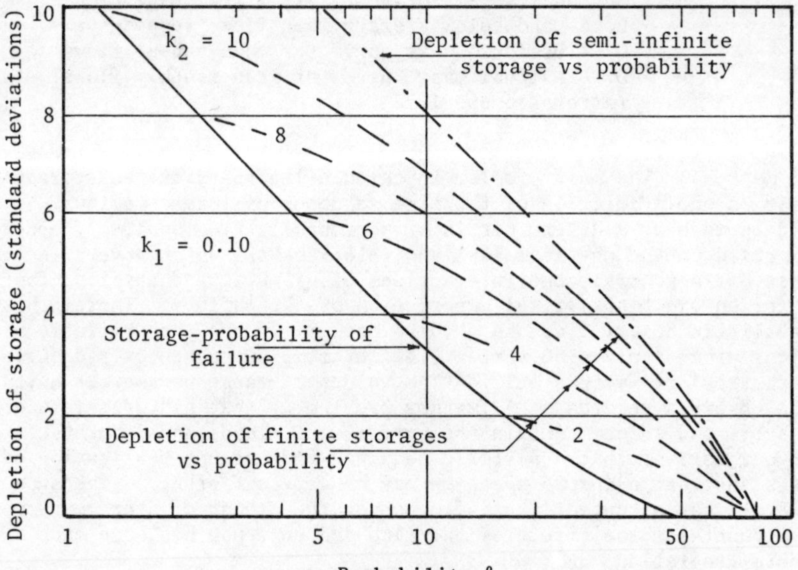

Fig. 10.11. Probability of failure and of depletion of storage from Monte Carlo simulation (Gould, 1960).

individual storage capacity curves, i.e., the points when the particular reservoir is empty and failure to supply the demand occurs. The curves to the right of it are called depletion-probability curves. These represent the relationship between the depletion of a reservoir and the proportion of years in which there is likely to be a greater depletion. Gould also proposed a formula for storage calculations. This is a relationship fitted to 240 sets of Monte Carlo simulations using various combinations of the draft D, storage ratios and skewness of the inflow distributions. The formula is

$$(k_1 + 0.15)(k_2 + b) = c \qquad 10.9$$

where k_1 is the draft ratio = $(\bar{Q} - D)/S$, k_2 is the storage ratio = V/S, V is the volume of the annual or carryover storage, $\bar{Q}$ is the mean (population) annual inflow, S is the standard deviation of the annual net inflow, and b and c are parameters given in Fig. 10.12 as a function of the skewness of the inflow distribution and the probability of failure. The length of the critical period in years is expressed as $L = (k_2 + b)^2/c = c/(k_1 + 0.15)^2$. The formulae are said to be suitable for utilization greater than about 80% of the coefficient of variation of annual streamflow volumes. Gould also discussed the problem of confidence limits and sampling errors.

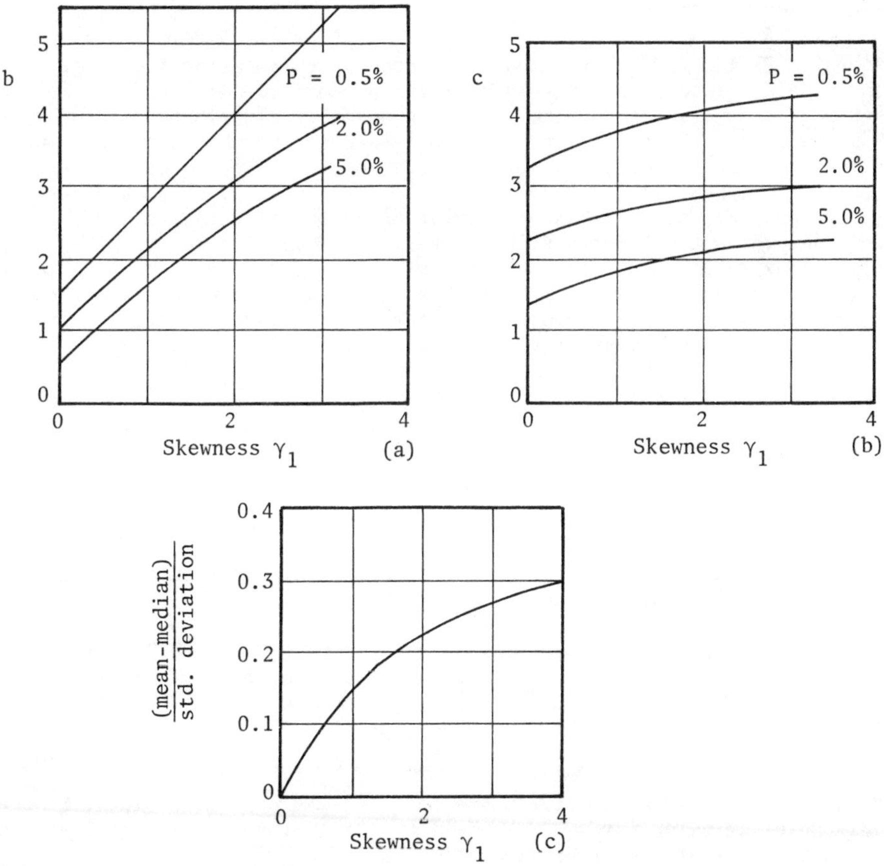

Fig. 10.12. (a) and (b) Values of b and c in eqn 10.9. (c) Graph for estimation of skewness from the mean, median and standard deviation (Gould, 1960).

Analytical methods: Exact analytical solutions of the storage problem exist only for a limited number of idealised models. A brief survey of these methods is given by Yevjevich (1972). The bulk of the analytical methods are of statistical nature.

Hurst (1951) showed that the asymptotic mean value of the adjusted range of independent identically distributed random variables with zero mean and unit variance is

$$E[aR_n] = \sqrt{\frac{n\pi}{2}} \simeq 1.253\sqrt{n} \qquad 10.10$$

He compared this result with a range of data from a wide variety of natural processes. The data included river discharges, rainfalls, varves, temperature and pressure records, tree rings and sunspot numbers. Additional data were presented by Hurst et al. (1965). Hurst found that the range of all these data could be described by

$$R = \sigma\left(\frac{n}{2}\right)^K \qquad 10.11$$

where K = 0.726 with a standard deviation of 0.082 and σ is the standard deviation of the data for which the range is calculated. The range of K values for all the data was 0.46 to 0.95. This simple expression has become known as the "Hurst phenomenon" and has been subject to a number of analytical studies which are reviewed by Klemes (1974).

Subsequently Feller (1951) attacked the storage problem using the theory of Brownian motion. He developed relationships for the asymptotic values of the mean of the range and of the variance for the case of a normally distributed independent random input variable X. The Gaussian variable X is described by the mean μ and standard deviation σ. The standardised normal variable $x = (X - \mu)/\sigma$ will have the maximum departures, S_n^+ and S_n^-, and a range R_n. The magnitudes of these are given by the respective products of the standard deviation σ and the values of S_n^+, S_n^- and R_n. Feller's development was for continuous variable X_t. The sum S_i of a discrete series is approximately equal to the integral value of the continuous variable $(I_j - Q_j)$. The range R_n is approximately equal to the range of the integral of the continuous variable for the same length of record. The expected asymptotic range, its variance and coefficient of variance are

$$E[R_n] = 2\sqrt{2n/\pi} \simeq 1.6\sqrt{n} \qquad 10.12$$

$$\text{Var } R_n = 4n(\ln 2 - 2/\pi) \simeq 0.226\,n \qquad 10.13$$

and

$$C_v(R_n) = \sqrt{\text{Var } R_n}/E[R_n] = \sqrt{\frac{1}{2}\pi \ln 2 - 1} \simeq 0.297$$

respectively. The expected range of a discrete variable should be slightly smaller than that of a continuous variable.

Since for a normal distribution $E[S_n^+] = -E[S_n^-]$ and $E[R_n] = 2E[S_n^+]$, the expected value of $E[S_n^+] = \frac{1}{2}E[R_n] \simeq 0.8\sqrt{n}$. The asymptotic variance of S_n^+ is

$$\text{Var } S_n^+ = \frac{\text{Var } R_n}{2(1 - \rho_n)} = \frac{2n(\ln 2 - 2/\pi)}{1 - \rho_n} \simeq \frac{0.113\,n}{1 - \rho_n} \qquad 10.14$$

where ρ_n is the correlation between the positive and negative values, S_n^+ and S_n^-. There is no exact value for ρ_n. Yevjevich (1972) proposed an empirical expression for ρ_n, obtained from numerical simulation, as

$$\rho = 1 - \frac{0.30\sqrt{n}}{\sqrt{n} - 0.37} \qquad 10.15$$

with this value of ρ_n eqn 10.14 yields

$$\text{Var } S_n^+ = 0.377\ n - 0.139\ \sqrt{n} \qquad 10.16$$

The values obtained using the above equation depart significantly from the exact values for $n < 10$.

The corresponding expressions for the asymptotic mean, variance and coefficient of variation of the adjusted range, and the asymptotic mean of the adjusted surplus (Hurst, 1951, 1956a & b, Feller, 1951) are

$$E[aR_n] = \sqrt{\pi n/2} \simeq 1.25\ \sqrt{n} \qquad 10.17$$

$$\text{Var } [aR_n] = (\pi^2/6 - \pi/2)n \simeq 0.074\ n \qquad 10.18$$

$$C_v(aR_n) = \sqrt{\pi n/3 - 1} \simeq 0.217 \qquad 10.19$$

$$E[aS_n^+] = \sqrt{\pi n/8} \simeq 0.625\ \sqrt{n} \qquad 10.20$$

The exact expected value of R_n for a given value of n (the above values apply when $n \to \infty$) was given by Anis and Lloyd (1953) for large n ($n < \infty$) as

$$E[R_n] = \sqrt{(2/\pi)} \sum_{i=1}^{n} i^{-\frac{1}{2}} \qquad 10.21$$

where

$$\sum_{i=1}^{n} i^{-\frac{1}{2}} = \int_{\frac{1}{2}}^{n+\frac{1}{2}} x^{-\frac{1}{2}}\, dx = 2[(n + \tfrac{1}{2})^{\frac{1}{2}} - (\tfrac{1}{2})^{\frac{1}{2}}] \simeq 2\sqrt{n}$$

A good approximation is given by

$$E[aR_n] \simeq \sqrt{8n/\pi} - 1 \simeq 1.6\ \sqrt{n} - 1 \qquad 10.22$$

The corresponding approximation for the expected value of the adjusted range is

$$E[aR_n] \simeq \sqrt{\pi n/2} - \sqrt{\pi/2} = 1.25\ (\sqrt{n} - 1) \qquad 10.23$$

For a normal variable $E[R_n] = 2E[S_n^+]$ and is half the value given by eqn 10.21. i.e., $\sqrt{2n/\pi} \simeq 0.8\ \sqrt{n}$. The variance

$$\text{Var } R_n = \frac{0.60 \text{ Var } S_n^+}{1 - 0.37/\sqrt{n}} \qquad 10.24$$

where the empirical value of ρ_n from eqn 10.15 has been used.

Generally, the greater the variability of the mean the more expensive is the storage. About three times more reservoir storage is required to maintain an outflow at 80% of the mean flow when the coefficient of variance $C_v = S_n(Q)/\bar{Q} = 0.5$ as compared with 0.3.

Yevjevich (1972) showed that the skewness of the distribution has little effect on these values but has a marked effect on the confidence limits of R_n when the coefficient of skewness of the random variable X is $C_s(X) > 1.0$.

Hurst (1951) was also the first to attempt the deficit analysis. Using his long sequences of data he attempted to find a relationship between the adjusted maximum accumulated deficit and the adjusted range for the condition that the demand (constant controlled outflow) is a percentage of the sample mean. He plotted observed values of the adjusted range versus adjusted maximum deficit, and found by simple regression analysis that either of the following relationships fitted the data equally well

$$\log(aF_n/aR_n) = -0.11 - 0.88(\bar{Q} - Q_d)/\sigma \qquad 10.25$$

and

$$(aF_n/aR_n) = 0.91 - 0.89[(\bar{Q} - Q_d)/\sigma]^{\frac{1}{2}} \qquad 10.26$$

where $\bar{Q}$ is the sample mean flow rate, Q_d is the constant demand which is smaller than $\bar{Q}$, σ is the standard deviation of the flow record Q_i, aF_n is the maximum adjusted deficit, and aR_n is the adjusted range. When $\bar{Q} - Q_d = 0$, $aF_n/aR_n = 1$ and neither equation fits. Similarly, when $Q_d \to 0$, $aF_n/aR_n \to 0$ because no storage is required. Generally, for $(\bar{Q} - Q_d)/\sigma > 1$ the relationship was found to be ill-defined. Note, the large reduction in the required storage when the demand is only a little below the mean flow rate. A reduction of 0.1 σ and 0.2 σ reduces the ratio (aF_n/aR_n) to 0.63 and 0.52, respectively.

Gomide (1975) expressed the level of regulation as

$$\alpha = \{1 - \frac{E[X_t]}{E[Y_t]}\} 100\% \qquad 10.27$$

where $E[X_t]$ is the mean net input and $E[Y_t]$ is the mean natural discharge. For full regulation $E[X_t] = 0$ and $\alpha = 100\%$. For the particular case of constant regulated output, X_t and Y_t have the same variance σ^2 and their mean values may be written as

$$E[X_t] = \mu\sigma \qquad 10.28$$

$$E[Y_t] = c\sigma \qquad 10.29$$

where $c = 1/C_v$, C_v is the coefficient of variation and μ is a number between zero and c. For this case eqn 10.27 becomes

$$\alpha = (1 - \mu/c)\ 100\% = (1 - \mu C_v)\ 100\% \qquad 10.30$$

An illustration of Gomide's results is shown in Fig. 10.13, which also enables evaluation of eqn 10.30.

Hurst et al. (1965) also extended the range analysis using eqn 10.11 to the case where two streams with ranges of R_1 and R_2 join. If the discharges of the streams are both normally distributed with standard deviations σ_1 and σ_2, then it can be shown that the standard deviation of their combined flow is

$$\sigma^2 = \sigma_1^2 + \sigma_2^2 + 2r\sigma_1\sigma_2 \qquad 10.31$$

where r is the correlation coefficient between the discharges of streams 1 and 2. For this case, with a given value of n, the relationship $R_i = \sigma_i(n/2)^{K_i}$, where $K_i = 0.5$, yields

$$R^2 = R_1^2 + R_2^2 + 2rR_1R_2 \qquad 10.32$$

Equation 10.32 shows that for perfect correlation (r = 1) the range is $R = R_1 + R_2$ and for zero correlation $R^2 = R_1^2 + R_2^2$, i.e., the maximum value of R is associated with perfect correlation.

The majority of the analytical solutions of the range and deficit problem are by the Markov chain technique, for example, Moran (1959), Lloyd (1963), Prabhu (1964), Yevjevich (1965), Klemes (1970), Salas-La Cruz (1972) and Gomide (1975). These range from models for independent normally distributed input variables to models where the inputs are dependent on preceding inputs and the storage is subject to prescribed

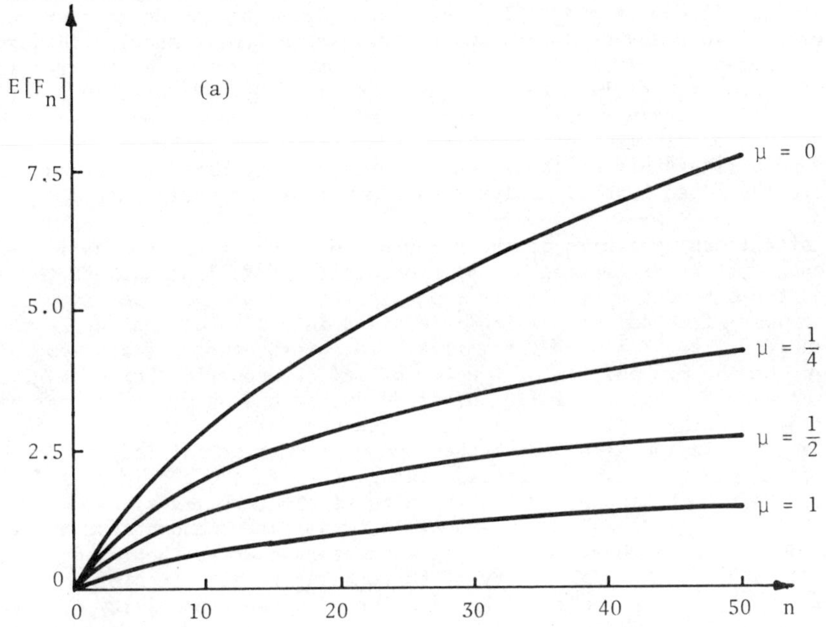

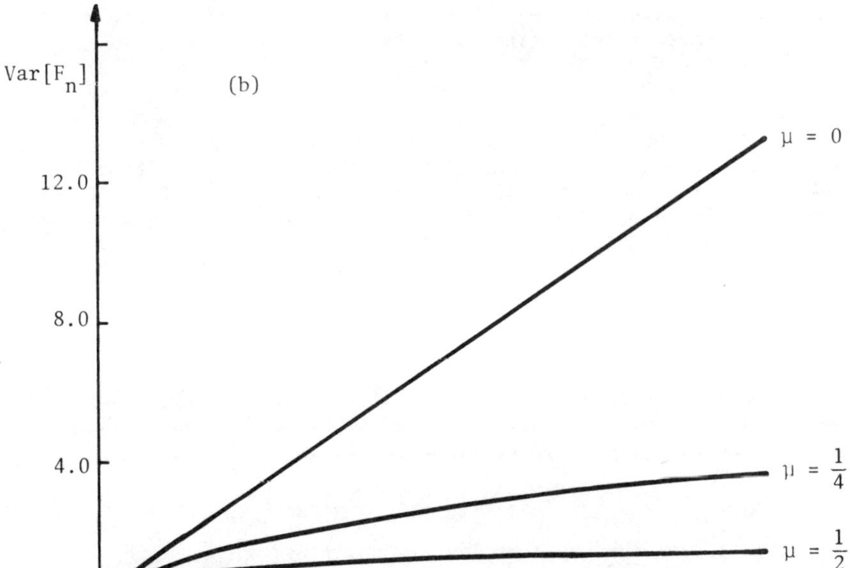

Fig. 10.13. (a) Expected value of deficit F_n for independent normal net inputs and (b) the variance of F_n (Gomide, 1975).

conditions. The range analysis or the infinite reservoir theory assumes a reservoir which is capable of storing all surplus and supplying any deficit of water. This type of reservoir analysis corresponds to a Markov chain model with absorbing boundaries. Moran's analysis is a finite reservoir theory since it works with a given value of storage, similar to the scheme in Fig. 10.7 or Table 10.2. It corresponds to a Markov chain model with reflecting boundaries. The deficit analysis is a Markov chain model with one absorbing and one reflecting boundary. Markovian models command a literature on their own. The ideas are briefly introduced in Chapter 12. The Moran type of analysis is a powerful and useful method.

One of the significant features of the Moran method, and of the probabilistic methods in general, is the estimation of the probability of failing to meet the demand. The logic of the method can be illustrated with the aid of a simple example. Consider a reservoir of capacity S_c which is divided into five convenient increments or storage units, i.e., K = 5. The inflow into the reservoir is assumed to be random. Let p_i be the probability of i units of inflow occurring during a given period. Moran assumed that the inflow occurred during one half of the year and the drawoff during the other half. Thus, if at the beginning of the inflow the storage is 2 units and the inflow is 4 units, then one unit must be spilled (or discharged) since the sum is greater than the storage capacity. This one unit does not satisfy a demand. After the inflow period, D units (including losses) are drawn off as demand. If D is greater than the volume in storage the demand cannot be met. Let P_i be the probability of there being i units in storage at the beginning of the inflow period, and P_i' be the probability of there being i units in storage at the end of the demand period. In general, when dealing with the probability of a series of events, the probability of an event A being followed by an *independent* event B is given by the product of their separate probabilities. The probability of occurrence of any one of a number of alternative events is obtained by addition of the probabilities of the respective events. Thus, the probability P_2' of having 2 units in storage after D = 2 units have been drawn off is the sum of the probabilities of having 4 units in storage at the beginning of the cycle

$$P_2' = P_0(p_4) + P_1(p_3) + P_2(p_2) + P_3(p_1) \qquad 10.33$$

The first term is the probability of the reservoir being empty and there being 4 units of inflow, the second is the probability of there being 1 unit in the reservoir and 3 units of inflow, etc. A similar formula applies for P_1'. For P_3', however, the probability involves drawoff from a full reservoir and this may involve exact filling or filling and some spilling, i.e.,

$$P_3' = P_0(p_5) + P_1(p_4 + p_5) + P_2(p_3 + p_4 + p_5) + P_3(p_2 + p_3 + p_4 + p_5) \qquad 10.34$$

where p_5 is the probability of 4 or more units of inflow. Similar complications arise for the probability of the reservoir being empty at the end of the demand period since the reservoir may be emptied without satisfying the demand

$$P_0' = P_0(p_0 + p_1 + p_2) + P_1(p_0 + p_1) + P_2(p_0) \qquad 10.35$$

Thus, this problem is described by four linear equations.

In general, the problem is described by a set of (K - D + 1) equations, which can be summarized in matrix form as shown in eqn 10.36. The coefficients on the left-hand side are the probabilities of inflow and each column represents the probability distribution of stored water at the end of the period (final state) given the amount of water at the beginning of the period (initial state). Hence, the sum of each column is unity. The set of (K - D + 1) linear equations in as many unknowns has

Initial States: 0, 1, 2, ...

Final States: 0, 1, 2, ...

$$\begin{bmatrix} \sum_{i=0}^{D} p_i & \sum_{i=0}^{D-1} p_i & \sum_{i=0}^{D-2} p_i & \cdots & p_0 + p_1 & p_0 \\ p_{D+1} & p_D & p_{D-1} & \cdots & p_2 & p_1 \\ p_{D+2} & p_{D+1} & p_D & \cdots & p_3 & p_2 \\ \vdots & \vdots & \vdots & & \vdots & \vdots \\ p_{K-1} & p_{K-2} & p_{K-3} & \cdots & \cdot & p_{D-1} \\ \sum_{i=K}^{\infty} p_i & \sum_{i=K-1}^{\infty} p_i & \sum_{i=K-2}^{\infty} p_i & & & \sum_{i=D}^{\infty} p_i \end{bmatrix} \begin{bmatrix} P_0 \\ P_1 \\ P_2 \\ \vdots \\ \\ P_{K-D} \end{bmatrix} = \begin{bmatrix} P'_0 \\ P'_1 \\ P'_2 \\ \vdots \\ \\ P'_{K-D} \end{bmatrix} \tag{10.36}$$

(K - D) independent equations. Moran omitted the last equation and replaced it by the requirement that the sum of the probabilities P_i at any time must be equal to unity, i.e.,

$$\Sigma P_i = 1 \tag{10.37}$$

or in the above example

$$P_0 + P_1 + P_2 + P_3 = 1$$

If the probabilities of inflow are $p_o = 0.1$, $p_1 = p_2 = 0.2$, $p_3 = 0.3$, $p_4 = p_5 = 0.1$, $(\Sigma p_i = 1)$, then the set of equations is as follows

$$P'_0 = 0.5P_0 + 0.3P_1 + 0.1P_2$$

$$P'_1 = 0.3P_0 + 0.2P_1 + 0.2P_2 + 0.1P_3$$

$$P'_2 = 0.1P_0 + 0.3P_1 + 0.2P_2 + 0.2P_3$$

$$P'_3 = 0.1P_0 + 0.2P_1 + 0.5P_2 + 0.7P_3$$

or

$$\begin{matrix} 0.5 & 0.3 & 0.1 & 0 \\ 0.3 & 0.2 & 0.2 & 0.1 \\ 0.1 & 0.3 & 0.2 & 0.2 \\ 0.1 & 0.2 & 0.5 & 0.7 \end{matrix} \quad \begin{matrix} P_0 \\ P_1 \\ P_2 \\ P_3 \end{matrix} \quad \begin{matrix} P'_0 \\ P'_1 \\ P'_2 \\ P'_3 \end{matrix}$$

If the reservoir is empty at t = 0 then $P_0 = 1$, $P_1 = P_2 = P_3 = 0$ and $P'_0 = 0.5$, $P'_1 = 0.3$, $P'_2 = 0.1$, $P'_3 = 0.1$ or there is a 50% chance of the reservoir remaining empty at the end of the period and a 10% chance of the reservoir having 3 units of storage. These P'_i values now become the P_i values for next step which yields the new values of $P'_0 = 0.35$, $P'_1 = 0.24$, $P'_2 = 0.18$, $P'_3 = 0.23$, $(\Sigma P'_i = 1)$. The computations can be continued and the results will soon (after about 10 steps) become independent of the initial storage conditions of step one, i.e., $P'_i \rightarrow P_i$. In the limit the distribution becomes stationary and the P_i values become the steady state probabilities. For example, by equating P'_i to P_i the steady state equations are

$$0 = -0.5P_o + 0.3P_1 + 0.1P_2$$
$$0 = 0.3P_o - 0.8P_1 + 0.2P_2 + 0.1P_3$$
$$0 = 0.1P_o + 0.3P_1 - 0.8P_2 + 0.2P_3$$
$$0 = 0.1P_o + 0.2P_1 + 0.5P_2 - 0.3P_3$$

and

$$P'_o + P'_1 + P'_2 + P'_3 = 1$$

which yield the steady state probabilities as

$$P_o = 0.139, \quad P_1 = 0.165, \quad P_2 = 0.202, \quad P_3 = 0.494$$

The stationary probabilities may also be obtained by finding the transition probability matrix $P = [p_{ij}]$ and raising it to the power of n to give P^n, which is the n-step transition matrix. By solving for steady state probabilities directly, rather than step by step, the information on how the reservoir approaches the steady state condition is lost. For the example used, starting with $P_o = 1$ the probability $P'_o = 0.35$ after one step and 0.139 is the steady state value.

A failure occurs if the reservoir has one unit of storage and no inflow or if the reservoir is empty and has zero or one unit of inflow, i.e.,

$$P_{failure} = P_o(p_1 + p_2) + P_1(p_o)$$
$$= 0.139(0.2 + 0.1) + 0.165(0.1) = 0.058 \qquad 10.38$$

There are a number of methods for estimating the transition matrix. The calculations are usually carried out in terms of the net addition to the storage X, which is the difference between the inflow and drawoff, i.e., $X = V_Q - D$. The observed records give the distribution of annual (or seasonal) runoff, Fig. 10.14a, which when superimposed on the drawoff distribution function yields the net flow distribution function F(X). Figure 10.14b illustrates F(X) for the case of a uniformly distributed drawoff. Figure 10.14c illustrates the association of F(X) with the storage states.

Although the demand in Fig. 10.14 has been shown constant, usually both the inflow and the demand vary seasonally. For example, the average monthly flow and monthly demand may look as shown in Fig. 10.15. The reservoir would have to store winter water for summer use and water from wet years to supplement the flows of dry years.

Alternatively, one could fit a probability distribution to the annual (or seasonal) net inflows. The reservoir capacity is again divided into a number of equal volume increments and the probabilities of increase in storage are estimated from the distribution function of the inflows, f(X) dX, Fig. 10.16. Assume that the values from Fig. 10.16 (the area elements under the curve) are as shown in Table 10.3. Thus, starting with an empty reservoir the probability of it remaining empty at the end of the period is 0.003 + 0.007 + 0.01 = 0.02 and that of it being in state 5 is 0.16. Likewise, starting at state 1, etc. The procedure for building the transition matrix is analogous to that of the graphical construction (Fig. 10.14). The resulting matrix is shown in Table 10.4.

In cases of distinct seasonal variation probability distributions are fitted to observed data on the net wet and net dry season flows. For each season the probability transition matrix is constructed as discussed above. The matrices for the two seasons are then multiplied together to give the yearly transition matrix. The p_{ij} term in the annual matrix is obtained by multiplying column i of one matrix with row j of the other matrix. The result depends on whether the wet season matrix is multiplied by the dry season matrix or vice versa. A wet season matrix times a dry season matrix gives the probabilities at the end of the dry season which are the critical values. The converse gives the probability values at the beginning of the dry season when the reservoir should be at its maximum storage.

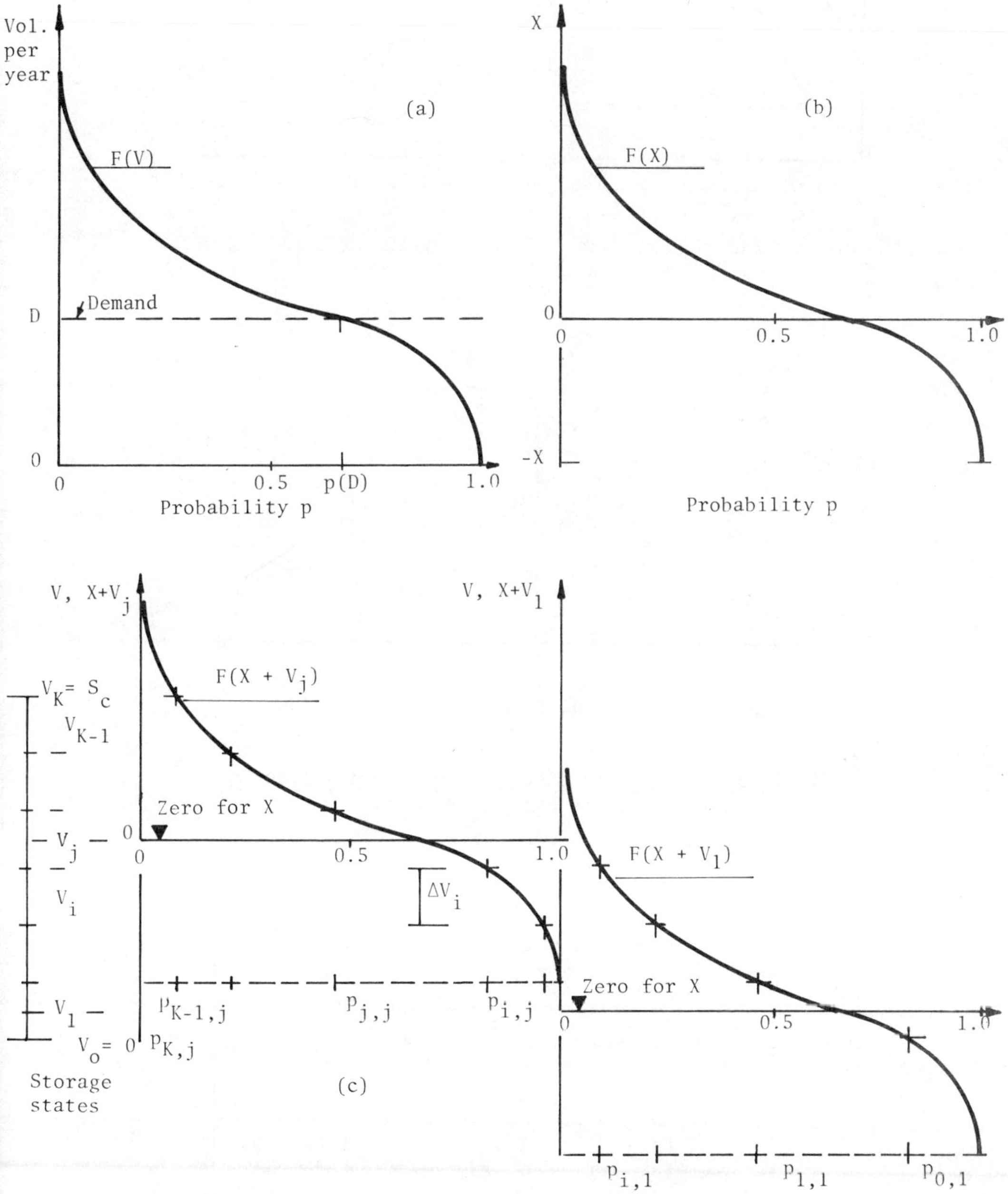

Fig. 10.14. Illustration of the distribution function of net inflows and storage states for Moran's method.

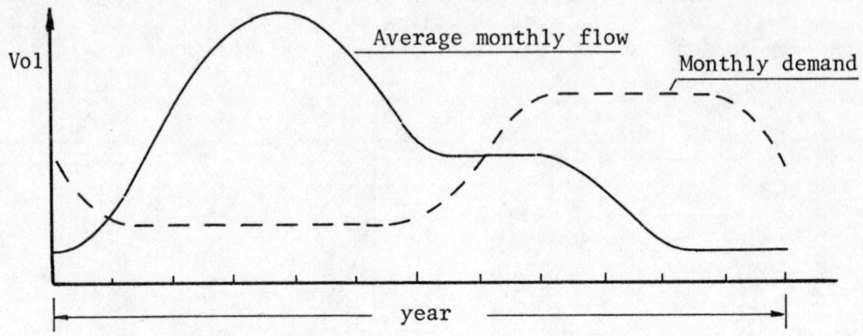

Fig. 10.15. Illustration of within year variation of inflow and demand.

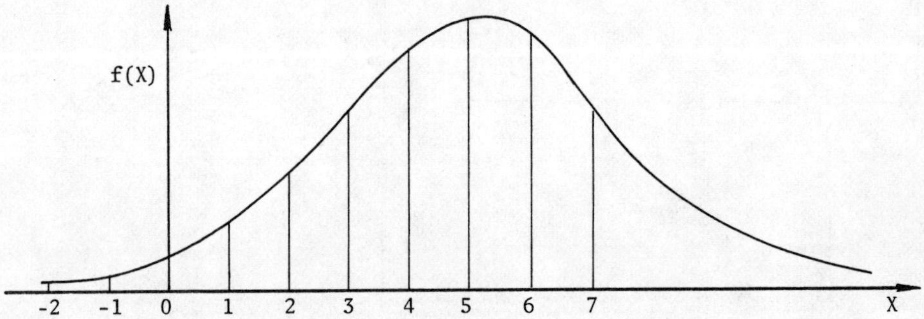

Fig. 10.16. Frequency distribution of net inflows.

TABLE 10.3 Probability Distribution of Net Inflows of Fig. 10.16

Storage increase (volume units)	f(X) dX
> 7	0.210
6 to 7	0.138
5 to 6	0.170
4 to 5	0.162
3 to 4	0.134
2 to 3	0.090
1 to 2	0.050
0 to 1	0.026
-1 to 0	0.010
-2 to -1	0.007
< -2	0.003

TABLE 10.4 An Example of a Transition Matrix Derived from the Distribution of Net Inflows Table 10.3

		Initial state								
	Storage in range	0	1	2	3	4	5	6	7	8
Final state 0	< 0	0.020	0.010	0.003	0	0	0	0	0	0
1	0 to 1	0.026	0.010	0.007	0.003	0	0	0	0	0
2	1 to 2	0.950	0.026	0.010	0.007	0.003	0	0	0	0
3	2 to 3	0.090	0.050	0.026	0.010	0.007	0.003	0	0	0
4	3 to 4	0.134	0.090	0.050	0.026	0.010	0.007	0.003	0	0
5	4 to 5	0.162	0.134	0.090	0.050	0.026	0.010	0.007	0.003	0
6	5 to 6	0.170	0.162	0.134	0.090	0.050	0.026	0.010	0.007	0.003
7	6 to 7	0.138	0.170	0.162	0.134	0.090	0.050	0.026	0.010	0.007
8	> 7	0.210	0.348	0.518	0.680	0.814	0.904	0.954	0.980	0.990

The yearly matrix has the same probability distribution of storage at the beginnings and ends of the years. One could develop transition matrices for each month. Then the probability distribution at the end of one month is the same as that at the beginning of the next and in this way a month to month correlation is introduced. However, the solution by this route tends to be rather long. The summer and winter (dry and wet) matrices assume no correlation between the seasonal flows. The effect of correlation is to underestimate the storage required. Extension of Moran's method to correlated flows has been discussed by Lloyd (1963).

Another alternative method for construction of the transition matrix is by "probability routing" introduced by Langbein (1958) and further developed by Gould (1961). The method developed by Gould is semi-graphical and uses overlay sheets on which the cumulative departures from the demand (adjusted for evaporation, seepage and rainfall) for a year have been drawn. The underlay sheet represents the initial and final storage states, Fig. 10.17. In the discussion of Moran's method it was shown that each matrix column represents the probability distribution of stored water at the end of the year, given a particular initial volume. The values are obtained by routing the flow, one year at a time, through the reservoir using the underlay and overlay sheets. The routing commences for each year with the given amount of storage corresponding to the column under consideration. The results are recorded on a tally sheet as shown on next page. The probabilities in the matrix are obtained by dividing the observed number in the box by the number of years of record. In addition the number of times a failure to supply the demand within a year occurs is recorded for each initial state. This information would not necessarily appear in final state because the reservoir could be filled up subsequently before the end of the year.

The effects of evaporation and rainfall for the given surface area can be allowed for by adjustment of the scales on the underlay. Gould suggested that the initial position of the overlay be lowered from the normal scale values by the appropriate amount for six months for each state and the final position be raised from the normal values by the same amount. Gould also incorporated a rationing rule for when storage levels fall below a certain level as shown in Fig. 10.18.

There are also numerous graphical procedures and precomputed solutions for storage calculations. Klemes (1973), for example, reports on a computed solution by Svanidze which is arranged in charts for an input described by a three parameter Gamma distribution of reservoir inflow.

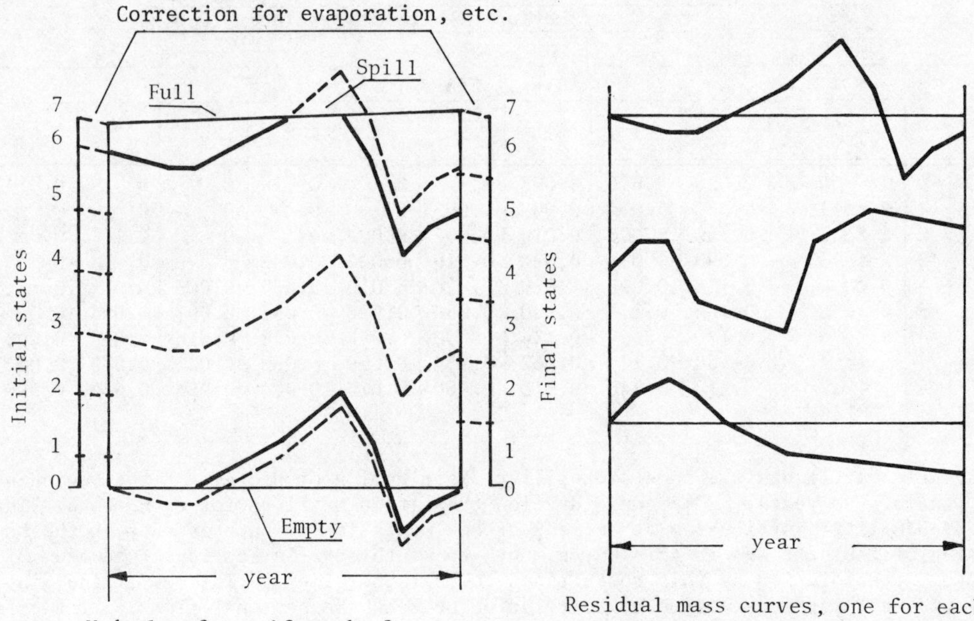

Underlay for uniform draft.

Residual mass curves, one for each year of record. Scale as for storage states on the underlay.

A portion of the overlay strip.

Fig. 10.17. Illustration of the construction of the transition matrix by probability routing using the underlay and overlay sheets.

TALLY SHEET: For.... for.... years of record (e.g. 40 years).

		Initial state								
		0	1	2	3	4	5	6	7	8
Final state	0	0.175	0.150	etc.						
	1	0.100								
	2	0.100								
	3	0.150								
	4	0.225								
	5	0.125								
	6	0.075								
	7	-								
	8	0.050								
No. of times failure, occurred within a year		14	6	4	2	1	1			
Prob. of failure		0.350	0.150	0.100	0.050	0.025	0.025			

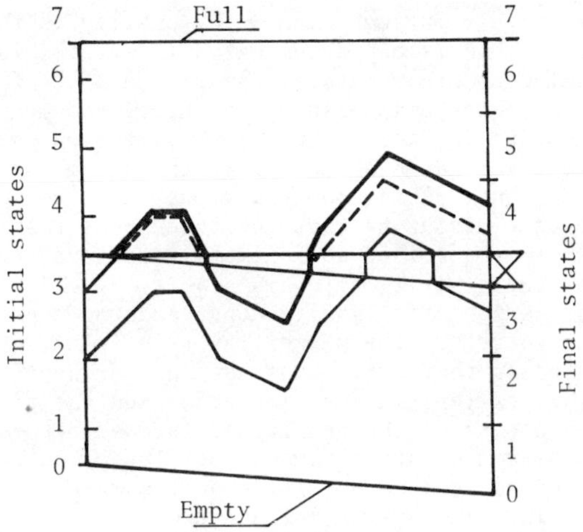

Fig. 10.18. Illustration of the use of a modified underlay and overlays when a rationing rule applies at levels below half-full reservoir. Overlay of a year is shown at two positions, initial states 2 and 3.

Table of failure probabilities prepared from the observed failures.

Starting state	Steady state probability starting in state i	Probability of failure at any time in a year	Product of probabilities
0	0.015	0.350	0.0053
1	0.021	0.150	0.0032
2	0.060	0.100	0.0060
3	0.140	0.050	0.0070
4	.	.	.
.	.	.	.
.	.	.	.
.	.	.	.
	Overall probability of failure		

It also needs to be stressed that the methods of storage calculation discussed assume a steady state condition over the life of the reservoir. In many reservoirs the storage capacity changes drastically over relatively short periods (less than 50 years). This means that the value of the dependability E changes with time, even if the starting value is 100%. The problem of practical importance is then to assess the constancy with time of the design value of the dependability. For example, if the design value is E = 90%, what is the chance that the ten failure years of the 100 year life of the reservoir occur during the first 20 years. Problems of this kind are discussed by Klemes (1969).

10.4 Stratification in Reservoirs

A phenomenon of considerable influence on water quality is the stratification of the water in a reservoir. This is a problem which is treated in texts on limnology. Briefly, the albedo of deap clear water is small and most of the solar energy received at the reservoir surface is absorbed as heat. Some of this heat is transmitted downwards by conduction and by turbulent diffusion caused by wave action. The remainder is lost to the atmosphere mainly through evaporation and by longwave radiation exchanges. Thus, during sunny periods and warm weather the surface tends to warm up, and turbulent mixing by wave action ensures that the reservoir is covered by a warmer layer of water of approximately constant temperature. Because this layer is lighter the stratification is stable, Fig. 10.19a. The absorption of solar energy decreases exponentially with depth and the depth over which mixing occurs is also small. Therefore the warmer layer is relatively thin. When the incident radiation decreases there will be a net loss of heat from the water, the thickness of the warm top layer (epilimnion) increases and the thermocline (metalimnion) becomes very marked and thin. The density difference between the epilimnion and the hypolimnion (the bottom constant temperature layer) decreases and if cooling continues the top layer may become heavier than the water underneath. This is an unstable condition. The top layer sinks and the bottom layer comes to the surface. This overturn of the lake or reservoir can be assisted by wind drag on the surface. Wind drag causes the reservoir surface to tilt and this leads to circulation, Fig. 10.19b. If the tilting exposes the hypolimnion on the windward side mixing of the two layers occurs even under conditions of stable stratification.

Stratification means that only the top layer is in contact with the air. The hypolimnion may be starved of oxygen for long periods and if, for example, there is decaying organic matter on the bed which consumes oxygen the water quality may rapidly deteriorate and fish may die, etc. Shallow reservoirs and lakes usually remain isothermal through mixing throughout the year.

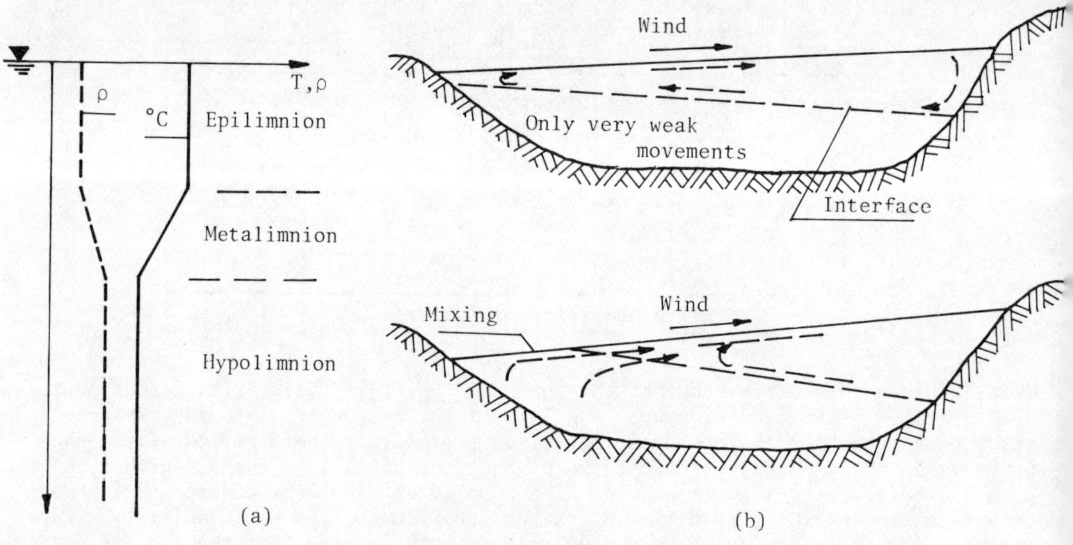

Fig. 10.19. Stratification of water in a reservoir. (a) Variation of density and temperature with depth. (b) Effect of wind on mixing between layers.

10.5 Siltation of Reservoirs

A major problem associated with the design and operation of reservoirs is siltation. Reservoirs fed by rivers, which carry a large sediment load, can lose most of their useful storage volume in a relatively short time. There are numerous examples of rapid siltation, e.g., the Anchicaya dam in Colombia lost 25% of its reservoir capacity in 21 months and Shihmen reservoir in Taiwan lost 45% in five years. Problems of sediment transport enter the reservoir design in several ways. The designer has to consider the response of the river to the reservoir upstream and downstrean from it, the siltation of the reservoir itself, and the handling of the river during the construction of the reservoir. Prediction of the rate of sediment supply to the reservoir and the amount retained in the reservoir is still a highly empirical art. The sediment carried by a river originates from sheet erosion of the catchment, various types of bank and gully erosion and landslides. Siltation of the reservoir depends on both the rate of erosion and the volume of reservoir relative to the mean annual flow. If the reservoir volume is small relative to the mean annual flow, say the ratio is less than one, then trouble from siltation is very likely. Estimates of the sediment yield of a catchment can be obtained from inspection of the catchment on the ground and from aerial photographs which give information on the vegetal cover, areas of erosion and landslides, etc. This information is correlated with data on observed soil losses from similar catchments. In many regions research stations collect and publish data on rates of soil erosion from specific catchments. A summarizing discussion and literature references on sediment sources and sediment yield can be found in Chapter IV of the ASCE Manual-Sedimentation Engineering, edited by Vanoni (1975). The finer sediments are, in general, carried in suspension. The coarser sands and gravels move as bedload. If a suspended sediment rating curve is available then, by multiplying the flow rates with the appropriate sediment concentration values, a sediment transport-duration curve is obtained. The area under this curve represents the volume of suspended sediment carried per annum by flows represented by the flow-duration curve. The sediment transport rates can also be calculated from flow data with the aid of sediment transport formulae (Raudkivi, 1976) and summed to annual totals. These calculations should be supported and checked by field measurements.

The distribution of sediment within the reservoir and the retention of sediment depend on sediment size and texture, reservoir inflow and outflow, and the size and shape of the reservoir. In principle, the settling of discrete non-flocculating particles is described by their fall velocity. In an oversimplified picture the particle is moving forward at the velocity of flow and down at its fall velocity. Correspondingly, the coarser sediments settle out first and form a delta at the upstream end of the reservoir. However, the settling of the finer particles is greatly affected by the flow pattern in the reservoir and an inflow, laden with fine suspended sediment, can flow as a well-defined stream through the reservoir. Such a flow is called a *density current*. A density current is the flow of a fluid of slightly different density under, through or over another fluid without loss of identity through mixing at the interfaces, although the fluid of the density current is miscible with the other fluid. The density differences may arise from differences in temperature, salinity, or sediment content. Figure 10.20 illustrates a typical reservoir delta. A sediment laden stream, when it flows into the nearly static water of the reservoir, has higher density and momentum than the water in the reservoir. Under its momentum the sediment laden stream penetrates a certain distance into the reservoir and then plunges under the lighter water of the reservoir. It continues to flow along the bed underneath the clearer water, Fig. 10.12. This flow is basically a hydraulic problem and exhibits most of the features of open channel flow. The major differences are the much greater surface drag at the interface, compared with that between air and water and the mixing with reservoir water across the interface. The flow may continue until it runs out of momentum (and has deposited its sediment load) or until it is stopped by the dam. It may even have enough momentum to climb the face of the dam and flow over the spillway. Such flows of

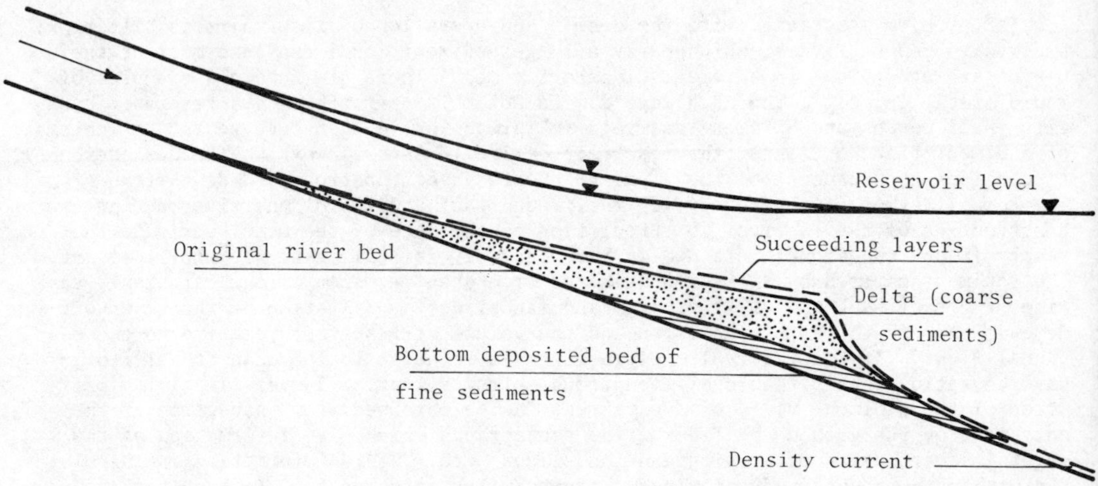

Fig. 10.20. Diagrammatic illustration of a reservoir delta.

muddy water over the spillway have been reported, for example, for Lake Mead in the USA and Lake Matahina in New Zealand.

Singh and Shah (1971) related from laboratory studies the depth at the plunge point, Y, to the critical depth $y_c = (q^2/g)^{1/3}$ as

$$Y = 1.85 + 1.3[q^2/(\Delta\rho/\rho)g]^{1/3} \tag{10.39}$$

where $\Delta\rho$ is the density difference, ρ is the density of the water in the reservoir, and q is the flow rate per unit width (cm is used as the unit). For large values of the second term 1.85 may be neglected and the densimetric Froude number $Fr_D = V/\sqrt{(\Delta\rho/\rho)gy}$ becomes $Fr_D \simeq 0.67$. No limiting density difference was noted in the experiments, where the limit was set by the available depth at the plunge point. A similar expression was derived from a model of energy conserving flow by Savage and Brimberg (1975), i.e., $Y = 1.587\{q^2/[(\Delta\rho/\rho)g]\}^{1/3}$. The authors also developed an equation for the interface and discussed results from numerical analysis. A model based on bedload transport was presented by Yücel and Graf (1973).

The actual accumulation of sediment in a reservoir depends on the fraction of the inflowing sediment retained. This ratio of deposited to total sediment inflow is called the *trap efficiency*, η_{tr}. Its value depends on the sediment fall velocity, rate of flow through the reservoir, the geometric features of the reservoir and outlet structure, reservoir operation as well as on the age of the reservoir. It can also depend on chemical properties if flocculation occurs. The methods for estimation of trap efficiency are empirical. Churchill (1947) presented a relationship between the percentage of incoming sediment passing the reservoir and the *sedimentation index* of the reservoir. The index is a retention period (obtained by dividing the volume of the reservoir at mean operating level by average daily inflow rate) divided by a velocity (obtained by dividing the inflow by the average cross-sectional area of the reservoir). Churchill's relationship was based on data from Tennessee Valley Authority reservoirs. Brune (1953) presented data from 44 normally ponded reservoirs as percentage of sediment trapped versus the ratio of reservoir volume to annual mean inflow. His graph has three lines, for coarse, medium and fine sediments, respectively. Borland (1971) concluded that Churchill's method

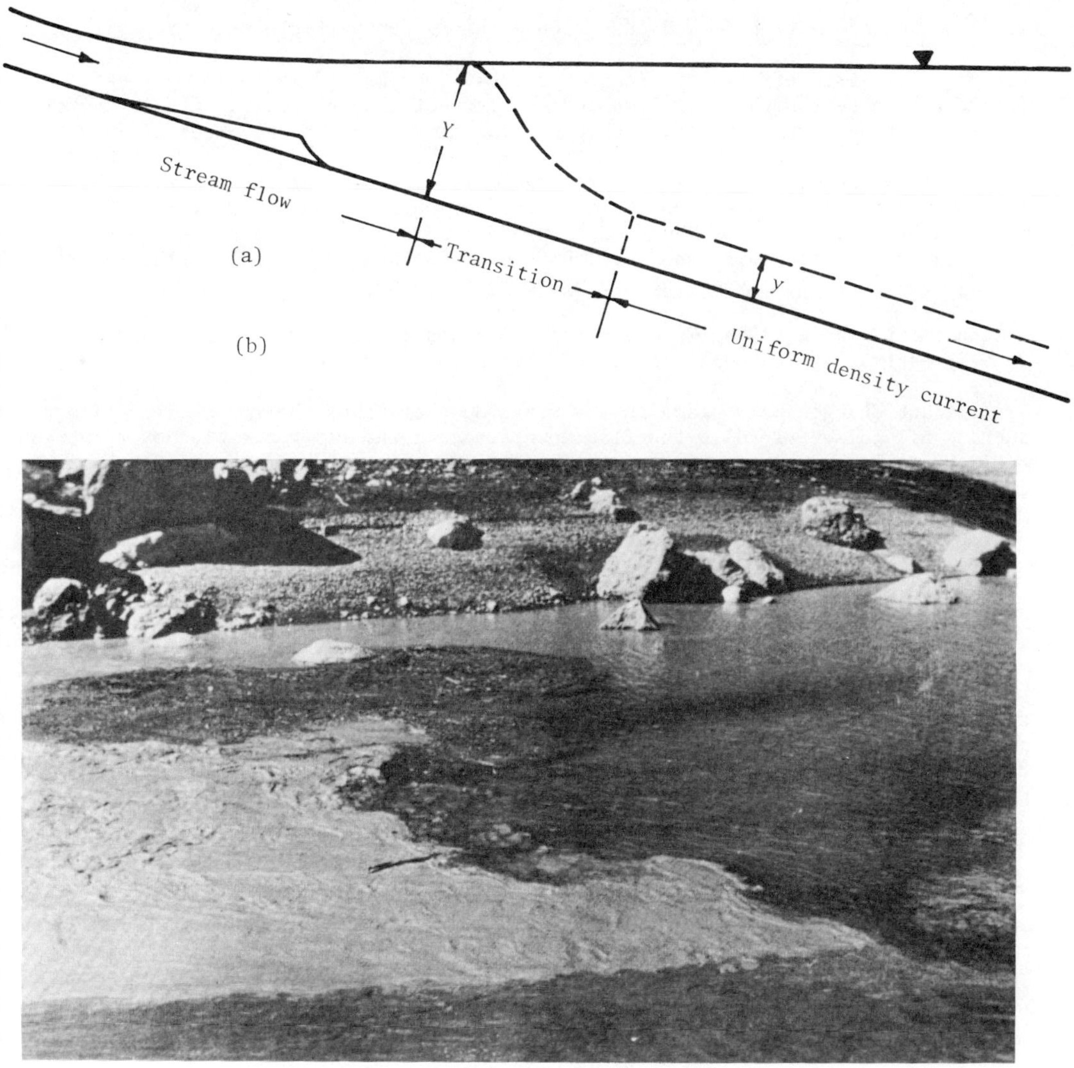

Fig. 10.21. Plunge point of a sediment laden stream at the head of a reservoir. (a) Definition sketch. (b) Photograph of the discontinuity at the plunge point (Duquennois, 1956).

gave better results than Brune's curves. Karaushev (1966) presented a phenomenological model based on distribution and decay of the "mixing of fluid" (turbulence and eddying) in the reservoir from which he expressed the trap efficiency as

$$\eta_{tr} = 1 - (1 - V_r)\exp[- \phi V_r/(1 - V_r)] \qquad 10.40$$

where V_r is the capacity-inflow ratio, $\phi = wT_s/D$, T_s is duration of spillover period in seconds and D is mean depth of the reservoir. This gives a family of η_{tr} versus V_r curves, with ϕ as a parameter. A $\phi \simeq 30$ value corresponds to Brune's medium sediment curve. A similar expression was proposed by Borland (1971). These expressions enable estimation of sedimentation within given particle size ranges.

The reservoir trap efficiency changes progressively as sedimentation proceeds. Sedimentation alters both the volume and geometry of the reservoir. It is therefore adviceable to analyse trap efficiency in time increments of the order of 10 years.

The distribution of sediments in a reservoir is estimated by the empirical area-reduction method, developed from field survey data by Borland and Miller (1958) and modified by Lara (1962). The data is presented in the form of design curves as a function of the reservoir type, U.S. Bureau of Reclamation (1973). The reservoirs are divided into four types, according to their shape.

The sediment load arriving into the reservoir has to be converted into sediment volume by estimating the bulk density of the deposited sediment.

Significant changes take place in the river downstream of the reservoir. The reduction of sediment load in the flow downstream of the reservoir without reduction of flow, as in hydropower plants, can lead to significant degradation of the river channel. For details of analysis of the degradation problem the reader is referred, for example, to Komura and Simons (1967) and discussions. Apart form the changes consequent upon the reduction of sediment load there are changes in river channel geometry that occur because of the modified magnitudes and frequencies of flow peaks downstream of the reservoir. The effects of the modified time distribution of flows could be significant over a long length of the river. Two contrasting forms of adjustment could occur. When flow is released from the reservoir in surges, with increased peak flow rates but decreased frequencies, the channel capacity could be increased. Where flood waters are impounded a decrease in the channel capacity occurs. The decrease in channel capacity can be quite marked with the reduction of peak discharges and it persists for a distance downstream until the catchment contributing to the stream reaches about four to five times the area draining to the reservoir.

10.6 Catchment Yield

Catchment yield refers to the volume of water available from a stream at a given location over a specified period of time. The yield depends on meteorological events, geology, vegetal cover and use of the catchment. In general, studies of catchment yield are concerned with the effects of land use and land use management on streamflow. In principle, catchment yield is a problem of water balance. The yield

$\int_{t_1}^{t_2} Q\,dt$ = the change in volume of water in storage (in the surface reservoirs and channels, on the ground and vegetation, and in the ground as groundwater and soil moisture)

+ $\int_{t_1}^{t_2} i\,dt$ minus the volume of water evaporated and transpired from the catchment,

where Q is the rate of discharge in the stream at a given location and i is the intensity of the precipitation. The evaluation of all the terms in this water balance relationship is seldom possible with any degree of accuracy. In particular, the changes in the volume of water in the ground are difficult to estimate. The boundaries of the groundwater reservoirs are frequently quite different from those of the catchment and the fluctuations in groundwater storage may depend on supply from several catchments.

The simplest expression for yield from a catchment, Y, in terms of the depth of water, are of the form

$$Y = aP - c$$

where P is the depth of precipitation for the period, c accounts for losses, and a is a reduction factor which accounts for interception. If the yield is expressed as $Y = P - E$, then the evaporation $E = c + (1 - a)P$. Such linear, or more complex, relationships can be fitted to local data.

In general, climatic and geological parameters are not subject to control or management procedures, except for the surface texture of soils, and the water yield can be influenced only by land use. Land use and land use management can, however, account for appreciable variation in the water yield from a given catchment. Land usage ranges from forestry and agriculture to urban and industrial use, all of which exert different influences on the hydrological conditions of the catchment. The primary influence of land use is the modification of the infiltration rate and the soil moisture potential. It was pointed out in the discussion of infiltration that a good cover of vegetation or a forest protects soil from compaction and creates conditions favourable for infiltration. At the same time vigorous plant cover increases evapotranspiration and depletion of soil moisture storage, and hence may reduce the volume of groundwater recharge. If less water goes to groundwater storage the stream flows will decrease. The classic texts dealing with the hydrological influences of forest cover and vegetation are those by Kittredge (1948) and Colman (1953).

The bulk of the research data on the effect of forest cover on catchment yield relates to the extremes of full forest cover or no cover. Appreciable increases in streamflow have been reported after clean cutting of the forest. For example, Pierce et al. (1970) report from a study in New Hampshire, U.S.A., streamflow increases form 240 to 346 mm per water year, with the greatest increase in late summer when the flows were normally very low. The total precipitation amounts were 1242 and 1304 mm per annum, respectively. The catchment was cleared of woody vegetation and treated for three years to minimize regrowth. They also report large changes in the water chemistry. The nitrate (NO_3^-) concentration increased by an average of fifty times, major cation levels (sulphate $SO_4^=$, calcium C_a^{++}, potassium K^+) increased from three- to twenty-fold and sediment concentration increased nine-fold. Hibbert (1967) summarizes the results of 39 studies of forest treatment designed to alter streamflow characteristics, as well as one study (1971) on conversion of a catchment to grass.

Analyses of streamflow records generally indicate an increase in the median value of the discharge-frequency curve after clean cutting. Reduction in evaporation from catchments on which all vegetation has been cut, but left on the ground, contributes to higher baseflow rates. Increases in the yield and peak flow rates have also been reported from catchments where the forest has been clear cut in alternate strips. In general, poor logging practices, fires, etc., tend to create conditions where the runoff peaks increase and low flow rates decrease, with little change in the total yield. The problem of runoff from forested areas is basically far more complex than that from catchments where the runoff is predominantly surface runoff. For runoff from forested catchments the hydrograph concept is of little value. On forested catchments overland flow is rare. These catcments have high infiltration capacities, which are seldom exceeded by rainfall intensities. The velocity of overland and channel flow is usually of the order of 0.1 m/s or more, whereas the subsurface velocities seldom exceed 5×10^{-5} m/s. Therefore, it requires only a few metres of subsurface flow to separate the subsurface flow clearly from the surface runoff. The concept of constant duration of baselength of the runoff hydrograph for a given duration of rainfall is no more an adequate approximation of the observed behaviour. The baselength of the hydrograph on a given catchment depends on the antecedent conditions in the soil, the swelling and shrinkage of the pores, etc. The storage-discharge relationship displays a hysteresis loop, which is related to the depth of the porous soil, the length of the hillside segments and their slopes. The contribution to streamflow is derived from direct precipitation and from discharge of subsurface flow through the wetted perimeter of the channel. Both the channel size and the exfiltration are functions of time. The overriding effect of the infiltration and the associated delay of runoff from forested catchments is clearly evident from the volumes of storm runoff hydrographs on headwater streams. These volumes account for only about ten percent of the annual precipitation.

The streamflow at a given gauging station of a forested catchment depends primarily on the permeability and layering of the soil. The soils can range from deep porous soils (and an ephemeral stream) to a thin layer of porous cover as illustrated in Fig. 10.22, and make the different responses and yields for a given rainfall depth obvious. In addition there is the major problem that no catchment is homogeneous. The enormous variability of the permeability, within layers and from layer to layer, and the variation of the geometry and slope of the layers make modelling of such a system extremely difficult. Some conceptual ideas are presented by Betson and

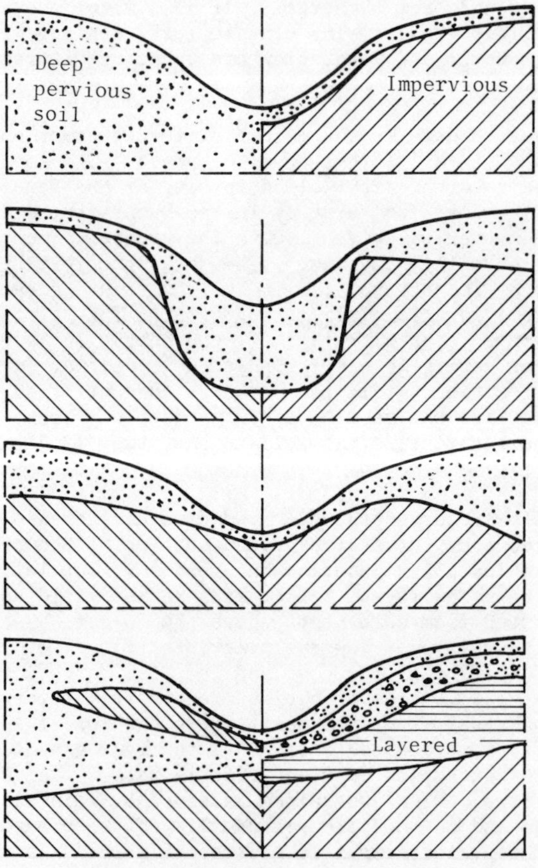

Fig. 10.22. Schematic presentation of eight different catchment soil profiles out of a very large number of possible combinations.

Marius (1969), Dickinson and Whiteley (1970), and Nutter and Hewlett (1971). Dooge (1960) modelled the streamflow or baseflow probelm by using linear storage elements (S = KQ) through which the infiltration water may be routed to other storage elements in series, directly to streamflow or back to the atmosphere by evapotranspiration. His model was aimed at modelling a baseflow which is maintained by depletion of groundwater storage (and snowmelt in some locations and seasons) during periods of no precipitation, but it could be extended to the forest runoff problem. In order to apply such a model observed data must be available. The discharge Q_n from a storage element during a time period can be represented by the routing equation

$$Q_n = C_o R_n + C_1 R_{n-1} + C_2 Q_{n-1} \qquad 10.41$$

where R_n and R_{n-1} are the recharges during the period and the period before, respectively, Q_{n-1} is the discharge from the element during the previous period, and the routing coefficients are

$$C_o = 1 - \frac{K}{T}\,(1 - e^{-T/K})$$

$$C_1 = \frac{K}{T}\,(1 - e^{-T/K}) - e^{-T/K}$$

$$C_2 = e^{-T/K}$$

where T is the duration of uniform recharge and K is the storage delay time of an element. The coefficients depend on the value of K/T only.

There are also numerous studies of the effects of management of agricultural lands on water yield. It has been shown that autumn ploughing decreases runoff and that sod condition increases winter water yield. High rates of runoff have been reported from areas of spring-planted grain and new pastures. Sartz (1970) reported from an 8-year study in Wisconsin, U.S.A., peak runoff rates from 64 mm/hr from tilled land to zero from undisturbed forested areas during a major storm. He found that the peak runoff rates from tilled land averaged two and half times those from meadows, which in turn were 1.4 times those from abandoned land. Peak runoff rates from heavily grazed pastures were three times those from lightly grazed pastures. However, Sharp et al. (1966), from a study of the Great Plains area of the United States, concluded: "No statistical approach was found that would consistently assess effects of land treatment on streamflow from river basins, or even prove conclusively that such effects do or do not exist. In a few cases, streamflow appeared to be increasing. In some, it appeared to be decreasing. In all cases, streamflow fluctuated considerably, due to climatic or other causes".

The effects of land usage have also been previously discussed in connection with the individual processes of interception, infiltration and evapotranspiration. Interception of rain or snow by vegetation and surface litter, compaction of ground by raindrops and animals, etc., all have pronounced effects on the water balance components and on the catchment yield. Forest, for example, can accumulate a large quantity of snow, and the infiltration into the litter covered ground is usually high. At the same time the evapotranspiration losses can be high. Not only can trees deplete the soil moisture, but through deep roots they can also lower the groundwater level. Some trees have roots extending more than 20 m into the ground. This extraction of water by the deep roots does not only affect the water balance but could also have a major effect on the slope stability of some hillsides.

10.7 Sediment Yield and Transport

Soil erosion decreases the potential productivity of land and is the largest source of pollutant of all surface waters. Problems are caused where the eroded material is deposited. Erosion of only 1 mm of soil over 1 km^2 is equivalent to 1000 m^3 in original volume and more in a freshly deposited form. On most arable land soil erosion is a very important land management problem.

The physical process of soil erosion is very complex. It depends on the amount and intensity of the rainfall, the physical and chemical properties of the soil, the geologic and topographic features of the catchment and the use of the land and land management practices. In some regions (and it need not be desert) the erosion can be entirely caused by wind. The problem of wind erosion of agricultural land in the Middle West of the United States in the 1930-s, the dust bowl, is well known. The severity of that particular wind erosion problem was due to inappropriate land management practices. In the present context, however, attention is centred on erosion by water rather than wind. For wind erosion the reader is referred to Bagnold (1941) and Chepil and Woodruff (1963).

Soil erosion by water begins with the impact of rain drops which detach soil particles by splash. The accumulated rain water transports these particles down the slope and may dislodge additional ones. The impact from the rainfall depends on the size and fall velocity of the raindrops, the drop size distribution and the total volume of rain. The impact energy of an annual rainfall of 750 mm is of the order of 15-60 kJ m^{-2}. How this energy is spent depends on the surface conditions of the catchment. Where the ground is covered by dense vegetation, litter and mulch, the energy of the raindrops is dissipated on this cover and very little, if any, soil erosion takes place. On bare soil the energy goes into detaching the soil particles and in compaction of the soil. The latter reduces the rate of infiltration and increases overland flow and soil erosion. In general, the erosion from well-forested areas and from good grassland is in very small quantities and mainly limited to scouring in the stream channels and to soil from land slides in steep country. The water in the streams from such catchments is clear. The rate of soil loss from agricultural land is much higher and depends strongly on land management practices and on the crops grown. The velocity of the overland flow, V, depends on the runoff rate q, the slope S, the hydraulic roughness n of the flow channels and on the concentration of flow into small grooves and rivulets (rilling). In the small grooves the velocity is approximately proportional to $(Sq/n^2)^{1/3}$ and, since the carrying capacity of the flow is approximately proportional to V^5, it is clear that small changes in any of these variables will have a large effect on the erosion rate.

Most of the methods in use to estimate the rate or quantity of soil eroded use an empirical "soil erodibility" factor for which there is no physical definition. The soil loss equation is usually written as

$$E = RK\ LS\ CP \qquad 10.42$$

where E is the rate of soil erosion per unit area, R is the rainfall factor describing the rainfall characteristics and erosive power, K is the soil erodibility factor, L is the length of the slope, S is the slope, C is the cropping-management factor, and P is an erosion control practice factor. Various investigators have proposed formulae which incorporate all or some of these terms, e.g., Musgrave (1947) proposed

$$E \propto KS^{1.35}\ L^{0.35}\ r_{30}^{1.75} \qquad 10.43$$

where r_{30} is the maximum annual 30 min. rainfall rate. Meyer and Monke (1965) suggested

$$E \propto L^{1.9}\ S^{3.5}\ d^{-0.5} \qquad 10.44$$

where d is the particle size of the eroding soil, and the formula is restricted to $[E(kg/m/min.)]^{1/2}/d(mm) > 2$. An example of procedures based on the regression analysis approach is given by Wischmeier and Smith (1965).

Probably the most versatile model available for study of soil erosion is the simulation model by Meyer and Wischmeier (1969) illustrated in Fig. 10.23. Each of the four processes is evaluated with the aid of semi-empirical relationships. For each increment the quantity of soil available for transport is the material detached in that increment by the rainfall and runoff plus material brought into it by the overland flow from upstream. If this quantity of soil is less than the total transport capacity, then the erosion is limited by the availability of eroded soil; if not, transport capacity is the limiting factor.

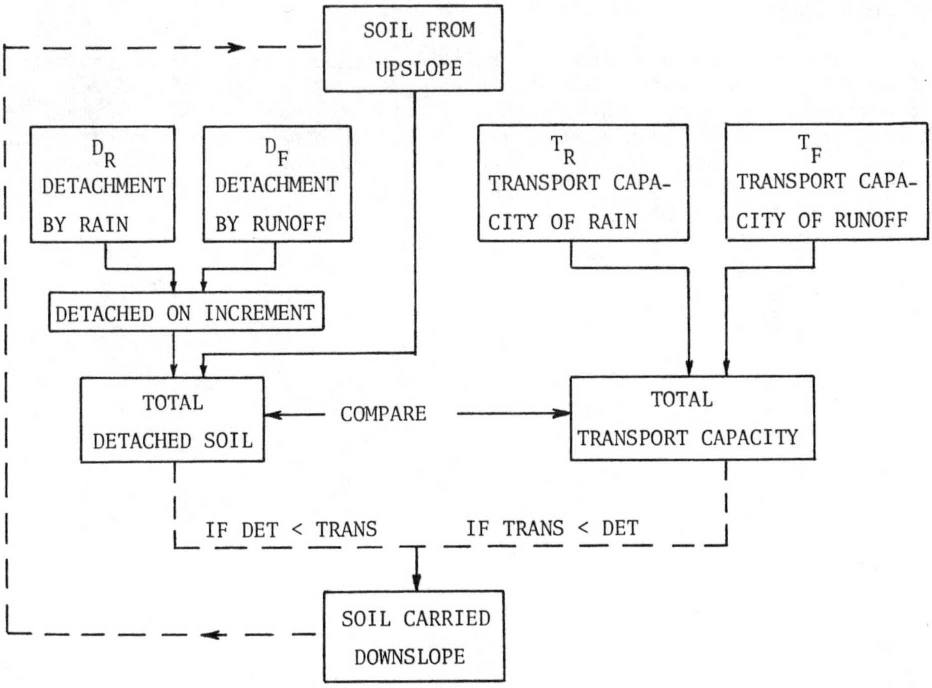

Fig. 10.23. The four component model of soil erosion by Meyer and Wischmeier.

The soil detachment by rainfall, D_R, was expressed as

$$D_R = S_{DR} A_i I^2 \qquad 10.45$$

where S_{DR} is the soil detachment coefficient, A_i is the area of the increment and I is the intensity of the rainfall. The detachment by runoff, D_F, was given as

$$D_F = S_{DF} A_i (qS)^{2/3} \qquad 10.46$$

The transport by rainfall, T_R, was evaluated from

$$T_R = S_{TR} SI \qquad 10.47$$

where S_{TR} is a soil transport coefficient.

The soil transport by runoff is approximated by

$$T_F = S_{TF}(qS)^{5/3} \tag{10.48}$$

The basic problem is the assessment of the values of the coefficients. However, since it is a computer simulation the results can be obtained for a large range of conditions and these can be checked against observations from the given locality.

The initial overland flow concentrates into rills, gullies and streams. Rill and gully erosion problems are treated with open channel flow methods. Major difficulties are encountered in estimating the resistance coefficients for these spatially varied flows.

On a larger scale one of the most reliable methods for obtaining the rate of erosion from a given catchment is to establish a correlation between the sediment load in the stream (including washload) and its water discharge. The sediment load can be obtained from measurements and this information can be supplemented by calculations of the total sediment load by methods described, for example, by Raudkivi (1976). For further reading Vanoni (1975) on erosion and Carson (1971) on erosion by landslides are recommended.

Chapter 11

HYDROLOGICAL MODELLING AND WATER RESOURCES SYSTEMS

The main problem of applied hydrology is the determination of river flows given certain physical parameters such as rainfall, temperature, wind and catchment parameters. These flows are not only required for flood forecasting but also for prediction of the effects of proposed changes of the catchment and, in general, for water resources management. The processes which link rainfall with river flows are essentially deterministic, governed by physical laws which are reasonably well-known, but the boundary conditions (i.e., the physical description of the catchment and the initial conditions and distributions) make a solution based on the direct application of the laws of physics impracticable. As a consequence the hydrologists have turned to empirical and analytical modelling of the catchments. Models are used to predict frequencies of events, make short-term forecasts, extend records, generate synthetic sequences of input data for more extensive models, predict the hydrological performance of a catchment, forecast the effects of physical changes of the catchment, and optimise design and operating procedures of water resources projects, etc.

Widely varying interpretations are attached to terms such as hydrological modelling, mathematical modelling, systems analysis and simulation. In the present context *modelling* refers to relationships between the real system, such as a catchment, and its mathematical (or physical) models. *Simulation* is concerned with the relationships between the model and computer, i.e., with the production of answers from the model. Furthermore, the various techniques will be subdivided into two broad groups. In the first group are the methods of *mathematical modelling of hydrological systems, simulation* and *systems analysis,* and in the second group is the analysis of *water resources systems.* The first group is concerned primarily with the performance of the catchment, prediction and data generation, that is, with the problems of input and output. The analysis of water resources systems is concerned with the multipurpose use of the resource and with obtaining the best possible solution. It is a decision making process and incorporates not only the physical factors involved but also economic, social, environmental and political considerations. Both groups need data. It is therefore useful to discuss briefly the nature of hydrological data.

11.1 The Nature of Hydrological Data

Observed hydrological data may be either deterministic or random. Deterministic data can be described by an explicit mathematical relationship. The bulk of hydrological data relating to the time history of the process is random or stochastic.

There is no explicit mathematical relationship for such data. Random data has the property that each particular record is unique. A set of time history records $[x_t]$ is called an ensemble and it is said to be a random process if it can be described by statistical properties. The classification of random data is illustrated in Fig. 11.1. A stationary random process is characterized by statistical properties which do not depend on time. This means that the values of the statistical properties of n-years of record remain the same, irrespective of where on the time scale the record was taken. Likewise, joint statistical properties at two or more times are dependent only on the time differences, and not on the actual times. Processes which do not satisfy these conditions are non-stationary. A stationary process is said to be *ergodic* if time averages on particular individual records are the same for all records.

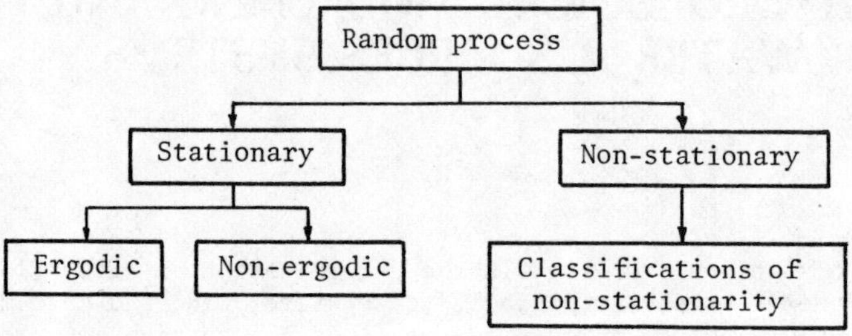

Fig. 11.1. Classification of random processes.

Hydrological data are generally neither purely deterministic nor purely random. Most records contain both deterministic (periodic) components and random components. Again the random component is not often normally distributed (i.e., pure chance dependent). Instead there may be a strong persistence or internal dependence effect, for example, a low flow corresponding to a very severe drought condition will never follow after a maximum flood observation.

A widely used assumption is that the hydrological variable X_t can be separated into components representing the trend x_T, oscillatory features x_p, and a random component x_t as follows

$$X_t = x_T + x_p + x_t \tag{11.1}$$

Figure 11.2 by Kisiel (1969) illustrates this concept of composition of time series. The trend, if it exists, is assumed to be deterministic and its future value may thus be predicted. The trend is identified in the record by moving mean (average) analysis. The oscillatory component is also a deterministic term, which can be identified by methods of harmonic analysis. The usual procedure is to find the auto-correlation function and transform it into a spectral density function. A periodic sine function has an auto-correlation function of cosine form and a single peak spectral density function at the frequency of the sine function, Fig. 11.3. Once the frequency has been identified, this component can be subtracted from the record. The process is repeated until all the periodic components have been removed. The need to separate the oscillatory component can at times be by-passed. For example, the annual records of monthly streamflows can be separated into twelve records of n-years, one for each month. The record for a given month from year to year is assumed to be free from seasonal oscillations. Thus, for any month $x_t = X_t - \bar{x}$, where the trend x_T has been separated from X_t before. For analysis, it is usually convenient to use a standardized variable defined as

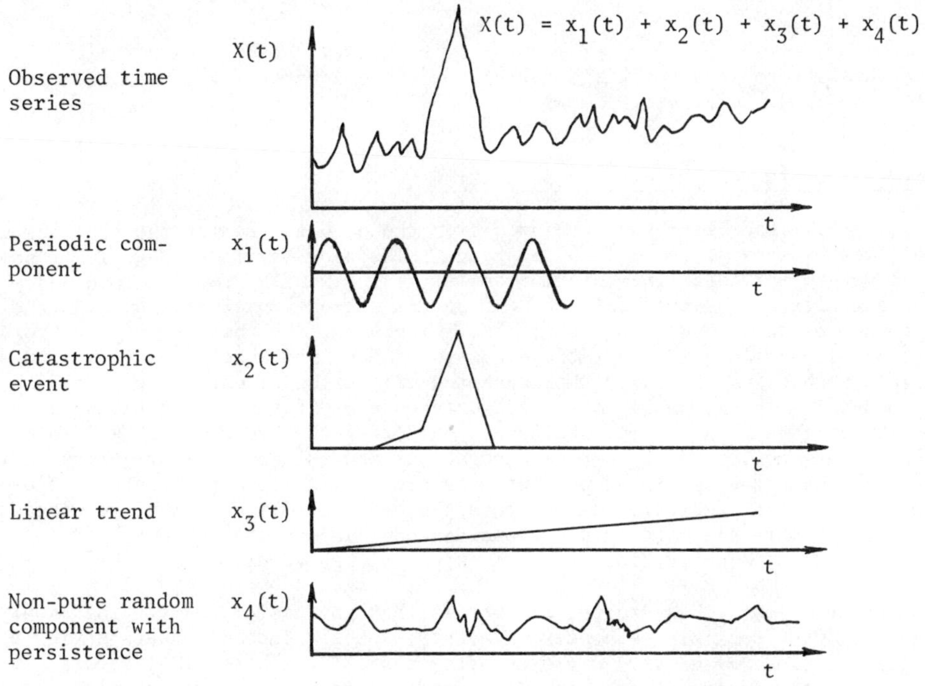

Fig. 11.2. Components of a time series after Kisiel (1969).

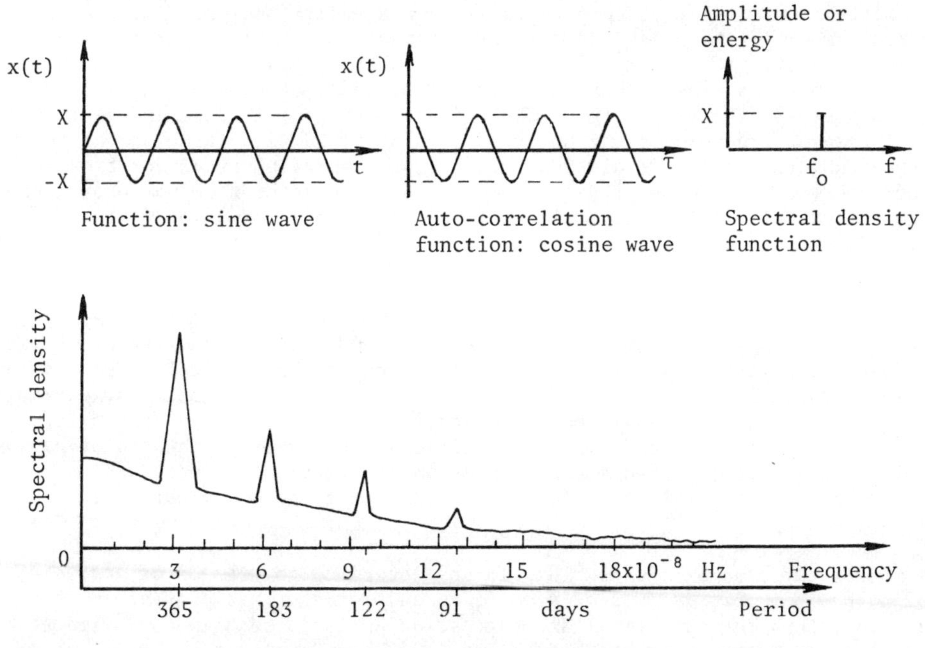

Fig. 11.3. Illustration of auto-correlation and spectral density functions.

$$y_t = \frac{X_t - \bar{x}_t}{\sigma_x} \qquad 11.2$$

The data can relate to a large variety of measurements, e.g., rainfall depth, streamflow, soil moisture, etc.

11.2 Hydrological Models and Simulation

Hydrological modelling has developed from the need to predict the hydrological output, for example, the runoff hydrograph. The need for modelling, i.e., some indirect method for obtaining information, arises primarily from the inadequacy of hydrological data. Observed data, of course, is extremely valuable but records seldom extend over an adequate length of time for probability analysis of the hydrological variable, particularly the extreme events. The inadequacy of data will always remain a problem regardless of how much the collection of data is expanded, because it is economically untenable to collect data on every catchment in the world. Furthermore, man-made changes are continuously altering the hydrological environment and hence, data collected may not be homogeneous and suitable for frequency analysis. Such non-homogeneous time-dependent data are very difficult to interpret and its usefulness for the prediction of future behaviour is limited. The hydrologist is, therefore, more and more turning towards modelling of the various hydrological processes and to computer simulation of catchment behaviour.

Another problem with hydrological data is that any historical record is only one of the very many possible sequences of events; the sequence of events during the prediction period may be very different from that of the historical record, even though the statistical parameters, e.g., mean and variance, etc., may remain constant. Therefore, an important group of models is concerned with data generation.

Future events may be predicted by a deterministic or a probabilistic model. The deterministic model is based on physical laws and relationships. It usually involves a large number of variables and complex interactions, which together make the use of the deterministic model very impractical. In the probabilistic model the actual physical processes play no part. Only statistical measures of the variables are used to generate future events. The object is to generate sequences of data which have the same structure as the observed record. The structure is measured in terms of the mean, variance, correlation, etc. The generation processes provide any desired number of sequences of data. The generation methods do not provide new information; they only enable the designer to use all the information contained in the historical sequence. If the historical record is of doubtful quality, then the generated data will be suspect as well.

Hydrological models are often referred to as hydrological systems. A *hydrological system*, according to the definition by Dooge (1968) and as modified by Clarke (1973), "is a set of physical, chemical and/or biological processes acting upon an input variable or variables, to convert it (them) into an output variable (or variables)". The word *model* has been assigned to a simplified representation of a system. Models of hydrological systems may be *physical* scale models of the prototype, *analog* models such as the resistance-capacitance analog (review of these was made by Diskin, 1967a), or *mathematical* models which represent the behaviour of the system by a set of equations.

The models may also be divided into groups according to methods of analysis, i.e., *parametric analysis, analytical techniques* and *mathematical simulation.* Parametric analysis refers to statistical and regression analysis techniques. Parameters are fitted to a regression or statistical model without explicit reference to the physical system. The lable "analytical techniques" includes models developed using systems analysis. In these models some attempt is made to describe the physical system, but only within the limits of the technique used. Restrictions arise from the need to have models for which economic numerical solutions can be obtained. Mathematical simulation refers to models which attempt to describe the physical system.

The various hydrological models may also be divided into *event models* and *continuous models*. The event model, for example, produces the hydrograph for a particular storm input, whereas the continuous model produces a continuous flow record.

The model (or system) has to generate an output from an input or inputs, which may also be interacting. A number of such models have been introduced earlier without emphasis on the modelling aspect, e.g., the hydrograph (runoff) models and routing models, etc. The output, in general, depends on the nature of the input, the physical laws involved and the system itself. In some studies the nature of the system may not be knwon or is ignored, as in unit hydrograph studies where, once the unit hydrograph has been determined from the input-output records, it is used without any reference to the nature of the catchment or the physical laws involved. The derivation of a synthetic unit hydrograph, the study of the adequacy of the unit hydrograph procedure, etc., however, call for examination of the relationships between the unit hydrograph, the characteristics of the catchment and the physical laws governing the catchment behaviour. In such problems the system itself has to be known or has to be synthesized. Dooge (1968) illustrates the problems which arise with systems as follows:

	Type of problem	Input	System	Output
Analysis	Prediction	✓	✓	?
	Identification	✓	?	✓
	Detection	?	✓	✓
Synthesis (simulation)		✓	?	✓

The hydrological system may be a closed system or an open system. The hydrological cycle discussed in Chapter 1 is a closed system which operates as a result of the radiation input. In this system the hydrologist is mainly concerned with the transfer and transformation of water in the cycle. Usually the problem is confined to a particular catchment. Frequently the atmospheric ocean and the lithospheric parts of the system are ignored. The resulting subsystem for the catchment cuts certain lines of moisture transport and is no longer a closed system. A simplified system of this kind for a catchment is shown in Fig. 11.4a. If the interflow Q_i is ignored as well, then a model of the catchment could be as shown in Fig. 11.4b. The subsystem involving soil moisture has a feedback loop for separation of rainfall into rainfall excess and infiltration. The central problem in all these models is the analytical formulation of each of these processes involved in the model. This may be achieved by an empirical relationship (e.g., curve fitting) or by a function derived from physical considerations. The ideal is a function based on physical laws and free of calibration and matching to the problem.

The solution of the prediction problem requires knowledge of the system. In practice this means that assumptions have to be made about the nature of the system. The more assumptions made, the easier is the solution and correspondingly the greater is the risk that the model will not characterize the catchment correctly. The most important assumptions usually made are those of time invariance and linearity. The unit hydrograph method is a linear and time invariant system

$$Q(t) = \int_0^t i(\tau)u(t - \tau)d\tau$$

Once the input $i(\tau)$ and the instantaneous unit hydrograph are known, the solution of the prediction problem is fairly straight forward.

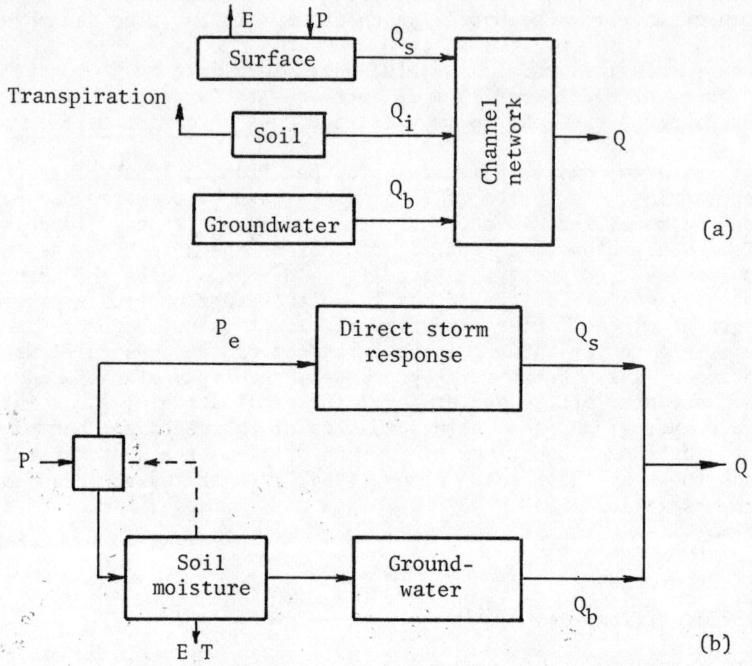

Fig. 11.4. The catchment as a system (a), and a simplified model of the catchment (b).

The solution of the problem of identification and detection is more involved. Detection, for example, may involve finding the unit hydrograph. The methods for doing this are the trial and error method by Collins (1939); the method of moments (Nash, 1959); Laplace transforms (Paynter, 1952, Diskin, 1967); Fourier transforms (Levi and Valdes, 1964); harmonic coefficients (O'Donnell, 1960); Laguerre coefficients (Dooge, 1965); least squares (Snyder, 1955, 1961); and time series analysis (Eagleson et al., 1966, Bayazit, 1966).

Whatever the required end result, the use of hydrological models (systems) involves several basic steps:
(1) Definition of the problem
(2) Choice of the class of the hydrological model
(3) Choice of the type of model in the chosen class
(4) Calibration of the model (optimization of parameters)
(5) Evaluation of model performance
(6) Use of the model

In general, a model is continuously changing. It may start as a rough empirical model, but as the study proceeds more and more of the empirical components may be replaced by functions more firmly based on theory. Usually different models have different uses, although some may have several uses. Of the various types of models the physical scale models of the catchment are the most restricted in their use. Eagleson (1969) writes: "While there appears to be a class of watershed problems for which scale modelling may be a legitimate method of study, the class is small and included in that larger class for which digital computer solution of the equations of motion shows at least equal promise". There is no clear-cut division between mathematical modelling, simulation, hydrological modelling and systems analysis;

neither is there such a division between these and the usual "every day" hydrological calculations, such as frequency analysis and unit hydrograph calculations, etc. The primary purpose of hydrological models is to simulate the catchment behaviour and to predict the effect of changes of conditions and variables on this behaviour. There is a subsection of modelling which is entirely concerned with generation of synthetic sequences of hydrological data. These data may subsequently serve as input data to a more extensive model.

It is important to realize that the model, however elaborate, is only an oversimplified description of the real catchment and climatic variables. The model should be looked upon as a useful tool, which provides guidance in solving a particular problem rather than the answer to all the hydrological problems. At present the entire subject of hydrological modelling is, to quote Dooge (1972), overshadowed by a proliferation of approaches and techniques, innumerous models based on the same technique, and a lack of adequate techniques to evaluate models in a particular situation or to choose between them.

11.2.1 Classification of mathematical models

In the mathematical or analytical models the functional relationships between the input, x(t), and the output, y(t), variables are represented as

$$f(x_t,\ y_t;\ \ \partial x/\partial t,\ \partial y/\partial t;\ \ \partial^2 x/\partial t^2,\ \partial^2 y/\partial t^2;\ \ldots\ \alpha_1, \alpha_2, \ldots) = 0 \qquad 11.3$$

where the α_i are parameters measured in the field or determined from data. When the variables are measured at discrete intervals, the partial derivatives are replaced by their finite difference approximations, $\partial x/\partial t \simeq \frac{1}{2}(x_{t+1} - x_{t-1})$, $\partial^2 x/\partial t^2 \simeq \frac{1}{2}(x_{t+1} - 2x_t + x_{t-1})$. Then eqn 11.3 becomes

$$f(x_t,\ y_t;\ \ x_{t-1},\ y_{t-1};\ \ x_{t-2},\ y_{t-2};\ \ldots\ \alpha_1,\ \alpha_2,\ \ldots) = 0 \qquad 11.4$$

This describes an input-output model. When the properties of a single hydrological variable are characterized by one or more equations, the models are referred to as *single-variable* models. It is usually necessary to use a simplified functional relationship and to accept some residual error ε, i.e.,

$$y_t = f(x_t, x_{t-1},\ x_{t-2},\ y_{t-1},\ y_{t-2},\ \ldots\ \alpha_1,\ \alpha_2,\ \ldots) + \varepsilon_t \qquad 11.5$$

Clarke (1973) orders the mathematical models into four main groups:
(1) stochastic-conceptual (SC)
(2) stochastic-empirical (SE)
(3) deterministic-conceptual (DC)
(4) deterministic-empirical (DE)
any of which may be classified as linear in the systems-theory sense (LST) or non-linear in the systems-theory sense (NLST). Groups 1 and 2 may be linear (LSR) or non-linear (NLSR) in the statistical regression sense. The models in any group may further be classed as lumped, probability-distributed, or geometrically distributed. A more detailed classification was given by Kisiel (1969).

If any of the variables is a random variable having a probability distribution, then the model is *stochastic* rather than statistical. This emphasizes the time-dependence of the hydrological variables in the model. Conversely, if all the variables are free from random variations, then the model is *deterministic*. The distinction between *conceptual* and *empirical* depends on whether or not the functional relationship between the variables is derived from consideration of the physical processes. This distinction is, however, artificial since many physical laws contain empirical constants. *Linearity* in the *systems analysis sense* means that solutions can be superimposed. The model is *linear* in the *statistical regression sense* if it is linear in the parameters to be estimated. For example, if the input x(t) and output y(t) are related by $y = \alpha + \beta x + \gamma x^2$, then the model would be linear in the statistical regression sense and non-linear in the systems-theory sense. The converse applies if $y = \alpha + x/\alpha$.

A lumped model ignores the spatial distribution of the input variables and parameters, which characterize the physical processes. There is also a lumping in time when daily average flows or rainfalls are substituted for the actual inputs. A *probability-distributed* model describes the spatial variation of the input variables without geometric reference to the points at which the input is measured or estimated, for example, a uniform probability distribution over the catchment. A *geometrically-distributed* model expresses the spatial variability in terms of the orientation of the network points relative to each other and their spacings.

An advanced and detailed treatment of stochastic modelling can be found in the book Dynamic Stochastic Models from Empirical Data by Kashyap and Rao (1976).

11.3 Generation of Sequences of a Single Hydrological Variable

The methods used to generate sequences of a hydrological variable at a given site depend both on the variable and the time period. One may be interested in the generation of rainfall sequences with a time increment of 5 or 10 minutes or of sequences of annual river discharges. The internal dependence (persistence) of data depends on the length of the time interval. Short period rainfall sequences are usually highly correlated whereas yearly rainfall sequences are not. Thus the first step is to investigate the observed record for persistence. A measure of persistence is the lag-one serial correlation coefficient, r_1, which gives a relation between the value at a given instant and that at one time interval earlier. For a purely random independently distributed variable r_1 differs from zero only by sampling variation and is close to one for sequences with strong internal dependence. Negative values of r_1, which imply that large values in the sequence tend to be followed by small values and vice-versa, are not common in hydrological sequences. The stationarity assumption implies that the joint probability distribution is the same for all times the same interval apart. The nature of this joint distribution is indicated by the scatter diagrams of pairs of values (x_t, x_{t+1}) separated by a constant lag of k time intervals, Fig. 11.5.

In the description of time series the first and second moment functions (i.e., the mean and variance) of the sample play the central role. The higher order moments may be required at times but, because of the third or higher power of the variable, their values tend to fluctuate extensively owing to the influence of the few extreme values of the sample.

If the data sequence is denoted by $[x_t]$, t = 1, 2, ... n, then the expected value of x or the mean, which is constant for a stationary process, is

$$\bar{x} = \frac{1}{n}\sum_{1}^{n} x_t \qquad 11.6$$

and is taken to be the estimate of the population mean. For a continuous variable the expected value can be expressed as

$$E[X_t] = \int_{-\infty}^{\infty} x_t \, dF(X_t) = \int_{-\infty}^{\infty} x_t \, f(x) \, dx \qquad 11.7$$

where F signifies the cumulative distribution function and f(x) the probability density function of x.

The variance, which is also constant for a stationary process, is given by

$$\sigma_x^2 = \mathrm{Var}[X_t] = \int_{-\infty}^{\infty} (x_t - \bar{x})f(x)\,dx = \frac{1}{n}\sum_1^n (x_t - \bar{x})^2 \qquad 11.8$$

The sample estimate of the variance is frequently designated by S^2. The best estimate of the population variance is $\sigma^2 = [1/(n - 1)]\Sigma(x_t - \bar{x})^2$.

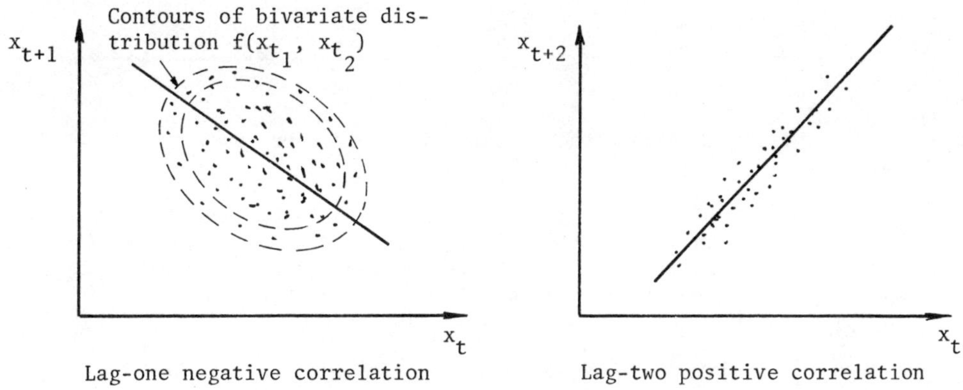

Fig. 11.5. Scatter diagrams showing correlation.

The covariance between x_t and x_{t+k} is called the *auto-covariance* at lag k and is given by

$$C(k) = \mathrm{Cov}[X_t, X_{t+k}] = \frac{1}{n}\Sigma(x_t - \bar{x})(x_{t+k} - \bar{x})$$
$$= \frac{1}{n}\Sigma x_t x_{t+k} - (\bar{x})^2 \qquad 11.9$$

When the expected value $\bar{x} = 0$, C(k) is called the auto-correlation function. When the lag is zero, C(0) is the variance of the sequence.

The serial or auto-correlation function is given by

$$\rho(k) = \frac{\mathrm{Cov}[X_t, X_{t+k}]}{\{\mathrm{Var}[X_t]\mathrm{Var}[X_{t+k}]\}^{1/2}} = \frac{C(k)}{C(0)} \qquad 11.10$$

for a stationary process. For a sample of n observations the best estimate is

$$r(k) = \frac{\sum_1^{n-k} (x_t - \bar{x})(x_{t+k} - \bar{x})}{(n - k)S^2} \qquad 11.11$$

For a purely random independent process r(k) = 0. The graphical representation of r(k) versus k is called the correlogram, Fig. 11.6.

The important question is what value of r(k) differs from zero only by sampling variation? For a strictly random sequence it has been shown that the serial corre-

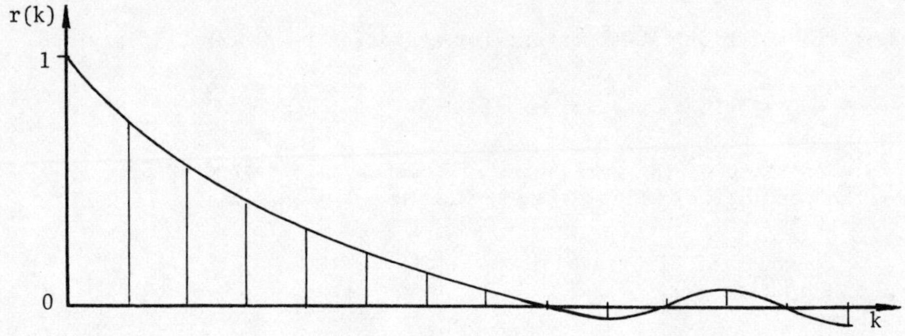

Fig. 11.6. Diagrammatic illustration of a correlogram.

lation coefficients of all lags other than zero have a large-sample standard error of $\pm 1/\sqrt{n}$. A simple approximate test of significance is thus given by comparison of r(k) with $\pm 2/\sqrt{n}$. Values of r(k) outside the range $\pm 2/\sqrt{n}$ indicate persistence in the data sequence. For lag-one an exact test yields that r_1 is significantly different from zero when it lies outside the range

$$-\frac{1}{n-1} \pm 1.96\,\frac{(n-2)}{(n-1)^{3/2}}$$

The distribution of r(k) = 0, which is asymptotically normally distributed with a mean

$$E[r(k)] = \frac{S_1^2 - S_2^2}{n-1} \qquad 11.12$$

and variance

$$\mathrm{Var}[r(k)] = \frac{S_2^2 - S_4}{n-1} + \frac{S_1^4 - 4S_1^2 S_2 + 4S_1 S_3 + S_2^2 - 2S_4}{(n-1)(n-2)} - E^2[r(k)] \qquad 11.13$$

where

$$S_j = \sum_{i=1}^{n} x_i^j \qquad 11.14$$

is used for testing the null hypothesis

$$H_o : r(k) = E[r(k)] = 0 \qquad 11.15$$

for zero correlation. A 5% level of significance is given by

$$t = \frac{r(k) - E[r(k)]}{\sqrt{\mathrm{Var}[r(k)]}} > t_{5\%} = 1.65 \qquad 11.16$$

i.e., if t > 1.65 the time series has a correlated generating structure at 5% significance level.

11.3.1 Internally independent random sequences

If the random component is an independent random variable, then a counting or Monte-Carlo process can be used to generate new sequences. The counting process is a sampling of simple events of a specific type. It rests on the unique relationship between the probability density function (p.d.f.), $f(x)$, and its cumulative distribution function (c.d.f.), i.e.,

$$F(x_t) = \int_{-\infty}^{x_t} f(x)\, dx \qquad 11.17$$

where $F(x_t)$ is defined over the interval (0, 1) and constitutes a probability scale. The p.d.f. for the process to be simulated has to be estimated from the observed record. The more common distributions are the log-normal, the gamma and the Pearson III distributions.

The cumulative distribution derived can now be entered with uniformly distributed random numbers r^*, distributed over the interval $0 \leq r^* \leq 1$. Such pseudo-random numbers can be generated by use of random number tables or by the many existing computer subroutines. (The use of sequences of pseudo-random numbers with distributions other than uniform is possible, for example, a normally distributed probability of x_t in a realization). The generation procedure is illustrated in Fig. 11.7.

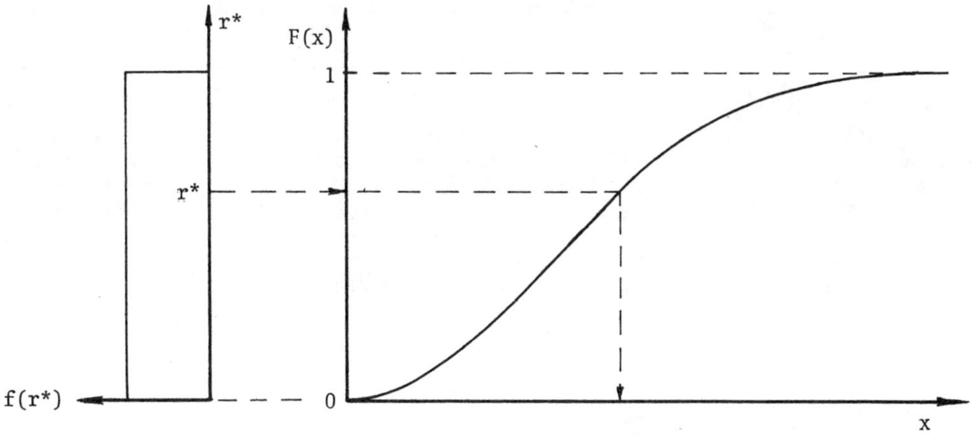

Fig. 11.7. Data generation by a counting process.

For a particular values of r^*, say r_t^*, the corresponding value of x_t is given by the inverse function of F, if known, that is

$$x_t = F^{-1}(r_t^*)$$

Thus, if

$$r_t^* = F(x_t) = \int_{-\infty}^{x_t} f(x)\, dx$$

then

$$\Pr[X \leq x_t] = F(x_t) = \Pr[r_t^* \leq F(x_t)] = \Pr[F^{-1}(r_t^*) \leq x_t]$$

and consequently $F^{-1}(r_t^*)$ is a variable that has $f(x)$ as its probability density function. The solution for the inverse function, except for simple forms of $f(x)$, is not easy and not required. The actual cumulative distribution function can be stored in the computer in tabular form and the sampling is carried out as illustrated in Fig. 11.7 for a sequence of as many data points as desired. The model generates internally independent purely random sequences.

11.3.2 Sequences with internal persistence

Most of the hydrological data sequences for monthly, daily, and hourly increments of time exhibit internal dependence. This means that the outcome of the next realization is not independent of what happened before but depends on the magnitudes of one or more preceding values. The system is said to have a memory. Several methods have been developed to deal with problems of this kind.

The simplest and most widely used method for generation of hydrological data sequences with internal dependence is the *auto-regressive scheme*. In this model the current value of the process is expressed as a function of the preceding values and a random term ε_t. Let $[z_t]$ represent the time series where each individual value of z_t is $x_t - \bar{z}$. Then

$$z_t = f(z_{t-1}, z_{t-2}, \ldots z_{t-k}) + \varepsilon_t \qquad 11.18$$

A special case occurs when the current value of z_t is expressed as a linear sum of the previous values, i.e.,

$$z_t = \alpha_1 z_{t-1} + \alpha_2 z_{t-2} + \ldots \alpha_k z_{t-k} + \varepsilon_t \qquad 11.19$$

The random variable ε_t is usually assumed to have zero mean, variance σ_ε^2, and to be normally and independently distributed in time and to be independent of z and other values, although different distributions can be allowed for (Raudkivi and Lawgun, 1972). The order of k is determined from serial correlation analysis and (k + 2) coefficients have to be determined, i.e., k values of α, $\bar{z}$ and σ_ε^2. The principles of estimation of the coefficients α have been outlined, e.g., by Jenkins and Watts (1968) and in hydrological context by Carlson et al. (1970). A practical estimate may be obtained from

$$\begin{bmatrix} 1 & r_1 & r_2 & \ldots & r_{k-1} \\ r_1 & 1 & r_1 & \ldots & r_{k-2} \\ r_2 & r_1 & 1 & \ldots & r_{k-3} \\ \cdot & \cdot & \cdot & & \\ \cdot & \cdot & \cdot & & \\ r_{k-1} & r_{k-2} & r_{k-3} & \ldots & 1 \end{bmatrix} \begin{bmatrix} \alpha_1 \\ \alpha_2 \\ \alpha_3 \\ \cdot \\ \cdot \\ \alpha_k \end{bmatrix} = \begin{bmatrix} r_1 \\ r_2 \\ r_3 \\ \cdot \\ \cdot \\ r_k \end{bmatrix} \qquad 11.20$$

where r_i is the empirical value of the i-th auto-correlation coefficient. This yields for a second order process (two steps of memory)

$$\alpha_2 = \frac{r_2 - r_1^2}{1 - r_1^2} \quad \text{and} \quad \alpha_1 = \frac{r_1 - r_1 r_2}{1 - r_1^2} \qquad 11.21$$

whence

$$z_t = \left(\frac{r_1 - r_1 r_2}{1 - r_1^2}\right) z_{t-1} + \left(\frac{r_2 - r_1^2}{1 - r_1^2}\right) z_{t-2} + \varepsilon_t \qquad 11.22$$

For a first order process

$$z_t = r_1 z_{t-1} + \varepsilon_t \qquad 11.23$$

For the independently distributed random component

$$\sigma_\varepsilon^2 = \frac{1}{1 - r_1^2} \qquad 11.24$$

and an estimate of this variance is given by

$$S_\varepsilon^2 = (\frac{n-1}{n})(1 - r_1^2)(\frac{1}{n-3}) \sum_1^n (x_t - \bar{x})^2 \qquad 11.25$$

The correlogram of the population of z_t is represented by $r_k = r_1^k$. For a second order process the estimate of σ_ε^2 is

$$S_\varepsilon^2 = \frac{n-2}{n-5} [C(0) - \alpha_1 C(1) - \alpha_2 C(2)] \qquad 11.26$$

In some records longer persistence effects are apparent than can be readily reproduced by the auto-regressive models. This long term persistence has become known as the Hurst phenomenon. Mandelbrot and Wallis (1969) showed that the fractional Gaussian noise model has an auto-correlation structure corresponding to a very long memory.

11.3.3 The Thomas-Fiering model

Thomas and Fiering (1962) developed a model to generate monthly streamflows. In this method the n-years of record are separated into twelve records, one for each month, and twelve linear regression equations are used. The record for a given month is regressed upon that for the preceding month. The seasonal variation (periodicity) is catered for by the use of the monthly regression relationships. The model assumes, in fact, lag-one (monthly) persistence. This type of persistence is caused by storage effects of water as soil moisture and groundwater. There may also be some carry-over effect in seasonal weather patterns.

For each month the regression relationship upon the month before has to be established, Fig. 11.8.

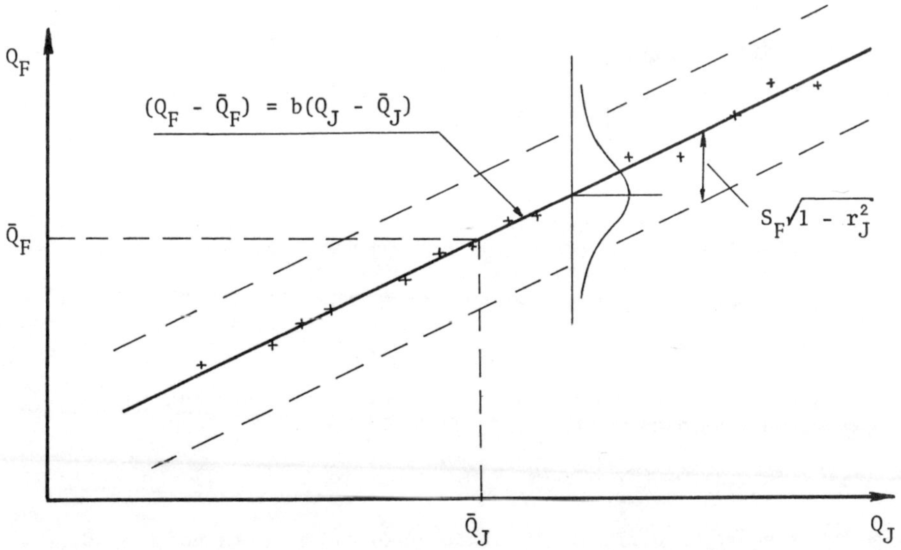

Fig.11.8. Regression of monthly flows of February upon January.

For linear regression the best estimate of Y given X_i is

$$\hat{Y}_i = \bar{Y} + b(X_i - \bar{X}) \tag{11.27}$$

where

$$b_{YX} = \frac{\sum_1^n X_i Y_i - n\bar{X}\bar{Y}}{\sum_1^n X_i^2 - n\bar{X}^2} \tag{11.28}$$

The standard error of the estimate of Y_i is given by $\sigma_Y\sqrt{1 - r^2}$, being the measure of the random variation or variation of Y_i not explained by regression. The variance can be expressed as

$$\sigma^2 = \sigma^2 r^2 + \sigma^2(1 - r^2) \tag{11.29}$$

where σ^2 is the total variance, $\sigma^2 r^2$ is the variance explained by regression and $\sigma^2(1 - r^2)$ is the variance not accounted for by regression. The correlation coefficient is

$$r = \frac{\sum_1^n X_i Y_i - n\bar{X}\bar{Y}}{\sqrt{(\sum_1^n X_i^2 - n\bar{X}^2)(\sum_1^n Y_i^2 - n\bar{Y}^2)}} \tag{11.30}$$

Thus,

$$(Q_F - \bar{Q}_F) = b_J(Q_J - \bar{Q}_J) + t_i S_F\sqrt{1 - r_J^2} \tag{11.31}$$

where t_i is a random, usually normally and independently distributed standardized variate with zero mean and unit variance, and S_F is the best estimate of $\sigma_F \equiv \sigma_Y$, where

$$S_Y^2 = \frac{1}{n - 1}\sum_1^n (Y_i - \bar{Y})^2$$

Generally,

$$Q_{i+1} = \bar{Q}_{i+1} + b_i(Q_i - \bar{Q}_i) + t_i S_{i+1}\sqrt{1 - r_i^2} \tag{11.32}$$

where i ranges from 0 to 12, and $b_i = r_i S_{i+1}/S_i$. For generation 48 parameters have to be established, 12 of $\bar{Q}$, 12 of $\bar{b}$, 12 of $\bar{S}$ and 12 of r. Twelve or more years of record are required to estimate these parameters with reasonable accuracy. For twelve years of record the correlation coefficients r_i are each based on ten degrees of freedom and the S_{i+1} values are based on eleven degrees of freedom. The constant spread of the random component (Fig. 11.8) leads to a too large a spread for small discharges and vice versa. It may also be found that values in the generated sequences are sometimes negative. It has been recommended that these negative values be retained and used to calculate subsequent values, but that they be replaced by zeros in the completed sequence. No negative values occur if the model is fitted to log Q instead of Q.

The model can be extended to higher orders of persistence by using multiple regression equations, such as

$$Q_{i+1} = \bar{Q}_{i+1} + b_{1,i}(Q_i - \bar{Q}_i) + b_{2,i}(Q_{i-1} - \bar{Q}_{i-1}) + t_i S_{i+1}\sqrt{1-r_i^2} \tag{11.33}$$

where r_i is now the multiple correlation coefficient obtained when Q_{i+1} is regressed upon Q_i and Q_{i-1}. Colston and Wiggert (1968) also included rainfall on the left hand side of the equation.

Special problems are encountered with application of this method to ephemeral streams. Clarke (1973) suggests that when the historical record is long enough to yield parameters with acceptable accuracy, the following procedure can be used:
(1) For each month i, record the number of months n_i with flow out of the total record of N years and define the probability $p_i = n_i/N$.
(2) Calculate Q_i and S_i and fit the Thomas-Fiering model to successive pairs of months for which flows have occurred. Generate sequences of monthly flow as follows:
(a) For month i, choose a pseudo-random number r* uniformly distributed between (0,1). If $p_i > r^* > 0$ flow will occur in this month; if $r^* > p_i$ no flow will occur in month i and the process is repeated for month i + 1.
(b) If month i is the first month of the year for which flow is to occur, then select a pseudo-random normal deviate (i.e., $y = x_i - \bar{x}$) for a distribution with mean and variance equal to the mean monthly flow and variance of the flows in this month, respectively.

Example 11.1. A certain river has discharges in excess of 10^6 m^3 per month ($\bar{Q} < 0.4$ m^3/s) only during the monsoon season. The available record of flows is as follows:

Year	Mean monthly flow $\bar{Q}$ in m^3/s			
	July	Aug.	Sept.	Oct.
1	49	154	114	-
2	56	161	115	34
3	27	103	126	-
4	25	107	27	36
5	29	137	78	-
6	42	113	47	-
7	37	130	-	-
8	48	140	109	44
9	42	85	63	38
10	55	120	59	-

Use the Thomas-Fiering model and generate monthly flows.

The probabilities $p_i = n_i/N$ are $p_J = 1$, $p_A = 1$, $p_S = 0.9$, $p_O = 0.4$. The mean $\bar{Q}$ and variance of the monthly flows are:

	July	Aug.	Sept.	Oct.
$\bar{Q}$	41	125	82	38
Variance	127.56	567.56	1241.75	20
S	11.29	23.82	35.24	4.47

where $\text{Var} = [1/(n-1)]\Sigma(Q - \bar{Q})^2$. Fit the Thomas-Fiering model to the flows of July to Oct., eqn. 11.31,

$$Q_A = 125 + 1.034(Q_J - 41) + t_1 \times 23.82\sqrt{1 - 0.490^2}$$

where the b_i are evaluated by eqn 11.28 and r by eqn 11.30

$$Q_S = 82 + 0.662(Q_A - 125) + t_2 \times 35.24\sqrt{1 - 0.474^2}$$

where n = 9 in eqn 11.28.

$$Q_O = 38 - 0.039(Q_S - 82) + t_3 \times 4.47\sqrt{1 - (-0.526)^2}$$

in which b and r values are evaluated with n = 4.

Since $p_J = 1$ and $p_A = 1$ generated flows will always occur in July and August. No flow will occur in September if the generated uniformly distributed random number $r^* > 0.9$ and none in October if $r^* > 0.4$. The first value of the *normally distributed* random number sequence starts the generating process with the flow for July. For example, $r^* = -1.070$ the for the first year

$$Q_J = 41 + 11.29(-1.070) = 28.92 \text{ m}^3/\text{s}.$$

The August flow of the first year is obtained using this July value and the second value of the normally distributed random number sequence, for example, $r^* = 0.765$

$$Q_A = 125 + 1.034(28.92 - 41) + 0.765 \times 23.82\sqrt{1 - 0.490^2} = 128.39 \text{ m}^3/\text{s}$$

and this allows Q_S for the first year to be calculated, that is, if the other sequence of random numbers shows that there is a flow in September, etc. The same procedure is followed for the subsequent years.

11.3.4 Generation of data sequences for daily and shorter time intervals

In general, the shorter the time interval the more difficult it is to find a suitable model for data generation. Also the cost to the user of these models (in terms of computer time) increases rapidly. The majority of models introduced for synthesis of data for short time intervals are of the Markov chain type. The central feature of this method is the probability transition matrix. To illustrate the idea consider a rainfall sequence of $[x_1, x_2, \ldots]$ mm of rain and assume that this sequence can be grouped into the intervals 0, 0-1 and 0-2 mm of rain (in practice there can be many more intervals), which are known as states 0, 1, 2 (using sequential numbering). From the observed record one can determine the number of occurrences of no rain, up to 1 mm and up to 2 mm of rain following an interval with no rain; likewise, following an interval with up to 1 mm of rain, and following one with up to 2 mm of rain. Designating with n_{oo} the number of instances where an interval with no rain follows one with no rain, n_{o1} the number of instances where 0-1 mm of rain occurs in the interval following one with no rain, etc., then the probability of an interval without rain following one without rain is

$$p_{oo} = n_{oo}/(n_{oo} + n_{o1} + n_{o2})$$

where the sum $(n_{oo} + n_{o1} + n_{o2})$ is the total number of increments with the starting state of no rain. Similarly

$$p_{o1} = n_{o1}/(n_{oo} + n_{o1} + n_{o2})$$

$$p_{1o} = n_{1o}/(n_{1o} + n_{11} + n_{12}) \quad \text{etc.}$$

Hence, a probability matrix can be established as shown

		Final state 0	1	2
Starting state	0	p_{oo}	p_{o1}	p_{o2}
	1	p_{1o}	p_{11}	p_{12}
	2	p_{2o}	p_{21}	p_{22}

The conditional probability $P(s_i|s_j)$ of the process, if in state s_i, of moving to state s_j is the one-step probability. If such probabilities are known for all pairs of states, then they may be arranged in a square matrix form, known as the *transition (probability) matrix*, $P = [p_{ij}]$, where p_{ij} is the element in the i-th row and j-th column. All p_{ij} values are non-negative. The sum of the p_{ij} values in any single row equals one.

The simplest procedure for generation of a synthetic sequence is then as follows: Generate a pseudo-random number r*, uniformly distributed between 0 and 1, and given the initial state, for example, state 0, then if $r^* < p_{oo}$ the final state is 0, if $p_{oo} < r^* < p_{oo} + p_{o1}$ the final state is 1, and if $r^* > p_{oo} + p_{o1}$ the final state is 2. For the second number the state given above is the initial condition, for example, state 2. Then generate a new random number and continue as above.

A Markov chain model for rainfall generation of this simple type was discussed by Raudkivi and Lawgun (1970). A serious limitation of such models is that the model will reproduce only transitions which have been observed. Consequently, extreme events may not be, or are inadequately, represented.

This basic model for generation by the Markov process can be combined with the memory effect. Thus, the lag-one Markov process for generation of synthetic sequences is

$$x_{t+1} - \mu = \rho_1(x_t - \mu) + \sigma_x\sqrt{1 - r_1^2}\,\varepsilon_{t+1} \tag{11.34}$$

where μ is the mean, σ_x is the standard deviation of x, ρ_1 is the lag-one serial correlation coefficient for x, and ε_{t+1} is a random component that has zero mean and unit variance and is independent of x. The sample estimates of $\bar{x}$, S_x, and r_1 replace the unknown population values of μ, σ_x and ρ_1. The lag-one Markov process is a third-order stationary process. The synthetic sequences generated by it will resemble the historical events with $\bar{x}$, S_x, r_1 and skewness $\gamma_x (= \mu_3/\sigma^3)$. Denoting $y_t = (x_t - \bar{x})/S_x$ eqn 11.34 becomes

$$y_{t+1} = r_{y1}y_t + \sqrt{1 - r_{y1}^2}\,\varepsilon_{t+1} \tag{11.35}$$

If the y_t values are normally distributed eqn 11.35 represents a strictly stationary process, because then the process is completely defined by the first and second statistical moments (μ and σ^2). In general, in terms of the first and second moments eqn 11.35 represents a weakly stationary process. Sequences of realizations of eqn 11.35 produce mutually independent sequences, but not always with the correct skewness. Matalas (1967) showed, for example, that for gamma distributed events, the skewness values are not matched. If the historical events follow a 3-parameter log-normal distribution, then matching skewness values is possible. When (x - a) is log-normally distributed

$$y = \ell n(x - a) \tag{11.36}$$

where a is the lower bound of the random variate x. For this distribution the best estimates are

$$\hat{\mu}_x = a + \exp[1/(2\sigma_y^2) + \mu_y] \tag{11.37}$$

$$\hat{\sigma}_x^2 = \exp[2(\sigma_y^2 + \mu_y)] - \exp(\sigma_y^2 + 2\mu_y) \tag{11.38}$$

$$\hat{\gamma}_x = \frac{\exp(3\sigma_y^2) - 3\exp(\sigma_y^2) + 2}{[\exp(\sigma_y^2) - 1]^{3/2}} \tag{11.39}$$

When y is generated by the lag-one Markov process

$$y_{t+1} - \mu_y = \rho_{y1}(y_t - \mu_y) + \sqrt{1 - \rho_{y1}^2}\,\sigma_y\varepsilon_{t+1} \tag{11.40}$$

and in terms of x, the generating process is

$$x_{t+1} = a + \{\exp[\mu_y(1 - \rho)]\}(x_t - a)^{\rho}\,\delta_{t+1} \qquad 11.41$$

where

$$\rho = \rho_{y1} \quad \text{and} \quad \delta_{t+1} = \exp[(1 - \rho_{y1}^2)^{1/2}\,\sigma_y \varepsilon_{t+1}]$$

When x_t and δ_{t+1} are assumed to be independent

$$\hat{\rho}_{x1} = r_1 = [\exp(\sigma_y^2 \rho_{y1}) - 1]/[\exp(\sigma_y^2) - 1] \qquad 11.42$$

The values of a, μ_y, σ_y and ρ_{y1} are obtained by equating the historical values of $\bar{x}$, S_x, γ_x and r_1 to eqns 11.37, 11.38, 11.39 and 11.42, respectively. With these values eqn 11.40 may be used to generate a sequence of y values. Adding a to each of the anti-logs of y leads to a sequence of x values that will resemble the historical sequence. The use of values of a, μ_y, σ_y and ρ_{y1} obtained from the logarithms of historical events, leads to a synthetic sequence which will not resemble the historical one.

A model for simulation of rainfall sequences with 10 minutes as the time increment was discussed by Raudkivi and Lawgun (1974). The duration of the rainfall events (periods of continuous rain) are modelled by an auto-regressive scheme, which takes into account the skewness of the random component. The rainfall depths within each duration are modelled with the aid of the probability transition matrix (extrapolated to produce extreme events), and the intervals with no rain between the periods of continuous rainfall are modelled by a Monte-Carlo process using the observed cumulative distribution function.

11.4 Generation of Sequences of Several Hydrological Variables

The Thomas-Fiering model: A frequent requirement is to generate simultaneous sequences of two or more variables, for example, flow sequences of two tributaries of river, or river discharge and irrigation water requirements, etc. The models here have to preserve the serial correlation structure for each variable and the cross-correlational structure between all pairs of variables.

For monthly flow data, the Thomas-Fiering model can be expanded (Bernier, 1971). Using standardized variable $[y_t = (x_t - \bar{x})/\sigma_x]$ for station 1 and $[z_t = (x_t - \bar{x})/\sigma_x]$ for station 2, the data sequences are

$$[y_1, y_2, \ldots y_t, \ldots y_n]$$

and

$$[z_1, z_2, \ldots z_t, \ldots z_n]$$

and the Thomas-Fiering model becomes

$$\begin{aligned} y_t &= b_{11} y_{t-1} + b_{12} z_{t-1} + \eta_t \\ z_t &= b_{21} y_{t-1} + b_{22} z_{t-1} + \zeta_t \end{aligned} \qquad 11.43$$

or

$$\begin{bmatrix} y_t \\ z_t \end{bmatrix} = \begin{bmatrix} b_{11} & b_{12} \\ b_{21} & b_{22} \end{bmatrix} \begin{bmatrix} y_{t-1} \\ z_{t-1} \end{bmatrix} + \begin{bmatrix} \eta_t \\ \zeta_t \end{bmatrix}$$

For each pair of months seven parameters must be estimated for the model: four b_{ij}, var η_t, var ζ_t and cov (η_t, ζ_t). Hence, 7 x 12 = 84 parameters are needed for the year. In addition 12 parameters are required for each of the monthly means, $\bar{y}_i$ and $\bar{z}_i$, and the standard deviations, S_{yi} and S_{zi}. Thus, a total of 132 parameters is required.

The sequence for estimation of the parameters is as follows:
(1) Tabulate data from the n-years of record:

	Jan.		Febr.		March	
	Stat. 1	Stat. 2	Stat. 1	Stat. 2	Stat. 1	Stat. 2
Year 1	y_1	z_1	y_2	z_2	y_3	z_3
Year 2	y_{13}	z_{13}	y_{14}	z_{14}	y_{15}	z_{15}
Year 3	y_{25}	z_{25}	y_{26}	z_{26}	y_{27}	z_{27}
.	.	.	.	.	.	.
.	.	.	.	.	.	.
.	.	.	.	.	.	.

(2) Calculate for each pair of months, as shown for January and February, the following values:

$$\Sigma_{11}(1, 1) = (y_1^2 + y_{13}^2 + \ldots)/n$$

$$\Sigma_{11}(2, 2) = (z_1^2 + z_{13}^2 + \ldots)/n$$

$$\Sigma_{11}(1, 2) = (y_1 z_1 + y_{13} z_{13} + \ldots)/n = \Sigma_{11}(2, 1)$$

$$\Sigma_{11} = \begin{bmatrix} \Sigma_{11}(1, 1) & \Sigma_{11}(1, 2) \\ \Sigma_{11}(2, 1) & \Sigma_{11}(2, 2) \end{bmatrix} \qquad 11.44$$

$$\Sigma_{22}(1, 1) = (y_2^2 + y_{14}^2 + \ldots)/n$$

$$\Sigma_{22}(2, 2) = (z_2^2 + z_{14}^2 + \ldots)/n$$

$$\Sigma_{22}(1, 2) = (y_2 z_2 + y_{14} z_{14} + \ldots)/n = \Sigma_{22}(2, 1)$$

$$\Sigma_{22} = \begin{bmatrix} \Sigma_{22}(1, 1) & \Sigma_{22}(1, 2) \\ \Sigma_{22}(2, 1) & \Sigma_{22}(2, 2) \end{bmatrix} \qquad 11.45$$

$$\Sigma_{21}(1, 1) = (y_2 y_1 + y_{14} y_{13} + \ldots)/n$$

$$\Sigma_{21}(1, 2) = (y_2 z_1 + y_{14} z_{13} + \ldots)/n$$

$$\Sigma_{21}(2, 1) = (z_2 y_1 + z_{14} y_{13} + \ldots)/n$$

$$\Sigma_{21}(2, 2) = (z_2 z_1 + z_{14} z_{13} + \ldots)/n$$

$$\Sigma_{21} = \begin{bmatrix} \Sigma_{21}(1, 1) & \Sigma_{21}(1, 2) \\ \Sigma_{21}(2, 1) & \Sigma_{21}(2, 2) \end{bmatrix} \qquad 11.46$$

(3) The estimates of b_{ij} are given by

$$\begin{bmatrix} b_{11} & b_{12} \\ b_{21} & b_{22} \end{bmatrix} = \begin{bmatrix} \Sigma_{21}(1, 1) & \Sigma_{21}(1, 2) \\ \Sigma_{21}(2, 1) & \Sigma_{21}(2, 2) \end{bmatrix} \begin{bmatrix} \Sigma_{11}(2, 2) & -\Sigma_{11}(1, 2) \\ -\Sigma_{11}(2, 1) & \Sigma_{11}(1, 1) \end{bmatrix} \Big/ \Delta \qquad 11.47$$

where $\Delta = \Sigma_{11}(1, 1)\ \Sigma_{11}(2, 2) - \Sigma_{11}(2, 1)\ \Sigma_{11}(1, 2)$. Equation 11.47 in expanded form is

$$b_{11} = [\Sigma_{21}(1, 1)\ \Sigma_{11}(2, 2) - \Sigma_{21}(1, 2)\ \Sigma_{11}(2, 1)]/\Delta$$

$$b_{12} = [\Sigma_{11}(1, 1)\ \Sigma_{21}(1, 2) - \Sigma_{21}(1, 1)\ \Sigma_{11}(1, 2)]/\Delta$$

$$b_{21} = [\Sigma_{21}(2, 1)\ \Sigma_{11}(2, 2) - \Sigma_{21}(2, 2)\ \Sigma_{11}(2, 1)]/\Delta$$

$$b_{22} = [\Sigma_{11}(1, 1)\ \Sigma_{21}(2, 2) - \Sigma_{21}(2, 1)\ \Sigma_{11}(1, 2)]/\Delta$$

(4) Calculate the variances, var η_t and var ζ_t, and covariance, cov (η_t, ζ_t):

$$\begin{aligned} \text{var } \eta_t &= \Sigma_{22}(1, 1) - b_{11}\ \Sigma_{21}(1, 1) - b_{12}\ \Sigma_{21}(1, 2) \\ \text{var } \zeta_t &= \Sigma_{22}(2, 2) - b_{21}\ \Sigma_{21}(2, 1) - b_{22}\ \Sigma_{21}(2, 2) \qquad 11.48 \\ \text{cov } (\eta_t, \zeta_t) &= \Sigma_{22}(1, 2) - b_{11}\ \Sigma_{21}(2, 1) - b_{12}\ \Sigma_{21}(2, 2) \end{aligned}$$

The chance dependent random terms, η^*_{i+1} and ζ^*_{i+1}, are generated with the aid of a bivariate normal distribution. For this the Box-Müller (1958) method of pseudo-random normally distributed deviates (zero mean and unit variance) may be used. By this method the pseudo-random numbers r^*_1 and r^*_2, uniformly distributed over the interval (0, 1) are transformed to values of u_1 and u_2, where

$$u_1 = (-2\ \ell n\ r^*_1)^{1/2} \cos(2\ \pi r^*_2)$$

and 11.49

$$u_2 = (-2\ \ell n\ r^*_1)^{1\ 2} \sin(2\ \pi r^*_2)$$

With these the values of η^*_{i+1} and ζ^*_{i+1} are

$$\eta^*_{i+1} = u_1 \sqrt{\text{var } \eta_t} \qquad 11.50$$

and

$$\zeta^*_{i+1} = u_1[\text{cov}(\eta_t, \zeta_t)/\sqrt{\text{var } \eta_t}] + u_2[\text{var } \zeta_t - \frac{\text{cov}^2(\eta_t, \zeta_t)}{\text{var } \eta_t}]^{1/2} \qquad 11.51$$

(5) The generation may be started with the mean values of y_{t-1} and z_{t-1} (e.g., with $\bar{y}_J$ and $\bar{z}_J$). Then $y_1 = 0$ and $z_1 = 0$, and

$$y_2 = b_{11} \times 0 + b_{12} \times 0 + \eta^*_2$$

$$z_2 = b_{21} \times 0 + b_{22} \times 0 + \zeta^*_2$$

$$y_3 = b_{11}y_2 + b_{12}z_2 + \eta^*_3$$

$$z_3 = b_{21}y_2 + b_{22}z_2 + \zeta^*_3$$

From the above the values at station 1 and 2 are

$$Y_J = y_1 S_{yJ} + \bar{y}_J \qquad \text{and} \qquad Z_J = z_1 S_{zJ} + \bar{z}_J$$

Matalas' multivariate model: Matalas (1967) presented a model which requires the estimation of fewer parameters than the multivariate Thomas-Fiering model. In the auto-regressive model it was assumed that the sequence was stationary. Matalas assumes, in addition, that the correlation between the sequences depends only on the time increment (lag) and not on the actual time. A weakly stationary multivariate generating process is defined as

$$\underset{\sim}{y}_{t+1} = \underset{\sim}{A}\underset{\sim}{y}_t + \underset{\sim}{B}\underset{\sim}{\varepsilon}_{t+1} \qquad 11.52$$

where y_t and y_{t+1} are standardized variates ($y_t = (x_t - \bar{x})/\sigma_x$) and (m x 1) matrices (m-dimensional vectors) at time points t and t + 1, respectively; $\underset{\sim}{A}$ and $\underset{\sim}{B}$ are (m x m) square matrices of coefficients; and $\underset{\sim}{\varepsilon}_{t+1}$ is a m-dimensional vector of values which are normally and independently distributed with zero mean and unit variance and independent of the elements of x_t. Matalas shows that the elements of the matrix $\underset{\sim}{A}$ are given by

$$\underset{\sim}{A} = \underset{\sim}{M}_1\underset{\sim}{M}_0^{-1} \qquad 11.53$$

where $\underset{\sim}{M}_0$ and $\underset{\sim}{M}_1$ are the matrices of zero and lag-one cross-covariances, respectively, and $\underset{\sim}{M}_0^{-1}$ is the inverse of $\underset{\sim}{M}_0$. For two stations the values of the two sequences, y_t and z_t, at time t are given by the vector

$$\begin{bmatrix} y_t \\ z_t \end{bmatrix} \qquad 11.54$$

and assuming that both sequences have n terms, then

$$\underset{\sim}{M}_0 = \begin{bmatrix} \frac{1}{n}\Sigma y_t^2 & \frac{1}{n}\Sigma y_t z_t \\ \frac{1}{n}\Sigma y_t z_t & \frac{1}{n}\Sigma y_t^2 \end{bmatrix} \qquad 11.55$$

$$\underset{\sim}{M}_1 = \begin{bmatrix} \frac{1}{n}\Sigma y_t y_{t-1} & \frac{1}{n}\Sigma y_t z_{t-1} \\ \frac{1}{n}\Sigma z_t y_{t-1} & \frac{1}{n}\Sigma z_t z_{t-1} \end{bmatrix} \qquad 11.56$$

If more than two sequences are involved, i.e., m-sequences, then the notation $y_t^{(1)}$, $y_t^{(2)}$, ... $y_t^{(m)}$ is used to distinguish the sequences. The elements of the matrix $\underset{\sim}{B}$ are given by the solution of

$$\underset{\sim}{B}\underset{\sim}{B}^T = \underset{\sim}{M}_0 - \underset{\sim}{M}_1\underset{\sim}{M}_0^{-1}\underset{\sim}{M}_1^T = \underset{\sim}{M}_0 - \underset{\sim}{A}\underset{\sim}{M}_1^T \qquad 11.57$$

where $\underset{\sim}{B}^T$ and $\underset{\sim}{M}_1^T$ are the tranposes of the matrices $\underset{\sim}{B}$ and $\underset{\sim}{M}_1$, respectively. Matrix $\underset{\sim}{B}\underset{\sim}{B}^T$ is symmetrical and Young (1968) showed that the matrix $\underset{\sim}{B}$ may be replaced by one of triangular form

$$\underset{\sim}{B} = \begin{bmatrix} b_{11} & 0 & 0 - - - - 0 \\ b_{21} & b_{22} & 0 - - - - 0 \\ b_{31} & b_{32} & b_{33} - - - 0 \\ \cdot & \cdot & \cdot \quad\quad\quad \cdot \\ \cdot & \cdot & \cdot \quad\quad\quad \cdot \\ b_{m1} & b_{m2} & b_{m3} - - - b_{mm} \end{bmatrix} \qquad 11.58$$

The inverse of a 2 x 2 symmetric matrix

$$\begin{bmatrix} a & c \\ c & b \end{bmatrix} \text{ is } \frac{1}{(ab - c^2)} \begin{bmatrix} b & -c \\ -c & a \end{bmatrix}$$

The elements in the diagonal are interchanged, signs of the off-diagonal terms are changed and all four terms are divided by $(ab - c^2)$. This procedure gives tha $\underset{\sim}{M}_0^{-1}$ matrix, and the $\underset{\sim}{A}$ matrix is obtained by multiplying $\underset{\sim}{M}_1$ by $\underset{\sim}{M}_0^{-1}$

$$\underset{\sim}{A} = \begin{bmatrix} a_{11} & a_{12} \\ a_{21} & a_{22} \end{bmatrix}$$

In general, the transpose of a matrix

$$\underset{\sim}{C} = \begin{bmatrix} c_{11} & c_{12} & \cdots & c_{1n} \\ c_{21} & c_{22} & \cdots & c_{2n} \\ \cdot & \cdot & & \cdot \\ \cdot & \cdot & & \cdot \\ \cdot & \cdot & & \cdot \\ c_{m1} & c_{m2} & & c_{mn} \end{bmatrix} \text{ is the matrix } \underset{\sim}{C}^T = \begin{bmatrix} c_{11} & c_{21} & \cdots & c_{m1} \\ c_{12} & c_{22} & \cdots & c_{m2} \\ \cdot & \cdot & & \cdot \\ \cdot & \cdot & & \cdot \\ \cdot & \cdot & & \cdot \\ c_{1n} & c_{2n} & & c_{mn} \end{bmatrix}$$

Thus,

$$\underset{\sim}{B}\underset{\sim}{B}^T = \begin{bmatrix} b_{11} & 0 \\ b_{21} & b_{22} \end{bmatrix} \begin{bmatrix} b_{11} & b_{21} \\ 0 & b_{22} \end{bmatrix} = \begin{bmatrix} b_{11}^2 & b_{11}b_{21} \\ b_{11}b_{21} & b_{21}^2 + b_{22}^2 \end{bmatrix} \qquad 11.59$$

Equating these elements with the right hand side of eqn 11.57 leads to the values of b_{11}, b_{21} and b_{22}. The two variable model is thus

$$y_t = a_{11}y_{t-1} + a_{12}z_{t-1} + b_{11}\underset{\sim}{\varepsilon}_{(t+1)y}$$

$$z_t = a_{21}y_{t-1} + a_{22}z_{t-1} + b_{21}\underset{\sim}{\varepsilon}_{(t+1)y} + b_{22}\varepsilon_{t+1)z} \qquad 11.60$$

where $\varepsilon_{(t+1)y}$ and $\varepsilon_{(t+1)z}$ are two pseudo-random normally distributed variates. The synthetic sequences may be started from their mean value, i.e., $y_0 = 0$, $z_0 = 0$.

For a second order auto-regression, eqn 11.52 is written as

$$y_{t+2} = \underset{\sim}{A}y_{t+1} + \underset{\sim}{B}y_t + \underset{\sim}{C}\varepsilon_{t+2}$$

or

$$y_t = \underset{\sim}{A}y_{t-1} + \underset{\sim}{B}y_{t-2} + \underset{\sim}{C}\varepsilon_t \qquad 11.61$$

where y_t is a m-dimensional vector, and

$$\underset{\sim}{A} = (\underset{\sim}{I} - \underset{\sim}{B})\underset{\sim}{M}_1\underset{\sim}{M}_0^{-1} \qquad 11.62$$

$$\underset{\sim}{B} = (\underset{\sim}{M}_2 - \underset{\sim}{M}_1\underset{\sim}{M}_0^{-1}\underset{\sim}{M}_1^T)(\underset{\sim}{M}_0 - \underset{\sim}{M}_1\underset{\sim}{M}_0^{-1}\underset{\sim}{M}_1^T)^{-1} \qquad 11.63$$

$$\underset{\sim}{C}\underset{\sim}{C}^T = \underset{\sim}{M}_0 - \underset{\sim}{A}\underset{\sim}{M}_1^T - \underset{\sim}{B}\underset{\sim}{M}_2^T \qquad 11.64$$

I is the (m x m) unit matrix (diagonal elements equal to one, all others zero), $\underset{\sim}{M}_0$

and $\underset{\sim}{M}_1$ are as defined before, and

$$\underset{\sim}{M}_2 = \begin{bmatrix} \frac{1}{n}\Sigma y_t^{(1)} y_{t-2}^{(1)} & \frac{1}{n}\Sigma y_t^{(1)} y_{t-2}^{(2)} & \cdots & \frac{1}{n}\Sigma y_t^{(1)} y_{t-2}^{(m)} \\ \frac{1}{n}\Sigma y_t^{(2)} y_{t-2}^{(1)} & \frac{1}{n}\Sigma y_t^{(2)} y_{t-2}^{(2)} & \cdots & \frac{1}{n}\Sigma y_t^{(2)} y_{t-2}^{(m)} \\ \vdots & \vdots & \cdots & \vdots \\ \frac{1}{n}\Sigma y_t^{(m)} y_{t-2}^{(1)} & \frac{1}{n}\Sigma y_t^{(m)} y_{t-2}^{(2)} & \cdots & \frac{1}{n}\Sigma y_t^{(m)} y_{t-2}^{(m)} \end{bmatrix} \qquad 11.65$$

is the lag-two serial and cross-correlation matrix.

ARMA and ARIMA models: The ARMA (auto-regressive-moving-average) and ARIMA (auto-regressive-integrated-moving-average) models are two powerful methods which combine the features of auto-regressive and moving average models. The moving average model itself has not been successful in hydrological applications, but the combined auto-regressive and moving average models are capable of representing stationary sequences as well as certain types of non-stationary sequences. For details, the reader is referred to Box and Jenkins (1970). The ARIMA model is also discussed by Clarke (1973) and Spolia and Chander (1974) showed that the Nash reservoir model is a special case of the ARMA formulation.

11.5 Hydrological Models

Under this heading an attempt will be made to outline the concept of the hydrological models designed to simulate catchment performance, for example, the discharge from the catchment as a function of time given the rainfall input and the catchment parameters. Since there are very many such models, no attempt will be made even to list them and only a few will be referred to by way of examples. These models are primarily aimed at forecasting the catchment response to various inputs and to changes in the catchment parameters.

The initial modelling attempts were aimed at forecasting volumes of runoff and the hydrograph past a gauging station. The coaxial graphical correlation method by Linsley et al. (1949) is one of the best examples of the volume forecast methods. The unit hydrograph method is an example of a model for the time distribution of runoff. Most of the models use linear elements in series and parallel or both, that is, linear storages and linear channels or translation. For example, Dooge (1960) proposed a model for groundwater discharge composed of linear storage elements. There are a few non-linear catchment models (e.g., Amorocho and Orlob, 1961; Bidwell, 1971) but, because of the greatly increased time required for computation and the almost universal lack of observed data to verify the model performance, their use is not widespread.

Hydrological modelling got a big boost with the advent of the electronic computer. A computer orientated model, which incorporates the ideas of the coaxial correlation method, is the API-Model (Sittner et al., 1969) for continuous synthesis of the hydrograph.

11.5.1 The API-Model

The API-Model is aimed at prediction of flow from individual catchments during periods when the flow consists of groundwater discharge and small amounts of direct runoff. The direct runoff consists of channel precipitation, surface runoff and subsurface runoff, and is computed from precipitation data by the use of the antecedent precipitation index (API) type rainfall-runoff relationship and a unit hydrograph. The groundwater discharge hydrograph is related to the direct runoff hydrograph. The outline of the model is shown in Fig. 11.9. It consists of four parts: the rainfall-runoff relationship, the unit hydrograph, the relationship between the groundwater and direct runoff hydrographs, and the relationship for evaluation of the groundwater recession coefficient.

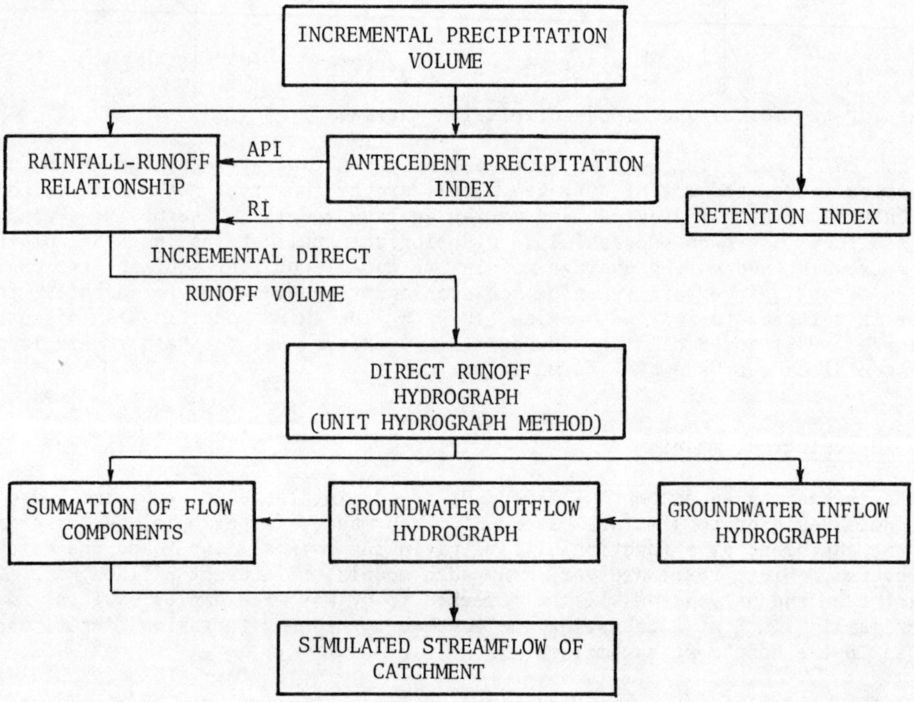

Fig. 11.9. Diagram of the API type hydrological model.

The antecedent precipitation index API reflects the soil moisture in the upper levels and is defined by an equation of the form

$$API = I = a_1P_1 + a_2P_2 + \dots a_tP_t \qquad 11.66$$

where P_t is the rainfall depth on the t-th day prior to the storm under consideration. The constant a_t is assumed to be given by a relationship, such as $a_t = 1/t$ or $a_t = k^t$, where k is a constant with a value in the range of 0.85-0.98 (Linsley et al., 1975). For continuous forecasting it is more convenient to use

$$I_t = I_o k^t$$

which assumes a logarithmically decreasing soil moisture content during periods of no rain. Putting t = 1 gives I_o as the value of the index the day before. If rain occurs on any day its amount is added to the index. The initial value of I_o is calculated from eqn 11.66 from the days of rain in the preceding 20 to 60 days (usually 30) with $a_t = k^t$. Thus, starting with this value 30 days before, the index decays exponentially with step increments at each day of rainfall. The value arrived at on the 30-th day is the API.

The rainfall, runoff and antecedent conditions are generally related to each other by the coaxial correlation method described by Linsley et al. (1949) and summarized in Fig. 11.10. In this form the API is not suitable for incremental type of calculations, mainly because it makes no allowance for interception and depression storage. Sittner et al. (1969) modified the approach by introducing a retention index RI, into the correlation, Fig. 11.11. The inflow I to the groundwater reser-

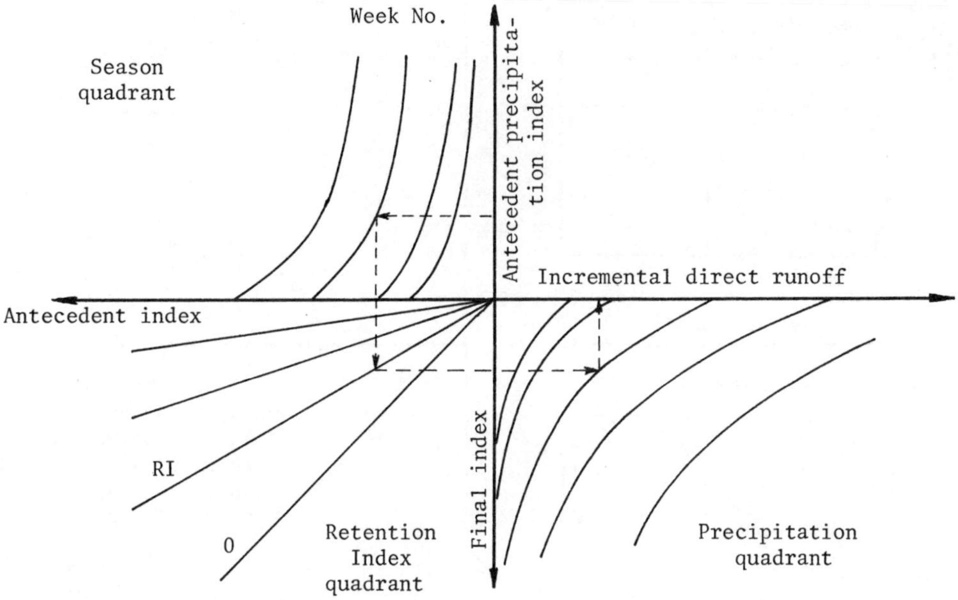

Fig. 11.11. Incremental rainfall-runoff relation, after Sittner et al. (1969).

voir is related to the direct runoff by a linear relationship I = Z(Q - G), where Q is the total discharge, G is the groundwater discharge and Z is the ratio of the instantaneous values of I and (Q - G). When the inflow is zero the groundwater discharge is assumed to follow a depletion pattern, $G_t = (K_g)^t G_0$, where G_0 is groundwater discharge at time zero and K_g is the groundwater recession factor.

The API-Model has been used on catchments from about 175 to 2115 km^2. It has to be calibrated for each catchment and this requires a fair amount of recorded data.

11.5.2 The USDAHL-70 Model

The U.S. Dept. of Agriculture Hydrograph Laboratory (1971) developed the "USDAHL-70 Model of watershed hydrology" for use on agricultural lands. The model consists of the "mainline" program and 14 subroutines. The subroutines model particular processes, such as evapotranspiration and infiltration, etc. The model is designed to follow what actually happens during the runoff process. The input to the model consists of a continuous record of rainfall, hydrological grouping of soils and land use, evapotranspiration parameters, infiltration parameters, routing relationships and geological parameters.

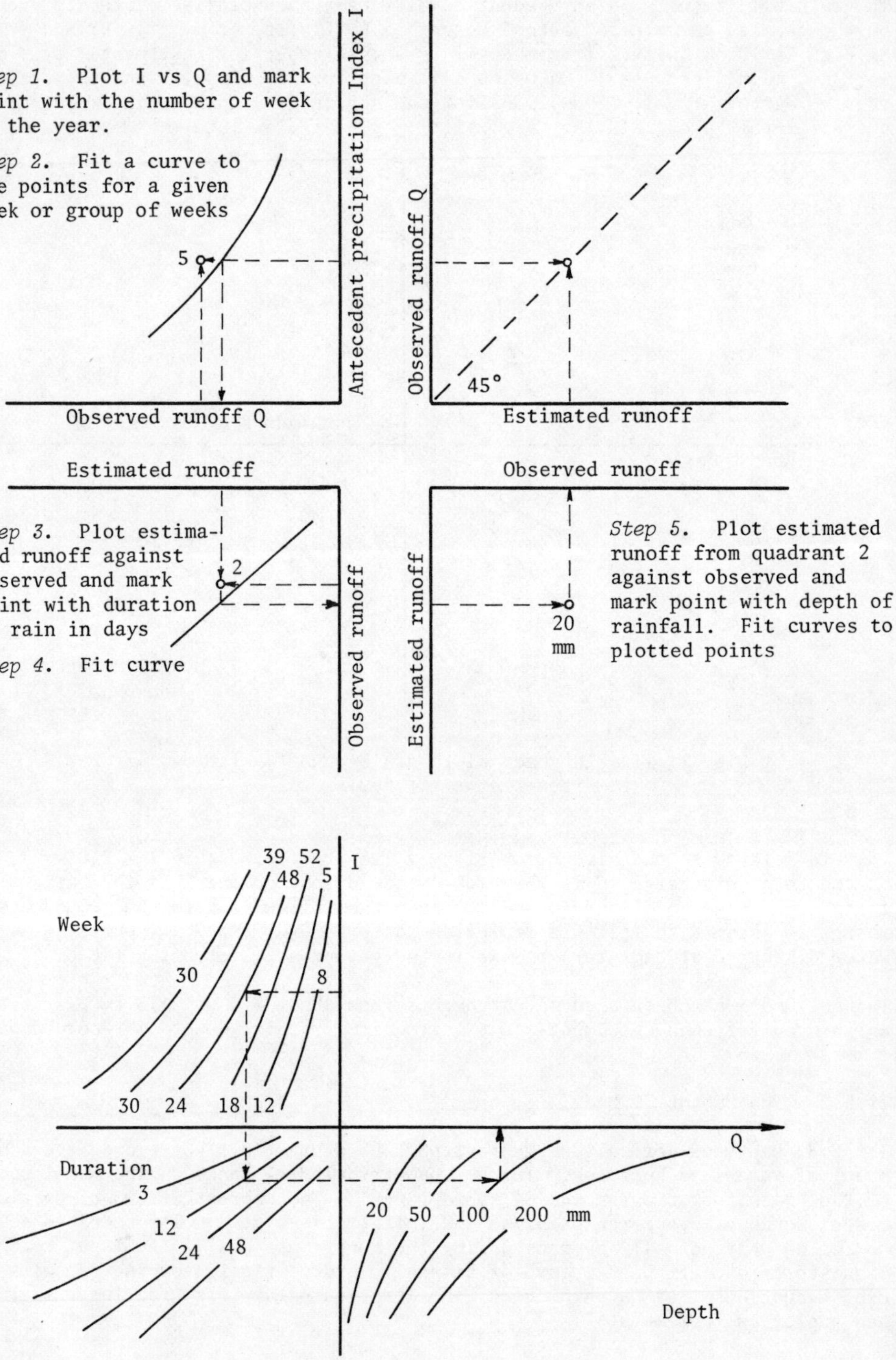

Step 1. Plot I vs Q and mark point with the number of week in the year.
Step 2. Fit a curve to the points for a given week or group of weeks
5
Antecedent precipitation Index I
Observed runoff Q
Observed runoff Q
45°
Estimated runoff
Estimated runoff
Observed runoff
Step 3. Plot estimated runoff against observed and mark point with duration of rain in days
Step 4. Fit curve
2
Observed runoff
Estimated runoff
20 mm
Step 5. Plot estimated runoff from quadrant 2 against observed and mark point with depth of rainfall. Fit curves to plotted points
39 52
48 5
I
Week
30
8
30 24 18 12
Q
Duration
3
12
24 48
20 50 100 200 mm
Depth

Fig. 11.10. Coaxial correlation model, Linsley et al. (1949).

11.5.3 The Monash Model

The Monash Model (Porter and McMahon, 1976; Porter, 1972, 1975) simulates the physical behaviour of the rainfall-runoff process. This is done by subdividing the catchment into a number of subcatchments, each of which is a complete unit of the drainage network. The subcatchments may be in series and parallel with up to four or five in series. Two versions of the model were prepared: the HYDROLOG which is based on a time unit of one day, and the HRCYCL which operates on a time unit of one hour. A schematic diagram of the model is shown in Fig. 11.12. The two versions are similar, but the daily time unit version involves necessary simplifications to the modelling of infiltration and catchment routing.

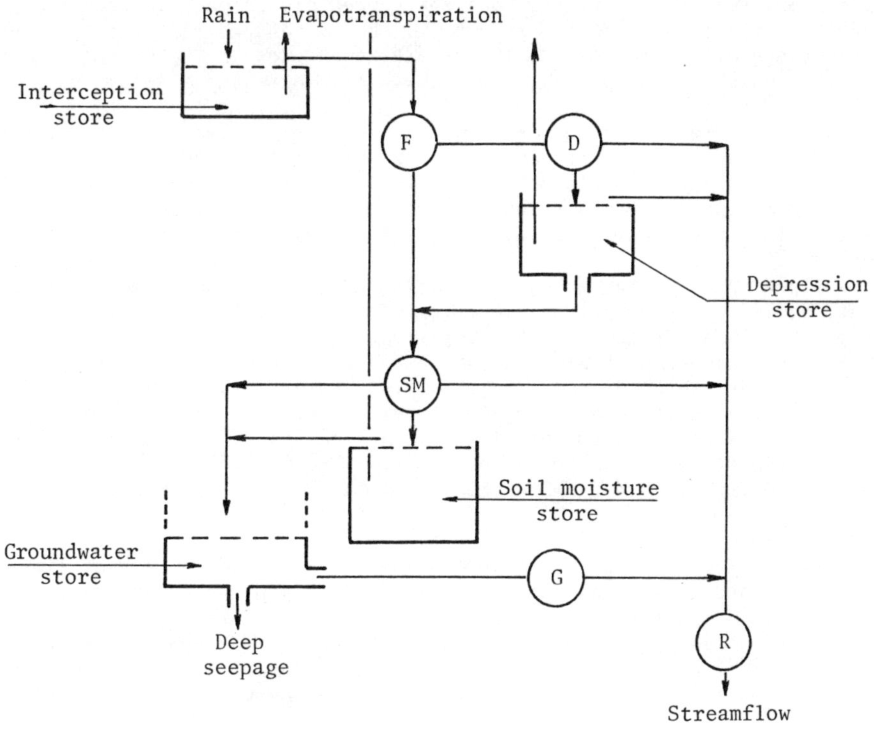

Fig. 11.12. Schematic diagram of the Monash Model. Legend: F = infiltration function, D = depression storage function, SM = soil moisture function, G = groundwater discharge function and R = routing function.

11.5.4 The Stanford Watershed Model IV

This catchment simulation model by Crawford and Linsley (1966), Linsley et al. (1975), is probably the most extensively used model, mainly because of its versatility in design. Figure 11.13 shows the flow chart of the model. The inputs required are the precipitation (hourly) and potential evapotranspiration (daily), as well as the physical description of the catchment and its hydraulic properties. The model is based on water-balance accounting and has to be calibrated for each catchment. For this records of rainfall and discharge from the given catchment must be available. Separate accounts are kept for segments of the catchment with differing soils, vegetation, land use, and precipitation. The physical features of each segment have to be described by a set of parameters. The amount of data and the computer time required increases with the number of segments.

The rainfall input to the segment must be the areal average for that segment. Hence the rainfall depth at the gauge has to be multiplied by an averaging factor (taken as the ratio of the catchment normal annual rainfall depth to the annual normal depth at the rain gauge). Interception loss is simulated by assuming an interception storage capacity (about 0-5 mm) according to the type of vegetation on the catchment. All the precipitation goes into interception storage until it is filled and the storage is emptied at the potential evaporation rate. The impervious area of the catchment is simulated by diverting a constant percentage of each increment of rainfall (impervious area runoff) to the streams. The percentage is taken to be equal to the percentage of continuous impervious area connected to the streams. The impervious area includes the area of stream surfaces, lakes and swamps.

Infiltration is one of the key features of the model. The authors use the concept of the cumulative frequency distribution of infiltration capacity. This distribution would be obtained from simultaneous infiltrometer measurements from a large number of points distributed over the catchment. Such measurements would apply for short time increments only, because infiltration capacity changes with time. This cumulative distribution is assumed to be linear. In Fig. 11.14a the position of the line is fixed by the point b, which is related to the ratio of moisture in the lower zone storage, S_L, to the nominal capacity of this zone L, Fig. 11.14b. This line and the net rainfall (rainfall less interception) line define the net infiltration which in due course contributes to the baseflow. Interflow is defined by line B (in Fig. 11.14a), which divides the rainfall excess triangle into surface runoff and interflow. The position of line B is fixed by (c.b.), where the factor c is again given in Fig. 11.14b in terms of (S_L/L). A fraction of the interflow storage is released to streamflow in each time increment. The interflow recession constant can be determined from flow data with the aid of the methods discussed in Chapter 7.3.

The surface runoff is accounted for by the upper-zone function, which directs a fraction of the surface runoff to the upper-zone storage. This storage simulates the soil moisture and depression storage. The volume ΔS_U. which enters the upper-zone storage S_U, is a function of the ratio of S_U to the upper-zone storage capacity U, Fig. 11.14c. When S_U/U is greater than S_L/L some of the upper-zone storage goes into the lower zone. This excess, together with the net infiltration, is divided between the lower-zone soil moisture storage and groundwater storage according to the ratio S_L/L. Fig. 11.14d.

The three zones of moisture regulate the soil-moisture profiles and groundwater. The upper zone gives the rapid response of small catchments while both upper and lower zones control overland flow, infiltration and groundwater storage. The runoff is modelled separately for channel flow and overland flow. The unit hydrograph is based on the area-time concept. All the individual processes are expressed algebraically and are linked together by the equation of continuity. The computed runoff is routed through the channels in the catchment to give the outflow hydrograph. For details of the procedure the references should be consulted.

An extension of the Stanford Model is the Kentucky Watershed Model (Liou, 1970; Ross, 1970). This model has the same structure as the Stanford Model, but in addition the catchment parameters are determined through the use of a self-calibrating model (OPSET). This program determines the optimum values of the parameters using a series of adjustment rules.

The experiences of conceptual runoff modelling (including snow) in Scandinavia are presented by Bergström (1976).

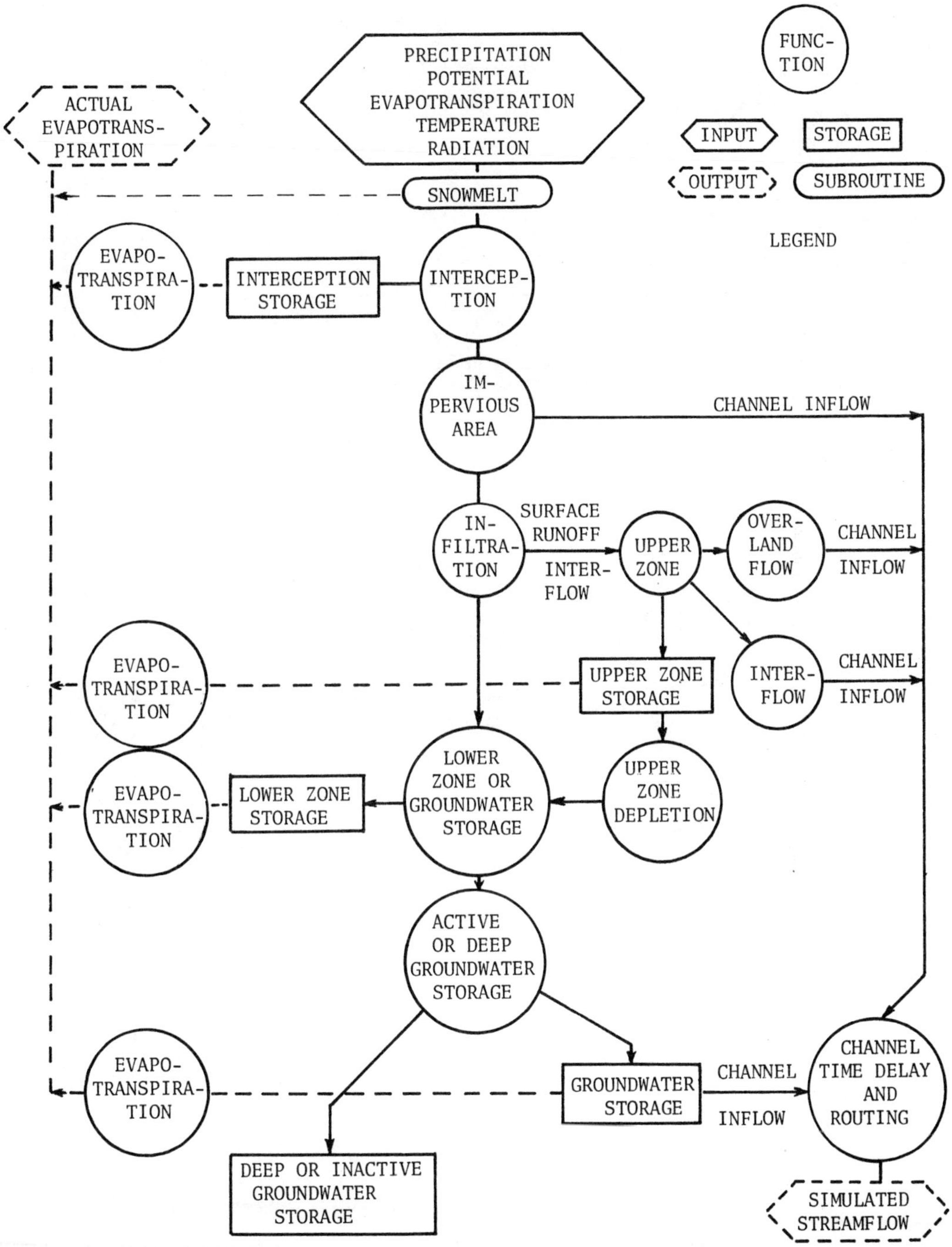

Fig. 11.13a. Flowchart of Stanford Watershed Model IV.

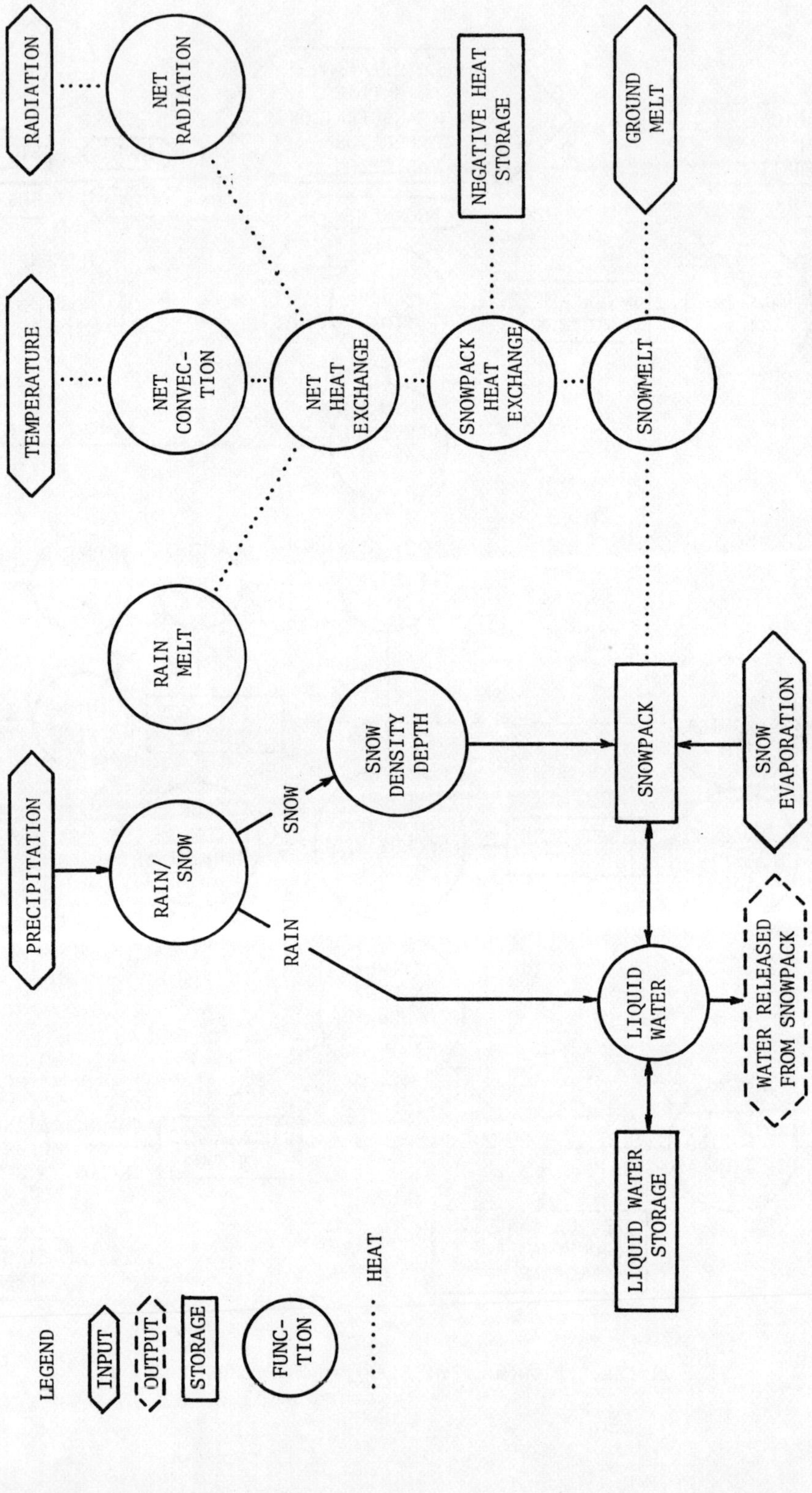

Fig. 11.13b. Stanford Watershed Model IV snowmelt subroutine.

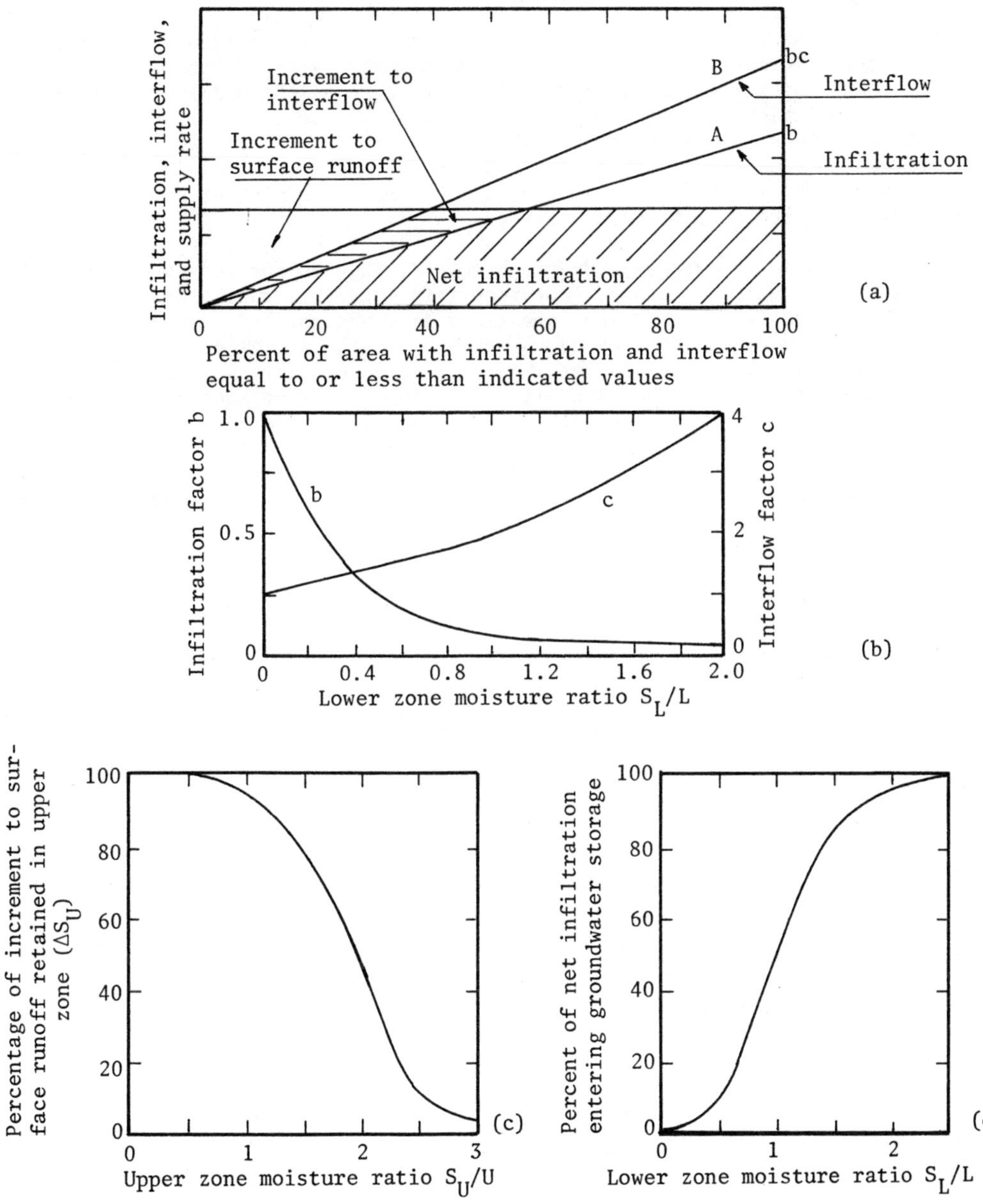

Fig. 11.14. (a) Cumulative infiltration-interflow function. (b) Variation of the infiltration factor b and interflow factor c as a function of S_L/L when INFILTRATION and INTERFLOW = 1.0. (c) Fraction of rainfall excess retained in upper-zone storage. (d) Division of infiltrated water between lower-zone moisture storage and groundwater. (Linsley et al., 1975).

11.5.5 Urban runoff

The ability of the various existing models to predict surface runoff, given the rainfall and estimated model parameters, was investigated by Dracup et al. (1973). They selected models of all types and concluded that "none of the models give totally satisfactory predictions of surface runoff. The flows predicted by the models have errors both in peak runoff and total hydrograph fit". (The report includes computer programs). However, a considerable amount of effort has been devoted to the development of special urban runoff models, both with and without reference to the water quality of the stormwater discharged.

Several versions of the simple hydrograph method, adapted for use on urban catchments are in use. These models assume 100% runoff from the hydraulically connected impervious areas. Runoff from pervious areas and from unconnected impervious areas, from which the runoff has to flow over pervious areas, is subject to infiltration losses. Infiltration may be assumed to occur at a constant rate or to depend on the antecedent rainfall. The rainfall excess for each time increment is multiplied by the catchment area and the resulting hydrograph is routed through an imaginary linear reservoir. The steps of the procedure are as follows:

(1) Select precipitation records from regional storms and express the depth of rainfall P(t) for each time increment Δt. On small catchments Δt may be 5 or 10 minutes On large urban catchments Δt may be up to an hour.

(2) Calculate the antecedent precipitation index (API) for the preceding 30 days. The apparent infiltration rate is then expressed as

$$f = f_c + e^{-fn(API)}, \qquad f < f_o$$

For example, with API in mm

$$f_{(mm/h)} = 5 + e^{(3.3 - API/25)}, \qquad f < 20 \text{ mm/h}$$

or with API in inches

$$f_{(in./h)} = 0.2 + e^{-API}, \qquad f < 0.7 \text{ in./h}$$

(3) Estimate the impervious fraction, I, of the total area which is hydraulically connected (i.e., through gutters, drains, channels) to the drainage system. If records are available, I can be estimated from the runoff depth of small storms where almost all the runoff is from the impervious areas. Such estimates tend to yield I values which are about half of the total fraction of impervious area.

(4) Measure the catchment area.

(5) Estimate the time of concentration (as for the Rational method).

(6) The runoff depth for each time increment is then calculated as follows:

Impervious area runoff $R0 = IP(t)$
Pervious area runoff $R1 = (1 - I)[P(t) - f] \geq 0$, where $f = f_c + e^{-fn(API)}$
Total runoff $R(t) = R0 + R1$

(7) The instantaneous hydrograph ordinate is defined as the product of the runoff depth R(t) and the drainage area A for each time increment, i.e.,

$$U(t)(m^3 s^{-1}) = \frac{R(t)(mm)\ A(m^2)}{\Delta t(sec)\ 10^3}$$

(8) The final hydrograph can be obtained by routing the instantaneous hydrograph through an imaginary reservoir with a time delay constant K equal to the time of concentration using

$$Q(t) = Q(t - 1) + K[U(t - 1) + U(t) - 2Q(t - 1)]$$

where

$$K = \frac{\Delta t}{2t_c + \Delta t}$$

These calculations are easily carried out on a programmable calculator or with a simple program on a small computer.

Chen and Shubinski (1971) designed a model to simulate the runoff from a catchment for any postulated pattern of rainfall. The catchment is subdivided into subcatchments with reasonably uniform characteristics, e.g., type of surface, slope, etc. The model inputs are the surface area, the widths of the subcatchments, the ground slope, Manning's n, the infiltration rate and the detention depth. The gutters and pipes are described by their geometric properties, slope and roughness.

11.6 Water Resources Systems

With the rapid increase in population and industrialization the demands on the available water resources have increased both in variety of uses and in the quantity of water required. Many of the demands are conflicting. The choices between the conflicting demands are influenced by economic, social and environmental considerations, by public opinion and by political pressure groups. Planning for such multi-purpose uses is primarily a process of decision making. In simple situations the consequences of one or the other choice can be readily predicted, but the analysis of complex problems, with a number of possible combinations of alternatives, is impossible without systems analysis techniques. These techniques provide the necessary rational for elimination of alternatives and for the reduction of the thousands of decisions to a few on the basis of input information at the various stages of interpretation. In the above sense the systems analysis or engineering is concerned only with decision making, regardless of whether it is in relation to planning, design, construction or operation of a project. According to Hall and Dracup (1970): "Systems engineering may be defined as the art and science of selecting from a large number of feasible alternatives, involving substantial engineering content, the particular set of actions which will best accomplish the overall objectives of the decision makers, within the constraints of law, morality, economics, resources, political and social pressures, and laws governing the physical, life, and other natural sciences".

The actual process of systems analysis consists of two parts. One part is the technique or science of systems analysis or operations research. The other is the art of accounting for the physical and social reality, i.e., judgement or whatever it may be called. The blind use of the technique alone in practical problems would fail in calculating the best combination because the number of combinations possible is frequently large. As Hall and Dracup point out, a system with 20 variables each with a range of 100 units, would involve 10^{40} combinations. Even at a speed of calculations of one operation per μs it would take over 3×10^{26} years to carry out the analysis. By the use of judgement the system is reduced to the principle interacting elements. The influences and processes not included in the system are considered through the specifications of the interactions with the environment in the form of inputs and outputs. The system in its environment will have inputs which may be controlled, partially controlled or uncontrolled. The outputs may be desirable, undesirable or neutral. Both the inputs and the outputs influence each other, i.e., there is a feedback.

The first step in systems analysis is the definition of the system. This includes its configuration, the network of surface and underground flows, and the points of water use and water control. A good deal of preliminary planning and screening occurs at this stage in order to reduce the number of alternatives. The elements of the system are conveniently classified as the *supply of water*, the *demand* for water, and the *design measures* for adjusting the supply and demand in space and time. The demand includes not only the industrial, domestic and irrigational requirements, but also the water required for fish, wildlife and recreation.

The next step is the organization and analysis of hydrological data so that these data can be used in the analysis of the system. This includes procedures for synthesis of hydrological events.

The third step is the definition of the system design variables and physical parameters. These include physical variables like the range of storage of the reservoirs, the range of probable aqueduct sizes, the likely reservoir sites, and the locations of water intakes, sewerage discharges and irrigation canals, etc., as well as the *system outputs* in the form of demands or targets and their distributions within the yearly cycle. The targets may be for production of electricity, supply of water for the many different purposes, levels of water quality and recreational demands. Operation policy variables also belong in this group. All these variables and parameters are called the *design or decision variables* which can be varied from one simulation to the next. In addition there are the *physical functions* and *constants* which relate the system components. These specify the physical realities and have to be satisfied, for example, friction factors, pump and turbine efficiencies, evaporation losses, infiltration rates, hydraulic head and reservoir storage relationships, and flood routing parameters, etc.

Based on these parameters the cost and benefit relationships which are associated with the design variables and outputs, have to be evaluated. The last step in the preparations is the evaluation of the economic consequences.

When each of the decision variables has been assigned a particular value, the resultant set of decisions is called a policy. A policy which does not contravene the set limits and constraints is a feasible policy and all feasible policies constitute a policy space. The policy space for the analysis generally contains a great many alternatives: engineering alternatives, management alternatives, timing and size alternatives, etc.

Space will not permit the discussion of specific details, such as the nature of the objective functions, optimal policy, constrained and unconstrained systems, gradient search procedures, linear programming, dynamic programming and simulation. For these reference should be first made to Hall and Dracup (1970), Hufschmidt and Fiering (1966), Biswas (1972, 1976), Haimes et al. (1975), Haimes (1977), Vansteenkiste (1975), and then to the specialized literature on systems analysis.

It is important to realize that the technique of analysis of water resources systems does not provide in itself the answer to all the water resources problems. Large scale water resources systems create problems, which make the application of optimization methods difficult, and unless applied by skilled personnel the results may be misleading. In these large systems there are many almost independent decision makers and controlling factors exercising their own different influences, as well as a large number of objectives to be optimized. These systems also involve a large element of uncertainity and risk due to the hydrological uncertainity, the inability to predict future events with reasonable accuracy and a high degree of irreversibility of the decisions carried out. It is appropriate to close with a quotation from Hall and Dracup:

"Systems analysis is a very powerful tool, capable of dealing with large-scale problems involving millions of people and billions of dollars. It can lead to decisions which (for better or for worse) may prove to be quite irreversible in an economic and social sense. As such, systems analysis is very dangerous tool in the hands of those lacking full understanding of the water resources systems and the multiplicity of objectives. It is equally dangerous when utilized by those who lack a reasonable appreciation of both the power and the limitations of the methods of systems analysis, however well they may be versed in the details and idiosyncrasies of the system".

Chapter 12

ANALYSIS OF INFORMATION

Truth has many faces and any one on its own is a lie.

An important part of hydrological work is concerned with analysis of information and decision making. The techniques of analysis are not strictly part of hydrology, but do form a very important part of the hydrologist's toolkit. In this Chapter a brief introduction to some of these techniques is attempted. Design and planning relate to future events whose magnitude and time of occurrence cannot be predicted with certainty. In general, hydrological phenomena such as rainfalls, streamflows, etc., are the result of the combination of many smaller events. A rainfall, for example, is composed of a number of processes which lead to the formation of the raindrop and involve the laws of thermodynamics and hydrodynamics in complex relationships. Few of such processes are subject to deterministic analytical solutions and it becomes necessary to resort to descriptions in terms of statistical parameters. The variables involved are functions of time and space and appear to fluctuate in a random manner. For practical application these variables have to be expressed in terms of probabilities and frequencies.

Selection of an appropriate level of probability is a decision on what level of risk is acceptable on economic, social or political grounds. Probability theory starts with certain axioms, generates theorems and is deductive in nature. It provides the basis for the prediction of the probability of an outcome, which is obtained from an exhaustive set of trials as the ratio of the number of times the event occur to the total number of trials. Statistics is concerned with decision making and is inductive in nature. The statistical techniques provide a procedure for decision making. Through the use of statistical methods masses of quantitative, and on its own nearly useless, information can often be reduced to a few parameters that convey clearly the nature of the raw data. Decision making, based on the data, requires inference and involves the risks and dangers associated with prediction and estimation. It also requires an understanding of the methods used and of their limitations. The usual problem is to infer, or to predict, the whole class of possible occurrences (the population) when only a few of them (the sample) have been observed The user of hydrological data seldom has any control over the selection of the sampl he gets what nature provides over the given period of time (see also Section 11.1). The analysis is further complicated by the fact that hydrological data are seldom truely random. Usually there is a tendency for persistence or a carry-over effect (e.g., a drought does not follow immediately after a flood) and this interferes with the requirement of randomness that each event is independent of all the other events

Difficulties arise also with the homogeneity of data. In order to draw any inferences from the sample about the population, the data in the sample must be homogeneous, that is, the conditions affecting the data should not have changed materially over that period. For example, changes in land usage could have an appreciable effect of streamflow data.

A classification of hydrological processes is shown in Fig. 12.1. The process may be deterministic (independent of chance) like flood routing through a reservoir, or probabilistic or stochastic where stochastic usually refers to time involvement in a random process, although Bernoulli in Ars Conjectandi in 1713 called it the science of conjecture or stochastic science, defined as "the art of estimating, as best one can, the probability of things so that in our judging and acting we may choose to follow the best, the safest, the surest, or the most soul-searching way". Box 3 in Fig. 12.1 represents probably the most common type of data, but at present the theory for analysis of non-stationary processes is inadequate. The stationary process (Box 2) is the usual time series analysis, the stochastic process, and Box 1 refers to the fitting of statistical distributions to data. For Boxes 1 and 2 the data must be homogeneous and when dealing with processes of nature it must be ascertained that the data actually depend on the physical processes and that all independent

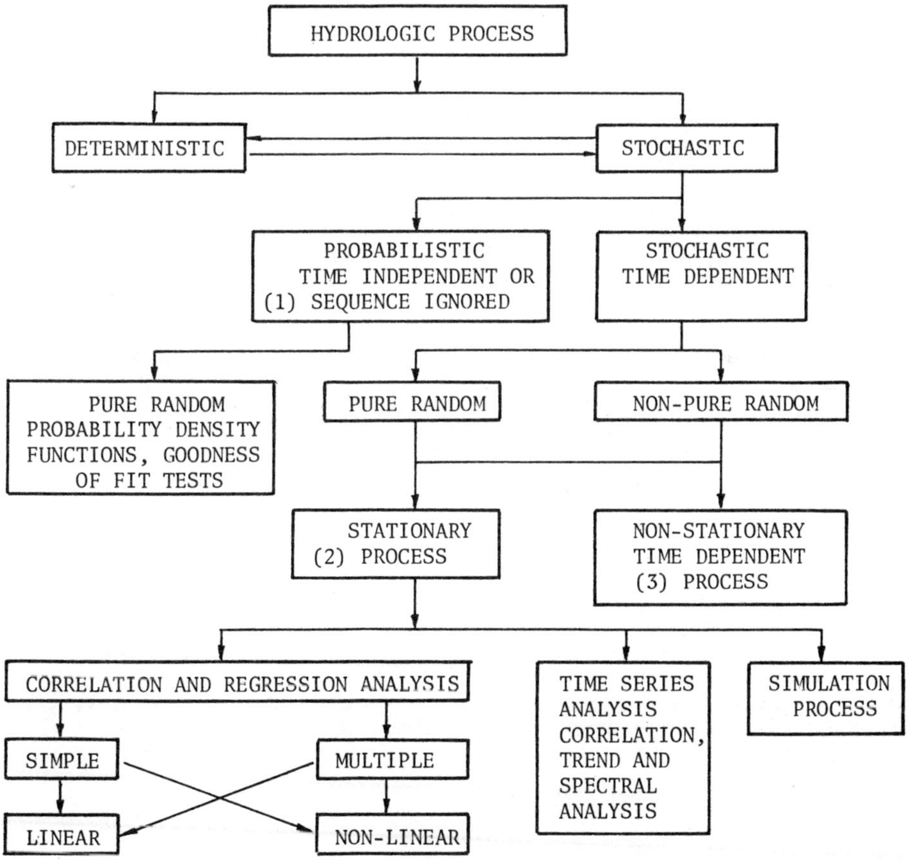

Fig. 12.1. Classification of statistical processes.

variables are included in the analysis. Otherwise the inferences drawn may be similar to that suggested by G.B. Shaw that "it can be shown by statistics, beyond doubt, that people who wear bowler hats and carry umbrellas have a much higher life expectancy than those who never dream of possessing such items. Hence, wearing bowler hats and carrying an umbrella contributes to longevity".

12.1 Frequency Distributions and Probability Paper

When the data are grouped into classes and the number of items in each class (the frequencies) are expressed as percentages one obtains the frequency distribution f(x). The area under the frequency distribution curve is unity. The summation of the f(x) yields the *cumulative distribution function* (c.d.f.) which can be interpreted as the probability of any value of x_i being greater than or equal to the given value of x, since the sum of the f(x) is equal to unity. Correspondingly, f(x) can then be looked upon as the probability distribution function or the *probability density function* (p.d.f.). The two are related by

$$F(x) = \int_{-\infty}^{x} f(x)\, dx \qquad 12.1$$

or

$$F(x) = \Sigma f(x_i)$$

where $f(x_i)$ are discrete values, and

$$\int_{-\infty}^{\infty} f(x)\, dx = 1$$

A probability density function must satisfy three criteria:
(a) the integral of f(x) over all possible values must be unity, (b) f(x) must have only positive values, and (c) f(x) must be a single-valued function.

If now the scale of the ordinate of the cumulative frequency plot, F(x), is so transformed that the curve becomes a straight line, then the resultant axes define a probability distribution paper for this particular type of distribution. The corollary is that if a set of data follows a straight line on probability paper then the data satisfy this type of distribution. Note, however, that every frequency distribution has its own probability paper.

The data are plotted on the probability paper in an increasing or decreasing order of probability, that is in an ascending or descending series, but there is no concurrence among statisticians about the *plotting position*. Commonly used values are m/(n + 1), m/n, and (m - ½)/n. For an *ascending* series the probability of x_i being *equal to or less than* x_m is

$$F(x) = \frac{m}{n+1} \qquad 12.2$$

where m is the sequence number of x in the series x_i, i.e., m = 1, 2, ..., n where n is the total number of items. The probability of the magnitude of x_m being exceeded is then

$$1 - F(x) = 1 - \frac{m}{n+1} \qquad 12.3$$

Hydrologists also use the recurrence interval or *return period*

$$T = \frac{1}{1 - F(x)} \qquad 12.4$$

i.e., T = (n + 1)/(n + 1 - m) for an ascending and T = (n + 1)/m for a descending series of data. It is the average interval between the occurrences of the given magnitudes of the data. If the interval between events is one year then a 100-year return period signifies an event which occurs on the average once in a hundred years or has probability of 1 in 100 of occurring in any given year.

It should be noted that in hydrologic application the statement that the probability of exceedance of a given flow rate is, for example, 5% may mean (a) a flow rate which is exceeded, on the average, five times in 100 years, or (b) a flow rate which is exceeded, on the average, in five years out of every hundred. In the latter case there may be more than five instances when this flow rate is exceeded but all are confined within the five years. The first case is strictly an annual frequency, not probability, since its value can exceed unity.

It should be evident that there are many functional forms (shapes) of the frequency distributions or p.d.f. These may be subdivided into *discrete variable* and *continuous variable* distributions. Wellknown examples of discrete distributions are the *binomial* distribution (tossing of the coin experiment), which give the probability of m successes in n trials as

$$P(x = m) = f(x) = \binom{n}{m} p^m q^{n-m} \qquad 12.5$$

where p is the probability of success in any one trial, $q = 1 - p$ is the probability of failure and

$$\binom{n}{m} = \frac{n!}{m!(n - m)!} \qquad 12.6$$

and the *Poisson distribution*

$$f(x, \lambda) = \frac{\lambda^x}{x!} e^{-\lambda}, \quad x = 0, 1, 2 \ldots; \quad \lambda > 0 \qquad 12.7$$

where λ is the rate of occurrence parameter.

The most widely known continuous distribution is the *normal* or *Gaussian* distribution

$$f(x_i; \mu,\sigma) = \frac{1}{\sigma\sqrt{2\pi}} \exp[- \frac{(x_i - \mu)^2}{2\sigma^2}]$$

where μ and σ are the location parameters, mean and standard deviation, respectively, $-\infty < x < \infty$, $-\infty < \mu < \infty$, and $\sigma > 0$. The variable x is called *standardized* when expresses as

$$z = \frac{x - \mu}{\sigma} \qquad 12.9$$

The p.d.f. then becomes

$$f(z) = \frac{1}{\sqrt{2\pi}} \exp(- \frac{1}{2} z^2), \quad -\infty < z < \infty \qquad 12.10$$

The variable z is normally distributed with zero mean and unit standard deviation. The cumulative distribution function, c.d.f., of x is

$$F(x_i; \mu, \sigma) = \frac{1}{\sigma\sqrt{2\pi}} \int_{-\infty}^{x} \exp[- \frac{(x_i - \mu)^2}{2\sigma^2}]dx \qquad 12.11$$

and of z

$$F(z) = \frac{1}{\sqrt{2\pi}} \int_{-\infty}^{z} e^{-\frac{1}{2}z^2} dz \qquad 12.12$$

The c.d.f of z for $\mu = 0$ and $\sigma = 1$ is tabulated in textbooks on statistics and mathematical tables. Table 12.1 shows values of z for given values of F(z) in %.

TABLE 12.1 Values of z for Given Values of F(z) for Normal Distribution

%	z(F)	%	z(F)	%	z(F)	%	z(F)	%	z(F)	%	z(F)
1	-2.326	21	-0.806	41	-0.228	61	0.279	81	0.878	99.1	2.366
2	-2.054	22	-0.772	42	-0.202	62	0.305	82	0.915	99.2	2.409
3	-1.881	23	-0.739	43	-0.176	63	0.332	83	0.954	99.3	2.457
4	-1.751	24	-0.706	44	-0.151	64	0.358	84	0.994	99.4	2.512
5	-1.645	25	-0.674	45	-0.126	65	0.385	85	1.036	99.5	2.576
6	-1.555	26	-0.643	46	-0.100	66	0.412	86	1.080	99.6	2.652
7	-1.476	27	-0.613	47	-0.075	67	0.440	87	1.126	99.7	2.748
8	-1.405	28	-0.583	48	-0.050	68	0.468	88	1.175	99.8	2.878
9	-1.341	29	-0.553	49	-0.025	69	0.496	89	1.227	99.9	3.090
10	-1.282	30	-0.524	50	0.000	70	0.524	90	1.282		
11	-1.227	31	-0.496	51	0.025	71	0.553	91	1.341	99.91	3.121
12	-1.175	32	-0.468	52	0.050	72	0.583	92	1.405	99.92	3.156
13	-1.126	33	-0.440	53	0.075	73	0.613	93	1.476	99.93	3.195
14	-1.080	34	-0.412	54	0.100	74	0.643	94	1.555	99.94	3.239
15	-1.036	35	-0.385	55	0.126	75	0.674	95	1.645	99.95	3.291
16	-0.994	36	-0.358	56	0.151	76	0.706	96	1.751	99.96	3.353
17	-0.954	37	-0.332	57	0.176	77	0.739	97	1.881	99.97	3.432
18	-0.915	38	-0.305	58	0.202	78	0.772	98	2.054	99.98	3.540
19	-0.878	39	-0.279	59	0.228	79	0.806	99	2.326	99.99	3.719
20	-0.842	40	-0.253								

Many of the distribution functions are not symmetrical and the peak does not coincide with the mean. Then the *median* divides the area under the curve into equal parts. The *mode* gives the value at which the peak occurs and the *mean* is the arithmetic average of the data. For a normal distribution these three coincide.

The distributions may be further subdivided into *single variable* and *multivariate* distributions. A single variable distribution is represented by the familiar p.d.f. curve, whereas that for a bivariate distribution is a surface.

12.2 Parameter Estimation

The statistical parameters, by which a distribution is defined, have to be estimated from the observed records, i.e., the sample data. Since sampling is subject to errors, it is desirable that the methods of parameter estimation keep the error in the estimated parameters to a minimum. The techniques used for parameter estimation are the *method of maximum likelihood*, the *method of moments* (central tendency), the *method of least squares*, and *graphical methods*.

The *method of maximum likelihood* is the most powerful method of parameter estimation but requires more work than other methods. In computer applications this is of little consequence and the maximum likelihood should be used, in preference to other methods, particularly when major decisions have to be based on a small sample of data or a sample with large scatter in the data. Detailed description of the method of maximum likelihood can be found, e.g., in Kendall and Stuart (1961). In principle, the idea is to choose as estimator of a parameter that function of the sample data which will on substitution make the probability of the sample a maximum. If a statistical population has a p.d.f. of $y = f(x, a)$ then the probability of obtaining a given value of x on random sampling is proportional to $f(x, a)$. The joint probability of a sample of n values of x, ($x = x_i$, i = 1, 2, ... n), is proportional to

$$L = f(x_1, a)f(x_2, a)\ldots f(x_n, a) \qquad 12.13$$

which is called the likelihood. The principle of maximum likelihood states that the estimate of a to choose is the one that maximizes the likelihood. Since f is always positive computation is simplified by use of the logarithm of L. For example, for the normal distribution the p.d.f. is given by eqn 12.8 and for a sample size of n

$$L = \frac{1}{(\sigma\sqrt{2n})^n} \exp[-\frac{1}{2\sigma^2}\sum_i (x_i - \mu)^2]$$

$$= (2\pi\sigma^2)^{-n/2} \exp[-\frac{1}{2\sigma^2}\sum_i (x_i - \mu)^2], \quad i = 1, 2, \ldots n$$

and

$$\ell n\ L = -\frac{n}{2}\ell n\ 2\pi - \frac{n}{2}\ell n\ \sigma^2 - \frac{1}{2\sigma^2}\sum_i (x_i - \mu)^2$$

Differentiation with respect to μ and σ^2 yields

$$\frac{\partial}{\partial\mu}\ell n\ L = \frac{1}{2\sigma^2}\Sigma - 2(x_i - \mu) = 0$$

$$\frac{\partial^2}{\partial\mu^2}\ell n\ L = (1/\sigma^2)\Sigma - 1 = \frac{n}{\sigma^2}$$

$$\frac{\partial}{\partial\sigma^2}\ell n\ L = -\frac{n}{2\sigma^2} + \frac{1}{2\sigma^4}\Sigma(x_i - \mu)^2 = 0$$

$$\frac{\partial^2}{\partial(\sigma^2)^2}\ell n\ L = \frac{n}{2\sigma^4} - (1/\sigma^6)\Sigma(x_i - \mu)^2 = -\frac{n}{2\sigma^4}$$

Thus, the maximum likelihood estimator for $\mu = \hat{\mu}$ is given by $\sum_i (x_i - \hat{\mu}) = 0$ or

$$\hat{\mu} = \frac{1}{n}\sum_i x_i = \mu \qquad 12.14$$

i.e., the sample mean and its value is independent of σ. The value of $\hat{\sigma}^2$, however, is not independent of μ. The solution for σ^2 with $\mu = \hat{\mu}$ yields

$$\hat{\sigma}^2 = \frac{1}{n}\Sigma(x_i - \mu)^2 \qquad 12.15$$

which is biassed because $x - \mu = (x - \bar{x}) + (\bar{x} - \mu)$ and $\Sigma(x - \mu)^2 = \Sigma(x - \bar{x})^2 + n(\bar{x} - \mu)^2 + 2(\bar{x} - \mu)\Sigma(x - \bar{x}) = \Sigma(x - \bar{x})^2 + n(\bar{x} - \mu)^2$, since $\Sigma(x - \bar{x}) = 0$. Over a large number of samples the average value of $\Sigma(x - \mu)^2 = n\sigma^2$, of $\Sigma(x - \bar{x})^2 = nS^2$ and of $(\bar{x} - \mu)^2$ will be the variance of $\bar{x}$, i.e., σ^2/n, where S^2 is the sample variance. The unbiassed maximum likelihood estimate is then

$$\hat{\sigma}^2 = \frac{1}{n - 1}S^2 = \frac{1}{n - 1}\sum_1^n (x_i - \bar{x})^2 \qquad 12.16$$

In general, n should be large enough to make the difference in $1/n$ and $1/(n - 1)$ insignificant.

The *method of moments* rests on the fact that distributions can be described in terms of the central tendency. This method is less rigorous than the method of maximum likelihood but is easier to apply. For a continuous distribution, Fig. 12.2, the first moment about an arbitrary axis $x = a$ is

$$\mu_1(a) = \int_{-\infty}^{\infty} (x - a)f(x)\, dx \qquad 12.17$$

If a = 0, the coordinate origin, the first moment is the mean of the distribution, i.e., the expected value E[x] or the familiar arithmetic mean $\mu_1(0) = \bar{x}$.

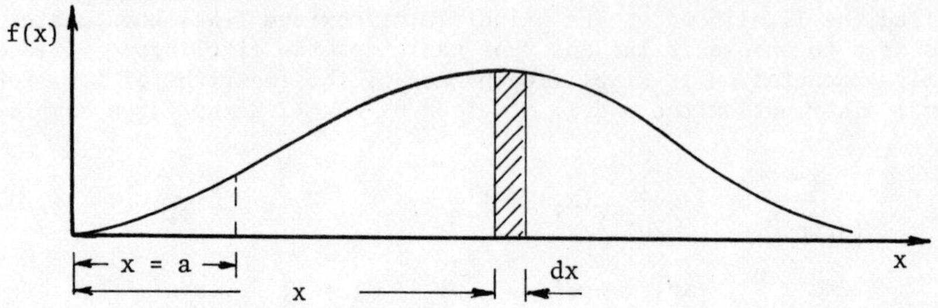

Fig. 12.2. Definition sketch for calculation of the moments.

In general, the r-th moment about the origin is

$$\mu_r(0) = \int_{-\infty}^{\infty} x^r f(x)\, dx \qquad 12.18$$

The first moment about the mean, $\mu_1(\bar{x}) = \mu_1 = 0$, and the r-th moment about the mean is

$$\mu_r = \int_{-\infty}^{\infty} (x - \bar{x})^r f(x)\, dx \qquad 12.19$$

In incremental form

$$\mu_1(0) = \bar{x} = \frac{1}{n}\sum_1^n f_i x_i \quad \text{and} \quad \mu_r = \frac{1}{n}\sum_1^n f_i (x_i - \bar{x})^r \qquad 12.20$$

where f_i is the frequency of x_i (the number of times x_i occurs) and Σf_i is the total number of values n. The moments μ_r about the mean and $\mu_r(0)$ are related by

$$\mu_r = \mu_r(a) - \frac{r}{1!}\bar{u}\,\mu_{r-1}(a) + \frac{r(r - 1)}{2!}\bar{u}^2\,\mu_{r-2}(a) - \ldots \qquad 12.21$$

where the general term is

$$(-1)^k \frac{r(r - 1)\ldots(r - k + 1)}{k!}\bar{u}^k\,\mu_{r-k}(a) \qquad 12.22$$

and $\bar{u}$ is from $x_i = a + u_i$ and $\bar{u} = \bar{x}$ when a = 0. When r = 0, $\mu_0 = 1$, when r = 1, $\mu_1 = 0$ and when r = 2

$$\mu_2 = \mu_2(a) - (\bar{x} - a)^2 = \sigma^2 \qquad 12.23$$

or

$$\mu_2 = \frac{1}{n}\Sigma f_i x_i^2 - \bar{x}^2, \quad a = 0 \qquad 12.24$$

When r = 3

$$\mu_3 = \mu_3(a) - 3\bar{u}\,\mu_2(a) + 2\bar{u}^3 \qquad 12.25$$

and when r = 4

$$\mu_4 = \mu_4(a) - 4\bar{u}\,\mu_3(a) + 6\bar{u}^2\,\mu_2(a) - 3\bar{u}^4 \qquad 12.26$$

The ratio $\sigma/\bar{x}$ is the coefficient of variation, i.e.,

$$C_v = \frac{\sigma}{\bar{x}} \quad 12.27$$

The third moment is used to indicate the value of skewness and its direction relative to the mean. The skewness is usually denoted by $\sqrt{\beta_1}$ or γ_1

$$\sqrt{\beta_1} = \gamma_1 = \mu_3/\sigma^3 \quad 12.28$$

where $\beta_1 = \mu_3^2/\mu_2^3$. A positive value of skewnwss implies that the distribution has a longer tail to the right.

The fourth moment is used to indicate the peakedness (kurtosis) of the distribution

$$\beta_2 = \mu_4/\mu_2^2 \quad 12.29$$

For the normal distribution $\beta_2 = 3.0$.

In application the higher moments become increasingly less reliable because of the strong influence of the few large values of the observations. Usually Greek letters are used to signify the moments of the population and Roman letters those of the sample. Thus, the population variance is σ^2 and the sample variance is S^2, the population mean is μ and the sample mean is m (or $\bar{x}$), likewise β and b or γ and g.

The *least squares* is a wellknown curve fitting technique and the *graphical methods* are used in conjunction with probability papers.

Example 12.1. For a given month the monthly mean rainfall depths for the 122 years of record at Auckland in 10 mm class intervals are as follows:

Class middle:	5	15	25	35	45	55	65	75	85	95	105	115	125
No. of items:	6	8	9	15	13	13	7	5	13	6	7	3	3
% :	4.92	6.56	7.38	12.30	10.66	10.66	5.76	4.10	10.66	4.92	5.76	2.46	2.46

Class middle:	135	145	155	165	175	185	195	205	215
No. of items:	2	3	2	1	1	3	-	1	1
% :	1.64	2.46	1.64	0.82	0.82	2.46	-	0.82	0.82

Calculate the first, second, third and fourth statistical moments, the skewness $\sqrt{\beta_1} = \gamma = \mu_3/\sigma^3$, peakedness $\beta_2 = \mu_4/\mu_2^2$ and fit the normal distribution to the data.

The calculation is shown in the tabular form. The results are as follows:
Mean $\bar{x} = a + \bar{u} = 105 - 35.41 = \underline{69.59}$ mm.
Variance $\mu_2 = \mu_2(a) - \bar{u}^2$

$$\mu_2(a) = \frac{1}{n} \Sigma f_i u_i^2 = 419\ 600/122 = 3\ 439.344$$

$$\mu_2 = \sigma^2 = 3\ 439.344 - 35.41^2 = \underline{2\ 185.476}$$

$$\mu_3 = \mu_3(a) - 3\bar{u}\ \mu_2(a) + 2\bar{u}^3$$

$$\mu_3(a) = -\ 21\ 762 \times 10^3/122 = -\ 178\ 377.049$$

$$\mu_3 = -\ 178\ 377 - 3 \times (-\ 35.41)3\ 439.344 + 2(-\ 35.41)$$

$$= -\ 178\ 377.049 + 365\ 361.513 - 88\ 798.939 = \underline{98\ 185.525}$$

$$\mu_4 = \mu_4(a) - 4u\ \mu_3(a) + 6\bar{u}^2\ \mu_2(a) - 3\bar{u}^4$$

$$\mu_4(a) = 255\ 608 \times 10^4/122 = 2\ 095.15$$

$$4 \times 35.41(-\ 178\ 377.05) = 25\ 265\ 325.22$$

$$6 \times 35,41^2 \times 3\ 439.34 = 25\ 874\ 902.36$$

$$3 \times 35.41^2 = -4\ 716\ 555.64$$

$$\mu_4 = 46\ 425\ 767$$

$$\sqrt{\beta_1} = \frac{\mu_3}{\sigma^3} = \frac{98\ 185.585}{46.794^3} = 0.961$$

$$\frac{\mu_4}{\mu_2^2} = \beta_2 = \frac{46\ 425\ 767}{2\ 185.5^2} = 9.72$$

x_i	f_i	$u_i = x_i - 105$	$f_i u_i$	$f_i u_i^2 \times 10^2$	$f_i u_i^3 \times 10^3$	$f_i u_i^4 \times 10^4$
5	6	-100	-600	600	-6000	60 000
15	8	- 90	-720	648	-5832	52 488
25	9	- 80	-720	576	-4608	36 864
35	15	- 70	-1050	735	-5145	36 015
45	13	- 60	-780	468	-2808	16 848
55	13	- 50	-650	325	-1625	8 125
65	7	- 40	-280	112	- 448	1 792
75	5	- 30	-150	45	- 135	405
85	13	- 20	-260	52	- 104	208
95	6	- 10	- 60	6	- 6	6
105	7	0	0	0	0	0
			-5270		-26711	
115	3	10	30	3	3	3
125	3	20	60	12	24	48
135	2	30	60	18	54	162
145	3	40	120	48	192	768
155	2	50	100	50	250	1 250
165	1	60	60	36	216	1 296
175	1	70	70	39	343	2 401
185	3	80	240	192	1536	12 288
195	0	90	0	0	0	0
205	1	100	100	100	1000	10 000
215	1	110	110	121	1331	14 641
	122		950	4196	4949	255 608
			-4320	419600	-21762×10^3	$255\ 608 \times 10^4$

$$\bar{u} = -\ 4320/122 = -\ 35.41$$

The p.d.f. of normal distribution is

$$f(x;\ \mu,\sigma)dx = \frac{1}{\sigma\sqrt{2\pi}} \exp[-\ \frac{(x - \mu)^2}{2\sigma^2}]dx$$

$$f(x;\ \bar{x},S)dx = \frac{1}{46.75 \times 2.51} \exp[-\ \frac{(x - 69.59)^2}{2 \times 2185.5}]10$$

$$= 8.53 \times 10^{-2} \exp[-\ 2.2878 \times 10^{-4}(x - 69.59)^2]$$

This evaluated for the midpoint values of each class yields the p.d.f. of a normal distribution with the above mean and standard deviation and superimposed on the histogram of the given data gives an indication on whether the distribution fits or not.

12.3 Notes on Selected Distributions

The *gamma distribution* is described by

$$f(x;\ \eta,\lambda) = \begin{cases} \dfrac{\lambda^\eta}{\Gamma(\eta)}\, x^{\eta-1}\, e^{-\lambda x}; & x \geq 0, \quad \lambda > 0, \quad \eta > 0 \\ 0 \quad ; & \text{elsewhere} \end{cases} \qquad 12.30$$

where $\Gamma(\eta)$ is the gamma function defined as

$$\Gamma(\eta) = \int_0^\infty x^{\eta-1}\, e^{-x}\, dx = (\eta - 1)!$$

where η is a positive integer. Figure 12.3 illustrates the shapes of this distribution and shows that η and λ are the shape and scale parameters, respectively. The cumulative gamma distribution is

$$F(x;\ \eta,\lambda) = \begin{cases} \dfrac{\lambda^\eta}{\Gamma(\eta)} \displaystyle\int_0^x x^{\eta-1}\, e^{-\lambda x}\, dx; & x \geq 0 \\ 0 \quad ; & x < 0 \end{cases} \qquad 12.31$$

It is known as the incomplete gamma function and is tabulated in mathematical tables. The probability that a random value from a gamma distribution with given λ and η is less than a given value x is obtained by evaluating first

$$u = \frac{\lambda x}{\sqrt{x}} \quad \text{and} \quad p = \eta - 1 \qquad 12.32$$

Then the tables give I(u, p) which is the desired cumulative probability $F(x;\ \eta,\lambda)$. The expected value (mean) is η/λ, variance is η/λ^2, $\sqrt{\beta_1} = 2/\sqrt{\eta}$ and $\beta_2 = 3(\eta + 2)/\eta$. The parameters, λ and η, can be estimated from data by the following approximate formulae

$$\hat{\lambda} = \frac{\bar{x}(n - 1)}{\sum_1^n (x_i - \bar{x})^2} = \frac{\bar{x}}{S^2} = (n - 1)\left[\frac{\Sigma x_i}{n\Sigma x_i^2 - (\Sigma x_i)^2}\right] \qquad 12.33$$

$$\hat{\eta} = \frac{\bar{x}^2 (n - 1)}{\Sigma (x_i - \bar{x})^2} \quad \text{or} \quad \hat{\eta} = \hat{\lambda}\bar{x} \qquad 12.34$$

A special case of the gamma distribution is the *chi-square distribution* for which $\lambda = ½$ and η is a multiple of ½.

The *exponential distribution* is a gamma distribution with $\eta = 1$. Then

$$f(x;\ \lambda) = \lambda e^{-\lambda x}; \quad x \geq 0, \quad \lambda > 0$$

and

$$F(x;\ \lambda) = \int_0^x \lambda e^{-\lambda x}\, dx = 1 - e^{-\lambda x} \qquad 12.35$$

The *log-normal distribution* has been extensively used in hydrology. It is a model for a random variable whose logarithm follows the normal distribution with parameters

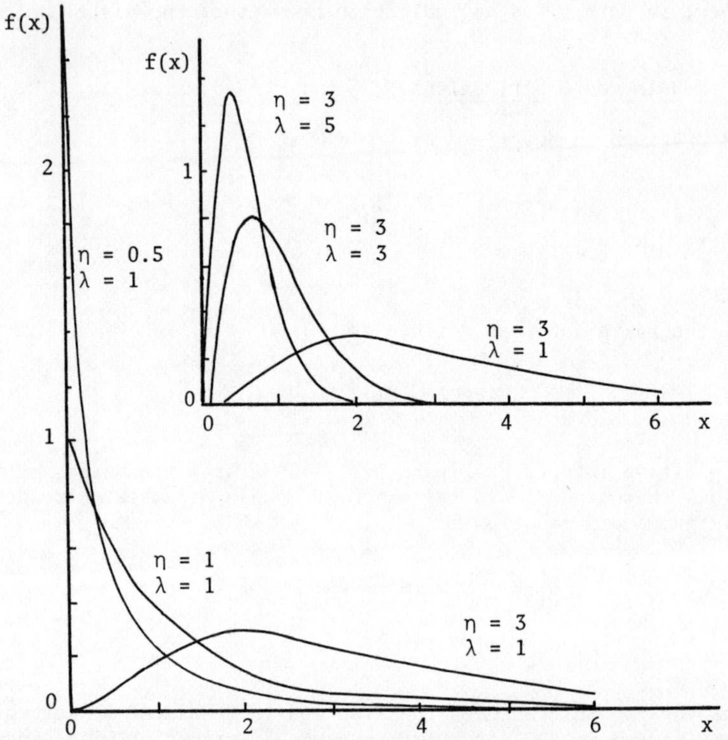

Fig. 12.3. Gamma distributions with $\eta = 3$ and various values of λ, and with $\lambda = 1$ and various values of η.

μ and σ. Hydrological data usually have a strongly skewed distribution and frequently the log transformation converts it to an approximately normal distribution. The log-normal distribution can be derived as a model for a process whose values result from the multiplication of many small errors, analogous to that of the normal distribution which is for addition of errors. A quick (rough) test is provided by plotting the data on log-normal probability paper.

If x is the initial variable, then

$$y = \ln x, \quad x = e^y \tag{12.36}$$

and the p.d.f. is

$$f(y) = \frac{1}{\sigma\sqrt{2\pi}} \exp[-(y-\mu)^2/2\sigma^2]$$

where $\mu \simeq \bar{y}$ and $\sigma = S_y$ are the mean and standard deviation of the y distribution, respectively. If f(x) and f(y) are the p.d.f. for the x and y distributions, respectively, then

$$f(x)\,dx = f(y)\,dy = f(y)\,\frac{1}{x}\,dx$$

since $y = \ln x$. Hence

$$f(x) = \frac{1}{x} f(y) = \frac{1}{x\sigma_y\sqrt{2\pi}} \exp[-(y - \bar{y})^2/2\sigma_y^2]$$

and the r-th moment about 0 of the variable x_i is

$$\mu_r(0) = \int_0^\infty x^r f(x)\, dx = \int_{-\infty}^\infty e^{ry} f(y)\, dy = \exp(r\bar{y} + \frac{1}{2} r^2\sigma_y^2) \qquad 12.37$$

Thus the mean of x_i is

$$\mu_1(0) = \bar{x} = \exp(\bar{y} + \frac{1}{2}\sigma_y^2) \qquad 12.38$$

and the variance of x_i (eqn 12.23) is

$$\mu_2 = \mu_2(0) - [\mu_1(0)]^2 = \sigma_x^2 = \exp(2\bar{y} + \sigma_y^2)(e^{\sigma_y^2} - 1)$$

$$= \bar{x}^2(e^{\sigma_y^2} - 1) = \bar{x}^2 C_v^2 \qquad 12.39$$

where $C_v = (e_y^{\sigma^2} - 1)^{1/2}$ is the coefficient of variance and

$$\sigma_y^2 = \ell n(1 + C_v^2) \qquad 12.40$$

The $C_v = \sigma/\bar{x}$ can be estimated from $C_v = [n/(n - 1)(\overline{x^2}/\bar{x}^2 - 1)]^{1/2}$, where $\bar{x} = (1/n)\Sigma x_i$ and $\overline{x^2} = (1/n)\Sigma x_i^2$ for large n. The third moment is

$$\mu_3 = \mu_3(0) - 3\bar{x}\mu_2(0) + 2\bar{x}^3 = e^{3\bar{y}}e^{(3/2)\sigma_y^2}(e^{3\sigma_y^2} - 3e^{\sigma_y^2} + 2) \qquad 12.41$$

and the skewness is

$$\gamma_1 = \frac{\mu_3}{\sigma^3} = \frac{\exp(3\sigma_y^2) - 3\exp(\sigma_y^2) + 2}{[\exp(\sigma_y^2) - 1]^{3/2}}$$

$$= [\exp(\sigma_y^2) - 1]^{1/2}[\exp(\sigma_y^2) + 2] = C_v[\exp(\sigma_y^2) + 2] = 3C_v + C_v^3 \qquad 12.42a$$

which may be approximated by

$$\gamma_1 \simeq 0.52 + 4.58\,\sigma_y^2 \qquad 12.42b$$

over the range $0.1 < \sigma_y^2 < 0.6$.

Since the y distribution is normal eqn 12.9 yields for the T-year return period event

$$\ell n\, x_T = y_T = \bar{y} + z\sigma_y \qquad 12.43a$$

where z is given in Table 12.1 for the probability level F(z) of T (z is here a frequency factor, a concept discussed later), e.g., for F(z) = 0.99 or T = 100, z = 2.326. Alternatively, eqn 12.43a can be written as

$$x_T = \bar{x} + K\sigma_x \qquad 12.43b$$

Substituting from eqn 12.38 and 12.39 yields

$$\exp(y_T) = \exp(\bar{y} + \frac{1}{2}\sigma_y^2)[1 + K(e^{\sigma_y^2} - 1)^{1/2}]$$

and

$$K = \frac{\exp(y_T - \bar{y} - \frac{1}{2}\sigma_y^2) - 1}{(e^{\sigma_y^2} - 1)^{1/2}} = \frac{\exp(z\sigma_y - \frac{1}{2}\sigma_y^2) - 1}{(e^{\sigma_y^2} - 1)^{1/2}} \qquad 12.44a$$

where the second form follows from eqn 12.43a. Substitution from eqn 12.40 leads to

$$K = \frac{1}{C_v}\left(\exp\{[\ln(1 + C_v^2)]^2 z - \frac{1}{2}\ln(1 + C_v^2)\} - 1\right) \quad 12.44b$$

Thus, x_T could also be calculated from eqn 12.43b with the aid of the K values, which can be tabulated for values of C_v and z.

The standard error of the estimate, S_T, in logarithmic form can be expressed as

$$S_T = \pm k\sigma_y/\sqrt{n} \quad 12.45$$

where

$$k = \sqrt{1 + \tfrac{1}{2} z^2} \quad 12.46$$

which leads to the prediction equation

$$\ln x_T = \bar{y} + z\sigma_y \pm k\sigma_y/\sqrt{n} \quad 12.43c$$

An alternative form of eqn 12.43c is

$$\ln x_T = \bar{y} + zS_y \pm \sqrt{E_{\bar{y}}^2 + (E_s z)^2} \quad 12.43d$$

where $E_{\bar{y}} = S_y/\sqrt{n}$ and $E_s = S_y/\sqrt{2n}$ for $n > 30$ are the standard errors of the mean and standard deviation, respectively. (When $n < 30$ use $\sqrt{n-1}$ and $\sqrt{2n-1}$, respectively). This gives the two-thirds confidence limits and if the square root is multiplied by 1.96 the 95% limits.

If the curve of log x against probability shows a curvature a three parameter log-normal distribution obtained through the transformation

$$y = \ln(x \pm x_o) \quad 12.47$$

may be used. If the curve is convex upwards, x_o is negative and signifies a lower bound of the distribution. If the curve is concave upwards x_o is positive. A rough estimate of x_o may be made from the graph. If x_1, x_2 and x_3 are the values of x corresponding to three equidistant points on the probability scale (e.g., $\pm\sigma$ from the mean) then $x_2 + x_o$ is the geometric mean of $x_1 + x_o$ and $x_3 + x_o$, so that

$$x_o = \frac{x_2^2 - x_1x_3}{x_1 + x_3 - 2x_2}$$

The value of x_o can be tested by plotting $\log(x + x_o)$. The mean of the distribution is then

$$\bar{x} = x_o + \exp(\bar{y} + \frac{1}{2}\sigma_y^2) \quad 12.48$$

and differs from the two parameter distribution value only by x_o. The variance is the same as for the two parameter distribution (eqn 12.39) and so is the skewness (eqn 12.42). Details of a graphical procedure for fitting data to a modified log-normal distribution ($y = a + b \ln x$) are given by McGuinness and Brakensiek (1964).

The 50% value on the log-normal probability plot is the mean and σ is estimated as either 2/5-th of the difference between the logarithms of the plotted 90-th and 10-th percentiles or the difference between the mean and either the value at 84.13% or 15.87%.

It is noted in passing that the cube-root transformation, $y = x^{1/3}$, is another transformation of data to "normality" which has been found to be useful for precipitation analysis (Stidd, 1953; Kendall, 1960 & 1966). Stidd noticed that when the precipitation data for different durations were plotted as cube root on the normal probability paper the straight lines were approximately parallel at spacing $P = at^n$, where t is the duration of the precipitation.

There is a family of distributions known as the (Karl) *Pearson distributions*. All of these can be generated from the differential equation

$$\frac{d\,f(x)}{dx} = \frac{(x - d)f(x)}{a + bx + cx^2}$$

where a, b, c and d are constants. These include the normal, beta (type I) and gamma (type III) distributions. The *log Pearson Type III* distribution is a three parameter distribution of the transformed variable $Z = \ell n\, x$, or more generally of $Z = \ell n(x - x_0)$ which is a four parameter distribution. The p.d.f. is

$$f(Z) = \frac{1}{\lambda^{\eta}\Gamma_{(\eta)}} e^{-(Z-Z_0)/\lambda} (Z - Z_o)^{\eta-1} \qquad 12.49$$

where Z_0, λ and η are the location, scale and shape parameters of the distribution, respectively, and $\Gamma_{(\eta)}$ is the gamma function. If $Z = \ell n\, x$, the p.d.f. is

$$f(x) = \frac{1}{x\lambda^{\eta}\Gamma_{(\eta)}} e^{-(\ell n\, x-Z_o)/\lambda} (\ell n\, x - Z_o)^{\eta-1} \qquad 12.50$$

Example 12.2. Fit the log-normal distribution to the following streamflow record: 4.05, 5.10, 6.10, 7.00, 7.95, 9.00, 10.10, 11.70, 13.40, 16.00, and 20.30 m^3/s. Find the 10 and 20 year return period flows.

m	Q m^3/s	Plotting position m/(n+1)	$y=\ell n\, Q$	$(y-\bar{y})^2$
1	4.05	0.083	1.399	0.642
2	5.10	0.167	1.629	0.326
3	6.10	0.250	1.808	0.154
4	7.00	0.373	1.946	0.065
5	7.95	0.417	2.073	0.016
6	9.00	0.500	2.197	0
7	10.10	0.583	2.313	0.013
8	11.70	0.667	2.460	0.068
9	13.40	0.750	2.595	0.156
10	16.00	0.833	2.773	0.328
11	20.30	0.917	3.011	0.658
			24.204	2.426

$$\bar{y} = 2.200 \qquad \sigma_y^2 = \frac{1}{n-1}\sum_1^n (y - \bar{y})^2 = 0.2426 \qquad \sigma_y = 0.4925$$

$$E_{\bar{y}} = \frac{\sigma_y}{\sqrt{n-1}} = 0.1557 \qquad E_s = \frac{\sigma_y}{\sqrt{2n-1}} = 0.1075 \qquad \bar{Q} = 10.189\ m^3/s$$

The fitted distribution is the given by $f(y)dy = (0.81/y)\exp[-(\ell n\, y - 2.200)^2/0.4852]dy$ and the prediction equation is $y = \bar{y} + z\sigma_y \pm [E_{\bar{y}}^2 \pm (E_s z)^2]^{\frac{1}{2}}dy = 2.200 + 0.4925z \pm (0.0242 + 0.0116z^2)^{\frac{1}{2}}$.

For a return period of 10 years, $F(x) = 1 - 1/T = 0.90$ for which, from Table 12.1, $z = 1.282$. Thus $y = 2.8314 \pm 0.2080$, or since $Q = e^y$

$$13.783 \le Q_{10} = 16.969 \le 20.893\ m^3/s$$

For T = 20 years, F(x) = 0.95, z = 1.645, and y = 3.0102 ± 0.2358 or

$$16.029 \le Q_{20} = 20.292 \le 25.667 \text{ m}^3/\text{s}$$

A better fit is obtained by using eqn 12.47.

Example 12.3. Let the mean of a record of 36 items be $\bar{x} = 15$, $\sigma_x = 5$ and $y_1 = 1.5$. Find the value of x which has a return period $T = 1/[1 - F(x)] = 100$, assuming that x has a log-normal distribution.

The skewness $\gamma = 1.5$ gives with the approximation of eqn 12.26

$$1.5 = 0.52 + 4.58\,\sigma_y^2 \qquad \sigma_y^2 = 0.214 \qquad \sigma_y = 0.463$$

and eqn 12.42a gives $\sigma_y^2 = 0.20$, $\sigma_y = 0.447$. Equation 12.39 gives

$$\bar{y} = \ell n\,\sigma_x - \frac{1}{2}\sigma_y^2 - \frac{1}{2}\ell n(e^{\sigma_y^2} - 1) = \ell n\,5 - \frac{1}{2}0.20 - \frac{1}{2}\ell n(1.22 - 1)$$

$$= 1.6094 - 0.10 + 0.7539 = 2.2633$$

Now from eqn 12.43d and F(x) = 0.99, z = 2.326

$$\ell n\,x_T = 2.2633 + 0.447 \times 2.326 \pm \sqrt{\frac{0.2}{36} + \frac{0.2}{72}2.326^2} = 3.3030 \pm 0.1435$$

or

$$23.56 \le x_{100} = 27.938 \le 31.39$$

12.4 Extreme Value Distributions

The importance of information is directly proportional to its improbability.

Hydrological studies are frequently concerned with extremes, such as the largest flood, the severest drought or the maximum precipitation. The basic statement of the extreme value theory is that the distribution of the largest (or smallest) values one taken from each of the N samples which contain n values, approaches a limiting form as n becomes large.

If the initial data has a p.f.d. of $f_0(x)$ and a c.d.f. of $F_0(x)$, then probability that all of the n observations will be equal to or less than x is

$$F(x) = [F_0(x)]^n \tag{12.51}$$

which is also the probability of x being the largest value among n observations. The p.d.f. is

$$f(x) = \frac{d[F_0(x)]^n}{dx} = n[F_0(x)^{n-1}]\frac{dF_0(x)}{dx} = n[F_0(x)]^{n-1}f_0(x) \tag{12.52}$$

It has been shown that when n becomes large the distribution of the extreme values, f(x), approaches a limiting distribution, independent of $f_0(x)$, which is of the same form as the parent one but has different parameters. If x is unlimited the probability F(x) can be shown to converge to

$$F(x) = \exp\{-\exp[-(a + x)/c]\} \tag{12.53}$$

and the p.d.f. is

$$f(x) = \frac{1}{c}\exp\{-(\frac{a + x}{c}) - \exp[-(\frac{a + x}{c})]\} \tag{12.54}$$

where a and c are statistical parameters for which Fisher and Tippett (1928) gave $a = \gamma c - \bar{x}$ and $c = (\sqrt{6}/\pi)\sigma$, where $\gamma = 0.5772157$ is the Euler constant. They also showed that eqn 12.51 has three possible forms of solution.

Jenkinson (1955) obtained a general solution for eqn 12.51 of the form

$$x - x_0 = \frac{\beta}{k}(1 - e^{-ky}) \qquad 12.55$$

where y is the dimensionless standardized variate expressed by the linear transformation

$$y = \frac{x - u}{c} = \alpha(x - u) \qquad 12.56$$

also known as the reduced variate (variable réduit), x_0 is the value of $x = u$ at $y = 0$ (the modal value of x) where $F(x) = e^{-1}$ and is a location parameter ($u = -a$). The values of u and α depend on the initial distribution and sample size and α has different values for the two tails of asymmetric distributions. The parameter β has the same units as x and is the slope of the $x - y$ curve at $y = 0$, $x = x_0$. The constant k is a curvature parameter. For $x_0 = 0$ eqn 12.55 yields

$$\frac{dy}{dx} = \frac{1}{\beta} e^{ky}$$

When k is positive

$y = -\infty$	$x = -\infty$	$dy/dx = 0$
$y = 0$	$x = 0$	$dy/dx = 1/\beta$
$y = \infty$	$x = \beta/k$	$dy/dx = \infty$

When k is negative

$y = -\infty$	$x = \beta/k$	$dy/dx = \infty$
$y = 0$	$x = 0$	$dy/dx = 1/\beta$
$y = \infty$	$x = \infty$	$dy/dx = 0$

When $k \to 0$, $x \to \beta y$ and this is the straight line known as the *Gumbel* or the *first extreme value distribution*, EV1. A negative value of k gives the second (EV2) and a positive value of k gives the third extreme value distribution (EV3). EV2 has a lower bound for the less extreme values and EV3 is bounded in the direction of the more extreme values ($y \to \infty$), Fig. 12.4.

The linear transformation, eqn 12.56, substituted in eqn 12.54 gives for EV1

$$f(x) = \alpha \exp[- y_1 - \exp(- y_1)]; \qquad y = \alpha(x - u) \qquad 12.57$$

and

$$F(x) = \exp[- \exp(- y_1)]; \qquad \alpha > 0 \qquad 12.58$$

where the scale parameter 1/c has been replaced by α. From eqn 12.58 and eqn 12.2 the reduced variate $y = - \ell n\{\ell n[T/(T - 1)]\}$. The parameters α and u estimated from sample values as

$$u \simeq \bar{x} - \frac{S_x}{\sigma_N}\bar{y}_N \qquad \text{and} \qquad \frac{1}{\alpha} = \frac{S_x}{\sigma_N} \qquad 12.59$$

in which the population values of σ_N and $\bar{y}_N$ are given in Table 12.3 as functions of N. For increasing values of N the mean $\bar{y}_N$ and the standard deviation $\sigma_N = \pi/(\alpha\sqrt{6})$ converge to the population mean $\gamma = 0.5772$ and the standard deviation $\pi/\sqrt{6}$, respectively.

The expected value prediction equation for EV1 (the Gumbel distribution) is then

$$x = u + y/\alpha \tag{12.60}$$

The skewness of this distribution $\sqrt{\beta_1} = C_s = 1.139$. When $C_v = 0.364$ and $C_s = 1.139$ the log-normal and Gumbel distributions are identical. This straight line on log-probability paper is a dividing line. For $C_v = 0.364$ all lines with $C_s > 1.139$ are concave and all lines with $C_s < 1.139$ are convex upwards.

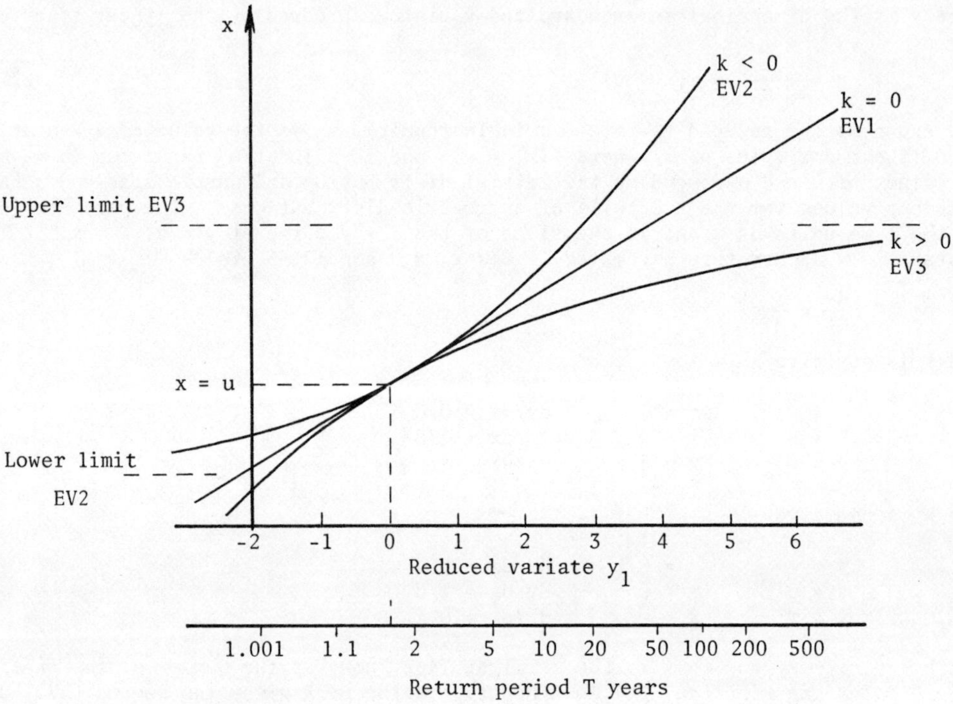

Fig. 12.4. The extreme value distributions as functions of the type 1 reduced variate: $x = u+(\beta/k)(1-e^{-ky_1})$

The asymptotic standard error theory gives that the m-th value is contained within the interval

$$\pm \Delta x_m = \frac{RSE}{\alpha\sqrt{N}} \tag{12.61}$$

with 67% probability, where the reduced standard error (RSE) is given in tabular form in Table 12.2 as a function of y. This control band is sufficiently accurate over the range $0.15 < m/(N + 1) < 0.85$. With increasing return period the standard error becomes independent of T, i.e., the control curves become parallel to the prediction line which is in the middle of the interval 0.32 T to 3.13 T. Alternatively, the spacing at x_N for 1 σ or 2/3 probability level is

$$\pm \Delta x_N = \frac{1.14071}{\alpha}$$

and the control curves are drawn parallel to the prediction line through the points $\hat{x}_N \pm \Delta x_N$. For 2 σ or 95.45% probability level the numerator is 3.06685. The transition can be plotted with the aid of the Table 12.3.

TABLE 12.2 Expected Means $\bar{y}_N$ and Standard Deviations σ_N of EV1

N	$\bar{y}_N$	σ_N	N	$\bar{y}_N$	σ_N	N	$\bar{y}_N$	σ_N
8	0.48430	0.90430	35	0.54034	1.12847	64	0.55330	1.17930
9	0.49020	0.92880	36	0.54100	1.13130	66	0.55380	1.18140
10	0.49520	0.94970	37	0.54180	1.13390	68	0.55430	1.18340
11	0.49960	0.96760	38	0.54240	1.13630	70	0.55477	1.18536
12	0.50350	0.98330	39	0.54300	1.13880	72	0.55520	1.18730
13	0.50700	0.99720	40	0.54362	1.14132	74	0.55570	1.18900
14	0.51000	1.00950	41	0.54420	1.14360	76	0.55610	1.19060
15	0.51280	1.02057	42	0.54480	1.14580	78	0.55650	1.19230
16	0.51570	1.03160	43	0.54530	1.14800	80	0.55688	1.19382
17	0.51810	1.04110	44	0.54580	1.14990	82	0.55720	1.19530
18	0.52020	1.04930	45	0.54630	1.15185	84	0.55760	1.19670
19	0.52200	1.05660	46	0.54680	1.15380	86	0.55800	1.19800
20	0.52355	1.06283	47	0.54730	1.15570	88	0.55830	1.19940
21	0.52520	1.06960	48	0.54770	1.15740	90	0.55860	1.20073
22	0.52680	1.07540	49	0.54810	1.15900	92	0.55890	1.20200
23	0.52830	1.08110	50	0.54854	1.16066	94	0.55920	1.20320
24	0.52960	1.08640	51	0.54890	1.16230	96	0.55950	1.20440
25	0.53086	1.09145	52	0.54930	1.16380	98	0.55980	1.20550
26	0.53200	1.09610	53	0.54970	1.16530	100	0.56002	1.20649
27	0.53320	1.10040	54	0.55010	1.16670	150	0.56461	1.22534
28	0.53430	1.10470	55	0.55040	1.16810	200	0.56715	1.23598
29	0.53530	1.10860	56	0.55080	1.16960	250	0.56878	1.23292
30	0.53622	1.11238	57	0.55110	1.17080	300	0.56993	1.24786
31	0.53710	1.11590	58	0.55150	1.17210	400	0.57144	1.25450
32	0.53800	1.11930	59	0.55180	1.17340	500	0.57240	1.25880
33	0.53880	1.12260	60	0.55208	1.17467	750	0.57377	1.26506
34	0.53960	1.12550	62	0.55270	1.17700	1000	0.57450	1.26851

Standard Errors of Reduced Central Values

y	$F_i(x)$	RSE	y	$F_i(x)$	RSE
-0.64034	0.15000	1.2548	1.00000	0.69220	1.8126
-0.50000	0.19230	1.2431	1.50000	0.80001	2.2408
0	0.36788	1.3108	1.81696	0.85000	2.5849
0.50000	0.54524	1.5057	2.00000	0.87342	2.8129

TABLE 12.3 Transition Values for EV1 Control Band

	Probability level			Probability level	
	68.269%	95.450%		68.269%	95.450%
Δx_N	$1.14078/\alpha$	$3.0669/\alpha$	Δx_{N-2}	$0.589/\alpha$	$1.35/\alpha$
Δx_{N-1}	$0.75409/\alpha$	$1.7820/\alpha$	Δx_{N-3}	$0.538/\alpha$	$1.17/\alpha$

The plotting of ordered data on extreme probability paper is usually done according to $P = m/(N + 1)$. However, Gringorten (1963) showed that with double exponential distribution improvement is obtained by using

$$P = \frac{m - a}{N + 1 - 2a} \qquad 12.62$$

as the plotting position. For N > 20, a = 0.44 and

$$P = \frac{m - 0.44}{N + 0.12}$$

(For N = 10, a = 0.448; N = 20, a = 0.443; N = 50, a = 440; N = 100, a = 0.439). Jenkinson (1969) proposed

$$P = \frac{m - 0.31}{N + 0.38}$$

The value form eqn 12.62 differs little from m/(N + 1).

The *second extreme value distribution*, EV2, has the c.d.f. of

$$F(x_2) = \exp\{- [1 - \alpha k(x_2 - u)]^{1/k}\}$$

$$k < 0 \qquad \alpha > 0 \qquad u + 1/\alpha k \le x_2 \le \infty \qquad 12.63$$

The lower bound of this distribution $u + 1/\alpha k$ is smaller than u since α and k are of opposite sign. The p.d.f. is

$$f(x_2) = \alpha[1 - \alpha k(x_2 - u)]^{(1/k)-1}\exp\{- [1 - \alpha k(x_2 - u)]^{1/k}\} \qquad 12.64$$

In terms of the reduced variate

$$y_2 = 1 - \alpha k(x_2 - u) \qquad 0 \le y_2 \le \infty \qquad 12.65$$

$$F(y_2) = \exp(- y_2^{1/k}) \qquad 12.66$$

and

$$f(y_2) = - \frac{y_2^{(1/k)-1}}{k} \exp(- y_2^{1/k}) \qquad 12.67$$

From eqn 12.65

$$x_2 = (u + \frac{1}{\alpha k}) - \frac{1}{\alpha k} y_2 \qquad 12.68$$

where $u + 1/\alpha k$ is the location parameter and $-1/\alpha k > 0$ is the scale parameter. The value of y_2 depends on k and the solution of eqns 12.65 and 12.68 for x_2 and y_2 for $F(x_2) = F(y_2)$ is aided by a table of $F(y_2)$ values, Table 12.4.

It has been shown that

$$y_2 = e^{-ky_1} \qquad 12.69$$

where y_1 is the reduced variate of the EV1 distribution. Equation 12.69 substituted in eqn 12.68 yields

$$x_2 = u + \frac{1}{\alpha k} (1 - e^{-ky_1}) = u + \frac{1}{\alpha} W(y_1; k) \qquad 12.70$$

which is the form proposed by Jenkinson (eqn 12.55) with $\beta = 1/\alpha$, and $W(y_1; k) = (1 - e^{-ky_1})/k$. Values of W are given in Table 12.5.

The EV1 and EV2 distributions are linked by $F_1(\ell n\, x_2) = F_2(x_2)$, where $x_1 = \ell n\, x_2$. Putting $u = \ell n\, v$ in eqn 12.58 yields

$$F_1(x_2) = \exp[-e^{-\alpha(\ell n\, x_2 - \ell n\, v)}] = \exp[- (\frac{v}{x_2})^{\alpha}] \qquad 12.71$$

By eqn 12.59, $\ell n\, v = \overline{\ell n\, x_2} - \bar{y}_N/\alpha$, where $1/\alpha = S(\ell n\, x_2)/\sigma_N$. A third parameter may be introduced by writing $x_1 = \ell n(x_2 - x_0)$, which yields

TABLE 12.4 Values of y_2 as a Function $F(y_2)$ and k

(Flood Studies Rep., 1975, Vol. 1:45)

$F(y_2)$	$k = -0.05$	$k = -0.10$	$k = -0.15$	$k = -0.20$	$k = -0.25$
0.00	0.000	0.000	0.000	0.000	0.000
0.05	0.947	0.896	0.848	0.803	0.760
0.10	0.959	0.920	0.882	0.846	0.812
0.20	0.976	0.954	0.931	0.909	0.888
0.30	0.991	0.982	0.973	0.964	0.955
0.40	1.004	1.009	1.013	1.018	1.022
0.50	1.018	1.037	1.057	1.076	1.096
0.60	1.034	1.069	1.106	1.143	1.183
0.70	1.053	1.109	1.167	1.229	1.294
0.80	1.078	1.162	1.252	1.350	1.455
0.90	1.119	1.252	1.401	1.568	1.755

Return period T	Nonexceedance probability F(y)	Normal $N(0,1)$ y	Exponential $y = \ell n\ T$	Gumbel or EV1 y_1	General extreme value $W(y_1;k)$									
					$k = -0.25$	$k = -0.20$	$k = -0.15$	$k = -0.10$	$k = -0.05$	$k = 0.05$	$k = 0.10$	$k = 0.15$	$k = 0.20$	$k = 0.25$
1.0101	0.01	−2.33	0.01	−1.53	−1.27	−1.32	−1.36	−1.42	−1.47	1.59	−1.65	−1.72	−1.79	−1.86
1.0526	0.05	−1.64	0.05	−1.10	−0.96	−0.99	−1.01	−1.04	−1.07	1.13	−1.16	1.19	1.23	−1.26
1.1111	0.10	−1.28	0.11	−0.83	−0.75	−0.77	−0.78	−0.80	−0.82	0.85	0.87	−0.89	−0.91	−0.93
1.250	0.20	−0.84	0.22	−0.48	−0.45	−0.45	−0.46	−0.46	−0.47	−0.48	−0.49	−0.49	−0.50	−0.51
1.428	0.30	−0.52	0.36	−0.19	−0.18	−0.18	−0.18	−0.18	−0.18	0.19	−0.19	0.19	−0.19	0.19
1.667	0.40	−0.25	0.51	0.09	0.09	0.09	0.09	0.09	0.09	0.09	0.09	0.09	0.09	0.09
2.00	0.50	0.00	0.69	0.37	0.38	0.38	0.38	0.37	0.37	0.36	0.36	0.36	0.35	0.35
2.50	0.60	0.25	0.92	0.67	0.73	0.72	0.71	0.69	0.68	0.66	0.65	0.64	0.63	0.62
3.33	0.70	0.52	1.20	1.03	1.18	1.14	1.11	1.09	1.06	1.00	0.98	0.96	0.93	0.91
5.00	0.80	0.84	1.61	1.50	1.82	1.75	1.68	1.62	1.56	1.45	1.39	1.34	1.30	1.25
10.00	0.90	1.28	2.30	2.25	3.02	2.84	2.68	2.52	2.38	2.13	2.02	1.91	1.81	1.72
20.00	0.95	1.64	3.00	2.97	4.41	4.06	3.74	3.46	3.20	2.76	2.57	2.40	2.24	2.10
25.00	0.96	1.75	3.22	3.20	4.90	4.48	4.10	3.77	3.47	2.96	2.74	2.54	2.36	2.20
40.00	0.975	1.96	3.69	3.68	6.03	5.43	4.90	4.44	4.04	3.36	3.08	2.83	2.60	2.40
50.00	0.98	2.05	3.91	3.90	6.61	5.91	5.30	4.77	4.31	3.54	3.23	2.95	2.71	2.49
75.00	0.9866	2.21	4.32	4.31	7.75	6.84	6.06	5.39	4.81	3.88	3.50	3.17	2.89	2.64
100.00	0.99	2.33	4.61	4.60	8.63	7.55	6.63	5.84	5.17	4.11	3.69	3.32	3.01	2.73
200.00	0.995	2.58	5.30	5.30	11.03	9.42	8.09	6.98	6.06	4.65	4.11	3.65	3.27	2.94
500.00	0.998	2.88	6.21	6.21	14.91	12.33	10.26	8.61	7.29	5.34	4.63	4.04	3.56	3.15
1000.00	0.999	3.09	6.91	6.91	18.49	14.90	12.12	9.95	8.25	5.84	4.99	4.30	3.74	3.29
5000.00	0.9998	3.54	8.52	8.52	29.64	22.46	17.25	13.44	10.62	6.94	5.73	4.81	4.09	3.52
10000.00	0.9999	3.72	9.21	9.21	36.00	26.55	19.87	15.12	11.70	7.38	6.02	4.99	4.21	3.60

$T = 1/[1 - F(y))]$ $\quad F(y) = 1 - 1/T$ $\quad y_1 = -\ell n\{- \ell n[(T - 1)/T]\}$ $\quad W(y_1; k) = [1 - \exp(-ky_1)]/k$

TABLE 12.5 Values of W as a Function of k, (Flood Studies Report, 1975, Vol. 1, p. 60).

$$F_1(x_2) = \exp[-(\frac{v}{x_2 - x_o})^{\alpha}] \qquad 12.72$$

The median reduced variate of EV2 is given from $F(y_2) = 0.5$ as $\tilde{y}_2 = (\ell n\ 2)^k$ and the mode as $y_2^* = (1 - k)^k$. The moments about $y_2 = 0$ are

$$\mu_r(0) = E[y_2^r] = \Gamma(1 + rk) \qquad k < 0 \qquad 12.73$$

Hence

$$\bar{y}_2 = \Gamma(1 + k) \qquad \text{Var } y_2 = \Gamma(1 + 2k) - \Gamma^2(1 + k)$$

and the skewness $\sqrt{\beta_1} = \mu_3/\mu_2^{3/2}$, where from eqn 12.73 and eqn 12.25 $\mu_3 = \Gamma(1 + 3k) - 3\Gamma(1 + 2k)\Gamma(1 + k) + 2\Gamma^3(1 + k)$. Table 12.6 is an abbreviated table of the values of the gamma function $\Gamma(x)$. These three parameters for a range of k values are shown in Table 12.7.

TABLE 12.6 Table of the Gamma Function for $0 < x \leq 2$

x	Γ(x)	x	Γ(x)	x	Γ(x)	x	Γ(x)
0.00	undefined	0.50	1.772 45	1.00	1.000 00	1.50	0.886 23
0.05	19.470 09	0.55	1.616 12	1.05	0.973 50	1.55	0.888 87
0.10	9.513 51	0.60	1.489 19	1.10	0.951 35	1.60	0.893 52
0.15	6.220 27	0.65	1.384 80	1.15	0.933 04	1.65	0.900 12
0.20	4.590 84	0.70	1.298 06	1.20	0.918 17	1.70	0.908 64
0.25	3.625 61	0.75	1.225 42	1.25	0.906 40	1.75	0.919 06
0.30	2.991 57	0.80	1.164 23	1.30	0.897 47	1.80	0.931 38
0.33	2.546 15	0.85	1.112 48	1.35	0.891 15	1.85	0.945 61
0.40	2.218 16	0.90	1.068 63	1.40	0.887 26	1.90	0.961 77
0.45	1.968 14	0.95	1.031 45	1.45	0.885 66	1.95	0.979 88
0.50	1.772 45	1.00	1.000 00	1.46	0.885 60 (minimum)	2.00	1.000 00

$$\Gamma(x) = \int_0^{\infty} e^{-t}\, t^{x-1}\, dt; \qquad x > 0$$

$$\Gamma(x + 1) = x\Gamma(x) \qquad \Gamma(n + 1) = n! \qquad n = 0, 1, 2, \ldots$$

For non-integer values of $x = \alpha$

$$\Gamma(\alpha) = \frac{\Gamma(\alpha + n + 1)}{\alpha(\alpha + 1)(\alpha + 2)\ldots(\alpha + n)}; \qquad \alpha \neq 0, -1, -2, \ldots$$

For negative $\alpha (\neq 0, -1, -2, \ldots)$ choose for n the smallest integer such that $\alpha + n + 1 > 0$.

TABLE 12.7 Moments $\bar{y}_2$ and Var y_2, and Skewness as a Function of k

k	$\bar{y}_2$	Var y_2	$\sqrt{\beta_1}$	k	$\bar{y}_2$	Var y_2	$\sqrt{\beta_1}$
0	0.366 513	0	1.139	-0.15	1.112482	0.060 426	2.532
-0.05	1.031 453	0.004 727	1.532	-0.20	1.164 225	0.133 763	3.535
-0.10	1.068 622	0.022 272	1.903	-0.25	1.225 413	0.270 803	5.606

The mean and variance of x_2 are obtained with the aid of the above values and eqn 12.68. Thus, $\bar{x}_2 = u + (1/\alpha k) - (1/\alpha k)\bar{y}_2$ and var $x_2 = (1/n)\Sigma(x_2 - \bar{x})^2 = (1/\alpha k)^2$ var y_2.

For data of good quality and adequate length several methods are available for estimation of the parameters, but for short records with appreciable scatter the method of maximum likelihood gives the best results. Jenkinson (1969) introduced a simple method which he found to give consistently close estimates to the maximum likelihood estimates. The variates, x_2 and W (eqn 12.70), have the same shape parameter. The W variate is considered to be divided into sextiles (six equal parts). The six mean values of the sextiles $\nu_1, \nu_2, \ldots, \nu_6$ are used to define the parameters of the W distribution:

$$\text{mean } \bar{\nu} = \frac{1}{6}\sum_1^6 \nu_i$$

$$\text{standard deviation } \sigma_\nu = [\frac{1}{6}\Sigma(\nu_i - \bar{\nu})^2]^{1/2} = (\frac{1}{6}\Sigma\nu^2 - \bar{\nu}^2)^{1/2}$$

$$\text{shape } \lambda = \frac{\nu_2 - \nu_1}{\nu_6 - \nu_5}$$

These quantities depend on k only and were tabulated by Jenkinson as shown in Table 12.8. The table is entered with the value of λ, estimated from the sample, yielding the values of σ_ν, $\bar{\nu}$ and k of the W distribution. The sextile means of the x_2 distribution are $\omega_1 . \omega_2, \ldots, \omega_6$ and these are related to ν_i by $\omega_i = u + \nu_i/\alpha$. The mean of $\bar{\omega}$ is $\bar{\omega} = u + \bar{\nu}/\alpha$, the variance of ω_i is $\sigma_\omega = \sigma_\nu/\alpha$, and the shape parameter

$$\lambda = \frac{\omega_2 - \omega_1}{\omega_6 - \omega_5} = \frac{u + \nu_2/\alpha - (u + \nu_1/\alpha)}{u + \nu_6/\alpha - (u + \nu_5/\alpha)} = \frac{\nu_2 - \nu_1}{\nu_6 - \nu_5}$$

Thus

$$\alpha = \frac{\sigma_\nu}{\sigma_\omega} \qquad u = \bar{\omega} - \bar{\nu}/\alpha \qquad 12.74$$

The values of λ, u and σ_ω are estimated from sample data which is divided into sextiles. The number of data items is seldom a multiple of six. One could arbitrarily decide, for example, that in a record of 46 items four sextiles have eight and two have seven items or proportion two-thirds of the eighth item into the first and one-third into the second sextile, etc., but either way, the effect on the final result is small. The recorded data yield $\lambda = (w_2 - w_1)/(w_6 - w_5)$, σ_W and $\bar{w}$ which serve as estimates of λ, σ_ω and $\bar{\omega}$, respectively.

TABLE 12.8 Values of k, $\bar{\nu}$, σ_ν and λ (Jenkinson, 1969).

k	$\bar{\nu}$	σ_ν	λ	k	$\bar{\nu}$	σ_ν	λ
-0.5	1.54	2.85	0.08	0.1	0.49	1.09	0.58
-0.4	1.22	2.24	0.11	0.2	0.41	1.01	0.70
-0.3	0.99	1.83	0.16	0.3	0.34	0.96	1.05
-0.2	0.82	1.55	0.23	0.4	0.28	0.92	1.39
-0.1	0.69	1.34	0.32	0.5	0.23	0.89	1.82
0	0.58	1.20	0.43	0.6	0.18	0.88	2.38
				0.7	0.13	0.87	3.13

Alternatively the parameters can be estimated by moments. For EV2 eqn 12.68 is $x_2 = A + By_2$. The moments depend only on y_2 as shown in Table 12.7. The variate x_2 is linearly related to y_2 and the skewness of x_2 and y_2 is the same. Thus, by calculating the skewness of the x_2 record and using Table 12.7, the k value, as well as $\bar{y}_2$ and var y_2 may be obtained. The mean is then $\bar{x}_2 = A + B\bar{y}_2$ and var $x_2 = B^2$ var y_2. With the aid of $\bar{x}_2$ and var x_2 as estimated from the sample, $B = (S_{x2}^2/\text{var } y_2)^{\frac{1}{2}}$ and $A = \bar{x} + B\bar{y}_2$. Since the skewness value of x_2 depends strongly on the quality of the data, this method is very sensitive to the quality of the data.

The *third extreme value distribution,* EV3, has the c.d.f.

$$F(x_3) = \exp\{- [1 - \alpha k(x_3 - u)]^{1/k}\}$$

$$k > 0 \qquad \alpha > 0 \qquad -\infty \le x_3 \le u + \frac{1}{\alpha k} \qquad 12.75$$

This distribution is of the same form as EV2, except that k > 0 and it has an upper limit which is greater than u. The p.d.f. is

$$f(x_3) = \alpha[1 - \alpha k(x_3 - u)]^{(1/k)-u} \exp\{- [1 - \alpha k(x_2 - u)]^{1/k}\} \qquad 12.76$$

Introducing the reduced variate

$$-y_3 = 1 - \alpha k(x_3 - u) \qquad -\infty \le y_3 \le 0 \qquad 12.77$$

yields

$$F(y_3) = \exp[- (-y_3)^{1/k}] \qquad 12.78$$

$$f(y_3) = - \frac{(-y_3)^{(1/k)-1}}{k} \exp [- (-y_3)^{1/k}] \qquad 12.79$$

and

$$x_3 = u + \frac{1}{\alpha k} + \frac{1}{\alpha k} y_3 = A + By_3 \qquad 12.80$$

The y_3 versus $F(y_3)$ relationship as a function of k is tabulated in Table 12.9, and values of x_3 and y_3 for $F(x_3) = F(y_3)$ are given by eqns 12.77 and 12.80.

TABLE 12.9 Values of y_3 as a Function of $F(y_3)$ and k. (Flood Studies Report, 1975, Vol. 1, p. 49)

$F(y_3)$	k=0.05	k=0.10	k=0.15	k=0.20	k=0.25
0.00	$-\infty$	$-\infty$	$-\infty$	$-\infty$	$-\infty$
0.05	-1.506	-1.116	-1.179	-1.245	-1.316
0.10	-1.043	-1.087	-1.133	-1.182	-1.232
0.20	-1.024	-1.049	-1.074	-1.100	-1.126
0.30	-1.009	-1.019	-1.028	-1.038	-1.048
0.40	-0.997	-0.991	-0.987	-0.983	-0.978
0.50	-0.982	-0.964	-0.947	-0.929	-0.912
0.60	-0.967	-0.935	-0.904	-0.874	-0.845
0.70	-0.950	-0.902	-0.857	-0.814	-0.773
0.80	-0.928	-0.861	-0.799	-0.799	-0.687
0.90	-0.894	-0.798	-0.714	-0.638	-0.570
1.00	0.000	0.000	0.000	0.000	0.000

It has been shown that if y_3 is the reduced variate of EV3 then

$$y_3 = - \exp(-ky_1) \qquad 12.81$$

Substituting this relationship in eqn 12.80 gives

$$x_3 = u + \frac{1}{\alpha k} (1 - e^{-ky}) = u + \frac{1}{\alpha} W(y_1; k) \qquad 12.82$$

where values of $W(y_1; k)$ are given in Table 12.5. Hence, the same methods of estimation of the parameters as those discussed for EV2 may be used.

The median of EV3 for $F(y_3) = 0.5$ is $\tilde{y}_3 = -(\ell n2)^k$ or

$$\tilde{x}_3 = u + \frac{1}{\alpha k} - \frac{1}{\alpha k} (\ell n 2)^k \quad 12.83$$

The modal value is

$$y_3^* = -(1 - k)^k$$

or

$$x_3^* = u + \frac{1}{\alpha k} - \frac{1}{\alpha k} (1 - k)^k \quad 12.84$$

The r-th moment $\mu_r(0) = E[y_3^r] = (-1)^r \Gamma(1 + rk)$. Hence

$$\text{mean } \bar{y}_3 = -\Gamma(1 + k)$$

$$\text{var } y_3 = \Gamma(1 + 2k) - \Gamma^2(1 + k)$$

$$\mu_3 = -\Gamma(1 + 3k) + 3\Gamma(1 + 2k)\Gamma(1 + k) - 2\Gamma^3(1 + k) \quad 12.85$$

These yield the mean $\bar{x}_3 = u + (1/\alpha k) + (1/\alpha k)\bar{y}_3$ and var $x_3 = [1/(\alpha k)^2]$var y_3. A few values of the moments and skewness are given in Table 12.10.

TABLE 12.10 Moments, Coefficients of Variation and Skewness for the Reduced Variate of EV3 as a Function of k. (Flood Studies Report, 1975, Vol. 1, p. 51).

k	$E[y_3]=\mu_1$	μ_2=var y_3	$\sigma=\sqrt{\mu_2}$	CV=$\lvert\frac{\sigma}{\mu_1}\rvert$	$\sqrt{\beta_1}=\mu_3/\mu_2^{3/2}$
0.05	-0.973 50	0.003 650	0.060 42	0.062 06	0.911 500
0.10	-0.951 35	0.013 093	0.114 42	0.120 28	0.623 041
0.15	-0.933 04	0.026 906	0.164 03	0.175 80	0.436 171
0.20	-0.918 16	0.044 242	0.210 34	0.229 09	0.255 755
0.25	-0.906 40	0.064 659	0.254 28	0.280 54	0.086 610

Gumbel (1954) applied EV3 to extremes of droughts. For the special case when all the droughts exceed zero discharge and the lower limit is zero

$$F(x_3) = \exp[-(x_3/u)^\alpha]$$

In this case the value of x = u, the characteristic drought, is exceeded 36.788% of the time, the median is $\tilde{x}_3 = u(\ell n\ 2)^{1/\alpha}$ and the modal value is $x_3^* = u(1 - 1/\alpha)^{1/\alpha} < u$. The mode exists only for $1/\alpha > 1$ and precedes the median if $1/\alpha > 0.30685$. With the transformation

$$x_3 = e^z \quad \text{and} \quad u = e^v$$

the c.d.f. becomes

$$F(z) = \exp[-e^{\alpha(z-v)}]$$

which is of the same form as eqn 12.71. The reduced variate

$$-y = \alpha(\ell n\ x - \ell n\ u) = \alpha'(\log x - \log u)$$

where $\alpha' = 2.30256\alpha$ and $1/\alpha' = 0.43429/\alpha$ if the logarithm to base 10 is used. Hence the simple EV1 methods can be used. The return period $T(x_3) \simeq (u/x)^\alpha$, which shows that the drought flows decrease asymptotically as a power of T, i.e., $x_3 \simeq u/[T(x_3)]^{1/\alpha}$.

Example 12.4. Annual maxima of the Manawatu River, N.Z., flows for a period of 45 years are as follows:
562, 600, 728, 852, 938, 949, 1019, 1061, 1070, 1096, 1110, 1121, 1121, 1237, 1254, 1337, 1362, 1382, 1390, 1467, 1487, 1557, 1614, 1671, 1671, 1763, 1784, 1787, 1787, 1855, 1933, 2010, 2067, 2112, 2166, 2236, 2379, 2400, 2582, 2633, 2633, 3186, 3256, 3384, 4432 m^3/s

$$\Sigma Q_i = 78\ 041, \qquad \Sigma Q_i^2 = 163\ 650\ 537$$

Determine the prediction equation for this data based on EV1 distribution together with the control curves.

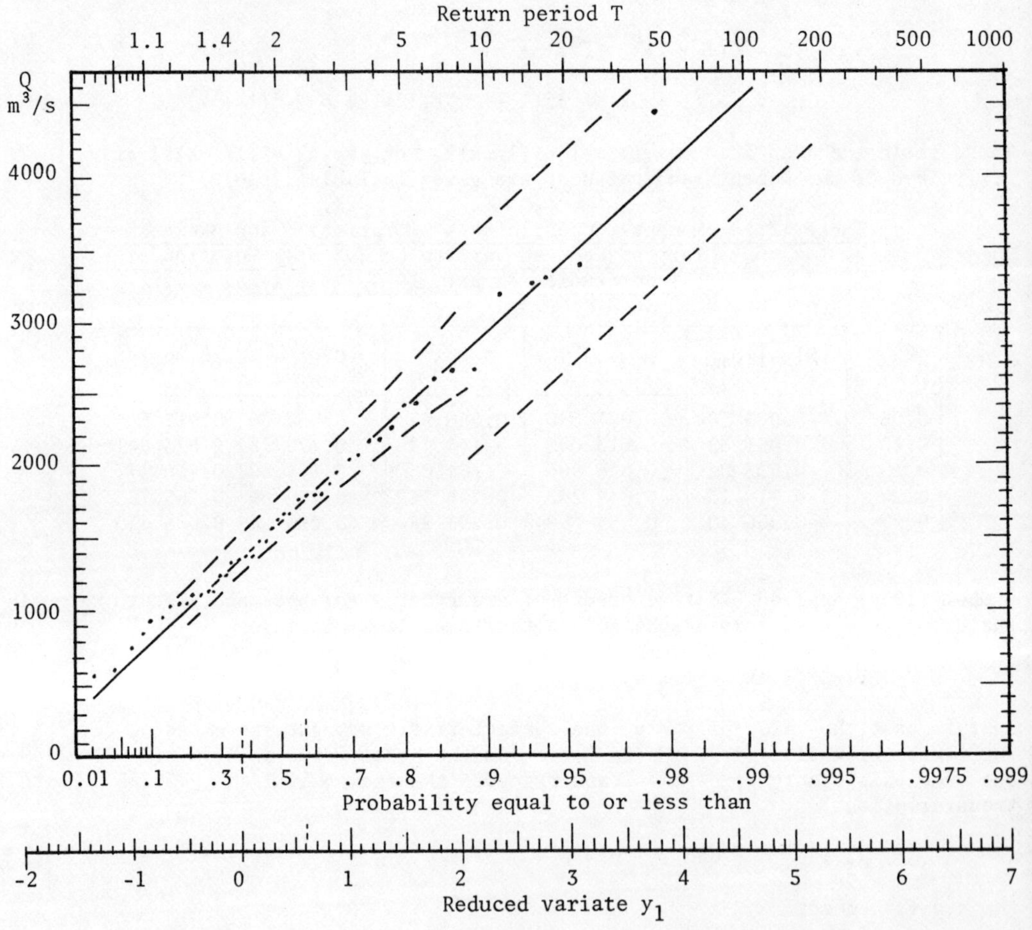

$$\bar{Q} = 78\ 041/45 = 1734\ m^3/s, \qquad \overline{Q^2} = 3\ 636\ 679$$

The prediction equation is $x = u + y/\alpha$, where $1/\alpha = s_x/\sigma_N$.

$$S_Q^2 = Q^2 - (\bar{Q})^2 = 629\ 075, \qquad S_Q = 793$$

For N = 45, Table 12.2 gives $\bar{y}_N = 0.5463$, $\sigma_N = 1.15185$. Hence $1/\alpha = 688.6$ and $u = \bar{Q} - \bar{y}_N/\alpha = 1734 - 0.5463 \times 688.6 = 1358$. Thus $Q = 1358 + 688.6y$. The control curves are calculated in the table below (eqn 12.58 and Table 12.2)

y	$f_1(Q)$	RES	ΔQ	Q	$Q + \Delta Q$	$Q - \Delta Q$
-0.5	0.19	1.2431	127.6	1014	1142	886
0	0.37	1.3108	134.6	1358	1493	1223
0.5	0.55	1.5057	154.6	1702	1857	1547
1.0	0.69	1.8126	186.1	2047	2233	1861
1.5	0.80	2.2408	230.0	2391	2621	2161
2.0	0.87	2.8129	288.7	2735	3024	2446

$$\Delta Q_N = \frac{1.14071}{\alpha} = 785.5 \qquad \frac{1}{\alpha\sqrt{N}} = \frac{688.6}{\sqrt{45}} = 102.65$$

Example 12.5. The records of annual maximum floods of the Waikato River at Ngaruawahia, New Zealand, for 46 years are as follows:
487.0, 533.5, 560.7, 560.7, 560.7, 560.7, 566.3, 566.3, 569.2, 574.8, 577.7, 580.5, 586.2, 594.7, 606.0, 606.0, 617.3, 617.3, 620.1, 623.0, 665.4, 679.6, 679.6, 679.6, 685.3, 693.8, 707.9, 716.4, 727.7, 727.7, 739.1, 747.6, 758.9, 781.5, 790.0, 798.5, 810.7, 855.2, 855.2, 877.8, 906.1, 928.8, 928.8, 947.2, 957.1, 1540.4 m^3/s

$\Sigma Q_i = 32\ 754.6$, $\Sigma Q_i^2 = 24\ 741\ 596.24$; $\Sigma(\ell n\ Q_i) = 301.0100175$, $\Sigma(\ell n\ Q_i)^2 = 1971.7799618$. Analyze this data by the EV2 distribution using (a) the transformation $x_1 = \ell n\ x_2$ and (b) the method of sextiles.

(a) The EV1 and EV2 distributions are linked by $F_1(\ell n\ x_2)$ where x_1 of EV1 is $x_1 = \ell n\ x_2$. Thus, $u(x_2) = \ell n(x_2) - \bar{y}_N/\alpha$, $1/\alpha = S(\ell n\ x_2)/\sigma_N$ and from Table 12.2 for N = 46, $\sigma_N = 1.1538$, $\bar{y}_N = 0.5468$, $S^2 = \overline{\ell n\ Q^2} - (\overline{\ell n\ Q})^2$

$$\overline{\ell n\ Q} = (1/N)\Sigma\ \ell n\ Q_i = 6.5436960, \qquad (\overline{\ell n\ Q})^2 = 42.8199573$$

$$\overline{(\ell n\ Q)^2} = 42.8647818$$

$$S^2 = 0.0448245 \qquad S = 0.2117180$$

$$1/\alpha = 0.211718/1.1538 = 0.1834963 \qquad \alpha = 5.4497020$$

$$u = \ell n\ v = \overline{\ell n\ x_2} - \bar{y}_N/\alpha = 6.5436960 - 0.5468x0.1834963 = 6.4433602,$$

$$v = 628.5152067$$

Thus, the cumulative distribution function is

$$F_L(x_2) = \exp[-(\frac{628.5152}{x_2})^{5.4497}]$$

For any given value of $Q = x_2$ this gives a value of probability and the prediction line can be plotted on the Gumbel paper. For $F(x_2) = 0.99$ the value of $x_2 = Q_{100} = 1461.9\ m^3/s$.

(b) The data is arbitrarily divided into sextiles with the first and last having seven items. The sextile means are

$$w_1 = 547.09\ m^3/s \qquad w_4 = 709.79\ m^3/s$$

$$w_2 = 581.94\ '' \qquad w_5 = 799.70\ ''$$

$$w_3 = 638.54\ '' \qquad w_6 = 1012.31\ ''$$

$$\bar{w} = 714.54\ '' \qquad \sigma_w = [(1/6)\Sigma w_i^2 - \bar{w}^2]^{\frac{1}{2}} = 158.24\ m^3/s$$

$$\lambda = (581.93 - 547.09)/(1012.31 - 799.70) = 0.16$$

For $\lambda = 0.16$ Table 12.8 yields $\sigma_\nu = 1.83$, $\bar{\nu} = 0.99$ and $k = -0.3$. Hence,

$\alpha = 1.83/158.24 = 0.011565$, $1/\alpha = 86.47$ and $u = 714.54 - 0.99\text{x}86.47 = 628.93\ m^3/s$ and $x_2 = [u + (1/\alpha k] - (1/\alpha k)\exp(-ky_1) = 628.93 - 288.23 + 288.23\exp(0.3y_1) = 340.70 + 288.23\exp(0.3y_1)$, e.g., $(x_2)_{100} = 340.7 + 288.23\exp(0.3\text{x}4.61) = 1489.83\ m^3/s$

Example 12.6. Estimate from the given 38 years of record of annual minimum streamflows the 100-year return period drought flow at the same gauging station by
(a) Gumbel transformation $x_3 = e^z$,
(b) equation 12.80 using the method of moments, and
(c) using the method of sextiles.

Frequency f	Discharge Q ℓ/s	Plotting position $\tilde{m} = \sqrt{m(m = f - 1)}$	$\frac{\tilde{m}}{N+1}$	log Q	f_i(10g Q)	$f_i(\log Q)^2$
3	870	$\sqrt{1(1+3-1)}$ = 1.732	0.044	2.9395192	8.8185576	25.9223194
2	750	$\sqrt{4(4+2-1)}$ = 4.479	0.115	2.8750613	5.7501226	16.5319550
7	650	$\sqrt{6(6+7-1)}$ = 8.485	0.218	2.8129134	19.6903938	55.3873726
7	520	$\sqrt{13(13+7-1)}$ = 15.716	0.403	2.7160033	19.0120231	51.6367175
2	500	$\sqrt{20(20+2-1)}$ = 20.494	0.525	2.6989700	5.3979400	14.5688781
2	475	$\sqrt{22(22+2-1)}$ = 22.494	0.577	2.6766936	5.3533872	14.3293773
1	430		0.615	2.6334685	2.6334685	6.9351563
5	410	$\sqrt{25(25+5-1)}$ = 26.926	0.690	2.6127839	13.0639195	34.1331985
6	320	$\sqrt{30(30+6-1)}$ = 32.404	0.831	2.5051500	15.0309000	37.6546591
1	225		0.927	2.3521825	2.3521825	5.5327625
1	210		0.945	2.3222193	2.3222193	5.3927025
1	175		0.974	2.2430380	2.2430380	5.0312195
38					101.6681521	273.0563183

f_iQ	f_iQ^2	f_iQ^3
2 610	2 270700	1 975509000
1 500	1 125000	843750000
4 550	2 957500	1 922375000
3 640	1 892800	984256000
1 000	500000	250000000
950	451250	214343750
430	184900	79507000
2 050	840500	344605000
1 920	614400	196608000
225	50635	11390625
210	44100	9261000
175	30625	5359375
19 260	10 962400	6 836964750

(a) <u>Gumbel method</u>: From the above table $\overline{\log Q} = 2.6754777$,
$\overline{(\log Q)^2} = 7.1856926 = \mu_2(0)$

$$S^2_{\log Q} = \overline{(\log Q)^2} - (\overline{\log Q})^2 = 0.0275116,\ S_{\log Q} = 0.1658664.$$

For N = 38 Table 12.2 gives $\bar{y}_N = 0.5424$ and $\sigma_N = 1.1363$. Hence, $1/\alpha' = S_{\log Q}/\sigma_N = 0.1459706$, $\log u = \overline{\log Q} + \bar{y}_N/\alpha' = 2.7546521$ and u = 568 ℓ/s, where u is the characteristic drought which is equalled or exceeded 36.79% of time (not the mode). The prediction equation is then $\log Q = \log u - y_1/\alpha' = 2.7546521 - 0.1459706\ y_1$. For a 100 year drought $y_1 = 4.60$ and log Q = 2.0831873, <u>Q = 121.1 ℓ/s</u>.

(b) <u>Method of moments</u>: For use of eqn 12.80

$$\bar{Q} = 506.84\ \ell/s,\ \overline{Q^2} = \frac{1}{n}\Sigma f_iQ_i^2 = \mu_2(0) = 288484.21$$

$\mu_3(0) = \frac{1}{n} \Sigma f_i Q_i^3 = 179920125$

$S_Q = \overline{Q^2} - \bar{Q}^2 = 288484.21 - 256886.78 = 31587.42, \; S_Q = 177.8 \; \ell/s$

$\mu_3 = \mu_3(0) - 3\bar{x}\,\mu_2(0) + 2\bar{x}^3$

$= 179290125 - 3 \times 506.84 \times 288484.21 + 2 \times 506.84^3 = 1675110.8$

$\gamma_1 = \sqrt{\beta_1} = \frac{\mu_3}{S^3} = \frac{1675110.8}{177.76^3} = 0.298223$

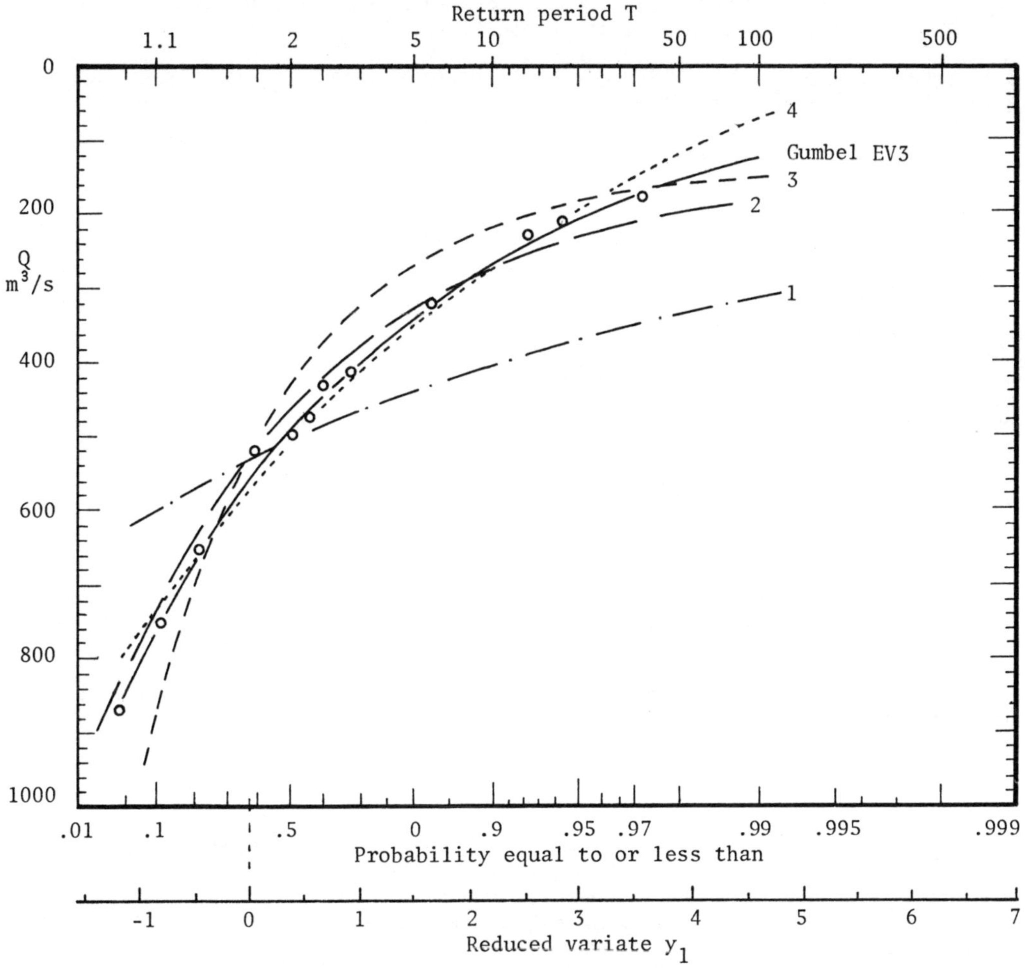

From Table 12.10 this yields $k = 0.188$, $\bar{y}_3 = -0.92167$, var $y_3 = 0.19941$. Since

$\text{var } x_3 = \frac{1}{(\alpha k)^2} \text{var } y_3$

$\frac{1}{\alpha k} = \left(\frac{\text{var } x_3}{\text{var } y_3}\right)^{\frac{1}{2}} = \left(\frac{31587.42}{0.19941}\right)^{\frac{1}{2}} = 398, \quad \frac{1}{\alpha} = 74.9$

In order to retain the EV3 shape when the numerical value of the increasing drought diminishes the data is to be plotted, as minus x-values. Then from

$$\bar{x} = u + \frac{1}{\alpha k} y_3, \qquad -506.84 = u + 398 + 398(-0.92167, \qquad u = -538 \qquad \text{and}$$

$$Q_3 = -140 + 398\, y_3 = -140 - 398\, e^{-ky_1}$$

For k = 0.188, $(Q_3)_{100}$ = - 168 or 168 ℓ/s and the prediction equation is superimposed on plotted data as line 1. Line 2 shows the prediction line for k = 0.5 and line 3 for k = 0.75. In general, k varies rapidly with skewness. The γ_1 versus k relationship for EV3 distribution starts at γ_1 = 1.139 and k = 0 (EV1) and approaches the k axis asymptotically as γ_1 goes to zero, i.e., for small values of γ_1 an accurate value of k is difficult to estimate.

(c) Method of sextiles: Let the first two sextiles with large flow rates have seven items then the sextile means are

$$w_1 = 772.86 \text{ ℓ/s} \qquad w_4 = 450.00 \text{ ℓ/s}$$

$$w_2 = 687.14 \text{ "} \qquad w_5 = 365.00 \text{ "}$$

$$w_3 = 516.67 \text{ "} \qquad w_6 = 261.67 \text{ "}$$

$$\bar{w} = 508.89 \text{ "}$$

$$\sigma_w = \left(\frac{1}{6}\,\Sigma w_i^2 - \bar{w}^2\right)^{\frac{1}{2}} = 176.45 \text{ ℓ/s}$$

$$\lambda = (w_2 - w_1)/(w_6 - w_5) = 0.83$$

For λ = 0.83 Table 12.8 yields $\sigma_\nu \simeq 1.0023$, ν = 0.39923 and k = 0.2154. Hence

$$\alpha = \sigma_\nu/\sigma_w = \frac{1.0023}{176.45} = 5.68 \times 10^{-3}$$

$$\frac{1}{\alpha} = 176.05, \qquad \frac{1}{\alpha k} = 817.32$$

$$u = \bar{w} - \bar{\nu}/\alpha = -508.89 - 176.05 \times 0.39923 = -579$$

and

$$x_3 = \left(u + \frac{1}{\alpha k}\right) + \frac{1}{\alpha k} y_3 = -579 + 817 - 817\, e^{-ky_1} = 238 - 817\, e^{-0.2154\, y_1}$$

The relationship $Q = x_3$ is shown as line 4. This example was selected to illustrate some of the difficulties met in practice.

12.5 Partial Duration Series

An objection raised against the annual maxima method is that some of the scarce extreme observations may be omitted, i.e., those which are less than the maximum in a particular year but still larger than the maxima of some of the other years. The partial duration series attempts to compromise between the use of the full record or only the annual maxima. By the partial duration method the lowest value of the annual maxima is selected as the cutting off point of the descending series. Various other criteria for truncation of the series are also in use. The top n observations retained from a N-year period of record form a sample called the annual exceedances. The number n is usually greater than N. The data do not form a complete distribution and any function which fits this portion of the data can be used. The greater number of observations in the partial duration series might be advantageous if the record is short, but usually most of the additional observations are of low magnitude and plot where the curve is already well-defined. The higher values are generally identical to those in annual maxima series. The method of exceedances introduces an ambiguity, because flood peaks are often not clearly defined, for example, in a multi-peak flood when a number of floods follow in close succession.

The ratio of the number of peaks, n, above a threshold value x_0 to the number of years of record, N, gives the rate of exceedance, r = n/N. In general, the number of peaks in successive years is not the same, nor is it the same from season to season wihtin a year. Several models have been proposed incorporating
(1) a constant rate of exceedance r from year to year,
(2) the number of exceedances above x_0 in a year as a Poisson variate, $p_i = e^{-r}r^i/i!$, and
(3) seasonal variation of the number of peaks.

When m is the order number of x_i in a descending sequence, then P' = m/N does not define a probability, but the average frequency, and is greater than unity for m > N. The frequency in years is given by 1/P' = N/m. If the events are randomly distributed in time, the relationship between the probabilities of exceedance P and the annual frequencies P' is

$$P = 1 - e^{-P'} = 1 - e^{-m/N} \qquad 12.86$$

The values of P and P' approach each other for increasing 1/P'. When 1/P' > 10 years, P ≃ P' for practical application. i.e., P = 0.0952 and P' = 0.10.

Chow (1953) wrote the prediction equation as

$$x = a \log T + b \qquad 12.87$$

where T ≃ N/m is the recurrence interval of the events. The constants a and b were fitted by the method of least squares

$$a = \frac{\overline{xy} - \bar{x}\,\bar{y}}{\overline{y^2} - \bar{y}^2} \qquad b = \bar{x} - a\bar{y} = \frac{\bar{x}\,\overline{y^2} - \overline{xy}\,\bar{y}}{\overline{y^2}\quad \bar{y}^2}$$

where $\bar{x} = (1/n)\Sigma x_i$, $\bar{y} = (1/n)\Sigma y_i$, $\overline{xy} = (1/n)\Sigma x_i y_i$ and $\overline{y^2} = (1/n)\Sigma y_i^2$. Using T = n/m as the plotting position Chow showed that at 68.3% probability the event will not occur in the return period

$$T_o = n(1.465^{1/m} - 1)$$

in N future trials, and will occur at least once in

$$T_1 = n(3.150^{1/m} - 1)$$

years. If n is not the number of years of record, then T is not one year but a fraction of a year depending on the ratio of n to the number of years of record.

The distribution function for exceedance models is usually assumed to be of exponential form

$$F(X) = 1 - e^{-(x-x_0)/c} \qquad 12.88$$

where X ≤ x given that $x \geq x_0$, x_0 is the base or threshold value, and c is a parameter given by $c = \bar{x} - x_0$, where $\bar{x} = (1/n)\Sigma x_i$. If the rate of exceedance is constant, then each period of T years contains rT peaks greater than x_0, i.e., the return period is rT or the probability is 1/rT. Thus, the event with a T-year return period is given by

$$F(X) = 1 - \frac{1}{rT}$$

These two expressions for F(X) yield

$$X(T) = x_o + c\, \ell n\, r + c\, \ell n\, T \qquad 12.89$$

The Poisson distribution of events leads to the same relationship. If seasonal variations are introduced, then with r_i as the number of peaks per season and $r = r_1 + r_2 + \ldots$,

$$X(T) = x_o + c\, \ell n\, (r_1 + r_2 + \ldots) + c\, \ell n\, T$$

If, instead of x_o, n is fixed beforehand, then $x_o = x_{min} - c/n$ and $c = n(\bar{x} - x_{min})/(n - 1)$. The variance of X(T) for x_o fixed in advance is

$$\text{Var } X(T) = \frac{c^2}{N}[1 + \frac{(\ell n\, T + \ell n\, r)^2}{r}]$$

and for n fixed in advance

$$\text{Var } X(T) = \frac{c^2}{rN}[\frac{(1 - \ell n\, r - \ell n\, T)^2}{rN - 1} + (\ell n\, r + \ell n\, T)^2]$$

12.6. Additional Remarks

A guide as to which of the distributions may be useful is given in Fig. 12.5 in terms of the kurtosis and skewness of the sample data. It must be borne in mind that the use of a distribution is simply a procedure by which a curve may be fitted to the sample data and frequently more than one distribution can be fitted equally well.

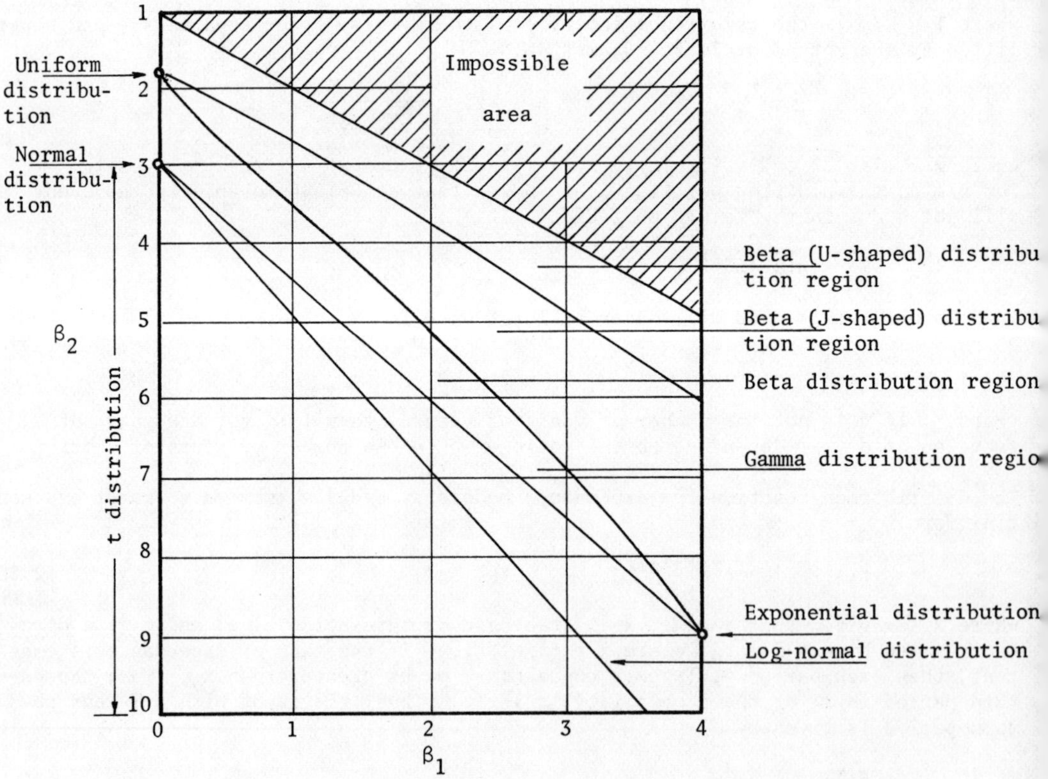

Fig. 12.5. Regions in (β_1, β_2) plane for various distributions, after E.S. Pearson. From Hahn and Shapiro (1967).

It is also important to remember that the single variable distributions are only a small part of all the possible distributions. If the process depends on more than one independent variable, it is said to have a multivariate distribution. A bivariate distribution, for example, has a frequency distribution surface, rather than a curve. Such a distribution is also known as a *joint distribution*, meaning that the outcome is the result of the simultaneous occurrence of two random variables x and y. The joint cumulative probability $P[X \le x, Y \le y] = F(x, y)$ in discrete form is

$$F(x_k, y_\ell) = \sum_1^k \sum_1^\ell P(x_i, y_j) \tag{12.90a}$$

or in continuous form

$$F(x, y) = \int_{-\infty}^{x} \int_{-\infty}^{y} f(x, y)\,dx\,dy \tag{12.90b}$$

where

$$\sum_1^m \sum_1^n P(x_i, y_j) = 1 \qquad \text{or} \qquad \int_{-\infty}^{\infty} \int_{-\infty}^{\infty} f(x, y)\,dx\,dy = 1$$

for $x_i (i = 1, 2, \ldots, m)$ and $y_j (j = 1, 2, \ldots, n)$. In general, F(x, y) cannot be determined from F(x) and F(y) but is related to these. Diagrammatically F(x, y) is the "volume" under the frequency distribution surface between given limits of x and y, or in terms of a sample space the total common area in Fig. 12.6. The joint pro-

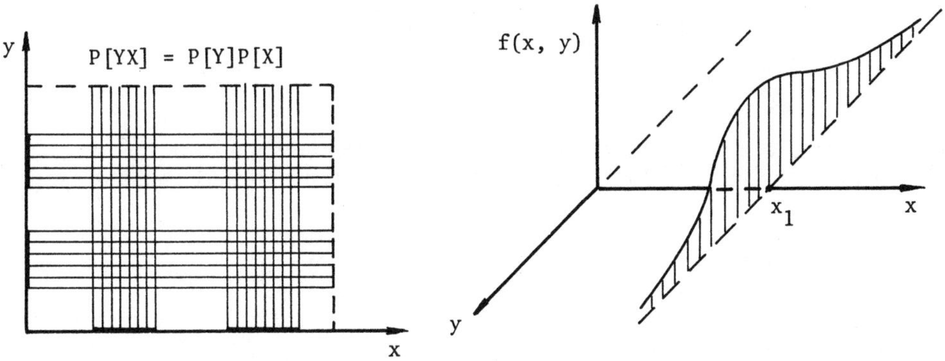

Fig. 12.6. Illustration of joint events in sample space.

bability distribution is also frequently illustrated by a double entry table as shown below. This table represents the 27 possible outcomes of placing 3 balls in 3 cells.

		0	1	2	3	Distribution of Y
	1	2q	0	0	q	3q
Y	2	6q	6q	6q	0	18q
	3	0	6q	0	0	6q
Distribution of X_i		8q	12q	6q	q	

Here Y denotes the number of cells occupied, X_i the number of balls in cell number i, and $q = 1/27$.

The joint density function is

$$F(x, y) = \frac{\partial^2 F(x, y)}{\partial x \, \partial y}.$$

In the study of several random variables the distribution of each variable is called marginal. Thus

$$F(x) = F(x, \infty) = \int_{-\infty}^{\infty}\int_{-\infty}^{\infty} f(x, y)dx\, dy \qquad 12.91a$$

is the marginal c.d.f and is the area of sample space to the left of the limit of x on the x - y plane (i.e., to the left of x_1 in Fig. 12.6). The *marginal probability density functions* are related to the joint density function as

$$f(x) = \int_{-\infty}^{\infty} f(x, y)dy \qquad 12.92a$$

$$f(y) = \int_{-\infty}^{\infty} f(x, y)dx \qquad 12.93a$$

In discrete form the c.d.f. is

$$F(x_k) = \sum_1^k \sum_1^n P(x_i, y_j) \qquad 12.91b$$

and the marginal distributions are

$$P(x_i) = \sum_{j=1}^{n} P(x_i, y_j) \qquad 12.92b$$

$$P(y_j) = \sum_{i=1}^{m} P(x_i, y_j) \qquad 12.93b$$

The vertical plane which intersects the probability surface at x_1 in Fig. 12.6 defines a profile f(x, y), the area of which equals the marginal density $f(x_1)$. From this

$$f(x_1)dx = P[x_1 < X \leq x_1 + dx] = dx \int_{-\infty}^{\infty} f(x_1, y)dy$$

which equals the "mass" in the vertical element $x_1 < x \leq x_1 + dx$. Similarly, $f(y_1)dy$ equals the "mass" in a strip $y_1 < y \leq y_1 + dy$. The marginal density functions are most useful for finding conditional probabilities.

The *conditional probability* P(B|A) of an event B with respect to some other event A is the probability that B will occur, given that A takes place. The distribution function is

$$F(y|A) = P[Y \leq y|A] = \frac{P[Y \leq y, A]}{P(A)}\ ; \qquad P(A) \neq 0$$

or in terms of the probability space, Fig. 12.7

$$P(B|A) = \frac{P(AB)}{P(A)} \qquad 12.94$$

Once the event A has taken place the set A becomes the sample space instead of the original identity set I and AB represents the joint space. It is seen that the conditional probability equals the ratio of F(x, y) to the marginal distribution F(x). The corresponding density function is found by differentiation with respect to y, i.e.

$$f(y|X \leq x) = \frac{\partial F(x, y)/\partial y}{F(x)} = \frac{\int_{-\infty}^{\infty} f(x, y)dx}{\int_{-\infty}^{\infty}\int_{-\infty}^{x} f(x, y)dx\, dy} \qquad 12.95$$

and $f(y|x_1 \leq X \leq x_2)dy$ is the ratio of the probability "mass" in the rectangle, x_1x_2dy, to the total mass in the strip between x_1 and x_2, Fig. 12.7.

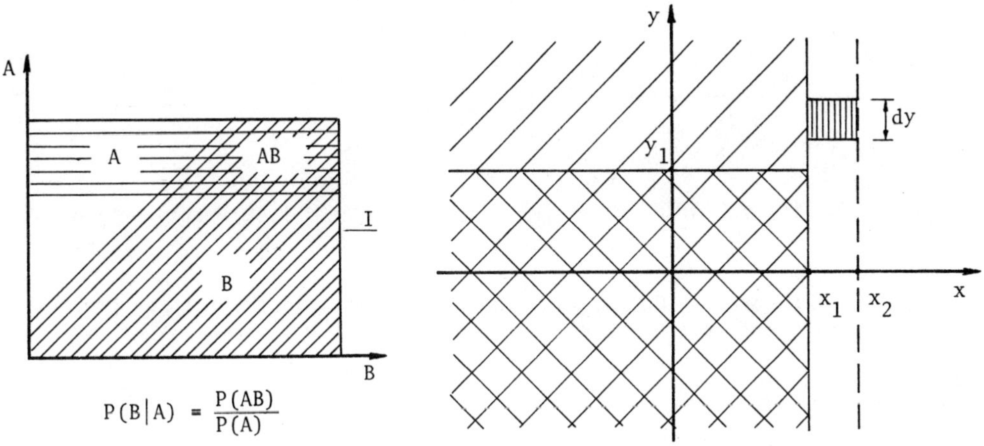

$$P(B|A) = \frac{P(AB)}{P(A)}$$

$$F[y|X \leq x_1] = \frac{P[X \leq x_1, Y \leq y_1]}{P[X \leq x_1]} = \frac{F(xy)}{F(x)}$$

Fig. 12.7. Illustration of conditional probability.

The joint probability concept as shown in Fig. 12.7 is based on the assumption that A and B are independent events. Then the probability that both A and B will occur is the joint probability given by

$$P(A \text{ and } B) = P(AB) = P(A)P(B)$$

This can be stated in a more general form, which removes the requirement of independence, as follows

$$P(A \text{ and } B) = P(AB) = P(A)P(B|A)$$

If A_1 and A_2 are two mutually exclusive events, then $P(A_1 \text{ or } A_2) = P(A_1 + A_2) = P(A_1) + P(A_2)$. Then

$$P(A_1B) = P(A_1)P(B|A_1)$$
$$P(A_2B) = P(A_2)P(B|A_2)$$

from which

$$P(B) = P(A_1B) + P(A_2B) = P(A_1)P(B|A_1) + P(A_2)P(B|A_2)$$

or in general

$$P(B) = \sum_{i=1}^{n} P(B|A_i)P(A_i)$$

where the A_i are mutually exclusive and $\sum^n P(A_i) = 1$. This leads to an important law known as Bayes' theorem. Bayes' theorem provides a method for combining the initial or prior probabilities concerning the occurrence of some event with the related experimental data to obtain a revised probability. The initial probabilities are obtained from knowledge of the physical situation or analysis of past data. Thus, given $P(A_i)$ and $P(B|A_i)$, and seeking $P(A_i|B)$ it follows from the joint probability

$$P(A_iB) = P(B)P(A_i|B) = P(A_i)P(B|A_i)$$

that

$$P(A_i|B) = \frac{P(A_iB)}{P(B)} = \frac{P(A_i)P(B|A_i)}{P(B)}$$

Substituting for P(B) gives

$$P(A_i|B) = P(A_i)\frac{P(B|A_i)}{\sum_{i=1}^{n} P(B|A_i)P(A_i)} \qquad i = 1, 2, \ldots, n$$

which is Bayes' theorem. On the right hand side $P(A_i)$ is the initial probability and the rest is the factor by which it is revised on the basis of experimental data.

Moments again provide the simplest description of distributions. The joint moment of order r and s is defined as

$$\mu_{r,s}(0) = \int_{-\infty}^{\infty}\int_{-\infty}^{\infty} x^r y^s f(x, y)\,dx\,dy \qquad 12.96$$

When r = 1 and s = 0

$$\mu_{1,0}(0) = \int_{-\infty}^{\infty} x\left[\int_{-\infty}^{\infty} f(x, y)\,dy\right]dx = \int_{-\infty}^{\infty} xf(x)\,dx = \bar{x}$$

since the expression in brackets is the marginal p.d.f. of x. When r = 0, s = 1, $\mu_{0,1}(0) = \bar{y}$. Written for moments about the mean

$$\mu_{r,s} = \int_{-\infty}^{\infty}\int_{-\infty}^{\infty} (x - \bar{x})^r (y - \bar{y})^s f(x, y)\,dx\,dy \qquad 12.97$$

When r = 2, s = 0 the result is σ_x^2 and for r = 0, s = 2 it is σ_y^2. When r = 1, s = 1 the moment is known as the covariance, cov(x, y) or $\sigma_{x,y}$. Then the correlation coefficient is

$$\rho = \frac{\text{cov}(x, y)}{\sigma_x\sigma_y} = \frac{\sigma_{xy}}{\sigma_x\sigma_y} \qquad 12.98$$

which varies between ±1 and is a measure of the dependence or association between the two variables.

A phenomenon may also depend on a number of independent factors, each of which contributes to the scatter of the results. The variance of such a system is then the sum of the component variances caused by the individual factors. By a technique, known as *analysis of variance*, the mean values of the variances can be tested for significance. The total variance is partitioned into component parts, each of which may be assigned a particular cause, and the relative importance of the various components is assessed.

For example, if there are k experimental drainage areas in a region where the regional runoff may be considered as homogeneous, is there an appreciable difference in runoff from these catchments? The measurements of runoff from the k lots form mutually exclusive categories or classes, each representing the population of all the possible results which could be obtained by a particular gauge. The n years of results for each gauge represent a sample of the population of runoffs resulting from common rainstorms. Thus, the data can be divided into k classes with n items in each. The total number of observations is $N = \Sigma n_i$ for i = 1 to k and, if x is the mean annual discharge per unit area, then the mean $\bar{x}_i$ for the i-th class (drainage area) is Σx_{ij} for j = 1 to n_i, divided by n_i. If all the areas have the same length of record, then n is common. The mean for all the areas is

$$x = \frac{1}{N}\sum_{i=1}^{n}\sum_{j=1}^{n_i} x_{ij}$$

Since each gauge is assumed to measure independently of the others, this is called the *one-way* or single variable classification. The variable of the classification is the drainage lot or its runoff and it has k values. The observations within any one class form a *random variable*, one for each class. Each observation is the sum of two components: the unknown mean of the population from which the observations are drawn and which is common to all the observations within the set, i.e., the systematic component, and the deviation from this mean considered to be a random variable. The k population means characterize the systematic (fixed) differences between the lots.

There may be, however, a significant variation between the individual values in each of the k classes, i.e., there might have been a significant difference in the yield from the rainstorms over the various drainage areas. In such a problem there are two random components and one speaks of a *two-way* classification. For details reference has to be made to statistical texts.

Example 12.7. The joint distribution of quarterly flows on river A and B are as shown.

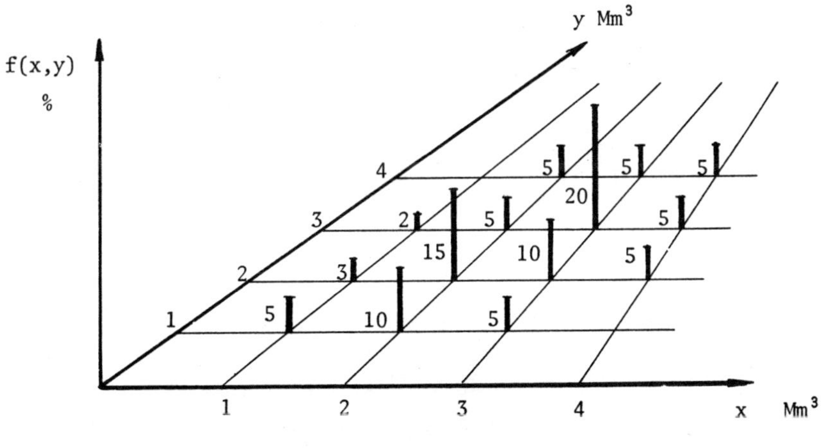

Find the marginal distributions of X and Y and compare the means, standard deviations, covariances and correlation coefficient.

The marginal distribution of X is obtained by summing over Y.

x :	1	2	3	4
f(x):	0.10	0.35	0.40	0.15

and of Y by summing over X

y :	1	2	3	4
f(y):	0.20	0.33	0.32	0.15

$$\bar{x} = \Sigma x\ f(x) = 1 \times 0.1 + 2 \times 0.35 + 3 \times 0.40 + 4 \times 0.15 = 2.6 \times 10^6\ m^3$$

$$S_x^2 = \Sigma(x - \bar{x})^2 f(x) = (1 - 2.6)^2 0.1 + (2 - 2.6)^2 0.35 + (3 - 2.6)^2 0.40$$

$$+ (4 - 2.6)^2 0.15 = 0.256 + 0.126 + 0.064 + 0.294 = 0.740\ (Mm^3)^2$$

$$S_x = 0.860\ Mm^3$$

$$\bar{y} = \Sigma y\ f(y) = 1 \times 0.20 + 2 \times 0.33 + 3 \times 0.32 + 3 \times 0.15 = 2.42 \times 10^6\ m^3$$

$$S^2_y = \Sigma(y - \bar{y})^2 f(y) = (1 - 2.42)^2 0.20 + (2 - 2.42)^2 0.33 + (3 - 2.42)^2 0.32$$

$$+ (4 - 2.42)^2 0.15 = 0.403 + 0.058 + 0.108 + 0.375 = 0.944\ (Mm^3)^2$$

$$S_y = 0.971\ Mm^3$$

$$\begin{aligned} Cov(x,y) &= \Sigma(x - \bar{x})(y - \bar{y}) f(x,y) \\ &= (1-2.6)(1-2.42)0.05 + (1-2.6)(2-2.42)0.03 + \\ &\quad (1-2.6)(3-2.42)0.02 + (2-2.6)(1-2.42)0.10 + \\ &\quad (2-2.6)(2-2.42)0.15 + (2-2.6)(3-2.42)0.05 + \\ &\quad (2-2.6)(4-2.42)0.05 + (3-2.6)(1-2.42)0.05 + \\ &\quad (3-2.6)(2-2.42)0.10 + (3-2.6)(3-2.42)0.20 + \\ &\quad (3-2.6)(4-2.42)0.05 + (4-2.6)(2-2.42)0.05 + \\ &\quad (4-2.6)(3-2.42)0.05 + (4-2.6)(4-2.42)0.05 = 0.644 \end{aligned}$$

$$\rho = \frac{0.644}{0.860 \times 0.971} = 0.771$$

Example 12.8. Two nurseries have supplied seedlings, 75% from nursery 1 and 25% from nursery 2. The success rate from previous planting was 90% form nursery 1 and 75% from nursery 2. What are the probabilities of picking a failing seedling from the mixed up supplies?

Let A_1 = seedling from nursery 1, A_2 = seedling from nursery 2, B_1 = good seedling, and B_2 = dying seedling, then

$$P(A_1) = 0.75 \qquad P(A_2) = 0.25$$

$$P(B_1|A_1) = 0.90 \qquad P(B_2|A_1) = 0.10$$

$$P(B_1|A_2) = 0.75 \qquad P(B_2|A_2) = 0.25$$

The probability that the failing seedling is from nursery 1 is

$$P(A_1B_2) = P(A_1)P(B_2|A_1) = 0.75 \times 0.10 = 0.075$$

or from nursery 2

$$P(A_2B_2) = P(A_2)P(B_2|A_2) = 0.25 \times 0.25 = 0.0625$$

and the probability of picking a failing seedling irrespective of supplier is

$$P(B_2) = P(A_1B_2) + P(A_2B_2) = P_1(A_1)P(B_2|A_1) + P(A_2)P(B_2|A_2)$$

$$= 0.075 + 0.0625 = 0.1375 \text{ or about one in seven.}$$

12.7 Frequency Factor

The variable may be thought of as $x = \bar{x} + \Delta x$, where the departure Δx depends on the dispersion characteristics of the distribution of x and the recurrence interval T (or probability). Chow (1951) suggested that this departure can be assumed to be proportional to σ and a frequency factor ϕ, so that

$$x = \bar{x} + \sigma\phi \qquad 12.99$$

or

$$\frac{x}{\bar{x}} = 1 + C_v\phi \qquad 12.100$$

where $C_v = \sigma/\bar{x}$ is the coefficient of variation and $\phi = f(\gamma_1, r^*)$, where γ_1 is the coefficient of skewness μ_3/σ^3 and r^* is a random number, $0 \leq r^* \leq 1$.

One of the earliest formulae for the analysis of the frequency of annual maximum daily flows is the empirical formula by Fuller

$$x = \bar{x}(1 + 0.8 \log T)$$

where

$$\phi = \frac{0.8}{C_v} \log T$$

For a *normal distribution*

$$\phi = \frac{x - \mu}{\sigma} \qquad 12.101$$

The cumulative probability of a value being equal to or less than x is

$$F[X \leq x] = \frac{1}{\sigma\sqrt{2\pi}} \int_{-\infty}^{x} \exp[-(x - \mu)^2/2\sigma^2]dx = \frac{1}{\sqrt{2\pi}} \int_{-\infty}^{\phi} \exp(-\phi^2/2)d\phi$$

and can be read from tables for a normal distribution. Likewise, for the log-normal distribution

$$y = \bar{y} + \sigma_y\phi \qquad 12.102$$

For the extreme value or Gumbel distribution the return period is

$$T = \frac{1}{1 - \exp\{-\exp[-(a + x)/c]\}}$$

and a rearrangement yields

$$x = -a - c\,\ell n[\ell n\,T - \ell n(T - 1)]$$

comparing this with eqn 12.60, $x = u + y_1/\alpha$, and observing that $y = -\ell n\{\ell n[T/(T - 1)]\}$ shows that $u = -a$ and $c = 1/\alpha$. Hence, from $y = \bar{y} + \sigma\phi = \gamma + (\pi/\sqrt{6})\phi$

$$\phi = -(\sqrt{6}/\pi)\{\gamma + \ell n[\ell n\,T - \ell n(T - 1)]\}$$

$$= -[1.1 + 1.795 \log(\log \frac{N + 1}{N + 1 - m})] \qquad 12.103$$

where $a = \gamma c - \bar{x}$, $c = (\sqrt{6}/\pi)\sigma$, the population mean $\gamma = 0.5772$, and N is the total number of years of record.

Harter (1969, 1971) published tables of ϕ for the Pearson Type III distribution in terms of the cumulative probability F(x) and the skewness γ_1, Table 12.11. Percentage points for negative γ_1 values corresponding to 1 - F(x) are found by reading the value for positive γ_1 and F(x) and changing the sign, e.g., the 95% point for γ_1 = -2.1 is obtained by reading the 5% value for γ_1 = 2.1, which is -0.91458, and changing the sign.

An abbreviated table of the frequency factor ϕ for the log-Pearson Type III distribution is shown in Table 12.12.

TABLE 12.12 K Values for the log-Pearson Type III Distribution, after Réménieras (1967)

Skew coefficient g	Recurrence interval, years							
	1.0101	1.2500	2	5	10	25	50	100
	Percent chance							
	99	80	50	20	10	4	2	1
3.0	−0.667	−0.636	−0.396	0.420	1.180	2.278	3.152	4.051
2.8	−0.714	−0.666	−0.384	0.460	1.210	2.275	3.114	3.973
2.6	−0.769	−0.696	−0.368	0.499	1.238	2.267	3.071	3.889
2.4	−0.832	−0.725	−0.351	0.537	1.262	2.256	3.023	3.800
2.2	−0.905	−0.752	−0.330	0.574	1.284	2.240	2.970	3.705
2.0	−0.990	−0.777	−0.307	0.609	1.302	2.219	2.912	3.605
1.8	−1.087	−0.799	−0.282	0.643	1.318	2.193	2.848	3.499
1.6	−1.197	−0.817	−0.254	0.675	1.329	2.163	2.780	3.388
1.4	−1.318	−0.832	−0.225	0.705	1.337	2.128	2.706	3.271
1.2	−1.449	−0.844	−0.195	0.732	1.340	2.087	2.626	3.149
1.0	−1.588	−0.852	−0.164	0.758	1.340	2.043	2.542	3.022
0.8	−1.733	−0.856	−0.132	0.780	1.336	1.993	2.453	2.891
0.6	−1.880	−0.857	−0.099	0.800	1.328	1.939	2.359	2.755
0.4	−2.029	−0.855	−0.066	0.816	1.317	1.880	2.261	2.615
0.2	−2.178	−0.850	−0.033	0.830	1.301	1.818	2.159	2.472
0	−2.326	−0.842	0	0.842	1.282	1.751	2.054	2.326
−0.2	−2.472	−0.830	0.033	0.850	1.258	1.680	1.945	2.178
−0.4	−2.615	−0.816	0.066	0.855	1.231	1.606	1.834	2.029
−0.6	−2.755	−0.800	0.099	0.857	1.200	1.528	1.720	1.880
−0.8	−2.891	−0.780	0.132	0.856	1.166	1.448	1.606	1.733
−1.0	−3.022	−0.758	0.164	0.852	1.128	1.366	1.492	1.588
−1.2	−3.149	−0.732	0.195	0.844	1.086	1.282	1.379	1.449
−1.4	−3.271	−0.705	0.225	0.832	1.041	1.198	1.270	1.318
−1.6	−3.388	−0.675	0.254	0.817	0.994	1.116	1.166	1.197
−1.8	−3.499	−0.643	0.282	0.799	0.945	1.035	1.069	1.087
−2.0	−3.605	−0.609	0.307	0.777	0.895	0.959	0.980	0.990
−2.2	−3.705	−0.574	0.330	0.752	0.844	0.888	0.900	0.905
−2.4	−3.800	−0.537	0.351	0.725	0.795	0.823	0.830	0.832
−2.6	−3.889	−0.499	0.368	0.696	0.747	0.764	0.768	0.769
−2.8	−3.973	−0.460	0.384	0.666	0.702	0.712	0.714	0.714
−3.0	−4.051	−0.420	0.396	0.636	0.660	0.666	0.666	0.667

TABLE 12.11 Percentage Points of Pearson Type III Distribution as a Function of Skewness γ_1

F(x)	γ_1 = 0	γ_1 = 0.1	γ_1 = 0.2	γ_1 = 0.3	γ_1 = 0.4	γ_1 = 0.5	γ_1 = 0.6	γ_1 = 0.7	γ_1 = 0.8	γ_1 = 0.9
0.0001	-3.71902	-3.50703	-3.29921	-3.09631	-2.89907	-2.70836	-2.52507	-2.35015	-2.18448	-2.02891
0.0005	-3.29053	-3.12767	-2.96698	-2.80889	-2.65390	-2.50257	-2.35549	-2.21328	-2.07661	-1.94611
0.0010	-3.09023	-2.94834	-2.80786	-2.66915	-2.53261	-2.39867	-2.26780	-2.14053	-2.01739	-1.89894
0.0020	-2.87816	-2.75706	-2.63672	-2.51741	-2.39942	-2.28311	-2.16884	-2.05701	-1.94806	-1.84244
0.0050	-2.57583	-2.48187	-2.38795	-2.29423	-2.20092	-2.10825	-2.01644	-1.92580	-1.83660	-1.74919
0.0100	-2.32635	-2.25258	-2.17840	-2.10394	-2.02933	-1.95472	-1.88029	-1.80621	-1.73271	-1.66001
0.0200	-2.05375	-1.99973	-1.94499	-1.88959	-1.83361	-1.77716	-1.72033	-1.66325	-1.60604	-1.54886
0.0250	-1.95996	-1.91219	-1.86360	-1.81427	-1.76427	-1.71366	-1.66253	-1.61099	-1.55914	-1.50712
0.0400	-1.75069	-1.71580	-1.67999	-1.64329	-1.60574	-1.56740	-1.52830	-1.48852	-1.44813	-1.40720
0.0500	-1.64485	-1.61594	-1.58607	-1.55527	-1.52357	-1.49101	-1.45762	-1.42345	-1.38855	-1.35299
0.1000	-1.28155	-1.27037	-1.25824	-1.24516	-1.23114	-1.21618	-1.20028	-1.18347	-1.16574	-1.14712
0.2000	-0.84612	-0.84611	-0.84986	-0.85285	-0.85508	-0.85653	-0.85718	-0.85703	-0.85607	-0.85426
0.3000	-0.52440	-0.53624	-0.54757	-0.55839	-0.56867	-0.57840	-0.58757	-0.59615	-0.60412	-0.61146
0.4000	-0.25335	-0.26882	-0.28403	-0.29897	-0.31362	-0.32796	-0.34198	-0.35565	-0.36889	-0.38186
0.4296	-0.17733	-0.19339	-0.20925	-0.22492	-0.24037	-0.25558	-0.27047	-0.28516	-0.29961	-0.31368
0.5000	0.00000	-0.01662	-0.03325	-0.04993	-0.06651	-0.08302	-0.09945	-0.11578	-0.13199	-0.14807
0.5704	0.17733	0.16111	0.14472	0.12820	0.11154	0.09478	0.07791	0.06097	0.04397	0.02693
0.6000	0.25335	0.23763	0.22168	0.20552	0.18916	0.17261	0.15589	0.13901	0.12199	0.10486
0.7000	0.52440	0.51207	0.49927	0.48600	0.47228	0.45812	0.44352	0.42851	0.41309	0.39729
0.8000	0.84162	0.83639	0.83044	0.82377	0.81638	0.80829	0.79950	0.79002	0.77986	0.76902
0.9000	1.28155	1.29178	1.30105	1.30936	1.31671	1.32309	1.32850	1.33294	1.33640	1.33889
0.9500	1.64485	1.67279	1.69971	1.72562	1.75048	1.77428	1.79701	1.81864	1.83916	1.85856
0.9600	1.75069	1.78462	1.81756	1.84949	1.88039	1.91022	1.93896	1.96660	1.99311	2.01848
0.9750	1.95996	2.00688	2.05290	2.09795	2.14202	2.18505	2.22702	2.26790	2.30764	2.34623
0.9800	2.05375	2.10697	2.15935	2.21081	2.26133	2.31084	2.35931	2.40670	2.45298	2.49811
0.9900	2.32635	2.39961	2.47226	2.54421	2.61539	2.68572	2.75514	2.82359	2.89101	2.95735
0.9950	2.57583	2.66965	2.76321	2.85636	2.94900	3.04102	3.13232	3.22281	3.31243	3.40109
0.9980	2.87816	2.99978	3.12169	3.24371	3.36566	3.48737	3.60872	3.72957	3.84981	3.96932
0.9990	3.09023	3.23322	3.37703	3.52139	3.66608	3.81090	3.95567	4.10022	4.24439	4.38807
0.9995	3.29053	3.45513	3.62113	3.78820	3.95605	4.12443	4.29311	4.46189	4.63057	4.79899
0.9999	3.71902	3.93453	4.15301	4.37394	4.59687	4.82141	5.04718	5.27389	5.50124	5.72899

F(x)	γ_1 = 1.0	γ_1 = 1.1	γ_1 = 1.2	γ_1 = 1.3	γ_1 = 1.4	γ_1 = 1.5	γ_1 = 1.6	γ_1 = 1.7	γ_1 = 1.8	γ_1 = 1.9
0.0001	-1.88410	-1.75053	-1.62838	-1.51752	-1.41753	-1.32774	-1.24728	-1.17520	-1.11054	-1.05239
0.0005	-1.82241	-1.70603	-1.59738	-1.49673	-1.40413	-1.31944	-1.24235	-1.17240	-1.10901	-1.05159
0.0010	-1.78572	-1.67825	-1.57695	-1.48216	-1.39408	-1.31275	-1.23805	-1.16974	-1.10743	-1.05068
0.0020	-1.74062	-1.64305	-1.55016	-1.46232	-1.37981	-1.30279	-1.23132	-1.16534	-1.10465	-1.04898
0.0050	-1.66390	-1.58110	-1.50114	-1.42439	-1.35114	-1.28167	-1.21618	-1.15477	-1.09749	-1.04427
0.0100	-1.58838	-1.51808	-1.44942	-1.38267	-1.31815	-1.25611	-1.19680	-1.14042	-1.08711	-1.03695
0.0200	-1.49188	-1.43529	-1.37929	-1.32412	-1.26999	-1.21716	-1.16584	-1.11628	-1.06864	-1.02311
0.0250	-1.45507	-1.40314	-1.35153	-1.30042	-1.25004	-1.20059	-1.15229	-1.10537	-1.06001	-1.01640
0.0400	-1.36584	-1.32414	-1.28225	-1.24028	-1.19842	-1.15682	-1.11566	-1.07513	-1.03543	-0.99672
0.0500	-1.31684	-1.28019	-1.24313	-1.20578	-1.16827	-1.13075	-1.09338	-1.05631	-1.01973	-0.98381
0.1000	-1.12762	-1.10726	-1.08608	-1.06413	-1.04144	-1.01810	-0.99418	-0.96977	-0.94496	-0.91988
0.2000	-0.85161	-0.84809	-0.84369	-0.83841	-0.83223	-0.82516	-0.81720	-0.80837	-0.79868	-0.78816
0.3000	-0.61815	-0.62415	-0.62944	-0.63400	-0.63779	-0.64080	-0.64380	-0.64436	-0.64488	-0.64453
0.4000	-0.39434	-0.40638	-0.41794	-0.42899	-0.43949	-0.44942	-0.45873	-0.46739	-0.47538	-0.48265
0.4296	-0.32740	-0.34075	-0.35370	-0.36620	-0.37824	-0.38977	-0.40075	-0.41116	-0.42095	-0.43008
0.5000	-0.16397	-0.17968	-0.19517	-0.21040	-0.22535	-0.23996	-0.25422	-0.26808	-0.28150	-0.29443
0.5704	0.00987	-0.00719	-0.02421	-0.04116	-0.05803	-0.07476	-0.09132	-0.10769	-0.12381	-0.13964
0.6000	0.08763	0.07032	0.05297	0.03560	0.01824	0.00092	-0.01631	-0.03344	-0.05040	-0.06718
0.7000	0.38111	0.36458	0.34772	0.33054	0.31307	0.29535	0.27740	0.25925	0.24094	0.22250
0.8000	0.75752	0.74537	0.73257	0.71915	0.70512	0.69050	0.67532	0.65959	0.64335	0.62662
0.9000	1.34039	1.34092	1.34047	1.33904	1.33665	1.33330	1.32900	1.32376	1.31760	1.31054
0.9500	1.87683	1.89395	1.90992	1.92472	1.93836	1.95083	1.96213	1.97227	1.98124	1.98906
0.9600	2.04269	2.06573	2.08758	2.10823	2.12768	2.14591	2.16293	2.17873	2.19332	2.20670
0.9750	2.38364	2.41984	2.45482	2.48855	2.52102	2.55222	2.58214	2.61076	2.63810	2.66413
0.9800	2.54206	2.58480	2.62631	2.66657	2.70556	2.74325	2.77964	2.81472	2.84848	2.88091
0.9900	3.02256	3.08660	3.14944	3.21103	3.27134	3.33035	3.38804	3.44438	3.49935	3.55295
0.9950	3.48874	3.57530	3.66073	3.74497	3.82798	3.90973	3.99016	4.06926	4.14700	4.22336
0.9980	4.08802	4.20582	4.32263	4.43839	4.55304	4.66651	4.77875	4.88971	4.99937	5.10768
0.9990	4.53112	4.67344	4.81492	4.95549	5.09505	5.23353	5.37087	5.50701	5.64190	5.77549
0.9995	4.96701	5.13449	5.30130	5.46735	5.63252	5.79673	5.95990	6.12196	6.28285	6.44251
0.9999	5.95691	6.18480	6.41249	6.63980	6.86661	7.09277	7.31818	7.54272	7.76632	7.98888

F(x)	γ_1 = 2.0	γ_1 = 2.1	γ_1 = 2.2	γ_1 = 2.3	γ_1 = 2.4	γ_1 = 2.5	γ_1 = 2.6	γ_1 = 2.7	γ_1 = 2.8	γ_1 = 2.9
0.0001	-0.99990	-0.95234	-0.90908	-0.86956	-0.83333	-0.80000	-0.76923	-0.74074	-0.71429	-0.68966
0.0005	-0.99950	-0.95215	-0.90899	-0.86952	-0.83331	-0.79999	-0.76923	-0.74074	-0.71429	-0.68966
0.0010	-0.99900	-0.95188	-0.90885	-0.86945	-0.83328	-0.79998	-0.76922	-0.74074	-0.71428	-0.68965
0.0020	-0.99800	-0.95131	-0.90854	-0.86929	-0.83320	-0.79994	-0.76920	-0.74073	-0.71428	-0.68965
0.0050	-0.99499	-0.94945	-0.90742	-0.86863	-0.83283	-0.79973	-0.76909	-0.74067	-0.71425	-0.68964
0.0100	-0.98995	-0.94607	-0.90521	-0.86723	-0.83196	-0.79921	-0.76878	-0.74049	-0.71415	-0.68959
0.0200	-0.97980	-0.93878	-0.90009	-0.86371	-0.82959	-0.79765	-0.76779	-0.73987	-0.71377	-0.68935
0.0250	-0.97468	-0.93495	-0.89728	-0.86169	-0.82817	-0.79667	-0.76712	-0.73943	-0.71348	-0.68917
0.0400	-0.95918	-0.92295	-0.88814	-0.85486	-0.82315	-0.79306	-0.76456	-0.73765	-0.71227	-0.68336
0.0500	-0.94871	-0.91458	-0.88156	-0.84976	-0.81927	-0.79015	-0.76242	-0.73610	-0.71116	-0.68759
0.1000	-0.89464	-0.86938	-0.84422	-0.81929	-0.79472	-0.77062	-0.74709	-0.72422	-0.70209	-0.68075
0.2000	-0.77686	-0.76482	-0.75211	-0.73880	-0.72495	-0.71067	-0.69602	-0.68111	-0.66603	-0.65086
0.3000	-0.64333	-0.64125	-0.63833	-0.63456	-0.62999	-0.62463	-0.61854	-0.61176	-0.60434	-0.59634
0.4000	-0.48917	-0.49494	-0.49991	-0.50409	-0.50744	-0.50999	-0.51171	-0.51263	-0.51276	-0.51212
0.4296	-0.43854	-0.44628	-0.45329	-0.45953	-0.46499	-0.46966	-0.47353	-0.47660	-0.47888	-0.48037
0.5000	-0.30685	-0.31872	-0.32999	-0.34063	-0.35062	-0.35992	-0.36852	-0.37640	-0.38353	-0.38991
0.5704	-0.15516	-0.17030	-0.18504	-0.19933	-0.21313	-0.22642	-0.23915	-0.25129	-0.26282	-0.27372
0.6000	-0.08371	-0.09997	-0.11590	-0.13148	-0.14665	-0.16138	-0.17564	-0.18939	-0.20259	-0.21523
0.7000	0.20397	0.18540	0.16682	0.14827	0.12979	0.11143	0.09323	0.07523	0.05746	0.03997
0.8000	0.60944	0.59183	0.57383	0.55549	0.53683	0.51789	0.49872	0.47934	0.45980	0.44015
0.9000	1.30259	1.29377	1.28412	1.27365	1.26240	1.25039	1.23766	1.22422	1.21013	1.19539
0.9500	1.99573	2.00128	2.00570	2.00903	2.01128	2.01247	2.01263	2.01177	2.00992	2.00710
0.9600	2.21888	2.22986	2.23967	2.24831	2.25581	2.26217	2.26743	2.27160	2.27470	2.27676
0.9750	2.68888	2.71234	2.73451	2.75541	2.77506	2.79345	2.81062	2.82658	2.84134	2.85492
0.9800	2.91202	2.94181	2.97028	2.99744	3.02330	3.04787	3.07116	3.09320	3.11399	3.13356
0.9900	3.60517	3.65600	3.70543	3.75347	3.80013	3.84540	3.88930	3.93183	3.97301	4.01286
0.9950	4.29832	4.37186	4.44398	4.51467	4.58393	4.65176	4.71815	4.78313	4.84669	4.90884
0.9980	5.21461	5.32014	5.42426	5.52694	5.62818	5.72796	5.82629	5.92316	6.01858	6.11254
0.9990	5.90776	6.03865	6.16816	6.29626	6.42292	6.54814	6.67191	6.79421	6.91505	7.03443
0.9995	6.60090	6.75798	6.91370	7.06804	7.22098	7.37250	7.52258	7.67121	7.81839	7.96411
0.9999	8.21034	8.43064	8.64971	8.86753	9.08403	9.29920	9.51301	9.72543	9.93643	10.14602

F(x)	γ_1 = 3.0	γ_1 = 3.1	γ_1 = 3.2	γ_1 = 3.3	γ_1 = 3.4	γ_1 = 3.5	γ_1 = 3.6	γ_1 = 3.7	γ_1 = 3.8	γ_1 = 3.9
0.0001	-0.66667	-0.64516	-0.62500	-0.60606	-0.58824	-0.57143	-0.55556	-0.54054	-0.52632	-0.51282
0.0005	-0.66667	-0.64516	-0.62500	-0.60606	-0.58824	-0.57143	-0.55556	-0.54054	-0.52632	-0.51282
0.0010	-0.66667	-0.64516	-0.62500	-0.60606	-0.58824	-0.57143	-0.55556	-0.54054	-0.52632	-0.51282
0.0020	-0.66667	-0.64516	-0.62500	-0.60606	-0.58824	-0.57143	-0.55556	-0.54054	-0.52632	-0.51282
0.0050	-0.66666	-0.64516	-0.62500	-0.60606	-0.58824	-0.57143	-0.55556	-0.54054	-0.52632	-0.51282
0.0100	-0.66663	-0.64514	-0.62499	-0.60606	-0.58823	-0.57143	-0.55556	-0.54054	-0.52632	-0.51282
0.0200	-0.66649	-0.64507	-0.62495	-0.60603	-0.58822	-0.57142	-0.55555	-0.54054	-0.52631	-0.51282
0.0250	-0.66638	-0.64500	-0.62491	-0.60601	-0.58821	-0.57141	-0.55555	-0.54054	-0.52631	-0.51282
0.0400	-0.66585	-0.64465	-0.62469	-0.60587	-0.58812	-0.57136	-0.55552	-0.54052	-0.52630	-0.51281
0.0500	-0.66532	-0.64429	-0.62445	-0.60572	-0.58802	-0.57130	-0.55548	-0.54050	-0.52629	-0.51281
0.1000	-0.66023	-0.64056	-0.62175	-0.60379	-0.58666	-0.57035	-0.55483	-0.54006	-0.52600	-0.51261
0.2000	-0.63569	-0.62060	-0.60567	-0.59096	-0.57652	-0.56242	-0.54867	-0.53533	-0.52240	-0.50990
0.3000	-0.58783	-0.57887	-0.56953	-0.55989	-0.55000	-0.53993	-0.52975	-0.51952	-0.50929	-0.49911
0.4000	-0.51073	-0.50863	-0.50585	-0.50244	-0.49844	-0.49391	-0.48888	-0.48342	-0.47758	-0.47141
0.4296	-0.48109	-0.48107	-0.48033	-0.47890	-0.47682	-0.47413	-0.47088	-0.46711	-0.46286	-0.45819
0.5000	-0.39554	-0.40041	-0.40454	-0.40792	-0.41058	-0.41253	-0.41381	-0.41442	-0.41441	-0.41381
0.5704	-0.28395	-0.29351	-0.30238	-0.31055	-0.31802	-0.32479	-0.33085	-0.33623	-0.34092	-0.34494
0.6000	-0.22726	-0.23868	-0.24946	-0.25958	-0.26904	-0.27782	-0.28592	-0.29335	-0.30010	-0.30617
0.7000	0.02279	0.00596	-0.01050	-0.02654	-0.04215	-0.05730	-0.07195	-0.08610	-0.09972	-0.11279
0.8000	0.42040	0.40061	0.38081	0.36104	0.34133	0.32171	0.30223	0.28290	0.26376	0.24484
0.9000	1.18006	1.16416	1.14772	1.13078	1.11337	1.09552	1.07726	1.05863	1.03965	1.02036
0.9500	2.00335	1.99869	1.99314	1.98674	1.97951	1.97147	1.96266	1.95311	1.94283	1.93186
0.9600	2.27780	2.27785	2.27693	2.27506	2.27229	2.26862	2.26409	2.25872	2.25254	2.24558
0.9750	2.86735	2.87865	2.88884	2.89795	2.90599	2.91299	2.91898	2.92397	2.92799	2.93107
0.9800	3.15193	3.16911	3.18512	3.20000	3.21375	3.22641	3.23800	3.24853	3.25803	3.26653
0.9900	4.05138	4.08859	4.12452	4.15917	4.19257	4.22473	4.25569	4.28545	4.31403	4.34147
0.9950	4.96959	5.02897	5.08697	5.14362	5.19892	5.25291	5.30559	5.35690	5.40711	5.45598
0.9980	6.20506	6.29613	6.38578	6.47401	6.56084	6.64627	6.73032	6.81301	6.89435	6.97435
0.9990	7.15235	7.26881	7.38382	7.49739	7.60953	7.72024	7.82954	7.93744	8.04395	8.14910
0.9995	8.10836	8.25115	8.39248	8.53236	8.67079	8.80779	8.94335	9.07750	9.21023	9.34158
0.9999	10.35418	10.56090	10.76618	10.97001	11.17239	11.37334	11.57284	11.77092	11.96757	12.16280

F(x)	γ_1 = 4.0	γ_1 = 4.1	γ_1 = 4.2	γ_1 = 4.3	γ_1 = 4.4	γ_1 = 4.5	γ_1 = 4.6	γ_1 = 4.7	γ_1 = 4.8	γ_1 = 5.0
0.0001	-0.50000	-0.48780	-0.47619	-0.46512	-0.45455	-0.44444	-0.43478	-0.42553	-0.41667	-0.40000
0.0005	-0.50000	-0.48780	-0.47619	-0.46512	-0.45455	-0.44444	-0.43478	-0.42553	-0.41667	-0.40000
0.0010	-0.50000	-0.48780	-0.47619	-0.46512	-0.45455	-0.44444	-0.43478	-0.42553	-0.41667	-0.40000
0.0020	-0.50000	-0.48780	-0.47619	-0.46512	-0.45455	-0.44444	-0.43478	-0.42553	-0.41667	-0.40000
0.0050	-0.50000	-0.48780	-0.47619	-0.46512	-0.45455	-0.44444	-0.43478	-0.42553	-0.41667	-0.40000
0.0100	-0.50000	-0.48780	-0.47619	-0.46512	-0.45455	-0.44444	-0.43478	-0.42553	-0.41667	-0.40000
0.0200	-0.50000	-0.48780	-0.47619	-0.46512	-0.45455	-0.44444	-0.43478	-0.42553	-0.41667	-0.40000
0.0250	-0.50000	-0.48780	-0.47619	-0.46512	-0.45455	-0.44444	-0.43478	-0.42553	-0.41667	-0.40000
0.0400	-0.50000	-0.48780	-0.47619	-0.46512	-0.45455	-0.44444	-0.43478	-0.42553	-0.41667	-0.40000
0.0500	-0.49999	-0.48780	-0.47619	-0.46511	-0.45454	-0.44444	-0.43478	-0.42553	-0.41667	-0.40000
0.1000	-0.49986	-0.48772	-0.47614	-0.46508	-0.45452	-0.44443	-0.43477	-0.42553	-0.41666	-0.40000
0.2000	-0.49784	-0.48622	-0.47504	-0.46428	-0.45395	-0.44402	-0.43448	-0.42532	-0.41652	-0.39993
0.3000	-0.48902	-0.47906	-0.46927	-0.45967	-0.45029	-0.44114	-0.43223	-0.42357	-0.41517	-0.39914
0.4000	-0.46496	-0.45828	-0.45142	-0.44442	-0.43734	-0.43020	-0.42304	-0.41590	-0.40880	-0.39482
0.4296	-0.45314	-0.44777	-0.44212	-0.43623	-0.43016	-0.42394	-0.41761	-0.41121	-0.40477	-0.39190
0.5000	-0.41265	-0.41097	-0.40881	-0.40621	-0.40321	-0.39985	-0.39617	-0.39221	-0.38800	-0.37901
0.5704	-0.34831	-0.35105	-0.35318	-0.35473	-0.35572	-0.35619	-0.35616	-0.35567	-0.35475	-0.35174
0.6000	-0.31159	-0.31635	-0.32049	-0.32400	-0.32693	-0.32928	-0.33108	-0.33236	-0.33315	-0.33336
0.7000	-0.12530	-0.13725	-0.14861	-0.15939	-0.16958	-0.17918	-0.18819	-0.19661	-0.20446	-0.21843
0.8000	0.22617	0.20777	0.18967	0.17189	0.15445	0.13737	0.12067	0.10436	0.08847	0.05798
0.9000	1.00079	0.98096	0.96090	0.94064	0.92022	0.89964	0.87895	0.85817	0.83731	0.79548
0.9500	1.92023	1.90796	1.89508	1.88160	1.86757	1.85300	1.83792	1.82234	1.80631	1.77292
0.9600	2.23786	2.22940	2.22024	2.21039	2.19988	2.18874	2.17699	2.16465	2.15174	2.12432
0.9750	2.93324	2.93450	2.93489	2.93443	2.93314	2.93105	2.92818	2.92455	2.92017	2.90930
0.9800	3.27404	3.28060	3.28622	3.29092	3.29473	3.29767	3.29976	3.30103	3.30149	3.30007
0.9900	4.36777	4.39296	4.41706	4.44009	4.46207	4.48303	4.50297	4.52192	4.53990	4.57304
0.9950	5.50362	5.55005	5.59528	5.63934	5.68224	5.72400	5.76464	5.80418	5.84265	5.91639
0.9980	7.05304	7.13043	7.20654	7.28138	7.35497	7.42733	7.49847	7.56842	7.63718	7.77124
0.9990	8.25289	8.35534	8.45646	8.55627	8.65479	8.75202	8.84800	8.94273	9.03623	9.21961
0.9995	9.47154	9.60013	9.72737	9.85326	9.97784	10.10110	10.22307	10.34375	10.46318	10.69829
0.9999	12.35663	12.54906	12.74010	12.92977	13.11808	13.30504	13.49066	13.67495	13.85794	14.22004

F(x)	γ_1 = 5.2	γ_1 = 5.4	γ_1 = 5.6	γ_1 = 5.8	γ_1 = 6.0	γ_1 = 6.2	γ_1 = 6.4	γ_1 = 6.6	γ_1 = 6.8	γ_1 = 7.0
0.0001	-0.38462	-0.37037	-0.35714	-0.34483	-0.33333	-0.32258	-0.31250	-0.30303	-0.29412	-0.28571
0.0005	-0.38462	-0.37037	-0.35714	-0.34483	-0.33333	-0.32258	-0.31250	-0.30303	-0.29412	-0.28571
0.0010	-0.38462	-0.37037	-0.35714	-0.34483	-0.33333	-0.32258	-0.31250	-0.30303	-0.29412	-0.28571
0.0020	-0.38462	-0.37037	-0.35714	-0.34483	-0.33333	-0.32258	-0.31250	-0.30303	-0.29412	-0.28571
0.0050	-0.38462	-0.37037	-0.35714	-0.34483	-0.33333	-0.32258	-0.31250	-0.30303	-0.29412	-0.28571
0.0100	-0.38462	-0.37037	-0.35714	-0.34483	-0.33333	-0.32258	-0.31250	-0.30303	-0.29412	-0.28571
0.0200	-0.38462	-0.37037	-0.35714	-0.34483	-0.33333	-0.32258	-0.31250	-0.30303	-0.29412	-0.28571
0.0250	-0.38462	-0.37037	-0.35714	-0.34483	-0.33333	-0.32258	-0.31250	-0.30303	-0.29412	-0.28571
0.0400	-0.38462	-0.37037	-0.35714	-0.34483	-0.33333	-0.32258	-0.31250	-0.30303	-0.29412	-0.28571
0.0500	-0.38462	-0.37037	-0.35714	-0.34483	-0.33333	-0.32258	-0.31250	-0.30303	-0.29412	-0.28571
0.1000	-0.38462	-0.37037	-0.35714	-0.34483	-0.33333	-0.32258	-0.31250	-0.30303	-0.29412	-0.28571
0.2000	-0.38458	-0.37036	-0.35714	-0.34483	-0.33333	-0.32258	-0.31250	-0.30303	-0.29412	-0.28571
0.3000	-0.38414	-0.37011	-0.35700	-0.34476	-0.33330	-0.32256	-0.31249	-0.30303	-0.29412	-0.28571
0.4000	-0.38127	-0.36825	-0.35583	-0.34402	-0.33285	-0.32230	-0.31234	-0.30294	-0.29407	-0.28569
0.4296	-0.37919	-0.36680	-0.35484	-0.34336	-0.33242	-0.32202	-0.31216	-0.30283	-0.29400	-0.28565
0.5000	-0.36945	-0.35956	-0.34955	-0.33957	-0.32974	-0.32016	-0.31090	-0.30198	-0.29344	-0.28528
0.5704	-0.34740	-0.34198	-0.33573	-0.32886	-0.32155	-0.31399	-0.30631	-0.29862	-0.29101	-0.28355
0.6000	-0.33194	-0.32914	-0.32519	-0.32031	-0.31472	-0.30859	-0.30209	-0.29537	-0.28854	-0.28169
0.7000	-0.23019	-0.23984	-0.24751	-0.25334	-0.25750	-0.26015	-0.26146	-0.26160	-0.26072	-0.25899
0.8000	0.02927	0.00243	-0.02252	-0.04553	-0.06662	-0.08580	-0.10311	-0.11859	-0.13231	-0.14434
0.9000	0.75364	0.71195	0.67058	0.62966	0.58933	0.54970	0.51089	0.47299	0.43608	0.40026
0.9500	1.73795	1.70155	1.66390	1.62513	1.58541	1.54487	1.50365	1.46186	1.41963	1.37708
0.9600	2.09490	2.06365	2.03073	1.99629	1.96048	1.92343	1.88528	1.84616	1.80618	1.76547
0.9750	2.89572	2.87959	2.86107	2.84030	2.81743	2.79259	2.76591	2.73751	2.70751	2.67603
0.9800	3.29567	3.28844	3.27854	3.26610	3.25128	3.23419	3.21497	3.19374	3.17062	3.14572
0.9900	4.60252	4.62850	4.65111	4.67050	4.68680	4.70013	4.71061	4.71836	4.72350	4.72613
0.9950	5.98602	6.05169	6.11351	6.17162	6.22616	6.27723	6.32497	6.36948	6.41086	6.44924
0.9980	7.90078	8.02594	8.14683	8.26359	8.37634	8.48519	8.59027	8.69167	8.78950	8.88387
0.9990	9.39827	9.57232	9.74190	9.90713	10.06812	10.22499	10.37785	10.52681	10.67197	10.81343
0.9995	10.92853	11.15402	11.37487	11.59122	11.80316	12.01082	12.21429	12.41370	12.60913	12.80069
0.9999	14.57706	14.92912	15.27632	15.61878	15.95660	16.28989	16.61875	16.94329	17.26361	17.57979

F(x)	γ_1 = 7.2	γ_1 = 7.4	γ_1 = 7.6	γ_1 = 7.8	γ_1 = 8.0	γ_1 = 8.2	γ_1 = 8.4	γ_1 = 8.6	γ_1 = 8.8	γ_1 = 9.0
0.0001	-0.27778	-0.27027	-0.26316	-0.25641	-0.25000	-0.24390	-0.23810	-0.23256	-0.22727	-0.22222
0.0005	-0.27778	-0.27027	-0.26316	-0.25641	-0.25000	-0.24390	-0.23810	-0.23256	-0.22727	-0.22222
0.0010	-0.27778	-0.27027	-0.26316	-0.25641	-0.25000	-0.24390	-0.23810	-0.23256	-0.22727	-0.22222
0.0020	-0.27778	-0.27027	-0.26316	-0.25641	-0.25000	-0.24390	-0.23810	-0.23256	-0.22727	-0.22222
0.0050	-0.27778	-0.27027	-0.26316	-0.25641	-0.25000	-0.24390	-0.23810	-0.23256	-0.22727	-0.22222
0.0100	-0.27778	-0.27027	-0.26316	-0.25641	-0.25000	-0.24390	-0.23810	-0.23256	-0.22727	-0.22222
0.0200	-0.27778	-0.27027	-0.26316	-0.25641	-0.25000	-0.24390	-0.23810	-0.23256	-0.22727	-0.22222
0.0250	-0.27778	-0.27027	-0.26316	-0.25641	-0.25000	-0.24390	-0.23810	-0.23256	-0.22727	-0.22222
0.0400	-0.27778	-0.27027	-0.26316	-0.25641	-0.25000	-0.24390	-0.23810	-0.23256	-0.22727	-0.22222
0.0500	-0.27778	-0.27027	-0.26316	-0.25641	-0.25000	-0.24390	-0.23810	-0.23256	-0.22727	-0.22222
0.1000	-0.27778	-0.27027	-0.26316	-0.25641	-0.25000	-0.24390	-0.23810	-0.23256	-0.22727	-0.22222
0.2000	-0.27778	-0.27027	-0.26316	-0.25641	-0.25000	-0.24390	-0.23810	-0.23256	-0.22727	-0.22222
0.3000	-0.27778	-0.27027	-0.26316	-0.25641	-0.25000	-0.24390	-0.23810	-0.23256	-0.22727	-0.22222
0.4000	-0.27776	-0.27026	-0.26315	-0.25641	-0.25000	-0.24390	-0.23810	-0.23256	-0.22727	-0.22222
0.4296	-0.27774	-0.27025	-0.26315	-0.25640	-0.25000	-0.24390	-0.23810	-0.23256	-0.22727	-0.22222
0.5000	-0.27751	-0.27010	-0.26306	-0.25635	-0.24996	-0.24388	-0.23808	-0.23255	-0.22727	-0.22222
0.5704	-0.27629	-0.26926	-0.26248	-0.25596	-0.24970	-0.24371	-0.23797	-0.23248	-0.22722	-0.22219
0.6000	-0.27491	-0.26825	-0.26175	-0.25544	-0.24933	-0.24345	-0.23779	-0.23236	-0.22714	-0.22214
0.7000	-0.25654	-0.25352	-0.25005	-0.24622	-0.24214	-0.23788	-0.23352	-0.22911	-0.22469	-0.22030
0.8000	-0.15478	-0.16371	-0.17123	-0.17746	-0.18249	-0.18643	-0.18939	-0.19147	-0.19277	-0.19338
0.9000	0.36557	0.33209	0.29986	0.26892	0.23929	0.21101	0.18408	0.15851	0.13431	0.11146
0.9500	1.33430	1.29141	1.24850	1.20565	1.16295	1.12048	1.07832	1.03654	0.99519	0.95435
0.9600	1.72412	1.68225	1.63995	1.59732	1.55444	1.51141	1.46829	1.42518	1.38213	1.33922
0.9750	2.64317	2.60905	2.57375	2.53737	2.50001	2.46175	2.42268	2.38288	2.34242	2.30138
0.9800	3.11914	3.09099	3.06137	3.03038	2.99810	2.96462	2.93002	2.89440	2.85782	2.82035
0.9900	4.72635	4.72427	4.71998	4.71358	4.70514	4.69476	4.68252	4.66850	4.65277	4.63541
0.9950	6.48470	6.51735	6.54727	6.57456	6.59931	6.62159	6.64148	6.65907	6.67443	6.68763
0.9980	8.97488	9.06261	9.14717	9.22863	9.30709	9.38262	9.45530	9.52521	9.59243	9.65701
0.9990	10.95129	11.08565	11.21658	11.34419	11.46855	11.58974	11.70785	11.82294	11.93509	12.04437
0.9995	12.98848	13.17258	13.35309	13.53009	13.70366	13.87389	14.04086	14.20463	14.36528	14.52288
0.9999	17.89193	18.20013	18.50447	18.80504	19.10191	19.39517	19.68489	19.97115	20.25402	20.53356

Harter (1969, 1971), by permission of Technometrics, American Statistical Association.

12.8 Reliability of Results from Frequency Analysis

The central limit theorem states that for a population with finite variance σ^2 and mean μ the distribution of the sample means from repeated sampling will be a normal distribution with mean μ and variance σ^2/n. This result does not depend on the type of distribution of the parent population. The standard deviation of the distribution of the sample means, $\sigma/\sqrt{n}$, is known as the *standard error* of the mean. The standard error of the standard deviation of the sample means is $\sigma/\sqrt{2n}$. The standard error of estimate accounts only for the errors in the parameters due to the shortness of the sample data. It does not account for errors due to the choice of a wrong distribution function.

The observed value of $\bar{x}$ of a random variable is thus within the range $\bar{x} \pm \sigma/\sqrt{n}$ with 67% probability and within the range $\bar{x} \pm 2\sigma/\sqrt{n}$ with 95% probability. The population variance, however, is usually unknown. Rough estimates can be obtained by using the sample standard deviation, i.e., $S/\sqrt{n}$ and $S/\sqrt{2n}$ or $S/\sqrt{n-1}$ and $S/\sqrt{2n-1}$ for small samples. An improved technique is to estimate the confidence limits for a sample with mean $\bar{x}$ and standard deviation S with the aid of the Student's t-distribution, which converges to the normal distribution as n grows large. A measure of the precision of $\bar{x}$ as an estimate of the unknown population mean μ is the *confidence interval*, $(1 - \alpha)100\%$. The values of $(1 - \alpha)$, e.g., 0.90, 0.99, etc., are known as confidence levels, i.e., in repeated sampling $(1 - \alpha)100\%$ of the sample means should fall within these limits. Thus, the estimate of the population mean is

$$\mu = \bar{x} \pm t_{[(1-\alpha),\ n-1]}\ S/\sqrt{n} \qquad 12.104$$

where t is given in statistical tables for a range of values of (n - 1) and confidence levels. The standard error of the mean as a percentage of the mean is shown as a function of the sample size and its coefficient of variation ($C_V = S_n/\bar{x}$) below:

	Sample size, n						Sample size, n				
C_V	5	10	20	50	100	C_V	5	10	20	50	100
0.10	4.5	3.2	2.2	1.4	1.0	0.60	26.8	19.0	13.4	8.5	6.0
0.20	8.9	6.3	4.5	2.8	2.0	0.80	35.8	25.3	17.9	11.3	8.0
0.40	17.9	12.7	8.9	5.7	4.0	1.00	44.7	31.6	22.4	14.1	10.0

The accuracy of C_V itself has a bearing on the results. The standard error of the coefficient of variation depends on the type of distribution of C_V, which is difficult to define. However, the variation of C_V with the type of distribution does not appear to be large.

It is of paramount importance to realize that the prediction line or equation obtained from a statistical analysis of an observed record is just one of an infinite number of such lines. The spread of these lines depends on the sample size. Figure 12.8 illustrates the spread of equally good prediction lines of best fit to data from samples drawn from an EV1 distribution (Benson, 1960) and demonstrates how difficult it is to estimate from a short record. The effect of the length of the record on the confidence limits was determined in terms of the length of record necessary to come within a given percentage of the correct value a certain percentage of the time as shown overleaf.

Interpretation of results can at times be made more difficult by grouping of data. For example, the data may consist of mainly low magnitude values and one or two very large values. The fitted distribution in this case would not match with either the low or high values.

Magnitude of flood T in years	Years of record necessary to come within 25% of the correct value 95% of the time	25% of the correct value 80% of the time	10% of the correct value 95% of the time	10% of the correct value 80% of the time
2.33	12	8	40	25
10	18	8	90	38
25	31	12	105	75
50	39	15	110	90
100	48	-	115	100

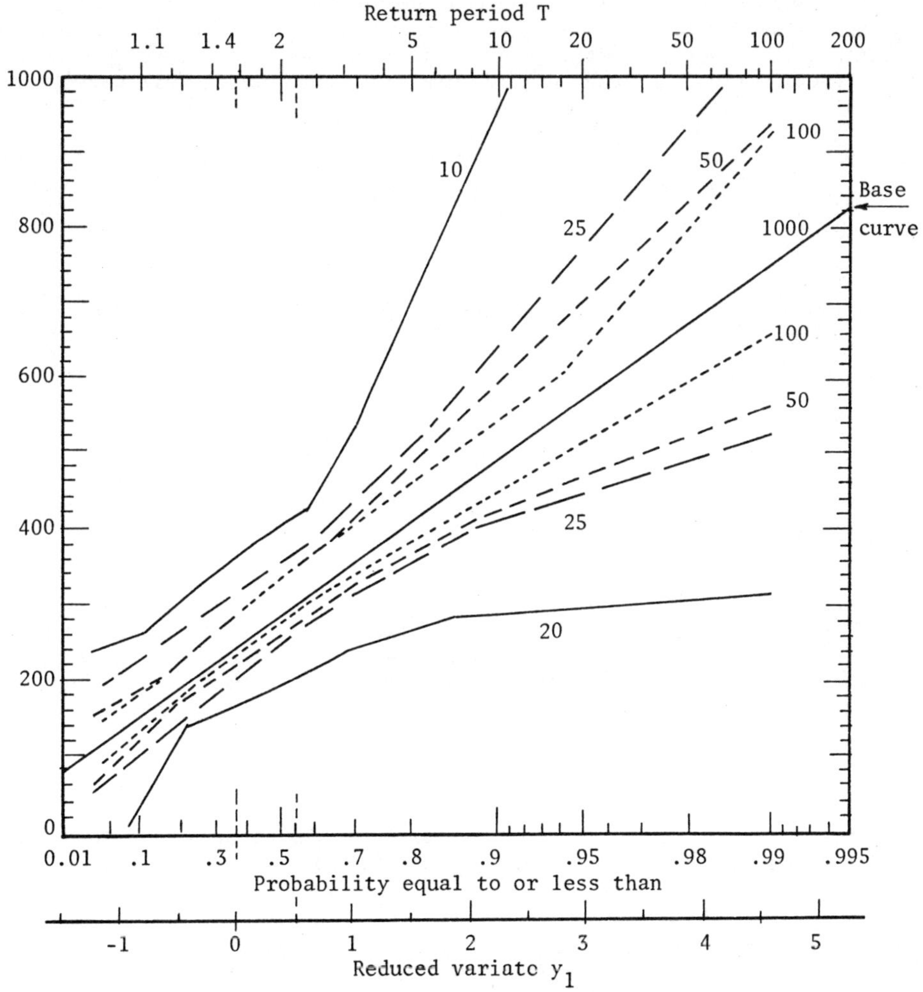

Fig. 12.8. Envelopes to frequency curves given by samples drawn from an EV1 distribution of 1000 items when the samples are 10, 25, 50 and 100 items, respectively. After Benson (1960).

In order to answer questions like "what is the chance of the annual runoff falling below half of the mean value, or exceeding five times the mean?" Or "what is the likelihood of five dry years in a row?", the probability distribution of the runoff must be estimated. The records are seldom long enough to provide adequate information to do this. Hydrological distributions are almost always asymmetric in shape. A measure of this asymmetry is the coefficient of skewness, C_s or $\sqrt{\beta_1}$, estimated from sample data by eqn 12.28, but C_s is strongly dependent on the distribution of x and subject to large errors. The standard errors of C_s can be several hundred percent of C_s for small n and C_v values but even for $n \sim 100$ and $C_v \sim 1.0$ the error is still in the 40-50% range.

In the preceding discussion of the various techniques, and errors in the estimates, it has been taken for granted that the "correct" distribution or relationship has been fitted to the observed data. This assumption has to be tested, because an incorrect distribution can lead to large errors in prediction, particularly at the extremes of the p.d.f.

The *goodness of fit test* shows whether or not the assumed distribution function is satisfactory and yields a measure of confidence in the results. The oldest, most commonly used, and perhaps most versatile procedure is the χ^2-test (chi-squared test) The basis of the test is the χ^2 distribution, but for the purposes of testing one does not need to know the distribution. The χ^2 distribution is derived from considering $X_1, X_2, \ldots, X_n$ to be n independent random variables with means $\mu_1, \mu_2, \ldots, \mu_n$ and variances $\sigma_1^2, \sigma_2^2, \ldots, \sigma_n^2$, respectively (each X could be looked upon as a sample). Then the standardized variate corresponding to X_i is

$$z_i = \frac{X_i - \mu_i}{\sigma_i}$$

and the χ^2 distribution is

$$f(\chi^2) = \left(\frac{\chi^2}{2}\right)^{\frac{1}{2}\nu-1} \frac{e^{-\chi^2/2}}{2\Gamma(\nu/2)} \qquad 12.105$$

where ν is the number of degrees of freedom, i.e., the number of independent normal variates whose squares are added to produce χ^2. In the χ^2-test, as in all tests of significance, it is postulated that there is no significant difference between the distributions being compared. This is the *null hypothesis*. The test measures the probability of the actual difference occurring due to chance alone and if this probability is low the null hypothesis is rejected, i.e., it is taken as proof that a real difference does exist.

To test a distribution the observed data are grouped into k classes. At least *five* observations should be in each class. Consequently, cells or classes which include the extremes have to be combined and this makes the test insensitive to deviations from the assumed model in the tail ends of the distribution. The test statistic is then

$$\chi_\alpha^{\prime 2} = \sum_{i=1}^{k} \frac{(f_{oi} - f_{ei})^2}{f_{ei}} \qquad 12.106$$

where f_{oi} is the observed frequency of the i-th class and f_{ei} is the expected theoretical frequency of the i-th class. χ^2 is distributed with $\nu = n - m - 1$ degrees of freedom (d.o.f.) and is tabulated as a function of the level of significance α,in statistical tables. Here the (-1) term means that one degree of freedom has been lost by the requirement that the frequencies must total n. The m value refers to the number of parameters estimated. If only the mean has been estimated, one is free to assign any values to the (n - 1) observations, but the last value must lead to the correct sample mean. If the standard deviation S has also been estimated, then another d.o.f. has been lost, etc. If χ_α^2 is smaller than the value from the table the fit satisfies the selected probability or confidence level.

The two common methods of procedure are:
Method 1: Here the data are initially arranged or naturally assigned into frequency classes, e.g., observations from discrete distributions. Each class should have at least five observations. If L_i and U_i are the lower and upper bounds of the i-th frequency class, then the model is used with the estimated (from sample) parameters to determine the probability of a random observation falling within each class, i.e.

$$P(L_i \leq x \leq U_i) \qquad i = 1, 2, \ldots, k$$

The estimated frequencies of the classes, f_{ei}, are obtained by multiplying each of the class probabilities by the sample size n,

$$f_{ei} = P(L_i \leq x \leq u_i)n$$

Finally the observed frequencies, f_{oi}, within each of the class intervals are counted. Then the estimate of χ^2 may be obtained.

Method 2: Here the data are not initially tabulated in classes and the choice of k is arbitrary. When n > 200 a rough rule is to take k as the integer closest to

$$k = 4[0.75(n - 1)^2]^{1/5} \qquad 12.107$$

For moderate values of n, k should be as large as possible with the limitation of not less than 5 per class.

The class boundaries $x_1, x_2, \ldots, x_k$ are determined from the cumulative distribution of the assumed model (using estimated parameters) as the values of $P(x \leq x_1) = 1/k$; $P(x \leq x_2) = 2/k$; ...; $P(x \leq x_{k-1}) = (k - 1)/k$. The probability of a random value falling within a given class is estimated to be 1/k for each class. The lower bound of the first class and the upper bound of the last class are the smallest and largest values that the random variable may take, e.g., 0 and ∞.

For example, for normal or log-normal distributions

$$P(x \leq x_1) = 0.5 - \frac{1}{\sqrt{2\pi}} \int_0^z e^{-z^2/2} dz = \frac{1}{k}$$

where $z = (x - \mu)/\sigma$, from which x_1 may be obtained, etc. For an exponential distributions, where $F(x, \lambda) = 1 - \exp(-\lambda x)$

$$P(x \leq x_1) = 1 - e^{-\lambda k_1} = \frac{1}{k}$$

Next the estimated frequencies, f_{ei}, are obtained by multiplying each of the class probabilities by the sample size n, i.e.

$$f_{ei} = \frac{n}{k} \qquad i = 1, 2, \ldots, k$$

Finally, the observed frequencies, f_{oi}, within each of the class intervals $x_i \leq x \leq x_{i+1}$ are counted. This yields the estimated value of χ^2.

For example, if the number of observations are divided into k = 20 classes and the two parameters $\bar{x}$ and S are calculated, then the number of d.o.f. $\nu = 20 - 2 - 1 = 17$. For this value the value χ_α^2 can be read from the table. If the calculated $\chi^2 > \chi_{0.05}^2$ the chances are less than 1/20 that the data could have originated from the assumed type distribution and the model has to be rejected.

Example 12.9. During a rainstorm the yield was recorded for each 10 min period as 0-0.99 mm, 1-1.99 mm, etc. The results were as follows:

x mm	0	1	2	3	4	5	6
obs. freq.	2	11	23	31	22	10	1

Could this sample be drawn from a normal population?

x	f_{oi}	u=x-3	fu	fu^2	$x-\bar{x}$	$\frac{x-\bar{x}}{S}$	F(z)	f_{ei} %	$f_{oi}-f_{ei}$	$\frac{(f_{oi}-f_{ei})^2}{f_{oi}}$
0	2	-3	- 6	18	-2.94	-2.36	0.0091			
	13				-2.44	-1.96	0.0250			
1	11	-2	-22	44	-1.94	-1.55	0.0606			
					-1.44	-1.15	0.1257	12.5	0.5	0.0192
2	23	-1	-23	23	-0.94	-0.75	0.2266			
					-0.44	-0.75	0.3632	23.8	0.8	0.0278
3	31	0	-51		0.06	0.05	0.5199			
					0.56	0.45	0.6736	31.0	0.0	
4	22	1	22	22	1.06	0.85	0.8023			
					1.56	1.25	0.8944	22.1	0.1	0.0005
5	10	2	20	40	2.06	1.65	0.9505			
	11									
6	1	3	3	9	3.06	2.45	0.9929	4.5	6.5	3.8404
	100		- 6	156						3.8884

$\bar{x} = a + (1/n)\Sigma f_i u_i = 3 - 0.06 = 2.94$; $\mu_2(a) = 1.56$; $\mu_2 = S^2 = 1.56 - (2.94 - 3)^2 = 1.556$; $S = 1.248$; d.o.f $\nu = n - k - 1 = 5 - 2 - 1 = 2$; $\chi^2 = 3.888 < \chi^2_{0.10,\ 2} = 4.605$

It can be assumed to 90% confidence level that the sample was drawn from a normal population, or there is 1 change in 10 of not being from normal population.

Example 12.10. A network of stations, spaced so that their recordings may be assumed to be independent, recorded the following distribution of hailstorms per station in a year.

Number of hailstorms per station:	0	1	2	3	4	5	6
Number of observed occurrences :	40	70	28	14	7	1	0

Hence, 40 stations recorded no hailstorms, 70 stations recorded one hailstorm, etc. Test the hypothesis that the distribution of hailstorms at a station is a Poisson distribution.

The Poisson distribution is

$$p(x) = \frac{e^{-\mu}\mu^x}{x!} \qquad x = 0, 1, 2, 3, 4, 5, 6$$

Calculate μ and $p(x)$.

x		f(x)	x f(x)	μ^x	x!	p(x)	Exp. number
0	40	0.2500	0	1	1	0.2846	45.5
1	70	0.4375	0.4375	1.2565	1	0.3577	57.2
2	28	0.1750	0.3500	1.5788	2	0.2247	36.0
3	14	0.0875	0.2625	1.9838	6	0.0941	15.1
4	7	0.0437	0.1750	2.4926	24	0.0296	4.7
5	1	0.0063	0.0315	3.1319	120	0.0074	1.2
6	0	0	0	3.9353	720	0.0016	0.3
	160	1.0000	1.2565 = μ			0.9997	160.0

The record shows that $\exp(-\mu) = \exp(-1.2565) = 0.2846$ and that the average number of hailstorms per station is $\mu \simeq 1.3$ per year. Calculate χ^2.

x	Observed		Expected		$\frac{(O - E)^2}{E}$
0	40		46		36/46 = 0.7826
1	70		57		169/57 = 2.9649
2	28		36		64/36 = 1.7778
3	14		15		1/15 = 0.0667
4	7	8	5	6	4/6 = 0.6667
5	1		1		
6	0		0		
					6.2587

Here one parameter, $\mu = \bar{x}$, has been estimated and hence $\nu = (5 - 1) - 1 = 3$. From tables of χ^2 distribution $6.2587 > \chi_{0.10,\ 3} = 6.251$. This means that there is slightly less than one chance in ten of the data originating from a Poisson distribution. At 10% level the difference is significant and the null hypothesis should be rejected. Had the computed value of $(O - E)^2/E$ been 16.3, then $16.3 > \chi_{0.001,\ 3} = 16.268$ and there would be less than one chance in a thousand of the data originating from the assumed distribution. The difference is highly significant and the null hypothesis of no difference should be rejected at 0.1% level.

12.9 Use of Historical Data

Frequently historical data on rare extremes are available which do not form part of a continuous record. Some historic data goes back to ancient times such as the flood levels on the Nile or the stone sculptures along the upper reaches of the Yangtze River, which show the flood levels of the great floods (A.D. 1153, 1227, 1560 and later). However, isolated observations of extremes over the last few centuries are in existence in most countries. Several tecniques have been proposed for inclusion of these historical data in frequency analyses. If a flood has been observed which is larger than any of the floods in the continuous record, then this may be regarded as the largest value in a period of $n_1 + n_2$ years, where n_1 is the length of the continuous record and n_2 is the number of years between the time of occurrence of this largest flood and the beginning of the continuous record. The continuous record remains unaltered.

When the historical flood which occurred n_2 years before the beginning of the continuous record is exceeded by a flood in the record, then the procedure is as illustrated by the following example:
A discharge of 1000 m^3/s occurred in 1876. The continuous record began in 1927 and in 1938 a discharge of 1500 m^3/s was recorded. The total number of years $n_1 + n_2 = 100$ (up to 1975 inclusive). The maximum flood in 1876-1975 has a sequence number $m = 100$ and a plotting position of $m/(100 + 1)$ or a return period of 101 years. The return period for the second largest flood in the same period is $(100 + 1)/2$ years. The second largest flood in 1926-1975 has a return period of $(49 + 1)/2$ years, the third largest a return period of $(49 + 1)/3$, etc.

For a larger number of historical floods Benson (1950) proposed the following formula for adjustment of the order numbers

$$m_1 = A + \frac{H - A}{T - A}(m - A) \qquad 12.108$$

where m = the sequence number of all the floods from the continuous and historical records arranged in descending order of magnitude; m_1 = the adjusted sequence number of the floods smaller than the smallest flood in the historical record; A = the number of annual floods equalling or exceeding the smallest historical flood; H = the length of the total record in years, and T = the total number of floods in the historical and continuous records.

Example 12.11. For the Waikato River, N.Z., there are historical records of floods for the years 1875, 1907, 1915 of 1540, 1869 and 872 m^3/s, respectively, and a continuous record from 1924 onwards. (728, 799, 929, 665, 748, 685, 561, 561, 561, 606, 623, 606, 694, 708, 561, 617, 580, 716, 739, 855, 569, 566, 680, 586, 617, 680, 575, 487, 680, 957, 595, 566, 782, 578, 1540, 620, 929, 906, 947, 728, 855, 790, 759, 878, 811, 533 m^3/s for the years 1924-1969 incl.). The total length of record is 95 years, in which 10 floods are greater than or equal to the lowest historical flood (1915) and 39 are smaller. These 10 floods are now considered to have occurred at annual intervals and the intervals between the remaining 39 floods are adjusted in the proportion of 85/39, where 85 is the total length of record less 10 years. The total number of floods is T = 49, A = 10, H = 95, $m_1 = A + [(H - A)/(T - A)](m - A) = 10 + 2.1795(m - 10)$, and the calculation is carried out as indicated below:

Year	Q m^3s^{-1}	Q^2	m	m-10	m_1	$T = \frac{95+1}{m}$
1907*	1 869	3 493 161	1		1	96
1958	1 540	2 371 600	2		2	48
1857*	1 540	2 371 600	3		3	32
1953	957	915 849	4		4	24
1962	947	896 809	5		5	19.20
1926	929	863 041	6		6	16
1960	929	863 041	7		7	13.71
1961	906	820 836	8		8	12
1967	878	770 884	9		9	10.67
1915*	872	760 384	10		10	9.60
1943	855	731 025	11	1	12.18	7.88
1964	855	731 025	12	2	14.36	6.69
1968	811	657 721	13	3	16.54	5.80
1925	799	638 410	14	4	18.72	5.13
1965	790	624 100	15	5	20.90	4.59
1956	782	611 524	16	6	23.08	4.16
1966	759	576 081	17	7	25 26	3.80
1928	748	559 504	18	8	27.44	3.50
1942	739	546 121	19	9	29.62	3.24
1924	728	529 984	20	10	31.80	3.02
	etc.					
	37 037	31 368 405				

$\bar{Q} = 755.86$, $Q^2 = 640171.53$, $(\bar{Q})^2 = 571320.02$, $S^2 = Q^2 = (Q^2) = 68851.51$, $S = 264.40$. The prediction equation by the Gumbel distribution is $Q = 633.93 + 217.98y$.

12.10 Correlation and Regression Analysis

Correlation analysis is used to determine the manner in which the independent variable or variables affect the dependent variable. Correlation of one dependent variable with one independent variable is known as simple correlation and that of two or more independent variables as multiple correlation. A plot of simultaneous measurements, the independent variable versus the dependent variable, results in a scatter diagram. The smaller the dispersion of the plotted data the better defined is the relationship between the variables. The pattern of the diagram also shows whether the relationship is linear or not. In hydrological work straight line approximations and fitting a curve by eye are at times satisfactory, but analytical correlation and curve fitting are to be preferred, because these provide also estimates of the dependability of the relationships.

The method of least squares is a popular curve-fitting technique. It makes the sum of the departures from the fitted line zero and the sum of the squares of the departures a minimum. If two series of observations x_i and y_i are assumed to satisfy a function $y = f(x,a,b,...)$, where a, b,... are parameters to be determined (e.g., $y = a + bx$), then in order to make $\Sigma(\Delta y_i)^2$ a minimum all the partial derivatives with respect to the parameters involved must be zero, i.e.,

$$\frac{\partial\Sigma(y_i - y)^2}{\partial a} = 0 \qquad \frac{\partial\Sigma(y_i - y)^2}{\partial b} = 0 \qquad \text{etc.}$$

For k parameters there are k simultaneous equations to solve. The important point to note is that the method fits the best line of the assumed form irrespective of the data. Thus, one could fit the best fitting parabola to a set of points which lie on a straight line or vice versa. For a straight line

$$y_i - y = y_i - (a + bx) = \varepsilon_i \tag{12.109}$$

and the two partial derivatives yield (omitting subscript i)

$$\Sigma y = na + b\Sigma x$$

$$\Sigma xy = a\Sigma x + b\Sigma x^2$$

from which

$$a = \frac{\Sigma x^2\Sigma y - \Sigma x\Sigma xy}{n\Sigma x^2 - (\Sigma x)^2} \tag{12.110}$$

and

$$b = \frac{n\Sigma xy - \Sigma x\Sigma y}{n\Sigma x^2 - (\Sigma x)^2} = \frac{\Sigma(x - \bar{x})(y - \bar{y})}{\Sigma(x - \bar{x})^2} = \frac{\Sigma xy - n\bar{x}\bar{y}}{\Sigma x^2 - n\bar{x}^2} \tag{12.111}$$

The scatter is assumed in y_i at given values of x_i, that is, the regression of y on x is being considered. Likewise, the scatter could be assumed to be in x_i, that is, the regression of x on y, or $x = A + By$

$$x = \frac{\Sigma y^2\Sigma x - \Sigma y\Sigma xy}{n\Sigma y^2 - (\Sigma y)^2} + \frac{n\Sigma xy - \Sigma x\Sigma y}{n\Sigma y^2 - (\Sigma y)^2}\,y \tag{12.112}$$

If y is independent of x, then the slope $b = 0$, i.e., there is zero correlation. Likewise, $B = 0$ if x is independent of y. Generally, if there is no correlation between the two variables the product $b \cdot B = 0$. For perfect correlation all points lie exactly on the line and the product equals one. A measure of the correlation is the correlation coefficient r

$$r = \frac{n\Sigma xy - \Sigma x\Sigma y}{\{[n\Sigma x^2 - (\Sigma x)^2][n\Sigma y^2 - (\Sigma y)^2]\}^{1/2}} = \frac{\Sigma\Delta x\ \Delta y}{\sqrt{\Sigma\Delta x^2\ \Sigma\Delta y^2}}$$

$$0 \le |r| \le 1 \tag{12.113}$$

where $\Delta x = x - \bar{x}$ and $\Delta y = y - \bar{y}$. For various significance levels the correlation coefficient r is given as a function of the number of d.o.f., $(n - 2)$, in statistical tables, (-2 because two constants fix the location of a straight line). In repeated sampling the distribution of r is a function of ρ, the correlation coefficient of the population, and n, the number of pairs of x_i, y_i. As ρ increases from zero the distribution becomes decidedly skew. However, a change of variable from r to

$$z = \tfrac{1}{2}\ell n[(1 + r)/(1 - r)]$$

converts the distribution to an approximately normal one and this may then be used to determine the accuracy of r as an estimate of ρ. The mean of this distribution is $\mu_z = \frac{1}{2}\ln[(1 + \rho)/(1 - \rho)]$ and the standard deviation $\sigma_z = (n - 3)^{-\frac{1}{2}}$. Thus, for example, the 95% probability level is given by

$$z = \frac{1}{2} \ln \frac{1 + \rho}{1 - \rho} \pm \frac{2}{\sqrt{n - 3}}$$

from which the two limiting values of r can be found.

For linear regression the best estimate of $\hat{y}$ is

$$\hat{y} = \bar{y} + b(x_i - \bar{x}) \qquad 12.114$$

On this a random component has to be added, namely the standard error of the estimate y_i. Without this component the regression analysis gives a unique value of y for a given value of x. The standard error is defined as $S_{y(x)}\sqrt{1 - r^2}$. The total variance

$$S^2_{y(x)} = S^2_{y(x)} r^2 + S^2_{y(x)}(1 - r^2) = \frac{1}{n - 2} \Sigma(y_i - \hat{y})^2 \qquad 12.115$$

where r is given by eqn 12.113. Thus, the regression equation for linear regression becomes

$$\hat{y}_i = \bar{y} + b(x_i - \bar{x}) \pm t_i S_{y(x)}\sqrt{1 - r^2} \qquad 12.116$$

where t_i is a random (usually normally) and independently distributed standardized variate with zero mean and unit variance. The standard error of the mean, $S_{\bar{y}} = S_{y(x)}/\sqrt{n - 2}$, is used to define $\bar{y} \pm tS_{\bar{y}}$, where t is the Student's t, read from tables for (n - 2) d.o.f. and the desired level of significance (2 d.o.f. have been lost in locating a straight line). The variance of the slope is

$$S^2_b = S^2_{y(x)}/\Sigma(x - \bar{x})^2 \qquad 12.117$$

and

$$b = \bar{b} \pm tS_b \qquad 12.118$$

For a specified value of $x_i = \xi$

$$S^2_{yi} = S^2_{y(x)}\left[\frac{1}{n} + \frac{(\xi - \bar{x})^2}{\Sigma(x_i - \bar{x})^2}\right] \qquad 12.119$$

and the variance of a is

$$S^2_a = S^2_{y(x)}\left[\frac{1}{n} + \frac{\bar{x}^2}{\Sigma(x_i - \bar{x})^2}\right] \qquad 12.120$$

If a variable depends on several independent variables, each of which varies randomly, the simple regression analysis still yields results, although with some loss in precision. However, if the independent variables vary according to some relationship, then multiple regression analysis must be used. For a linear relationship

$$y = b_o x_o + b_1 x_1 + b_2 x_2 + \ldots b_k x_k$$

or

$$\varepsilon = \Delta y - (b_1 \Delta x_1 + b_2 \Delta x_2 + \ldots b_k \Delta x_k)$$

where $b_o = a$, $x_o = 1$, b_i are the partial regression constants, $\Delta y = y - \bar{y}$, and $\Delta x = x - \bar{x}$. The criterion

$$\frac{\partial(\Sigma\varepsilon^2)}{\partial b_i} = 0$$

leads to a set of linear simultaneous equations. For two independent variables the result is a regression surface.

When applying correlation techniques it is necessary to appreciate that every value of x or y has a frequency function f(x) or f(y), which may not necessarily be the same one throughout the whole range of interest. If some statistic of the y distribution, e.g., the mean $\bar{y}$ for each value of x is plotted against x, then x and $\bar{y}$ are said to be correlated if a curve can be drawn through the points. Such a correlation ignores all other dependencies and should be used with caution. For example, one can find from laboratory experiments that evaporation is a function of the air and water temperatures. From these results the evaporation could be correlated with temperatures, but this could be misleading since evaporation from, say, a lake or field also depends on other variables, e.g., wind speed. Consideration of wind speed, as well as temperatures in the correlation analysis, would lead to a family of correlation surfaces, etc.

Correlation must be self-evident before it warrants analysis. In many situations a linear correlation is the easiest to apply, but one should not place much weight on the linearity. Nature, in general, is seldom linear. In particular, logarithmic presentation of data, unless based on physical reasoning, is suspect because of demagnification of inaccuracies by taking the logarithms. For $y = f(x)$, $d(\log y)/dx = [1/f(x)]f'(x)$. Hence, each $f'(x)$ is reduced by $1/f(x)$ and for large values of $f(x)$ the errors are hidden.

Great care must be exercised with multi-variate analysis. With the aid of computers pseudo-analytical equations of regression can be established for almost anything. Unless based on either very complete data or a physical model or both, these are nothing more than contrived nonsense. Anything can be correlated with any group of variables if the number of variables is large enough.

The *rank correlation* method is useful in the study of the significance of the dependence of data in sets of observations. When the sample is large the data may first be ranked in order of size, for example, the percentage of rainfall, x, infiltrating on a given catchment and the amount of rain in the preceding fortnight, y, are as follows:

x:	5	3	4	7	6	2	8	1	10	9
y:	3	7	4	2	9	1	6	5	8	10

If these numbers were written down in a purely random manner a certain amount of correlation can be expected to occur by chance. The question is whether the correlation present is more than might be expected from random sets of numbers. There are many rank correlation coefficients but here only Kendall's τ and Spearman's ρ are discussed.

The logic behind Kendall's correlation coefficient τ can be explained with the aid of the above data. Consider the pair of observations, (5, 3) and (3, 7), for example, taken from the x and y rows, respectively. The magnitudes of the two numbers in each observation increase in opposite directions and hence this is recorded as a negative correlation, which is scored as -1 or N. Likewise, (5, 6) and (3, 9) score as +1 or P. Pairs, such as (5, 5) and (3, 3), are recorded as zero correlation. Generally, the theory of combinations shows that the number of unordered samples of n different objects taken r at a time without repetition is

$$N_r = \binom{n}{r} = \frac{n!}{r!(n-r)!}$$

In the example there are $n(n - 1)/2 = 45$ possible pairs, and the sum $P + N = 45$

since there are no zero correlations. The difference P - N = S is a measure of the correlation. Kendall's correlation coefficient ($1 > \tau > -1$) is defined as

$$\tau = \frac{S}{\frac{1}{2}n(n-1)} \qquad 12.121$$

To evaluate P, rearrange the (x, y) pairs of data so that the top row is in natural ascending order and then take each number in the second row in turn and count how many numbers greater than it are to its right (e.g., for 5 in row 2 there will be 5 larger numbers to the right). The sum of these results gives P. Thus, here P = 30, N = 15 (since P + N = 45) and $\tau = 15/45 = 1/3$.

Spearman's rank correlation coefficient is defined as

$$\rho = 1 - \frac{6\Sigma d^2}{n(n^2-1)} \qquad 12.122$$

where $d = x_i - y_i$ and n is the number of x_i, y_i pairs.

Example 12.12. A frequent operation is the relating of rainfall data at site A to the rainfall data at site B. The latter site may have long term record whereas the former has only a short term record collected for the purpose of establishing a correlation relationship. Once such a relationship has been established the observations at site A may be discontinued.

Let the monthly rainfalls at A be y and at B be x.

	J	F	M	A	M	J	J	A	S	O	N	D
x	2	1	2	4	4	5	6	5	4	4	3	2
y	2	1	4	3	5	5	7	6	5	4	3	3

Relate y to x, and determine the regression line, correlation coefficient and the goodness of fit.

x	y	x^2	xy	$\Delta x = x-\bar{x}$	$\Delta y = y-\bar{y}$	$\Delta x\,\Delta y$	Δx^2	Δy^2	$\hat{y}$	$\Delta\hat{y}_i = y_i - \hat{y}_i$	$\Delta\hat{y}_i^2$
2	2	4	4	-1.5	-2	3.0	2.25	4	2.5	-0.5	0.25
1	1	1	1	-2.5	-3	7.5	6.25	9	1.5	-0.5	0.25
2	4	4	8	-1.5	0	0	2.25	0	2.5	1.5	2.25
4	3	16	12	0.5	-1	-0.5	0.25	1	4.5	-1.5	2.25
4	5	16	20	0.5	1	0.5	0.25	1	4.5	0.5	0.25
5	5	25	25	1.5	1	1.5	2.25	1	5.5	-0.5	0.25
6	7	36	42	2.5	3	7.5	6.25	9	6.5	0.5	0.25
5	6	25	30	1.5	2	3.0	2.25	4	5.5	0.5	0.25
4	5	16	20	0.5	1	0.5	0.25	1	4.5	0.5	0.25
4	4	16	16	0.5	0	0	0.25	0	4.5	-0.5	0.25
3	3	9	9	-0.5	-1	0.5	0.25	1	3.5	-0.5	0.25
2	3	4	6	-1.5	-1	1.5	2.25	1	2.5	0.5	0.25
42	48	172	193			25.0	25.00	32			7.00

$$\bar{x} = 3.5 \qquad \bar{y} = 4 \qquad S_{y(x)} = \sqrt{[7.00/(12-2)]} = 0.837$$

$$a = \frac{\Sigma x^2\Sigma y - \Sigma x\Sigma xy}{n\Sigma x^2 - (\Sigma x)^2} = \frac{172 \times 48 - 42 \times 193}{12 \times 172 - (42)^2} = \frac{8256 - 8106}{2064 - 1764} = \frac{150}{300} = 0.5$$

$$b = \frac{n\Sigma xy - \Sigma x\Sigma y}{n\Sigma x^2 - (\Sigma x)^2} = \frac{12 \times 193 - 42 \times 48}{300} = \frac{2316 - 2016}{300} = 1.0$$

Hence, the estimates are

$$\hat{y} = 0.5 + 1.0x$$

$$S_b = \frac{S_{y(x)}}{\sqrt{\Sigma(x - \bar{x})^2}} = \frac{0.837}{\sqrt{25}} = \frac{0.837}{5} = 0.167$$

$$S_a = S_{y(x)} \sqrt{\frac{1}{n} + \frac{\bar{x}^2}{\Sigma(x - \bar{x})^2}} = 0.837 \sqrt{\frac{1}{12} + \frac{3.5^2}{25}}$$

$$= 0.837 \sqrt{0 \cdot 573} = 0.837 \times 0.757 = 0.634$$

The significance of the slope is given by

$$t = \frac{b}{S_b} = \frac{1.0}{0.167} = 6$$

and for $\nu = 12 - 2 = 10$ the table for t-distribution shows that it is significant at better than 0.1% level (t = 4.587). For a

$$t = \frac{a}{S_a} = \frac{0.5}{0.634} = 0.79$$

and is not significant even at 10% level (1.812). Hence, not much reliance can be placed on a. The 95% confidence intervals for $\nu = 10$ are $a \pm tS_a$ and $b \pm tS_b$, i.e., $0.5 \pm 2.228 \times 0.634$ or $-0.913 < a < 1.913$, $1.0 \pm 2.228 \times 0.167$ or $0.628 < b < 1.372$. The correlation coefficient

$$r = \frac{\Sigma\Delta x\Delta y}{\sqrt{\Sigma\Delta x^2 \Sigma\Delta y^2}} = \frac{25}{\sqrt{25 \times 32}} = 0.884$$

and for $\nu = 12 - 2$ is better than 1% level of significance.

12.11 Risk

In the preceding chapters the words "decision making" and "risk" have been frequently used. The designer of any hydraulic structure, or a bridge across the river, must somehow find the answer to "what is the risk of failure?" For major structures the risk must be minimized but the risk does not stem from hydraulic causes alone. For large dams the most frequent cause of failure relates to foundation conditions, well ahead of failures due to inadequate spillway design, i.e., inadequate estimation of design flood. Inadequate spillway design accounts for about a third of the failures. Risk does not relate to floods alone, droughts too can cause major economic disasters.

A central aspect of risk analysis is the prediction of the probability of an event occurring during a specified period in future. For example, what is the probability M of a flood with an average probability of occurrence P being exceeded exactly k times during an N-year period? This probability is given by

$$M = 1 - \binom{N}{k}(1 - P)^{N-k} P^k \qquad 12.123$$

where

$$\binom{N}{k} = \frac{N!}{k!(N - k)!}$$

The probability of the largest event not occurring is given by k = 0 as $M = 1 - (1 - P)^N$. The relationship assumes that the average annual probability of occurrence is known exactly. In practice, this is not the case, and one has to use the value corresponding to the plotting position. The value of M is strongly affected by the accuracy of P. Equation 12.123 can also be used to estimate the return period required for a specified risk of occurrence within the stipulated life-span of the project, e.g., if a 10% chance is accepted of a flood being equalled or exceeded during the next 25 years, then the return period of the flood is $0.1 = 1(1 - 1/T)^{25}$ or T = 238 years.

The spillway designs are usually based on some selected return period of the flood and take no account of the increase of risk with increasing age of the structure or of the economically optimum conditions.

An alternative procedure is to relate the design not only to the magnitude and frequency of the flood (or drought) but also to the monetary value of the structure. In the case of a spillway it includes the unit cost of the spillway and cost of damage to property downstream, cost of lives, cost of lost production, etc. If ΔL is the incremental average loss for a particular design flood Q, then the average annual loss can be expressed as

$$C_1 = \Sigma \Delta L P$$

where P is the probability of the design flood being equalled or exceeded. Similarly, the average annual cost of the spillway can be expressed as

$$C_2 = \Delta c Q$$

where Δc is the incremental cost per unit discharge of providing spillway capacity.

The optimum design is then given by the minimum of

$$C = C_1 + C_2$$

A particular flow and structure will give a particular value of C. By repeated cal culations a function of C versus Q can be developed. Its minimum will indicate the optimum design conditions. A method of this type is described by the Task Committee (1973).

More detailed methods of risk analysis brake the risk down into the risks of component failures, such as hydrologic, hydraulic, structural, foundations. The calculation of the component risk is based on an assumed probability distribution. For a stationary hydraulic system the probability of occurrence of an event, X, greater than the design value, X_0, during the period of n years is $M(X > X_0)$ and that of non occurrence is (1 - M). If the design event has a return period of T years, then the annual probability of exceedence is $P(X \geq X_0) = 1/T$ and of non-exceedence (1 - P Thus, the probability that the design value will be exceeded at least once in n years is

$$M(X > X_0) = 1 - (1 - \frac{1}{T})^n \qquad 12.124$$

i.e., if the risk of failure is due to *independent annual events*. As $n \to \infty$, the probability $M \to 1 - 1/e$ and when T is much larger than n then $M(X > X_0) \simeq n/T$. Equation 12.124 is the same statement as eqn 12.123. For a partial duration series with an average of m observations per year

$$M(X > X_0) = 1 - (1 - \frac{1}{mT})^{nm} \qquad 12.125$$

Prasad (1971) showed that a project, which has been designed for a flood with T-year return period, has, after completion of n' years of the expected life of n years, a risk of failure given by

$$M(X > X_0) = 1 - (1 - \frac{1}{T})^{n-n'} \qquad 12.126$$

Writing $1 - M = [(1 - 1/T)^T]^{n/T} \to (1/e)^{n/T}$ as $n \to \infty$ shows that

$$n = T\,\ln[1/(1 - M)] \tag{12.127}$$

The average probability of occurrence of all events greater than X_0 (not just one) is given by

$$\bar{M} = \frac{1}{n'}\left(\frac{1}{n+1} + \frac{2}{n+1} + \frac{3}{n+1} + \cdots \frac{n'}{n+1}\right) = \frac{n+T}{2T(n+1} \to \frac{1}{2T} \text{ as } n \to \infty \tag{12.128}$$

i.e., the average probability of the occurrence of all events greater than X_0, of T-year return period, approaches the probability of the 2T-year event.

All the above expressions are for events which do not depend on a distribution. If one wants, for example, to calculate from eqn 12.124 the magnitude of the event corresponding to the design return period T, then a probability distribution must be used. Gumbel (1955) showed that this can lead to substantial variation in the value of the calculated standardized variable, $z = (x - \mu)/\sigma$. For example, for T = 1000 years z varied from 3.1 for normal distribution to 7.4 for a log-normal distribution with skewness of 2.94. The EV1 distribution yielded $z \simeq 5$. The discrepancy increases rapidly with increasing T.

For details on the decision theory the reader is referred to books, such as Chernoff and Moses (1959), Raiffa and Schlaifer (1961), Weiss (1961), and Blackwell and Girshick (1954).

12.12 Time Series

A time series is a set of sequential events that may be either continuous or discrete. Discrete series in hydrology mostly arise from the method of sampling or measurement. Since ancient times man has attempted to master the mechanisms of sequential events and to predict events to come. The earliest success was in the field of astronomy. The highly regular movements of heavenly bodies have been predicted for centuries. The understanding of time series, which represent a combination of random and periodic sequences, is quite recent and not complete as yet.

The nature of hydrological data was briefly discussed in Section 11.1 and the expression "stochastic process" has been repeatedly used. It is difficult to find a universal definition of a stochastic process in the literature. The definition used here is that a stochastic process is an ordered or indexed set of random variables. In hydrology the indexing is with reference to the time axis. Thus, particular importance is attached to the order of the random variables, which means that there will generally be a dependence between the random variables a certain time interval part.

The last two decades have seen the growth of and extensive literature on stochastic processes and modelling. Reviews of the literature are given by Kisiel (1969) and by Dawdy and Kalinin (1971). Attention is also drawn to the textbooks, e.g., Jenkins and Watts (1968), Box and Jenkins (1970), Bendat and Piersol (1971).

The object of time series analysis is to evaluate the statistical parameters of the population from the sample series, to predict the form of all the expected time series and to assess the confidence limits of the results obtained. The stationary time series is represented as a sum

$$X_t = x_T + x_p + x_t$$

where x_T, x_p and x_t are the trend, periodic and random components, respectively.

The trend, x_T, if it exists, is assumed to be deterministic and may be represented by a polynomial as

$$x_T(t) = \sum_{i=0}^{m} \alpha_n t^n \qquad n = 0, 1, 2, ...$$

where the coefficients, α_n, as a rule, are unknown. For a linear trend n = 1 and $x_T(t) = \alpha_1 t$. In order to determine the constants, data could be expressed as a Fourier series and differentiated m-times with respect to time. Then the coefficients a_n and b_n become $a_n(n\omega_0)^m$ and $b_n(n\omega_0)^m$, where $\omega_0 = 2\pi f_0$ is the frequency. However, hydrological data is usually not accurate enough for this method. The most widely used method for estimation of the trend is the *moving mean analysis*. Assume, for simplicity, that the data is in the form of a graphical record at intervals Δt. Then, assume that the record is covered by a sheet of paper which has a window $k\Delta t$ time units long, where k is an odd number, for example, $-2\Delta t$, $-1\Delta t$, 0, Δt, $2\Delta t$. The cover sheet is placed so that in the first instance the first k values of the record are visible. From these k values the mean is calculated and recorded at the centre of the window. Next the window is moved by Δt along the t-axis and a new average is calculated and recorded. This process is repeated until the entire record has been covered. The result is a new record $(N - k + 1)\Delta t$ long with substantially reduced oscillations. This averaging procedure may be repeated. The first passage eliminates linear trend, the second a parabolic trend. The method is easy to apply but, like differentiation, it introduces errors when the trend is non-linear. It can be shown that the average value of d^2x_t/dt^2 depends only on the first and last four values of the series and is inversely proportional to $4\Delta t^2(N - 4)$. For large N this is of no importance because the denominator will bring the value close enough to zero. The dependence on the end values also decreases with increasing k. However, since each passage reduces the data by (k - 1) values, the window size has to be limited. As a working rule $k < N/10$ for linear trend (one passage) and less for two or more smoothing operations. The trend can also be determined by the method of least squares. For this, the functional form of trend has to be assumed before computations start. It is a simple method if the trend is linear, but not with higher order trends. Once the form of the trend has been determined it can be separated from the data sequence. The remainder then contains the periodic or oscillatory component and the random component.

The periodic component is also a deterministic term and can be identified by the methods of harmonic analysis. This can at times be done indirectly, as was shown in Section 11.3.4, but, in general, it is helpful to separate the periodic component because its origin is different from that of the random component. The harmonic or Fourier analysis is based on the concept that any continuous oscillatory function can be described by superposition of elementary functions in the form

$$x_p(t) = \sum_{n=1}^{\infty} a_n \sin n\omega_0 t + \sum_{n=0}^{\infty} b_n \cos n\omega_0 t$$

where $\omega_0 = 2\pi/T$ is the basic frequency of the function of period T, i.e., $\omega_0 = 2\pi f_0$, $f_0 = 1/T$

$$a_n = \frac{2}{mT}\int_{-mT/2}^{mT/2} x_p(t) \sin n\omega_0 t \, dt \qquad n = 1, 2, 3, ...$$

$$b_n = \frac{2}{mT}\int_{-mT/2}^{mT/2} x_p(t) \cos n\omega_0 t \, dt$$

and the integral is over m periods. The periodic function can be described as soon as the coefficients a_n and b_n are known. The plot of these coefficients is the *line spectrum*. In practice xp(t) is unknown. When the expressions for a_n and b_n are written with $x = x_p + x_t$ then both can be written as the sum of two integrals.

The second integral is the cross-correlation of x_t with sin $n\omega_o t$ or cos $n\omega_o t$, respectively. If m is large enough then the product of a random function with a periodic one approaches zero.

There are several methods for estimation of the periods of periodic components. One of the simplest is the method based on the auto-correlation function. The auto-correlation function is an important characteristic of the stochastic process, in addition to the statistical parameters mean, variance, etc. The expected value of the cross products

$$E[x(t_i)x(t_i + \tau)] = \frac{1}{N}\sum_{j=1}^{N} x_j(t_i)x_j(t_i + \tau) \qquad 12.129$$

is, in general, for $0 \leq \tau \leq \tau_{max}$ different from zero. The above expected value is known as the *auto-correlation* $R_{xx}(\tau)$ or *auto-covariance* C(k) of the stochastic process, where k refers to the lag of k time units. When the expected value (mean) of x(t) is $\bar{x} = 0$, C(k) is called the auto-correlation function and when the lag is zero it becomes the variance. It is a symmetric function with a maximum at $\tau = 0$ and the auto-correlation of X(t), comprising trend, periodic and random components, is

$$R_{xx}(\tau) = R_{x_T x_T}(\tau) + R_{x_p x_p}(\tau) + R_{x_t x_t}(\tau) \qquad 12.130$$

A feature of the random function is that for τ larger than a certain value the last term in eqn 12.130 goes to zero. The auto-correlation of a periodic function is also periodic with the same period but the phase information is lost. Thus, once the trend has been eliminated the auto-correlation for $\tau > \tau_{max}$ is a periodic function and the period T of x_p can be read off from the auto-correlation diagram. This method runs into difficulties when the periodic component is composed of components from different unrelated causes and where $T_1 = kT_2$ or $T_2 = kT_1$ is not given by an integer value of k. The correlation for $\tau > \tau_{max}$ is then a superposition of two (or more) periodic functions, which leads to a modulated signal of varying amplitude and nodal values. Further complications arise when, in addition to frequencies, also the amplitudes of the individual periodic functions vary. Hydrologic data can have several periodic components, but usually some of these are known, e.g., the seasonal variation. When several periodic components are present it is simpler to estimate the periods from the spectrum.

The *spectrum* $S_{xx}(\omega)$ is the distribution function of the variance of the stochastic process, X(t), with frequency ω. The concept of spectrum assumes that X(t) is the result of superposition of an infinite number of oscillations. These give in any given frequency band a contribution $S_{xx}(\omega)d\omega$ to the variance. Thus, the area under the spectrum function is the variance of the process, i.e.,

$$\sigma_x^2 = \int_0^\infty S_{xx}(\omega)\,d\omega \qquad 12.131$$

The spectrum function is a Fourier transform of the auto-correlation function. The transforms are extensively discussed in the literature and they form part of standard computer programs. The spikes on the spectrum diagrams show dominant frequencies (periodic functions) which can then be removed one by one.

The *cross-correlation* $R_{xy}(\tau)$ of two variables X(t) and Y(t), measured at two locations, is a space-time-correlation. It can be subdivided into a symmetrical, $R_s(\tau)$, and unsymmetrical, $R_u(\tau)$, part and it goes to zero when $\tau \to \infty$, but does not usually have a maximum when $\tau = 0$. The subdivision has to satisfy

$$R_{xy}(\tau) - R_{xy}(-\tau) = 2R_u(\tau) \qquad 12.132$$

and

$$R_{xy}(\tau) + R_{xy}(-\tau) = 2R_s(\tau) \qquad 12.133$$

The cross-correlation also depends on whether X(t) or Y(t) is displaced by τ, i.e.,

$$R_{xy}(\tau) = R_{yx}(-\tau)$$

As from auto-correlation one can form a spectrum from cross-correlation, a *cross-spectrum*

$$S_{xy}(\omega) = \frac{2}{\pi}\int_0^\infty R_s(\tau)\cos\omega\tau d\tau - i\,\frac{2}{\pi}\int_0^\infty R_u(\tau)\sin\omega\tau d\tau \qquad 12.134$$

where the first integral is the covariance spectrum. The cross-spectrum gives the relationship between the equal frequency components of two stochastic processes and its main use is as a measure of *coherence*

$$Coh_{xy}(\omega) = \frac{|S_{xy}(\omega)|^2}{S_{xx}(\omega)S_{yy}(\omega)} \qquad 12.135$$

It is the percentage of the variance due to the process X(t) in the variance of the process Y(t), at any given frequency, and is not affected by a shift in time between X(t) and Y(t).

It should also be noted that the time interval Δt of data points determines the maximum frequency described by the data as

$$\Delta t = \frac{\pi}{\omega_{max}} = \frac{1}{2f_{max}}$$

since $\omega = 2\pi f$. This frequency, f_{max}, is the Nyquist or folding frequency and gives the length of the record $T = N\Delta t$. If the analysis is carried past this frequency it leads to the *aliasing error*. For example, data points at $\Delta t = 0.05$ s give a $f_{max} = 10$ Hz.

In time series modelling the Markov process and Markov chain are extensively used. The *Markov process* (for continuous data) and the *Markov chain* (for discrete data) are named after the Russian mathematician, A.A. Markov (1856-1922). In modern literature the Markov chain is also known as the AR (auto-regressive) model. The concept of the Markov chain can be illustrated with the aid of a very simple example involving two experiments:

Experiment 1: A coin is tossed, for example, six times and the sequence of heads, H, and tails, T, is recorded.

Experiment 2: A marble is placed into the middle urn C of a row of five urns marked A, B, C, D, E. The marble is moved to the next urn on the right when the coin shows H and to the left when the coin shows T. The experiment can be further specified, for example, by the rule that once the marble arrives into an end urn it must stay there, irrespective of what the coin shows. This is known as a process with *absorbing boundaries*. Alternatively, the rule could be that the marble must stay there when the coin shows H, but must move back into urn B or D, as the case may be, when the coin shows T. Such a process has *reflecting boundaries*. It is also possible for only one boundary to be absorbing or reflecting.

If experiment 1 yields H T H H T H, then the result of experiment 2, assuming absorbing boundaries, is

```
  H T H H T H
C D C D E E E
```

Another sequence may be

H T T H T H
C D C B C B C

In the first sequence the marble gets trapped in E. In the second sequence it does not. Usually all that can be said about which way the marble will move is that it will move to the right with 50% probability (in this example).

The state space for experiment 1 is [H, T]. The set of urns [A, B, C, D, E] forms the state space for experiment 2. The conditional probability of the process, if in state s_i, of moving to state s_j is $P(s_j|s_i)$, which is the *one-step* probability. If such probabilities are known for all pairs of states, they may be arranged in a square matrix, called transition (probability) matrix, $P = [p_{ij}]$, where p_{ij} is the element in the i-th row and the j-th column, e.g., for experiment 1

$$P = \begin{array}{c} \\ H \\ T \end{array} \begin{array}{c} \begin{array}{cc} H & T \end{array} \\ \left[\begin{array}{cc} \frac{1}{2} & \frac{1}{2} \\ \frac{1}{2} & \frac{1}{2} \end{array}\right] \end{array}$$

and for experiment 2

$$P = \begin{array}{c} \\ A \\ B \\ C \\ D \\ E \end{array} \begin{array}{c} \begin{array}{ccccc} A & B & C & D & E \end{array} \\ \left[\begin{array}{ccccc} 1 & 0 & 0 & 0 & 0 \\ \frac{1}{2} & 0 & \frac{1}{2} & 0 & 0 \\ 0 & \frac{1}{2} & 0 & \frac{1}{2} & 0 \\ 0 & 0 & \frac{1}{2} & 0 & \frac{1}{2} \\ 0 & 0 & 0 & 0 & 1 \end{array}\right] \end{array}$$

The matrices have non-negative elements and the sum of the elements in each row is equal to one. The probability of the marble moving to urn A (or E) initially is unity. The probability of the marble moving to B or D from state C is ½, etc.

The first state in which the process begins may be known, as in experiment 2, or it may be decided by a probability rule. In every case the initial probability (vector) must be known, e.g.,
Experiment 1. $\underset{\sim}{a} = (\frac{1}{2}, \frac{1}{2})$
Experiment 2. $\underset{\sim}{a} = (0, 0, 1, 0, 0)$

The probability of the process passing from state s_i to state s_j in n steps is called the n-step transition probability, $p_{ij}^{(n)}$, and is a square matrix. If P is the one-step transition matrix of a stationary finite Markov chain, then P^n is the n-step transition matrix. This statement can be proved in a step by step analysis. Clearly $p_{ij}^{(1)} = p_{ij}$. The two-step probability could be visualized as illustrated in Fig. 12.9. There are m mutually exclusive possible paths. The probabilities for each path are $p_{i1} \cdot p_{1j}$, $p_{i2} \cdot p_{2j}$, etc. Thus, the probability of the process

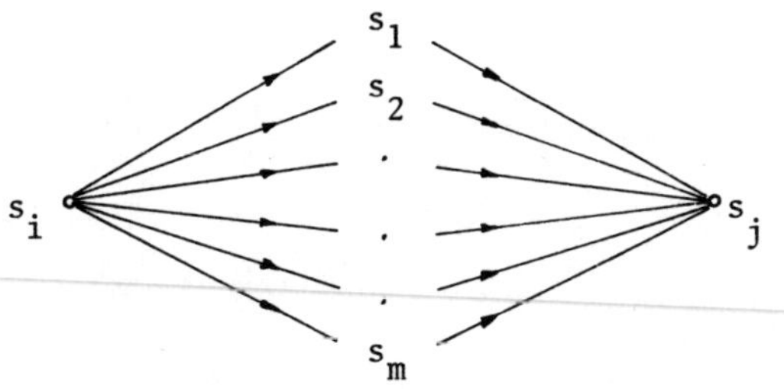

Fig. 12.9. Illustration of two-step probability.

passing in two steps from state s_i to state s_j is the sum of the probabilities

$$p_{ij}^{(2)} = \sum_{r=1}^{m} p_{ir} \cdot p_{rj}$$

By defintion of matrix multiplication the sum is the ij-th element of P^2, where P^2 is the matrix $p_{ij}^{(2)}$. The argument can be extended in the same manner. The calculation of the n-step probability matrix involves multiplication of matrices, e.g., if

$$E = \begin{bmatrix} A & B \\ C & D \end{bmatrix} \quad \text{and} \quad F = \begin{bmatrix} a & b \\ c & d \end{bmatrix}$$

then

$$EF = \begin{bmatrix} A & B \\ C & D \end{bmatrix}\begin{bmatrix} a & b \\ c & d \end{bmatrix} = \begin{bmatrix} Aa + Bc & Ab + Bd \\ Ca + Dc & Cb + Dd \end{bmatrix}$$

where the ij-th element is the scalar product of the i-th row vector of E and the j-th column vector of F. The product is defined only if the number of columns in E is the same as the number of rows in F. In the example the two-step transition probability matrix from eqn 12.125 is

$$P^2 = \begin{bmatrix} 1 & 0 & 0 & 0 & 0 \\ \frac{1}{2} & \frac{1}{4} & 0 & \frac{1}{4} & 0 \\ \frac{1}{4} & 0 & \frac{1}{2} & 0 & \frac{1}{4} \\ 0 & \frac{1}{4} & 0 & \frac{1}{4} & \frac{1}{2} \\ 0 & 0 & 0 & 0 & 1 \end{bmatrix}$$

Assuming the process begins in state C, the required probabilities are found in row 3 as $P(C \to A) = p_{CB} \cdot p_{BA} = \frac{1}{4}$, $P(C \to B) = p_{CB} \cdot p_{BB} = 0$, $P(C \to C) = p_{CB} \cdot p_{BC} + p_{CD} \cdot p_{DC} = \frac{1}{4} + \frac{1}{4} = \frac{1}{2}$, $P(C \to D) = p_{CD} \cdot p_{DD} = 0$, and $P(C \to E) = p_{CD} \cdot p_{DE} = \frac{1}{4}$

This type of Markov chain has a one-step memory and hence is called a first-order Markov chain. It is possible to extend the argument to two or more steps of memory. For a second-order Markov chain the probability transition matrix is three-dimensional. The practical difficulties (i.e., lack of data) with establishment of such transition matrices severely hinder the use of the higher order chains.

There are many physical processes which display a persistence or memory effect and can be modelled with the aid of the Markov chain technique. Some examples of use were introduced earlier in Chapters 10 and 11. The literature contains numerous examples of applications of the Markov chain problems to hydrology, as well as modelling of rainfalls, reservoir storage, streamflows, etc. Quimpo (1968), for example, used this technique to model daily river flows.

REFERENCES

Ackermann, W.C. (1964) Application of severe rainstorm data in engineering design, *Bull. Am. Meteorol. Soc.* 45, 204-206

Aitken, A.P. (1968) The application of storage routing methods to urban hydrology, *J. Inst. Eng. Australia* 40(1-2), 5-11.

Aitken, A.P., Ribeny, F.M.J. and Brown, J.A.H. (1972) The estimation of mean annual rainfall and runoff over the territory of Papua, New Guinea, *Civil Trans. Inst. Eng. Australia* 14, 49-56.

Aitken, A.P. (1973) Hydraulic design in urban areas - a review, Australian Water Resources Council, Techn. Paper, No.5.

Alekhin, Y.M. (1964) Short range forecasting of lowland river runoff, Transl. from Russian by Israel Program for Scientific, Transl. U.S. Dept. of Commerce and the Nat. Sci. Foundation, Washington, D.C., pp. 229.

Alexander, G.N. (1963) Using the probability of storm transposition for estimating the frequency of rare floods, *J. Hydrology* 1, 46-57.

Amorocho, J. and Orlob, G.T. (1961) Non-linear analysis of hydrologic systems, Water Resources Center, Contribution 40, Univ. of Calif. Davis.

Anis, A.A. and Lloyd, E.H. (1953) On the range of partial sums of a finite number of independent normal variables, *Biometrika* 40, 35-42.

Ardis, C.V., Ducker, K.J. and Lenz, A.T. (1969) Storm drainage practices of thirty-two cities, *Proc. ASCE*, 95, HY1, 383-408.

Askew, A.J. (1970) Derivation of formulae for variable lag time, *J. Hydrology* 10, 225-242.

Aslyng, H.C. and Jensen, S.V.E. (1965) Radiation and energy balance at Copenhagen 1955-1964, Hydrotechn. Lab. Roy. Veterinary and Agric. College, Copenhagen.

Bagnold, R.A. (1941) *The Physics of Blown Sand and Desert Dunes*, Methuen.

Baumgarten, A. (1956) Untersuchungen über den Wärme- und Wasserhaushalt eines jungen Waldes, *Bericht Deutschen Wetterdienstes*, No.28, Bad Kissingen.

Bayazit, M. (1966) Instantaneous unit hydrograph by spectral analysis and its numerical application, Proc. CENTO Symp. Hydrol. Water Resources Develop. Ankara, Turkey.

Bear, J. (1972) *Dynamics of Fluids in Porous Media*, American Elsevier, New York.

Beard, L.R. (1963) Flood control operation of reservoirs, *Proc. ASCE*, 89, HY1, 1-23.

Bell, F.C. (1969) Generalized rainfall-duration-frequency relationships, *Proc. ASCE*, 95, HY1, 311-327.

Bendat, J.S. and Piersol, A.G. (1971) *Random Data: Analysis and Measurement Procedures*, Wiley-Intersience.

Benson, M.A. (1950) Use of historical data in flood-frequency analysis, *Am. Geophys. Union Trans.* 31, 419-424.

Benson, M.A. (1960) Flood-frequency analyses, Manual of Hydrology, Part 3, Flood-Flow techniques, U.S. Geol. Survey Water-supply Paper, 1543-A, 51-74.

Benson, M.A. (1964) Discussion: Hydrology of spillway design: Large structures-adequate data, *Proc. ASCE*, 90, HY5, 297-300.

Bergström, S. (1976) Development and application of a conceptual runoff model for Scandinavian catchments, Dept. Water Resources Eng. Univ. of Lund, Bull. series A, No.52.

Bernier, J. (1971) Modèles probabilistes à variables hydrologiques multiples et hydrologie synthetique, IAHS-UNESCO-WMO, Proc. Warsaw Symp. 1971 on Mathematical Models in Hydrology, 1, 33-342.

Berry, F.A. jr., Bollay, E. and Beers, N.R. (Editors) (1945) *Handbook of Meteorology*, McGraw-Hill.

Betson, R.P. and Marius, J.B. (1969) Source area of storm runoff, *Water Resources Res.* 5, 574-582.

Bidwell, V.J. (1970) Multivariate analysis of non-linear hydrologic systems, Ph.D thesis, Univ. of Auckland, N.Z.

Bidwell, V.J. (1971) Regression analysis of non-linear catchment systems, *Water Resources Res.* 7, 1118-1126.

Biot, M.A. (1941) General theory of three-dimensional consolidation, *J. Appl. Phys.* 12, 155-164.

Biswas, A.K. (Editor) (1972) *Modelling of Water Resources Systems* (2 Vol.), Harvest House.

Biswas, Asist K. (1970) *History of Hydrology*, North-Holland Publishing Co.

Biswas, A.K. (Editor) (1976) *Systems Approach to Water Management*, McGraw-Hill.

Blackwell, D. and Girshick, M.A. (1954) *Theory of Games and Statistical Decisions*, Wiley.

Blaney, H.F. and Criddle, W.D. (1950) Determining water requirements in irrigated areas from climatological and irrigation data, U.S. Dept. Agric. SCS TP 96, Washington, D.C.

Borland, W.M. and Miller, C.R. (1958) Distribution of sediment in large reservoirs, *Proc. ASCE*, 84, HY2.

Borland, W.M. (1971) Reservoir sedimentation Ch. 29 in *River Mechanics*, edited by H.W. Shen, Colorado.

Box, G.E.P. and Müller, M.E. (1958) A note on the generation of random normal deviates, *Ann. Math. Statist.* 29, 610-611.

Box, J.E. and Taylor, S.A. (1962) Influence of soil bulk density on matrix potential, *Soil Sci. Soc. Am. Proc.* 26, 119-122.

Box, G.E.P. and Jenkins, G.M. (1970) *Time Series Analysis Forecasting and Control*, Holden-Day.

Boyer, M.C. (1957) A correlation of the characteristics of great storms, *Trans. Am.*

Geophys. Union, 38, 233-238.

Brandstetter, A. and Amorocho, J. (1970) Generalized analysis of small watershed responses, Univ. of Calif., Davis, Water Sci. Eng. Papers 1035.

Briggs, G.E. (1967) *Movement of Water in Plants*, Blackwell Scientific Publications, Oxford.

Brooks, R.H. and Gorey, A.T. (1966) Properties of porous media affecting fluid flow, *Proc. ASCE,* 92, IR2, 61-88.

Brune, G.M. (1953) Trap efficiency of reservoirs, *Trans. Am. Geophys. Union,* 34, 407-418.

Brunt, D. (1939) *Physical and Dynamical Meteorology*, Cambridge Univ. Press.

Buckingham, E. (1907) Studies in the movement of soil moisture, *U.S. Dept. Agric. Bur. Soils Bull.* 38, 29-61.

Budyko, M.I. (1956) Teplovoi balans zemnoi poverkhnosti, Gidrometerologicheskoe izdatel' stvo, Leningrad, Transl. U.S. Dept. of Commerce, Weather Bureau "*The Heat Balance of the Earth's Surface*", P1B 131692, Office of Techn. Services Dept. of Commerce, Washington, D.C. 1958.

Budyko, M.I., Efimova, N.A., Zubenok, L.I. and Strokina, L.A. (1962) The heat balance of the Earth's surface, *Akad. Nauk. USSR, Izv. Ser. Geogr.* 1, 6-16, Transl. No.153M, U.S. Dept. of Commerce, Office of Techn. Services, Document No.63-19851.

Budyko, M.I. and Gerasimov, I.P. (1961) The heat and water balance of the Earth's surface, the general theory of physical geography and the problem of the transformation of nature, *Soviet Geogr.* 2(2), 3-12.

Carlson, R.F., McCormick, A.J.A. and Watts, D.G. (1970) Application of linear random models to four annual streamflow series, *Water Resources Res.* 6, 1070-1078.

Carslaw, H.S. and Jaeger, J.C. (1959) *Conduction of Heat in Solids,* Clarendon Press.

Carrigan, P.H. Jr. (1971) A flood-frequency relation based on regional record maxima, U.S. Geol. Survey Prof. Paper 434-F.

Carson, M.A. (1971) *The Mechanics of Erosion*, Pion Ltd.

Carter, R.W. (1961) Magnitude and frequency of floods in suburban areas, U.S. Geol. Survey Prof. Paper 424-B.

Chamberlain, A.C. (1966) Transport of gases to and from grass and grass-like surfaces, *Proc. Roy. Soc. London (A)* 290, 236-265.

Chen, C.W. and Shubinski, R.P. (1971) Computer simulation of urban stormwater runoff, *Proc. ASCE,* 97, HY2, 289-301.

Chepil, W.S. and Woodruff, N.P. (1963) The physics of wind erosion and its control, *Adv. Agronomy* 15, 211-302.

Chernoff, H. and Moses, L.E. (1959) *Elementary Decision Theory*, Wiley.

Chery, D.L. Jr. (1967) A review of rainfall-runoff, physical models as developed by diemsional analysis and other methods, *Water Resources Res.* 3, 881-889.

Childs, E.C. and Collis-George, N. (1950) Permeability of porous materials, *Proc. Roy. Soc. London (A)* 201, 392-405.

Childs, E.C. (1969) *The Physical Basis of Soil Water Phenomena*, Wiley-Interscience.

Chow, Ven Te (1951) A general formula for hydrologic frequency analysis, *Trans. Am. Geophys. Union* 32, 231-237.

Chow, Van Te (1962) Hydrologic determination of waterway areas for the design of drainage structures in small drainage basins, Univ. Illinois Eng. Exp.

Station, Bull. 462, also in Chow (1964), Ch. 25, 22-25.

Chow, Ven Te (1964) *Handbook of Applied Hydrology* (section 25, part II), McGraw-Hill.

Churchill, M.A. (1947) Discussion of "Analysis and use of reservoir sedimentation data" by L.C. Gottschalk, Proc. Federal Interagency Sedimentation Conf. Denver, Colo. 139-140.

Clark, C.O. (1945) Storage and the unit hydrograph, *Trans. ASCE*, 110, 1419-1488.

Clarke, R.T. (1973) Mathematical models in hydrology, Irrigation and Drainage Paper No.19, Food and Agric. Org. of UN, Rome.

Collins, W.T. (1939) Runoff distribution graphs from precipitation occurring in more than one time unit, *Civil Eng.* (ASCE) 9, 559.

Colman, E.A. (1953) *Vegetation and Watershed Management*, Roland Press.

Colston, N.V. and Wiggert, J.M. (1970) A technique of generating a synthetic flow record to estimate the variability of dependable flows for a fixed flow capacity, *Water Resources Res.* 6, 310-315.

Conover, W.J. and Benson, M.A. (1963) Long-term flood frequencies based on extremes of short-term records, U.S. Geol. Survey Prof. Paper 450-E, E159-160.

Costin, A.B. et al. (1961) Studies in catchment hydrology in the Australian Alps III, Preliminary snow investigations, Div. of Plant Industry Techn. Paper No.15 CSIRO, Melbourne, Austr. pp. 31.

Court, A. (1961) Area-depth rainfall formulas, *J. Geophys. Res.* 66, 1823-1831.

Covey, W., Halstead, M.H., Hillman, S., Merryman, J.D., Richman, R.L. and York, A.H. (1958) Micrometeorological data collected by Texas A & M, *Project Prairie Grass, A Field Program in Diffusion* 2, 53-96, Geophys. Res. Papers No.59, Air Force Cambridge Res. Center, Bedford.

Crank, J. (1956) *Mathematics of Diffusion*, Oxford Univ. Press.

Crawford, N.H. and Linsley, R.K. (1966) Digital simulation in hydrology: Stanford Watershed Model IV, Techn. Rep. No.39, Stanford Univ. Calif.

Creager, W.P. and Justin, J.D. (1950) *Hydroelectric Handbook*, Wiley.

Cunge, J.A. (1969) On the subject of a flood propagation method, *J. Hydr. Res.* 7, 205-230.

Dalrymple, T. (1960) Flood-frequency analyses, U.S. Geol. Survey Water-supply Paper, 1543-A.

Davis, S.N. and De Wiest, R.J.M. (1966) *Hydrogeology*, Wiley.

Dawdy, D.R. and Kalinin, G.I. (1971) Mathematical modelling in hydrology, *Bull. Int. Assoc. Sci. Hydrol.* 16(4), 25-35.

Deacon, E.L. (1953) Vertical profiles of mean wind in the surface layers of the atmosphere, Geophys. Mem. No.91, Meteorol. Office, Air Ministry, London.

De Lisle, J.F. (1966) Mean daily insolation in New Zealand, *N.Z. J. of Sci.* 9, 992-1005.

de Quervain, M. (1951) Zur Verdunstung der Schneedecke, *Archiv für Meteorologie, Geophys. und Bioklimatologie Series B*, 3, 47-64.

De Wiest, R.J.M. (1965) *Geohydrology*, Wiley.

Dickinson, W.T. and Whiteley, H. (1970) Watershed areas contributing to runoff, Symp. on the Results of Res. on Representative and Exp. Basins, IASH Publ. No.96, 112-26.

Di Silvio, G. (1969) Flood wave modification along channels, *Proc. ASCE*, 95, HY9, 1589-1614.

Diskin, M.H. (1967) A Laplace transform proof of the theorem of moments for the instantaneous unit hydrograph, *Water Resources Res.* 3, 385-388.

Diskin, M.H. (1967a) A dispersion analog model for watershed systems, Int. Hydrol. Symp. Fort Collins.

Diskin, M.H. (1969) Thiessen coefficients by a Monte Carlo procedure, *J. Hydrol.* 8, 323-335.

Diskin, M.H. (1970) On the computer evaluation of Thiessen weights, *J. Hydrol.* 11, 69-78.

Dixon, R.M. (1966) Water infiltration responses to soil management practices, Ph. D thesis Univ. Wisconsin, pp. 175, Univ. Microfilms, Ann Arbor, Michg. (Diss. Abstr. 27;4).

Dixon, R.M. and Peterson, A.E. (1971) Water infiltration control: a channel system concept, *Soil Sci. Am. Proc.* 35, 968-973.

Dooge, J.C.I. (1959) A general theory of the unit hydrograph, *J. Geophys. Res.* 64, 241-256.

Dooge, J.C.I. (1960) The routing of groundwater recharge through typical elements of linear storage, Int. Assoc. Sci. Hydrol., General Assembly of Helsinki Publ. No.52, 286-300.

Dooge, J.C.I. (1965) Analysis of linear systems by means of Laguerre functions, *J. Soc. for Industr. and Appl. Math. (control) Series A*, 2(3), 396-408.

Dooge, J.C.I. (1968) The hydrologic cycle as a closed system, *Bull. Int. Assoc. Sci. Hydrol.* 13(1), 58-68.

Dooge, J.C.I. (1972) Mathematical models and hydrologic systems, In "*Modelling of Water Resources Systems*" Edited by A.K. Biswas, Harvest House, 1, 170-188.

Dracup, J.A., Fogarty, T.J. and Grant, S.G. (1973) Synthesis and evaluation of urban-regional rainfall-runoff criteria, Environmental Dynamics Inc. Los Angeles, Rep. for Office of Water Resources Res. U.S.A.

Duquennois, H. (1956) New methods of sediment control in reservoirs, *Water Power* 8, 174-180.

Eagleson, P.S. (1962) Unit hydrograph characteristics for sewered areas, *Proc. ASCE*, 88, HY2, 1-25.

Eagleson, P.S., Meija, R. and March, F. (1966) Computation of optimum realizable unit hydrographs, *Water Resources Res.* 2, 755-764.

Eagleson, P.S. (1967) Optimum density of rainfall-network, *Water Resources Res.* 3, 1021-1033.

Eagleson, P.S. (1969) Potential of physical models for achieving better understanding and evaluation of watershed changes, Effects of watershed changes on streamflow, Water Resources Symp. No.2, Univ. Texas, Austin, 12-25.

Eagleson, P.S. (1970) *Dynamic Hydrology*, McGraw-Hill.

Edlefsen, N.E. and Anderson, A.B.C. (1943) Thermodynamics of soil moisture, *Hilgardia* 15, 31-298.

Edson, C.G. (1951) Parameters for relating unit hydrographs to watershed characteristics, *Trans. Am. Geophys. Union* 32, 591-596.

Ekman, V.W. (1905) On the influence of the Earth's rotation on ocean currents, *Arkiv. Mat. Astron. Fysik, Stockholm*, 2(1-2).

Eliasson, J., St.Arnalds, S., Johannson, S. and Kjaran, S.P. (1973) Reservoir mechanism in an aquifer of arbitrary boundary shape, *Nordic Hydrol.* 4, 129-146.

Espey, W.H. Jr., Winslow, D.E. and Morgan, C.W. (1969) The effects of urbanization on peak discharge, Water Resources Symp. No.2, Univ. Texas, Austin.

Espey, W.H. Jr. and Winslow, D.E. (1974) Urban flood frequency characteristics, *Proc. ASCE*, 100, HY2, 179-293.

Fairbridge, R.W. (1961) Eustatic changes in sea level, *Physics and chemistry of the earth*, 4, 99-185.

Feller, W. (1951) The asymptotic distributions of the range of series of independent random variables, *Ann. Math. Stat.* 22, 427-432.

Fisher, R.A. and Tippett, L.H.C. (1928) Limiting forms of the frequency distribution of the largest or smallest member of a sample, *Proc. Cambridge Phil. Soc.* 24, 180-190.

Fletcher, N.H. (1962) *The Physics of Rainclouds*, Cambridge Univ. Press.

Floods and their computation (1967) Proc. Leningrad Symp. IASH-UNESCO-WMO, 2.

Flood Studies Report (1975) Hydrological studies 1, Meteorological studies 2, Flood routing studies 3, Hydrological data 4, Maps 5, Natural Environment Res. Council, London.

Folse, J.A. (1929) A new method of estimating stream flow based upon a new evaporation formula, Carnegie Inst. Washington, D.C. Publ. 400.

Frankenberger, E. (1960) Beiträge zum internationalen geophysikalischen Jahr, 1957-58 *Bericht Deuts. Wetterdienstes* 73(10), 84.

Franklin Institute Research Laboratories (1972) Investigation of porous pavements for urban runoff control, U.S. Environmental Prot. Agency, U.S. Gov. Print. Office, Washington, D.C.

Fried, J.J. (1975) *Groundwater Pollution Theory, Methodology, Modelling and Practical Rules*, Elsevier Scientific.

Frühling, A. (1894) Über Regen- und Abflussmengen für städtische Entwässerungskanäle, *Der Civilingenieur (Leipzig), Ser. 2*, 40, 541-558, 623-643.

Gardner, W. and Widstoe, J.A. (1921) The movement of soil moisture, *Soil Sci.* 11, 215-232.

Gardner, W.R. (1959) Solutions of the flow equations for drying of soils and other media, *Soil Sci. Soc. Am. Proc.* 23(3), 183-185.

Garstka, W.U., Love, L.D., Goodell, B.C. and Bertle, F.A. (1959) Factors affecting snowmelt and streamflow, U.S. Bureau of Recl. and U.S. Forest Service, 187 pp. Washington, D.C.

Gash, J.H. and Stewart, J.B. (1975) The average surface resistance of pine forest derived from Bowen Ratio measurements, *Boundary-Layer Meteorol.* 8, 453-464.

Goldberg, L. (1954) The absorption spectrum of the atmosphere, Ch. 9 in Vol II of *The Solar System, The Earth as a Planet*, Editor G.P. Kuiper, The Univ. of Chicago Press.

Gomide, F.L.S. (1975) Range and deficit analysis using Markov chains, Hydrology Paper No.79, Colorado State Univ. Fort Collins.

Gould, B.W. (1960) Water supply headworks storage estimation, *Austr. Civil Eng. and Constr.* 1(9), 46-52.

Gould, B.W. (1961) Statistical methods for estimating the design capacity of dams, *J. Inst. Eng. Australia* 33, 405-416.

Grace, R.A. and Eagleson, P.S. (1966) The modelling of overland flow, *Water Resources Res.* 2, 393-403.

Gray, D.M. (1961) Interrelationships of watershed characteristics, *J. Geophys. Res.* 66, 1215-1223.

Gray, D.M. (1961) Synthetic unit hydrographs for small drainage areas, *Proc. ASCE,* 87, HY4, 33-54.

Gray, D.M. (1970) *Handbook on the Principles of Hydrology*, Water Infor. Center Publ. Port Washington, N.Y.

Gringorten, I.I. (1963) A plotting rule for extreme probability paper. *J. Geophys. Res.* 68, 813-814.

Gumbel, E.J. (1954) Statistical theory of droughts, *Proc. ASCE,* 80; sep. No.439.

Gumbel, E.J. (1955) The calculated risk in flood control, *Appl. Sci. Res. Sec. A,* 5, 273-280.

Hack, J.T. (1957) Studies of longitudinal stream profiles in Virginia and Maryland, U.S. Geol. Survey Prof. Paper 294-B

Hadley, G. (1735) Concerning the cause of the general trade-winds, *Phil. Trans. Roy. Soc.* 39, 58-62.

Hahn, G.J. and Shapiro, S.S. (1967) *Statistical Models in Engineering,* Wiley.

Haimes, Y.Y., Hall, W.A. and Freedman, H.T. (1975) *Multiobjective Optimization in Water Resources Systems*, Elsevier Scientific.

Haimes, Y.Y. (1977) *Hierarchical Analysis of Water Resources Systems*, McGraw-Hill.

Hall, W.A. and Dracup, J.A. (1970) *Water Resources Systems Engineering*, McGraw-Hill.

Hanks, R.J. and Bowers, S.A. (1962) Numerical solutions of the moisture flow equations into layered soils, *Soil Sci. Soc. Am. Proc.* 26, 530-534.

Harbeck, G.E. (1962) A practical field technique for measuring reservoir evaporation utilizing mass transfer theory, U.S. Geol. Survey Prof. Paper 272-E.

Harter, H.L. (1969) A new table of percentage points of the Pearson Type III distribution, *Technometrics* 11(1), 177-187.

Harter, H.L. (1971) More percentage points of the Pearson Type III distribution, *Technometrics* 13(1) 203-204.

Haurwitz, B. (1941) *Dynamic Meteorology*, McGraw-Hill.

Hayami, S. (1951) On the propagation of flood waves, Bull. No.1 Disaster Prevention Res. Inst. Kyoto Univ. Japan.

Hayashi, T. (1965) Propagation and deformation of flood waves in natural channels, Anniversary Bull. Chuo Univ. Japan, 67-80.

Helvey, J.D. and Patric, J.H. (1965) Design criteria for interception studies, Symp. on Design of Hydrometeorol. Networks, WMO/IASH, Laval Univ. Quebec City.

Henderson, F.M. and Wooding, R.A. (1964) Overland flow and groundwater flow from a steady rainfall of finite duration, *J. Geophys. Res.* 69, 1531-1540.

Henderson, F.M. (1966) *Open Channel Flow*, Macmillan.

Herbst, P.H., Brendenkamp, D.B. and Barker, H.M. (1966) A technique for the evaluation of drought from rainfall data, *J. Hydrol.* 4, 264-272.

Hershfield, D.M., Weiss, L.L. and Wilson, W.T. (1955) Synthesis of rainfall-intensity-frequency regime, *Proc. ASCE,* 81, Paper 744.

Hershfield, D.M. and Wilson, W.T. (1957) Generalizing of rainfall-intensity-frequency data, Proc. Int. Assoc. of Sci. Hydrol. General Assembly of Toronto, 1, 499-506.

Hershfield, D.M. and Wilson, W.T. (1960) A comparison of extreme rainfall depths from tropical and non-tropical storms, *J. Geophys. Res.* 65, 969-982.

Hershfield, D.M. (1961) Rainfall frequency atlas of the United States, Weather Bureau Techn. Paper No.40.

Hershfield. D.M. (1962) Extreme rainfall relationships, *Proc. ASCE,* 88, HY6, 73-92.

Hibbert, A.R. (1967) Forest treatment effects on water yield, Proc. Int. Symp. Forest Hydrol. 527-543, Pergamon Press.

Hibbert, A.R. (1971) Increases in streamflow after converting chaparral to grass, *Water Resources Res.* 7, 71-80.

Holtan, H.N. (1961) A concept for infiltration estimates in watershed engineering, U.S. Dept. Agric. Agric. Res. Station, 41-51.

Holtan, H.N. (1971) A formulation for quantifying the influence of soil porosity and vegetation on infiltration, Third Int. Sem. for Hydrol. Professors, Purdue Univ. Lafayette, Indiana.

Horton, R.E. (1919) Rainfall interception, *U.S. Monthly Weather Rev.* 47.

Horton, R.E. (1924) Discussion of "The distribution of intense rainfall and some other factors in the design of stormwater drains", *Proc. ASCE,* 50, 660-667

Horton, R.E. (1931) The field, scope, and status of the science of hydrology, *Trans. Am. Geophys. Union* 12, 189-202.

Horton, R.E. (1940) An approach to the physical interpretation of infiltration capacity, *Soil Sci. Soc. Am. Proc.* 5, 399-417.

Horton, R.E. (1945) Erosional development of streams and their drainage basins; hydro logical approach to quantitative morphology, *Bull. Geol. Soc. Am.* 56, 275-370.

Howe, R.H.L. (1960) The application of aerial photographic interpretation to the investigation of hydrologic problems, *Photogrammetric Eng.* 26, 85-95.

Hufschmidt, M.M. and Fiering, M.B. (1966) *Simulation Techniques for Design of Water-Resources Systems*, Harvard Univ. Press (Macmillan, 1967).

Hurst, H.E. (1951) Long-term storage capacity of reservoirs, *Trans. ASCE,* 116, 776.

Hurst, H.E. (1956a) Methods of using long-term storage in reservoirs, *Proc. I.C.E.* 5, Part 1, 519.

Hurst, H.E. (1956b) The problem of long-term storage in reservoirs, *Int. Union Geophys. and Geodesy Inf. Bull.* London No.15, 463.

Hurst, H.E., Black, R.P. and Simaika, Y.M. (1965) *Long-term Storage, An Experimental Study*, Constable.

Irmay, S. (1954) On the hydraulic conductivity of unsaturated soils, *Trans. Am. Geophys. Union* 35, 463-467.

Iwagaki, Y. (1955) Fundamental studies on the runoff analysis by characteristics, Disaster Prevention Res. Inst. Bull. 10, Kyoto Univ. Japan.

Jackson, R.D., Reginato, R.J. and van Bavel, S.H.M. (1965) Comparison of measured and hydraulic conductivities of unsaturated soils, *Water Resources Res.* 1, 375-380.

Jackson, T.J. and Ragan, R.M. (1974) Hydrology of porous pavement parking lots, *Proc. ASCE,* 100, HY12, 1739-1752.

Jacob, C.E. (1950) Flow of groundwater in *Engineering Hydraulics,* edited by H. Rouse, Wiley.

Jenkins, G.M. and Watts, D.G. (1968) *Spectral Analysis and its Applications,* Holden-Day.

Jenkinson, A.F. (1955) The frequency distribution of the annual maximum (or minimum) values of meteorological elements, *Q.J. Roy. Meteorol. Soc.* 87, 158-171.

Jenkinson, A.F. (1969) Estimation of maximum floods (Ch. 5), World Meteorol. Org. Techn. Note No.98, WMO No.233, TP 126, Geneva.

Jennings, A.H. (1950) World's greatest observed point rainfalls, *Monthly Weather Rev.* 78, 4-5.

Joseph, E.S. (1970) Probability distribution of annual droughts, *Proc. ASCE,* 96, IR4, 461-474.

Kalinin, G.P. and Milyukov, P.I. (1957)"O raschete neustanovivshegosya dvizheniya vody v otkrytykh ruslakh" (On the computation of unsteady flow in open channels), *Meteorologiya i Gidrologiya (USSR)* 10, 10-18.

Karaushev, A.V. (1966) The silting of small reservoirs and ponds - theory and calculation method, *Am. Geophys. Union, Soviet Hydrol.* 35-46.

Kashyap, R.L. and Rao, A. R. (1976) *Dynamic Stochastic Models from Empirical Data,* Academic Press.

Kendall, G.R. (1960) The cube-root normal distribution applied to Canadian monthly rainfall totals, IASH Commission of Land Erosion, Helsinki, Publ. No.53, 250-260.

Kendall, G.R. (1966) Probability distribution of a single variable, Proc. Hydrol. Symp. No.5 "Statistical Methods in Hydrology", McGill Univ. 37-51.

Kendall, M.G. and Stuart, A. (1961) *The Advanced Theory of Statistics* (2), Charles Griffin.

Kepner, R.A., Boelter, L.M.K. and Brooks, F.A. (1942) Nocturnal wind velocity, eddy stability and eddy diffusion above a citrus orchard, *Trans. Am. Geophys. Union* 23, 239-249.

Kiefer, P.J. (1941) The thermodynamic properties of water vapour, *Monthly Weather Rev.* 69(11), also in *Handbook of Meteorology,* Edited by Berry, F.A. Jr. Bollay, E. and Beers, N.R. p. 392, McGraw-Hill (1945).

Kimball, H.H. (1914) The total radiation received on a horizontal surface from the sun and sky at Mt. Weather, Va. *Monthly Weather Rev.* 42, 474-487.

Kirkham, D. and Feng, C.L. (1949) Some tests of the diffusion theory and laws of capillary flow in soils, *Soil Sci.* 67, 29-40.

Kirpich, Z.P. (1940) Time of concentration of small agricultural watersheds, *Civil Eng. (ASCE)* 10(6), 362.

Kisiel, C.C. (1969) Time series analysis of hydrologic data, *Adv. in Hydrosci.* 5, 1-119.

Kittredge, J. (1948) *Forest Influences,* McGraw-Hill.

Kjaran, S.P. (1976) Theoretical and numerical models of groundwater reservoir mechanism, Inst. of Hydraulics and Hydraulic Eng. Techn. Univ. of Denmark, Series Paper No.13, pp.196.

Klemes, V. (1969) Reliability estimates for a storage reservoir with seasonal input, *J. Hydrol.* 7, 198-216.

Klemes, V. (L970) A two-step probabilistic model of storage reservoir with correlated inputs, *Water Resources Res.* 6, 756-767.

Klemes, V. (1973) Applications of hydrology to water resources management, World Meteorol. Org. Operational Hydrol. Rep. No.4, WMO-No.356.

Klemes, V. (1974) The Hurst phenomenon: A puzzle? *Water Resources Res.* 10, 675-688.

Kohler, M.A., Nordenson, T.J. and Fox, W.E. (1955) Evaporation from pans and lakes,

U.S. Weather Bureau Res. Paper 38.

Komura, S. and Simons, D.B. (1967) River-bed degradation below dams, *Proc. ASCE*, 93, HY4, 1-14, Disc. 94, HY1, 2, 3 & 5, 95, HY3.

Kozlowski, T.T. (1964) *Water Metabolism in Plants*, Harper and Row.

Kramer, P.J. (1969) *Plant and Soil Water Relationships*, McGraw-Hill.

Kulandaiswamy, V.C. (1964) A basic study of the rainfall excess-surface runoff relationship in a basin system, Ph.D thesis Univ. of Illinois, Urbana, Synopsis in Ch. 14, Chow (1964).

Lally, V.A. and Lichfield, E.W. (1969) Summary of status and plans for the Ghoast Ballon Project, *Bull. Am. Meteorol. Soc.* 50, 867-868.

Langbein, W.B. et al. (1947) Topographic characteristics of drainage basins, U.S. Geol. Survey Water-supply Paper 968-C, pp. 125-155.

Langbein, W.B. (1958) Queuing theory and water storage, *Proc. ASCE*, 84, HY3.

Lara, J.M. (1962) Revision of procedures to compute sediment distribution in large reservoirs, U.S. Bureau of Reclamation, Denver, Colorado.

Laurenson, E.M. (1964) A catchment storage model for runoff routing, *J. Hydrol.* 2, 141-163.

Laurenson, E.M. (1962) Hydrograph synthesis by runoff routing, Univ. of New South Wales, Water Res. Lab. Rep. No.66.

Laurenson, E.M. (1965) Storage routing methods of flood estimation, Inst. of Eng. Austr. *Civil Eng. Trans.* CE-7, 39-47.

Laurenson, E.M. and O'Donnell, T. (1969) Data error effects in unit hydrograph derivation, *Pro. ASCE*, 95, HY6, 1899-1917.

Leclerc, G. and Schaake, J.C. (1972) Derivation of hydrologic frequency curves, Massachusetts Inst. Tech. Rep. 142.

Lee, P.S., Lynn, P.P. and Shaw, E.M. (1974) Comparison of multiquadric surfaces for the estimation of areal rainfall, *Hydrol. Sci. Bull.* 19, 303-317.

Leopold, L.B. and Maddock, T. Jr. (1953) The hydraulic geometry of stream channels and some physiographic implications, U.S. Geol. Survey Prof. Paper 252, Washington, D.C.

Leopold, L.B. (1968) Hydrology for urban land planning - a guidebook on hydrologic effects of urban land use, U.S. Geol. Survey Circular 554.

Levi, E. and Valdés, R. (1964) A method for direct analysis of hydrographs, *J. Hydrol.* 2, 182-190.

Leyton, L. and Carlisle, A. (1959) Measurement and interpretation of interception by forest stands, AIHS, Hannoversch-Munden 1, 111-119.

Lighthill, M.J. and Whitham, G.B. (1955) On kinematic waves: Flood movement in long rivers, *Proc. Roy. Soc. London* 229, 281-316.

Linsley, R.K. and Ackermann, W.C. (1942) A method of predicting the runoff from rainfall, *Trans. ASCE*, 107, 825-846.

Linsley, R.K. Jr., Kohler, M.A. and Paulhus, J.L.H. (1975) *Hydrology for Engineers*, Second Ed. McGraw-Hill.

Liou, E.Y. (1970) OPSET: Program for computerized selection of watershed parameter values for the Stanford Watershed Model, Lexington, Univ. of Kentucky Water Resources Inst. Res. Rep. No.34.

Lloyd, E.H. (1963) Reservoirs with serially correlated inflows, *Technometrics* 5, 85.

Lloyd, E.H. (1963) A probability theory of reservoirs with serially correlate inputs,

J. Hydrol. 1(2), 99-128.

Lvovich, M.I. (1970) World water balance (general report), Proc. Reading Symp. July 1970, IASH-UNESCO-WMO 2, 401-415.

Lvovich, M.I. (1973) The water balance of the world's continents and a balance estimate of the world's freshwater resources, *Soviet Geography: Rev. and Transl. Am. Geogr. Soc.* 14(3), 135-152.

McAdams, W.H. (1954) *Heat Transmission*, 3rd Ed. McGraw-Hill.

McCarthy, G.T. (1938) The unit hydrograph and flood routing, Unpublished paper presented at a conference of the North Atlantic Div. U.S. Army Corps of Eng. 24(6).

McCuen, R.H. (1973) The role of sensitivity analysis in hydrologic modelling, *J. Hydrol.* 18, 37-53.

McDonald, W.F. (Ed.)(1938) Atlas of climatic charts of the oceans, U.S. Dept. of Agric. Weather Bureau, Washington, D.C.

McGauhey, P.H. (1968) *Engineering Management of Water Quality*, McGraw-Hill.

McGuinness, J.L. and Brakensiek, D.L. (1964) Simplified techniques for fitting frequency distributions to hydrological data, U.S. Dept. Agric. Handbook No.259, April 1964.

McMillan, W.D. and Burgy, R.H. (1960) Interception loss from grass, *J. Geophys. Res.* 65, 2389-94.

Maksimov, V.A. (1964) An outstanding rainstorm in the Donbas, *Am. Geophys. Union, Soviet Hydrology* (1), 68-69.

Mandelbrot, B.B. and Wallis, J.R. (1969) Computer experiments with fractional Gaussian noises, *Water Resources Res.* 5, 228-267.

Mandeville, A.N. and Rodda, J.C. (1970) A contribution to the objective assessment of areal rainfall amounts, *J. Hydrol. (N.Z.)* 9, 281-291.

Margules, M. (1906) Über Temperaturschichtung in stationär bewegter und in ruhender Luft, *Meteorol. Zeits. Hann-Band*, pp.243.

Mark, D.M. (1974) Line intersection method for estimating drainage density, *Geology*, 2(5), 235-236.

Martinec, J. (1975a) Snowmelt-runoff model for stream flow forecasts, *Nordic Hydrol.* 6, 145-154.

Martinec, J. (1975b) New methods in snowmelt-runoff studies in representative basins, Symp. Tokyo, IASH Publ. No.117.

Martinec, J. (1976) Snow and ice, Ch. 4 in *Facets of Hydrology*, Ed. by J.C. Rodda, Wiley, England.

Matalas, N.C. (1967) Mathematical assessment of synthetic hydrology, *Water Resources Res.* 3, 937-945.

Matern, B. (1960) Spatial variation, Medd. Statens Skogsforsknings Inst. Sweden, 49(5).

Melton, M.A. (1958) Geometric properties of mature drainage systems and their representation in an E_4 phase space, *J. Geol.* 66, 35-54.

Meriam, R.A. (1960) A note on the interception loss equation, *J. Geophys. Res.* 65(11), 3850-51.

Meyer, L.D. and Monke, E.J. (1965) Mechanics of soil erosion by rainfall and overland flow, *Trans. ASAE*, 8, 572-580.

Meyer, L.D. and Wischmeier, W.H. (1969) Mathematical simulation of the process of soil erosion by water, *Trans. ASAE,* 12(6).

Milankovitch, M. (1930) Mathematische Klimalehre in Köppen-Geiger, *Handbuch der Klimatologie,* 1, Berlin.

Miller, D.H. (1965) The heat and water budget of the Earth's surface, *Adv. in Geophys.* 2, 175-302.

Miller, C.F. (1968) Evaluation of runoff coefficients for small natural drainage areas, Univ. of Kentucky, Lexington, Water Res. Inst. Rep. No.14.

Mintz, Y. and Dean, G. (1952) The observed mean field of motion of the atmosphere, Geophys. Res. Paper 17, U.S. Air Force, Cambridge Res. Center, Cambridge Mass.

Mitchell, W.D. (1972) Model hydrographs, U.S. Geol. Survey Water-Supply Paper 2005.

Mockus, V. (1957) Use of storm and watershed characteristics in synthetic hydrograph analysis and application, U.S. Soil Conser. Service.

Monteith, J.L. (1965) Evaporation and environment, *Symp. Soc. Exp. Biol.* 19, 205-234.

Moore, R.E. (1939) Water conduction from shallow water tables, *Hilgardia,* 12, 383-426

Moran, P.A.P. (1959) *The Theory of Storage,* Methuen, London.

Murphy, C.E. and Knoerr, K.R. (1972) Modelling the energy balance of natural ecosystems, East. Deciduous Forest Biome-IBP Res. Rep. 72-10, 164 pp., Oak Ridge Nat. Lab. Oak Ridge, Tenn. and Duke Univ. Durham, N.C.

Murphy, C.E. and Knoerr, K.R. (1975) The evaporation of intercepted rainfall from a forest stand: An analysis by simulation, *Water Resources Res.* 11(2), 273-280.

Musgrave, G.W. (1947) Quantitative evaluation of factors in water erosion, a first approximation, *J. Soil and Water Conser.* 2(3), 133-138.

Nash, J.E. (1957) The form of the instantaneous unit hydrograph, Proc. IASH Assemblée Générale de Toronto, 3, 114-121.

Nash, J.E. (1959) Systematic determination of unit hydrograph parameters, *J. Geophys. Res.* 64, 111-115.

Nash, J.E. (1959a) A note on Muskingum flood-routing method, *J. Geophys. Res.* 64(8), 1053-1056.

Nash, J.E. (1960) A unit hydrograph study with particular reference to British catchments, *Proc. I.C.E.* 17, 249-282.

Neff, E.L. (1965) Principles of precipitation network design for intensive hydrologic investigations, WMO-IASH Symp. on Design of Hydrometeorol. Networks, Quebec.

Nutter, W.L. and Hewlett, J.D. (1971) Stormflow production from permeable upland basins, Proc. Third Int. Seminar for Hydrol. Professors, Dept. Agric. Eng. Purdue Univ. 248-258.

O'Donnell, T. (1960) Instantaneous hydrograph derivation by harmonic analysis, IASH Publ. No.51, 546-557.

Parlange, J-Y. (1971) Theory of water-movement in soils: 2. one-dimensional infiltration, *Soil Science,* 111, 170-174.

Paulhus, J.L.H. (1965) Indian Ocean and Taiwan rainfall set new records, *Monthly Weather Review,* 93(5), 331-335.

Paynter, H.M. (1952) Methods and results from MIT studies in unsteady flow, *J. Boston Soc. Civ. Eng.* 39, 120-165.

Penman, H.L. (1948) Natural evaporation from open water, bare soil, and grass, *Proc.*

Roy. Soc. London, A193, 120-145.

Penman, H.L. and Long, I.F. (1960) Weather in wheat: An essay in micro-meteorology, *Quart. J. Roy. Meteorol. Soc.* 86, 16-50.

Penman, H.L (1963) Vegetation and hydrology, Tech. Commun. 53, 40 pp. Commonwealth Bur. of Soils, Harpenden, England.

Pentland, R.L. and Cuthbert, D.R. (1971) Operational hydrology for ungauged streams by the grid square technique, *Water Resources Res.* 7, 283-291.

Pereira, H.C. (1973) *Land Use and Water Resources,* Cambridge Univ. Press.

Philip, J.R. (1957a) Evaporation and moisture and heat fields in the soil, *J. Meteorol.* 14, 354-366.

Philip, J.R. (1957b) The theory of infiltration, Parts I&II, *Soil Sci.* 83, 345-357, 435-448, Parts III-V, *Soil Sci.* 84, 163-178, 257-264, 329-339, Parts VI& VII, *Soil Sci.* 85(1958), 278-286, 33-337.

Philip, J.R. (1960) General method of exact solution of the concentration-dependent diffusion equation, *Austr. J. Phys.* 13(1), 1-12.

Pierce, R.S., Hornbeck, J.W., Likens, C.E. and Bormann, F.H. (1970) Effect of elimination of vegetation on stream water quantity and quality, Symp. on the results of research on representative and experimental basins, Wellinton, N.Z. IASH-UNESCO Publ. No.96, 311-328.

Porter, J.W. (1972) The synthesis of continuous streamflow from climatic data by modelling with a digital computer, Ph.D thesis, Dept. of Civil Eng. Monash Univ.

Porter, J.W. (1975) A comparison of hydrologic and hydraulic catchment routing procedures, *J. Hydrol.* 24, 33-349.

Porter, J.W. and McMahon, T.A. (1976) The Monash Model: User manual for daily program HYDROLOG, Monash Univ. Civil Eng. Res. Rep. 2/76, Melbourne.

Prabhu, N.U. (1964) *Time-dependent Results in Storage Theory,* Methuen, London.

Prasard, T. (1971) Discussion of "Risks in hydrologic design of engineering projects" *Proc. ASCE,* 97, HY1, 201-202.

Prescott, J.A. (1940) Evaporation from water surface in relation to solar radiation, *Trans. Roy. Soc. S. Austr.* 64, 114-118.

Price, R.K. (1973a) Flood routing methods for British rivers, Hydr. Res. Station, Wallingford, U.K. Rep. INT 111, *Proc. Inst. Civil Eng. London,* 55(12), 913-930.

Price, R.K. (1973b) Variable parameter diffusion method for flood routing, Hydr. Res. Station, Wallingflord, U.K. Rep. INT 115.

Priestly, C.H.B. (1959) *Turbulent Transfer in the Lower Atmosphere,* Univ. of Chicago Press.

Pruitt, W.O., Morgan, D.L. and Lourence, F.J. (1968) Energy, momentum and mass transfers above vegetative surfaces, Tech. Rep. ECOM-0447(E)-F, Univ. of Calif. Davis, 74 pp.

Pruitt, W.O. (1971) Factors affecting potential and actual evaporation and the prediction and measurement thereof, Proc. Third Int. Seminar for Hydrol. Professors, Dept. Agric. Eng. Purdue Univ. USA, 82-102.

Quimpo, R.G. (1968) Stochastic analysis of daily river flows, *Proc. ASCE,* 94, HY1, 43-57.

Ragan, R.M. and Duru, J.O. (1972) Kinematic wave nomograph for times of concentration, *Proc. ASCE,* 98, HY10, 1765- 1771.

Raiffa, H. and Schlaifer, R. (1961) Applied statistical decision theory, Graduate School of Business Administration, Harvard Univ.

Rantz, S.E. (1971) Suggested criteria for hydrologic design of storm drainage facilities in the San Francisco Bay Region, California, U.S. Dept. of the Interior, Geol. Survey, Water Resources Div. Prepared in cooperation with the U.S. Dept. of Housing and Urban Devel. Open-file Rep. Menlo Park, Calif. Nov 24.

Rao, R.A., Delleur, J.W. and Sarma, B.S.P. (1972) Conceptual hydrologic models for urbanizing basins, *Proc. ASCE,* 89, HY7, 1205-1220.

Raphael, J.M. (1962) Prediction of temperature in rivers and reservoirs, *Proc. ASCE,* 88, PO2, 157-181.

Raudkivi, A.J. and Lawgun, N. (1970) A Markov chain model for rainfall generation, Symposium on the results of research on representative and experimental basins, IASH-UNESCO Publ. No.96, 269-278.

Raudkivi, A.J. and Lawgun, N. (1972) Generation of serially correlated non-normally distributed rainfall durations, *Water Resources Res.* 8, 398-409.

Raudkivi, A.J. and Lawgun, N. (1974) Simulation of rainfall sequences, *J. Hydrol.* 22, 271-294.

Raudkivi, A.J. (1976) *Loose Boundary Hydraulics,* 2nd Edition, Pergamon Press.

Raudkivi, A.J. and Callander, R.A. (1976) *Analysis of Groundwater Flow,* Arnold.

Raudkivi, A.J. and Nguyen, Van U'u (1976) Soil moisture movement by temperature gradient, *Proc. ASCE,* 102, GT12, 1225-1244.

Reich, B.M. (1963) Short-duration rainfall-intensity estimates and other design aids for regions of sparse data, *J. Hydrol.* 1, 3-28.

Réméniéras, G. (1967) Statistical methods of flood frequency analysis in assessment of magnitude and frequency of flood flows, UN-WMO Water Resources Ser. 30, 50-108.

Richards, B.D. (1955) *Flood Estimation and Control,* Chapman & Hall, London.

Richards, L.A. (1931) Capillary conduction of liquids through porous medium, *Physics,* 1, 318-333.

Riehl, H. (1972) *Introduction to the Atmosphere,* McGraw-Hill.

Riehl, H. (1954) *Tropical Meteorology,* McGraw-Hill.

Rider, N.E. (1954) Eddy diffusion of momentum, water vapour, and heat near the ground *Phil. Trans. Roy. Soc. London,* A246, 481-501.

Rider, N.E., Philip, J.R. and Bradley, E.F. (1963) The horizontal transport of heat and moisture - a micro-meteorological study, *Quart. J. Roy. Meteorol. Soc.* 89, 507-531.

Rippl, W. (1883) The capacity of storage reservoirs for water supply, *Proc. Inst. Civil Eng.* 71, 270-278.

Road Research Laboratory (1963) A guide for engineers to the design of storm sewer systems, Road Note No.35, DSIR, HMSO, London.

Roche, M. (1963) *Hydrologie de Surface,* Gauthier-Villars, Paris.

Rodda, J.C. et al. (1969) Hydrologic network design - needs, problems and approaches, Rep. 12, World Meteorol. Org. Geneva.

Rodriguez-Iturbe, I. and Mejia, J.M. (1974) The design of rainfall networks in time and space, *Water Resources Res.* 10, 713-728.

Rodriguez-Iturbe, I. and Mejia, J.M. (1974a) On the transformation of point rainfall to areal rainfall, *Water Resources Res.* 10, 729-735

Rogers, J.S. and Klute, A. (1971) The hydraulic conductivity - water content relationship during non-steady flow through a sand column, *Soil Sci. Am. Soc. Proc.* 35, 695-700.

Ross, G.A. (1970) The Stanford Watershed Model: The correlation of parameter values selected by a computerized procedure with measurable physical characteristics of the watershed, Univ. of Kentucky Water Resources Inst. Res. Rep. No.35.

Runoff from snowmelt (1960) U.S. Army Corps of Engineers, EM 1110-2-1406, Washington, D.C.

Rutter, A.J. (1959) Evaporation from a plantation of Pinus Sylvestris in relation to meteorological and soil conditions, *Ass. Int. Hydr. Sci. Hannoversch-Munden,* 48, 101-110.

Rutter, A.J. (1972) *Transpiration,* Oxford Biology Readers, edited by J.J. Head and O.E. Lowenstein, Oxford Univ. Press.

Sala-La Cruz, J.D. (1972) Range analysis for storage problems of periodic-stochastic processes, Hydrology Paper No.57, Colorado State Univ. Fort Collins.

Salter, P.M. (1972) Areal rainfall analysis by computer, Proc. WMO-IASH Symposium: Distribution of precipitation in mountainous areas, 2, 497-509.

Sariahmed, A. and Kisiel, C.C. (1968) Synthesis of sequences of summer thunderstorms volumes for the Atterburg watershed in the Tucson area, Proc. IASH Symposium: Use of analog digital computers in hydrology, 2, 439-447.

Sartz, R.S. (1970) Effect of land use on the hydrology of small watersheds in southeastern Wisconsin, Symposium on the results of research on representative and experimental basins, IASH-UNESCO Publ. No.96, 286-295.

Satterlund, D.R. and Haupt, H.F. (1970) The disposition of snow caught by conifer crown, *Water Resources Res.* 6(2), 649-652.

Savage, S.B. and Brimberg, J. (1975) Analysis of plunging phenomena in water reservoirs, *J. Hydr. Res.* 13(2), 187-205.

Schaake, J.C. Jr., Geyer, J.C. and Knapp, J.W. (1967) Experimental examination of the rational method, *Proc. ASCE,* 93, HY6, 353-370.

Schulz, E.F. and Lopez, O.G. (1974) Determination of urban watershed response time, Colorado State Univ. Hydrol. Papers No.71.

Schumm, S.A. (1956) Evolution of drainage systems and slopes in badlands at Perth Amboy, New Jersey, *Bull. Geol. Soc. Am.* 67, 597-646.

Sellers, W.D. (1965) *Physical Climatology,* Univ. of Chicago Press.

Sharp, A.I., Gibbs, A.E., Owens, W.J. and Harris, B. (1966) Development of a procedure for estimating the effects of land and watershed treatment on streamflow, Tech. Bull. 1352, U.S. Dept. of Agric. and U.S. Dept. of Interior.

Sharp, A.L. and Holtan, H.N. (1940) A graphical method of analysis of sprinkled-plot hydrographs, *Trans. Am. Geophys. Union,* 21, 558-570.

Sharp, A.L. and Holtan, H.N. (1942) Extension of graphic methods of analysis of sprinkled-plot hydrographs to the analysis of hydrographs of control-plots and small homogeneous watersheds, *Trans. Am. Geophys. Union,* 23, 578-593.

Shaw, E.M. and Lynn, P.P. (1972) Areal rainfall evaluation using two surface fitting techniques, *Bull. Int. Assoc. Hydrol. Sci.* 17, 419-433.

Shen, J. (1963) A method of determining the storage-outflow characteristics on nonlinear reservoirs, U.S. Geol. Survey Prof. Paper 450-E, 167-168.

Sherman, L.K. (1932) Streamflow from rainfall by the unit-graph method, *Eng. News-Record* 108, 501-505.

Shidei, T. (1954) Studies on the damage to forest trees by snow pressures, Bull. Forest Exp. Stat. Meguro, Tokyo 73, 89 pp. (in Japanese with English summary).

Shreve, R.L. (1966) Statistical law of stream numbers, *J. Geol.* 74, 17-37.

Singh, B. and Shah, C.R. (1971) Plunging phenomenon of density currents in reservoirs *La Houille Blanche,* 26, 59-64.

Sittner, W.T., Schauss, C.E. and Monro, J.C. (1969) Continuous hydrograph synthesis with an API-type hydrologic model, *Water Resources Res.* 5, 1007-1022.

Slatyer, R.O. and McIlroy, I.C. (1961) *Practical Microclimatology*, CSIRO-UNSECO, Australia.

Slatyer, R.O. (1967) *Plant-water Relationships* (Experimental botany, an international series of monographs, 2), Academic Press.

Snow Hydrology (1956) Summary Report of the Snow Investigations, U.S. Army Corps of Engineers, North Pacific Div. Portland, Oregon.

Snyder, F.F. (1938) Synthetic unit graphs, *Trans. Am. Geophys. Union,* 19, 447-454.

Snyder, W.M. (1955) Hydrograph analysis by the method of least squares, *Proc. ASCE,* 81(9), No.793.

Snyder, W.M. (1961) Continuous parabolic interpolation, *Proc. ASCE,* 87, HY4, 99-111.

Solomon, S.I. et al. (1968) The use of a square grid system for computer estimation of precipitation, temperature, and runoff, *Water Resources Res.* 4, 919-929

Spolia, S.K. and Chander, S. (1974) Modelling of surface runoff systems by ARMA model, *J. Hydrol.* 22, 317-322.

Starr, V.P. (1948) An essay on the general circulation of the Earth's atmosphere, *J. Meteorol.* 5(2), 39-43.

Starr, V.P. and White, R.M. (1954) Balance requirements of the general circulation, Geophys. Res. Paper 35, U.S. Air Force Cambridge Res. Center, Cambridge, Mass.

Starr, V.P., Peixoto, J.P. and Crisi, A.R. (1965) Hemispheric water balance for the IGY, *Tellus* 17(4), 463-472.

Stewart, J.B. and Thom, A.S. (1973) Energy budgets in pine forests, *Quartl. J. Roy. Meteorol. Soc.* 99, 154-170.

Stidd, C.K. (1953) Cube root normal precipitation distributions, *Trans. Am. Geophys. Union,* 34(1), 31-35.

Stol, P.T. (1972) The relative efficiency of the density of rain-gauge networks, *J. Hydrol.* 15, 193-208.

Strahler, A.N. (1952) Hypsometric (area-altitude) analysis of erosional topography, *Bull. Geol. Soc. Am.* 63, 1117-1142.

Strahler, A.N.(1957) Quantitative analysis of watershed geomorphology, *Trans. Am. Geophys. Union* 38, 913-920.

Strahler, A.N. and Strahler A.H. (1974) *Introduction to Environmental Science,* Hamilton Publ. Co.

Subrahmanyam, V.P. (1967) Incidence and spread of continental drought, WMO-IHD Projects, Rep. No.2.

Sverdrup, H.U. (1946) The humidity gradient over the sea surface, *J. Meteorol.* 3, 1-8.

Swank, W.T. and Helvey, J.D. (1970) Reduction of streamflow increases following regrowth of clearcut hardwood forests, Symp. on the results of research on representative and experimental basins, IASH-UNESCO, Publ. No.96, 346-360.

Swinnerton, C.J., Hall, M.J. and O'Donnell, T. (1972) Dimensionless hydrograph design method for motorway stormwater drainage systems, *J. Inst. Highway Eng.* 19(11), 2-10.

Swinnerton, C.J., Hall, M.J. and O'Donnell, T. (1973) Conceptual model design for motorway stormwater drainage, *Civil Eng.* 68, 3-8.

Tanner, C.B. and Pelton, W.L. (1960) Potential evapotranspiration estimates by approximate energy balance method of Penman, *J. Geophys. Res.* 65, 3391-3413.

Task Committee (1973) Reevaluation of spillway adequacy of existing dams, *Proc. ASCE*, 99, HY2, 337-371.

Taylor, A.B. and Schwarz, H.E. (1952) Unit hydrograph lag and peak flow related to basin characteristics, *Trans. Am. Geophys. Union*, 33, 235-246.

Tholin, A.L. and Keifer, C.J. (1960) Hydrology of urban runoff, *Trans. ASCE*, 125, 1308-1379.

Thomas, M.A. and Fiering, M.B. (1962) Mathematical synthesis of streamflow sequences for the analysis of river basins by simulation, In *Design of Water Resources Systems*, edited by A.Maass et al. 459-493, Harvard Univ. Press.

Thornthwaite, C.W. (1948) An approach towards a rational classification of climate, *Am. Geogr. Review*, 38.

Thornthwaite, C.M. and Holzman, B. (1939) The determination of evaporation from land and water surfaces, *Monthly Weather Review*, 67, 4-11.

U.S. Bureau of Public Roads (1961) Hydraulic design series, No.2(April), by W.D.Potter.

U.S. Bureau of Reclamation (1973) *Design of Small Dams*.

U.S. Dept. of Agric. Soil Conservation Service (1972) *Engineering Handbook*, Sec. 4, Hydrol. Suppl. A, Washington, D.C.

U.S. Dept. of Agriculture (1971) USDAHL-70 Model of watershed hydrology, Agric. Res. Service, Techn. Bull. No.1435.

U.S. Weather Bureau (1960) Generalized estimates of probable maximum precipitation for the United States west of the 105th meridian for areas to 400 square miles and durations to 24 hours, Techn. Paper No.38.

van Bavel, C.H.M. and Fritschen, L.J. (1964) Energy balance studies over Sudan grass, 1962, Interim Rep. U.S. Water Conservation Laboratory, Tempe.

van Bavel, C.H.M. (1966) Potential evaporation: The combination concept and its experimental verification, *Water Resources Res.* 2, 455-467.

van de Hulst, H.C. (1949) Scattering in the atmospheres of the Earth and the Planets, Ch. 3 in *The Atmospheres of the Earth and Planets*, Editor G.P. Kuiper, Univ. of Chicago Press.

Vanoni, V. (Editor)(1975) *Sedimentation Engineering*, ASCE-Manual No.54, Ch. IV - Sediment sources and sediment yields.

Vansteenkiste, G.C. (Editor)(1975) *Computer Simulation of Water Resources Systems*, North-Holland Publishing Co.

Vehrencamp, J.E. (1953) Experimental investigation of heat transfer at an air-earth interface, *Trans. Am. Geophys. Union*, 34, 22-30.

Vehrencamp, J.E. (1951) An experimental investigation of heat and momentum transfer at a smooth air-earth interface, Dept. of Eng. Univ. of California, Los

Angeles.

Waggoner, P.E. and Reifsnyder, W.E. (1968) Simulation of the temperature, humidity and evaporation profiles in a leaf canopy, *J. Appl. Meteorol.* 7(3), 400-409.

Waggoner, P.E., Begg, J.E. and Turner, N.C. (1969) Evaporation of dew, *Agric. Meteorol.* 6, 227-230.

Walton, W.C. (1970) *Groundwater Resource Evaluation*, McGraw-Hill.

Watkins, L.H. (1962) The design of urban sewer systems, Road Res. Lab. U.K. Tech. Paper No.55, DSIR, HMSO, London.

Weiss, L. (1961) *Statistical Decision Theory*, McGraw-Hill.

Whitmore, J.S., van Eeden, F.J. and Harvey, K.J. (1960) Assessment of average annual rainfall over large catchments, S. Africa Dept. of Water Affairs Tech. Rep. No.14.

Wiesner, C.J. (1970) *Hydrometeorology*, Chapman and Hall Ltd.

Wilson, L.G. and Luthin, J.N. (1963) Effect of air flow ahead of wetting front on infiltration, *Soil Sci.* 96, 136-143.

Wischmeier, W.H. and Smith, D.D. (1958) Rainfall energy and its relationship to soil loss, *Trans. Am. Geophys. Union*, 39, 285-291.

Wischmeier, W.H. and Smith, D.D. (1965) Predicting rainfall erosion losses from cropland east of the Rocky Mountains, Agric. Handbook No.282, U.S. Dept. of Agric.

Wittenberg, H. (1975) A model to predict the effects of urbanization on watershed response, Proc. Nat. Symp. on Urban Hydrol. and Sediment Control, Univ. of Kentucky, Lexington, 161-167.

Wooding, R.A. (1965, 1966) A hydraulic model for the catchment-stream problem: I Kinematic-wave theory, *J. Hydrol.* 3, 254-267, II Numerical solutions, *J. Hydrol.* 3, 268-282, III Comparison with runoff observations, *J. Hydrol.* 4(1966), 21-37.

Woolhiser, D.A. and Ligget, J.A. (1967) Unsteady, one-dimensional flow over a plane - the rising hydrograph, *Water Resources Res.* 3, 753-771.

World Meteorological Organization (1969) Manual for depth-area-duration analysis of storm precipitation, WMO-No.237, TP.129.

World Meteorological Organization (1969) Estimation of maximum floods, Tech. Note No.98, WMO-No.233, TP.126.

World Meteorological Orgainzation (1973) Manual for estimation of probable maximum precipitation, WMO-No.332.

Wright, J.L. and Lemon, E.R. (1962) Estimation of turbulent exchange within a corn crop canopy at Ellis Hollow, Interim Rep. 62-7, N.Y.S. College of Agric. Itacha.

Yamamoto, G. and Shimanuki, A. (1964) Profiles of wind and temperature in the lowest 250 meters in Tokyo, Tohoku Univ. Sci. Rep. Series 5, *Geophys.* 15, 111-14.

Yevjevich, V.M. (1959) Analytical integration of the differential equation for water storage, *J. Res. Nat. Bureau of Standards*, 63B(1).

Yevjevich, V.M. (1965) The application of surplus, deficit and range in hydrology, Hydrol. Paper No.10, Colorado State Univ. Fort Collins.

Yevjevich, V.M. (1967) An objective approach to definitions and investigations of continental hydrologic droughts, Hydrol. Paper No.23, Colorado State Univ. Fort Collins.

Yevjevich, V.M. (1968) Misconceptions in hydrology and their consequences, *Water Resources Res.* 4(2), 225-232.

Yevjevich, V. (1972) *Stochastic Processes in Hydrology*, Water Resources Publications, Fort Collins.

Young, G.K. (1968) Discussion of "Mathematical assessment of synthetic hydrology" by N.C. Matalas, *Water Resources Res.* 4, 681-682.

Yücel, Ö. and Graf, W.H. (1973) Bed load deposition in reservoirs, Proc. 15th Congress IAHR, Istanbul, 1, 271-2781

Zheleznyakov, G.V. (1971) Interaction of channel and flood plain streams, Proc. 14th Congress IAHR, Paris, 5, 145-148.

Zoch, R.T. (1934) On the relation between rainfall and stream flow, *Monthly Weater Review*, Part I 62, 315-322, Part II 64, 105-121 (1936) and Part III 65, 135-147 (1937).

AUTHOR INDEX

SUBJECT INDEX